Python
第三方库开发应用实战

张有菊◎编著

人民邮电出版社

北京

图书在版编目（CIP）数据

Python第三方库开发应用实战 / 张有菊编著. -- 北京：人民邮电出版社，2020.5
 ISBN 978-7-115-51495-0

Ⅰ. ①P… Ⅱ. ①张… Ⅲ. ①软件工具－程序设计 Ⅳ. ①TP311.561

中国版本图书馆CIP数据核字(2020)第055031号

内 容 提 要

本书循序渐进地讲解了Python中常用第三方库的核心知识，并通过具体实例的实现过程演练了各个库的使用流程。全书共12章，分别讲解了Tornado框架、Django框架、Flask框架、数据库存储框架、数据库驱动框架、使用ORM操作数据库、特殊文本格式处理、图像处理、图形用户界面、数据可视化、第三方多媒体库、第三方网络开发库。

本书适用于已经了解了Python基础语法的读者，也适用于希望进一步提高自己Python开发水平的读者，还可以作为大专院校相关专业的师生用书和培训学校的教材。

◆ 编　　著　张有菊
　责任编辑　张　涛
　责任印制　王　郁　焦志炜

◆ 人民邮电出版社出版发行　北京市丰台区成寿寺路11号
　邮编　100164　电子邮件　315@ptpress.com.cn
　网址　https://www.ptpress.com.cn
　固安县铭成印刷有限公司印刷

◆ 开本：787×1092　1/16

　印张：30.75　　　　　　　　　　2020年5月第1版

　字数：830千字　　　　　　　　　2024年7月河北第3次印刷

定价：99.00 元

读者服务热线：(010)81055410　印装质量热线：(010)81055316
反盗版热线：(010)81055315
广告经营许可证：京东市监广登字20170147号

前　言

Python 的应用越来越广，这种语言的成功得益于第三方库的支持，第三方库能够帮助程序员迅速实现需要的功能，提高程序员的开发效率。为了让更多的读者了解 Python 中常用的框架和第三方库的使用方法，作者专门编写了本书。

本书内容

本书精心挑选了 12 类常用的库进行讲解，正是这些功能强大的第三方库吸引了众多程序员纷纷加入到 Python 开发者行列中。本书主要内容如下。

第 1 章讲解了 Tornado 框架的使用，主要包括 Tornado 开发基础、表单和模板操作、数据库操作等。

第 2 章讲解 Django 框架的应用，包括搭建 Django 开发环境、使用 Django 后台系统开发一个博客系统、开发一个在线商城系统、使用 Mezzanine 库、使用 Cartridge 库、使用 django-oscar 库等。

第 3 章讲解 Flask 框架，包括 Flask 开发基础、基于 Flask 开发 Web 程序、表单操作、使用数据库、收发电子邮件、开发图书借阅管理系统等。

第 4 章讲解数据库存储框架，包括 pickleDB、TinyDB 和 ZODB，并通过开发个人日志系统，讲解这些数据库存储框架的应用。

第 5 章介绍数据库驱动框架，包括连接 MySQL 数据库、连接 PostgreSQL 数据库、连接 SQLite3 数据库、连接 NoSQL 数据库等。

第 6 章介绍如何使用 ORM 操作数据库，如使用 mysqlclient 连接数据库、使用 Peewee 连接数据库、使用 Pony 连接数据库、使用 mongoengine 连接 MongoDB 数据库等。

第 7 章介绍特殊文本格式处理的框架知识，如 Tablib 模块、Office 模块/库、PDF 模块/库等。

第 8 章介绍图像处理的框架，包括使用 Pillow 库、使用 hmap 库，以及使用 pyBarcode 库创建条形码、使用 qrcode 库创建二维码、使用 face_recognition 库实现人脸识别等。

第 9 章讲解和图形用户界面有关的库，包括 PyQt 库、pyglet 库、toga 库、wxPython 库等。

第 10 章介绍与数据可视化有关的库，如 Matplotlib 库、pygal 库、csvkit 库、NumPy 库等。

第 11 章介绍多媒体库，如使用 audiolazy 库处理数字信号、使用 audioread 库实现音频解码、使用 eyeD3 库处理音频，以及 m3u8 库、mutagen 库、pydub 库、tinytag 库、moviepy 库、scikit-video 库的用法等。

第 12 章介绍网络开发库，包括 HTML、XML、HTTP 和 URL 的处理等。

本书特色

● 内容全面。

本书内容涵盖了常用的 Python 第三方库，通过案例，循序渐进地讲解了这些库中函数的使用方法，帮助读者快速掌握和应用这些库。

前言

- 通过实例驱动学习。

本书采用理论加实例的讲解方式,通过实例展示知识点的应用,达到了学以致用的目的。

- 提供更广泛的解决方案。

通过对本书的学习,读者可以构建自己的Python工具箱。借助该工具箱,读者能够使用Python开发各种类型的应用程序。

本书读者对象

本书适用于已经了解了Python基础语法的读者,也适用于希望进一步提高自己Python开发水平的读者,还可以作为大专院校相关专业的师生用书和培训学校的教材。

致谢

在编写过程中,本书得到了人民邮电出版社编辑的大力支持,正是各位编辑的高效工作,才使得本书能够顺利出版。另外,也十分感谢我的家人给予的巨大支持。本人水平毕竟有限,书中纰漏之处在所难免,诚请读者提出意见或建议,以便修订并使之更臻完善。编辑联系邮箱是 zhangtao@ptpress.com.cn。

最后感谢您购买本书,希望本书能成为您编程路上的挚友,祝您阅读快乐!

作者

资源与支持

本书由异步社区出品，社区（https://www.epubit.com/）为您提供相关资源和后续服务。

配套资源

本书配套资源包括书中示例的源代码。

要获得以上配套资源，请在异步社区本书页面中单击 配套资源 ，跳转到下载界面，按提示进行操作即可。注意，为保证购书读者的权益，该操作会给出相关提示，要求输入提取码进行验证。

如果您是教师，希望获得教学配套资源，请在社区本书页面中直接联系本书的责任编辑。

提交勘误

作者和编辑尽最大努力来确保书中内容的准确性，但难免会存在疏漏。欢迎您将发现的问题反馈给我们，帮助我们提升图书的质量。

当您发现错误时，请登录异步社区，按书名搜索，进入本书页面，单击"提交勘误"，输入勘误信息，单击"提交"按钮即可（见下图）。本书的作者和编辑会对您提交的勘误进行审核，确认并接受后，您将获赠异步社区的 100 积分。积分可用于在异步社区兑换优惠券、样书或奖品。

扫码关注本书

扫描下方二维码,您将会在异步社区微信服务号中看到本书信息及相关的服务提示。

与我们联系

我们的联系邮箱是 contact@epubit.com.cn。

如果您对本书有任何疑问或建议,请您发邮件给我们,并请在邮件标题中注明本书书名,以便我们更高效地做出反馈。

如果您有兴趣出版图书、录制教学视频,或者参与图书翻译、技术审校等工作,可以发邮件给我们;有意出版图书的作者也可以到异步社区在线提交投稿(直接访问 www.epubit.com/selfpublish/submission 即可)。

如果您所在的学校、培训机构或企业,想批量购买本书或异步社区出版的其他图书,也可以发邮件给我们。

如果您在网上发现有针对异步社区出品图书的各种形式的盗版行为,包括对图书全部或部分内容的非授权传播,请您将怀疑有侵权行为的链接发邮件给我们。您的这一举动是对作者权益的保护,也是我们持续为您提供有价值的内容的动力之源。

关于异步社区和异步图书

"异步社区"是人民邮电出版社旗下IT专业图书社区,致力于出版精品IT技术图书和相关学习产品,为作译者提供优质出版服务。异步社区创办于 2015 年 8 月,提供大量精品 IT技术图书和电子书,以及高品质技术文章和视频课程。更多详情请访问异步社区官网 https://www.epubit.com。

"异步图书"是由异步社区编辑团队策划出版的精品 IT 专业图书的品牌,依托于人民邮电出版社近 30 年的计算机图书出版积累和专业编辑团队,相关图书在封面上印有异步图书的LOGO。异步图书的出版领域包括软件开发、大数据、AI、测试、前端、网络技术等。

异步社区

微信服务号

目　录

第 1 章　Tornado 框架 ·············· 1
1.1　Tornado 框架简介 ············ 2
1.2　Tornado 开发基础 ············ 2
　　1.2.1　编写第一个 Tornado 程序 ···· 2
　　1.2.2　获取请求参数 ············ 3
　　1.2.3　使用 cookie ············ 5
　　1.2.4　URL 转向 ·············· 6
　　1.2.5　使用静态资源文件 ········ 7
1.3　表单和模板操作 ·············· 7
　　1.3.1　一个基本的注册表单 ······ 7
　　1.3.2　在模板中使用函数 ········ 9
1.4　数据库操作 ················ 12
　　1.4.1　实现持久化 Web 服务 ···· 12
　　1.4.2　图书管理系统 ············ 15

第 2 章　使用 Django ·············· 20
2.1　Django 简介 ················ 21
2.2　Django 开发基础 ············ 21
　　2.2.1　搭建 Django 环境 ········ 21
　　2.2.2　常用的 Django 命令 ······ 22
　　2.2.3　第一个 Django 项目 ······ 23
　　2.2.4　在 URL 中传递参数 ······ 25
　　2.2.5　使用模板 ·············· 27
　　2.2.6　使用表单 ·············· 31
　　2.2.7　实现基本的数据库操作 ···· 32
2.3　使用 Django 后台系统开发一个博客系统 ···················· 33
2.4　开发一个新闻聚合系统 ······ 36
　　2.4.1　基本设置 ·············· 36
　　2.4.2　获取聚合信息 ············ 37
　　2.4.3　视图处理 ·············· 40

　　2.4.4　模板文件 ·············· 42
2.5　开发一个在线商城系统 ······ 45
　　2.5.1　系统设置 ·············· 45
　　2.5.2　前台商城展示模块 ········ 46
　　2.5.3　购物车模块 ············ 49
　　2.5.4　订单模块 ·············· 53
2.6　使用 Mezzanine 库 ·········· 56
2.7　使用 Cartridge 库 ············ 58
2.8　使用 django-oscar 库 ········ 61

第 3 章　Flask 框架 ················ 62
3.1　Flask 开发基础 ·············· 63
　　3.1.1　Flask 框架简介 ·········· 63
　　3.1.2　Django 和 Flask 的对比 ···· 65
　　3.1.3　安装 Flask ·············· 65
　　3.1.4　第一个 Flask Web 程序 ···· 66
3.2　基于 Flask 开发 Web 程序 ···· 68
　　3.2.1　传递 URL 参数 ·········· 68
　　3.2.2　使用会话和 cookie ······ 70
　　3.2.3　使用 Flask-Script 扩展 ···· 71
　　3.2.4　使用模板 ·············· 72
　　3.2.5　使用 Flask-Bootstrap
　　　　　扩展 ·················· 74
　　3.2.6　使用 Flask-Moment 扩展
　　　　　本地化日期和时间 ·········· 76
3.3　表单操作 ···················· 78
　　3.3.1　使用 Flask-WTF 扩展 ···· 78
　　3.3.2　文件上传 ·············· 81
3.4　使用数据库 ·················· 82
　　3.4.1　Python 数据库框架 ········ 82
　　3.4.2　会员注册和登录 ·········· 82

目 录

    3.4.3 使用 Flask-SQLAlchemy 管理数据库 ·············· 84
3.5 收发电子邮件 ····················· 87
    3.5.1 使用 Flask-Mail 扩展 ······ 87
    3.5.2 使用 SendGrid 发送邮件 ························ 90
3.6 Flask+MySQL+ SqlAlchemy 信息发布系统 ··············· 92
    3.6.1 使用 Virtualenv 创建虚拟环境 ···················· 92
    3.6.2 使用 Flask 实现数据库迁移 ······················ 92
    3.6.3 具体实现 ················ 93
3.7 图书借阅管理系统 ·············· 100
    3.7.1 数据库设置 ············ 100
    3.7.2 登录验证与管理 ········ 101
    3.7.3 安全检查与页面跳转管理 ························ 102
    3.7.4 后台用户管理 ·········· 102
    3.7.5 图书管理 ·············· 103
    3.7.6 前台用户管理 ·········· 104

第 4 章 数据库存储框架 ············ 107

4.1 安装与使用 pickleDB ·········· 108
    4.1.1 安装 pickleDB ·········· 108
    4.1.2 使用 pickleDB ·········· 108
4.2 安装与使用 TinyDB ············ 108
    4.2.1 安装 TinyDB ············ 109
    4.2.2 使用 TinyDB ············ 109
4.3 如何使用 ZODB ················ 110
    4.3.1 安装并使用 ZODB ······ 110
    4.3.2 模拟银行存取款系统 ···· 112
4.4 个人日志系统(使用 Flask 与 TinyDB 实现) ··············· 114
    4.4.1 系统设置 ·············· 114
    4.4.2 后台管理 ·············· 117
    4.4.3 登录认证管理 ·········· 118
    4.4.4 前台日志展示 ·········· 121
    4.4.5 系统模板 ·············· 124

第 5 章 数据库驱动框架 ············ 131

5.1 连接 MySQL 数据库 ············ 132
    5.1.1 使用 mysqlclient ········ 132
    5.1.2 使用 PyMySQL ·········· 134
5.2 连接 PostgreSQL 数据库 ········ 146
    5.2.1 下载并安装 PostgreSQL ·············· 146
    5.2.2 使用 psycopg2 模块 ···· 148
    5.2.3 使用 queries 模块 ······ 155
5.3 连接 SQLite3 数据库 ············ 156
5.4 连接 SQL Server 数据库 ········ 158
5.5 连接 NoSQL 数据库 ············ 161
    5.5.1 使用 cassandra-driver 连接 Cassandra 数据库 ······ 161
    5.5.2 使用 PyMongo 驱动连接 MongoDB 数据库 ········ 162
    5.5.3 使用 redis-py 连接 Redis ·················· 165

第 6 章 使用 ORM 操作数据库 ······ 169

6.1 ORM 的背景 ···················· 170
6.2 使用 mysqlclient 连接数据库 ······················ 170
6.3 使用 Peewee 连接数据库 ········ 174
    6.3.1 Peewee 的基本用法 ······ 174
    6.3.2 使用 Peewee、Flask 与 MySQL 开发一个在线留言系统 ·············· 178
6.4 使用 Pony 连接数据库 ·········· 181
    6.4.1 Pony 的基础知识 ········ 181
    6.4.2 操作 SQLite 数据库 ······ 181
    6.4.3 操作 MySQL 数据库 ······ 186
6.5 使用 mongoengine 连接 MongoDB 数据库 ·············· 188

第 7 章 特殊文本格式处理 ·········· 191

7.1 使用 Tablib 模块 ················ 192
    7.1.1 基本用法 ·············· 192
    7.1.2 操作数据集中指定的行和列 ···················· 193

	7.1.3	删除并导出不同格式的数据 ············ 193
	7.1.4	生成一个 Excel 文件 ······ 194
	7.1.5	处理多个数据集 ············ 195
	7.1.6	使用标签过滤数据 ······ 197
	7.1.7	分离表格中的数据 ········· 197
7.2	使用 Office 模块/库 ············ 198
	7.2.1	使用 openpyxl 模块 ········ 198
	7.2.2	使用 pyexcel 模块 ········ 202
	7.2.3	使用 python-docx 模块 ····· 208
	7.2.4	使用 xlrd 和 xlwt 库读写 Excel ······ 219
	7.2.5	使用 xlsxwriter 库 ········· 221
7.3	使用 PDF 模块/库 ············ 229
	7.3.1	使用 PDFMiner 模块 ······ 229
	7.3.2	使用 PyPDF2 ············ 233
	7.3.3	使用 Reportlab 库 ········· 237

第 8 章 图像处理 ············ 242

8.1	使用 Pillow 库 ············ 243
	8.1.1	安装 Pillow 库 ············ 243
	8.1.2	使用 Image 模块 ············ 243
	8.1.3	绘制随机漫步图 ············ 249
	8.1.4	使用 ImageChops 模块合成图片 ········ 251
	8.1.5	使用 ImageEnhance 模块增强图像 ········ 253
	8.1.6	使用 ImageFilter 模块实现滤镜功能 ········ 255
	8.1.7	使用 ImageDraw 模块绘制图像 ········ 256
	8.1.8	使用 ImageFont 模块设置字体 ············ 258
	8.1.9	绘制指定年份的日历 ····· 259
8.2	使用 hmap 库 ············ 263
8.3	使用 pyBarcode 库创建条形码 ············ 264
8.4	使用 qrcode 库创建二维码 ········ 266
8.5	使用 scikit-image 库 ············ 269
	8.5.1	读取和显示 ············ 270
	8.5.2	像素操作 ············ 272
	8.5.3	转换操作 ············ 274
	8.5.4	绘制图像 ············ 276
	8.5.5	图像批处理 ············ 280
	8.5.6	缩放和旋转 ············ 281
8.6	使用 face_recognition 库实现人脸识别 ········ 283
	8.6.1	搭建开发环境 ············ 283
	8.6.2	面部特征 ············ 284
	8.6.3	识别人脸 ············ 286
	8.6.4	摄像头实时识别 ········ 289

第 9 章 图形用户界面 ············ 290

9.1	使用 PyQt 库 ············ 291
	9.1.1	第一个 GUI 程序 ············ 291
	9.1.2	菜单和工具栏 ············ 293
	9.1.3	界面布局 ············ 295
	9.1.4	事件处理 ············ 299
	9.1.5	对话框 ············ 303
	9.1.6	组件 ············ 307
	9.1.7	使用 Eric6 提高开发效率 ············ 316
9.2	使用 pyglet 库 ············ 317
	9.2.1	安装并尝试使用 pyglet ············ 317
	9.2.2	实现 OpenGL 操作 ········ 319
	9.2.3	开发一个 pyglet 游戏 ······ 320
9.3	使用 toga 库 ············ 330
	9.3.1	安装 toga 库并创建第一个 toga 示例 ············ 330
	9.3.2	使用基本组件 ············ 331
	9.3.3	使用布局组件 ············ 332
	9.3.4	使用绘图组件 ············ 333
9.4	使用 wxPython 库 ············ 334
	9.4.1	安装并使用 wxPython 库 ············ 334
	9.4.2	基本组件 ············ 335

第 10 章 数据可视化 ············ 340

| 10.1 | 使用 Matplotlib 库 ············ 341 |

目　录

10.1.1　搭建 Matplotlib 库的
　　　　使用环境……………341
10.1.2　初级绘图……………342
10.1.3　自定义散点图样式……345
10.1.4　绘制柱状图……………346
10.1.5　绘制多幅子图…………350
10.1.6　绘制曲线………………352
10.1.7　绘制随机漫步图………357
10.1.8　大数据分析某年的最
　　　　高温度和最低温度……360
10.1.9　在 Tkinter 中使用
　　　　Matplotlib 库绘制
　　　　图表……………………361
10.2　使用 pygal 库…………………362
10.2.1　安装 pygal 库…………362
10.2.2　使用 pygal 库模拟
　　　　掷骰子…………………363
10.3　使用 csvkit 库处理 CSV
　　　文件…………………………365
10.4　使用 Pandas 库………………371
10.4.1　安装 Pandas 库………371
10.4.2　从 CSV 文件读取
　　　　数据……………………371
10.4.3　选择指定数据…………376
10.4.4　与日期相关的操作……379
10.5　使用 NumPy 库………………383
10.5.1　安装 NumPy 库………383
10.5.2　数组对象………………384
10.5.3　使用通用函数…………388
10.5.4　使用 Matplotlib 库……393

第 11 章　第三方多媒体库……………395
11.1　使用 audiolazy 库处理数字
　　　信号…………………………396
11.1.1　安装并尝试使用
　　　　audiolazy 库……………396
11.1.2　实现巴特沃斯滤波器……396
11.2　使用 audioread 库实现音频
　　　解码…………………………397
11.3　使用 eyeD3 库处理音频……398

11.3.1　安装并尝试使用
　　　　eyeD3 库………………398
11.3.2　使用 eyeD3 库编程……398
11.3.3　MP3 文件编辑器………399
11.4　使用 m3u8 库…………………405
11.4.1　m3u8 库的介绍和
　　　　安装……………………406
11.4.2　下载 m3u8 视频并转换为
　　　　MP4 文件………………406
11.5　使用 mutagen 库………………407
11.5.1　安装并尝试使用
　　　　mutagen 库……………407
11.5.2　获取指定音频文件的
　　　　标签信息………………408
11.5.3　批量设置视频文件的
　　　　封面图片………………409
11.6　使用 pydub 库…………………414
11.6.1　安装并尝试使用
　　　　pydub 库………………414
11.6.2　使用 AudioSegment……415
11.6.3　截取指定的 MP3
　　　　文件……………………419
11.7　使用 tinytag 库…………………420
11.7.1　安装并尝试使用
　　　　tinytag 库………………420
11.7.2　开发一个 MP3
　　　　播放器…………………420
11.8　使用 moviepy 库………………428
11.8.1　安装 moviepy 库………429
11.8.2　剪切一段视频…………429
11.8.3　视频合成………………429
11.8.4　多屏显示………………430
11.8.5　设置视频属性…………431
11.8.6　使用 moviepy 库和
　　　　Matplotlib 库实现
　　　　数据的动态可视化……432
11.8.7　动画合成………………433
11.8.8　使用 moviepy 库和
　　　　numpy 库实现文本
　　　　动态化…………………434

11.9 使用 scikit-video 库……………435
 11.9.1 安装并尝试使用
   scikit-video 库……………435
 11.9.2 写入视频……………………436
 11.9.3 视频基准测试………………438
 11.9.4 图像的读取和写入…………438
 11.9.5 视频的读取和写入…………439

第 12 章 第三方网络开发库……………442
12.1 处理 HTML 和 XML……………443
 12.1.1 使用 Beautiful Soup 库……443
 12.1.2 使用 bleach 库………………449
 12.1.3 使用 cssutils 库………………454
 12.1.4 使用 html5lib 库……………455
 12.1.5 使用 MarkupSafe 库…………456
 12.1.6 使用 PyQuery 库……………457
12.2 处理 HTTP………………………461
 12.2.1 使用 aiohttp 库………………461
 12.2.2 使用 requests 库……………463
 12.2.3 使用 httplib2 库……………466
 12.2.4 使用 urllib3 库………………469
12.3 电子邮件…………………………472
 12.3.1 使用 envelopes 库……………472
 12.3.2 使用 Inbox 库………………473
12.4 处理 URL…………………………476
 12.4.1 使用 furl 库…………………476
 12.4.2 使用 purl 库…………………477
 12.4.3 使用 webargs 库……………479

第 1 章

Tornado 框架

Tornado 是 FriendFeed 使用的可扩展的非阻塞式 Web 服务器及其相关工具的开源版本。这个 Web 框架看起来有些像 web.py 或者 Google 的 WebApp。不过为了能有效地利用非阻塞式服务器环境，这个 Web 框架还包含了一些相关的有用的工具和优化措施。本章将详细讲解使用 Tornado 框架开发 Web 应用程序的核心知识。

1.1 Tornado 框架简介

Tornado 是 Python 中一种比较流行的、强大的、可扩展的 Web 非阻塞式开源服务器框架,也是一个异步的网络库,能够帮助开发者快速简单地编写出高速运行的 Web 应用程序。

Tornado 基于 Bret Taylor 和其他人员为 FriendFeed 所开发的网络服务框架,在 FriendFeed 被 Facebook 收购后得以开源。Tornado 在设计之初就考虑到了性能因素,旨在解决 C10K 问题。这样的设计使得它成为一个高性能的框架。此外,它还拥有用户验证、社交网络以及与外部服务(如数据库和网站 API)进行异步交互的工具。

自 2009 年第一个版本发布以来,Tornado 已经获得了很多社区的支持,并且在不同的场合得以应用。除了 FriendFeed 和 Facebook 外,还有很多公司在生产上转向 Tornado,包括 Quora、Turntable.fm、Bit.ly、Hipmunk 及 MyYearbook 等公司。

Tornado 框架的主要特点如下。
- 它是一种非阻塞式服务器框架。
- 运行速度快。
- 可以并发打开数千连接。
- 支持 WebSocket 连接。

在现实应用中,通常将 Tornado 框架分为如下 4 个部分。
- tornado.Web:创建 Web 应用程序的 Web 框架。
- HTTPServer 和 AsyncHTTPClient:HTTP 服务器与异步客户端。
- IOLoop 和 IOStream:异步网络功能库。
- tornado.gen:协程库。

1.2 Tornado 开发基础

在 Python 程序中使用 Tornado 框架之前,首先需要搭建 Tornado 框架环境。Tornado 框架可以通过 pip 或者 easy_install 命令进行安装。pip 命令如下。

```
pip install tornado
```

easy_install 命令如下。

```
easy_install tornado
```

在控制台中使用 easy_install 命令的安装界面如图 1-1 所示。

图 1-1 使用 "easy_install" 命令安装 Tornado 框架

1.2.1 编写第一个 Tornado 程序

在 Tornado 框架中,是通过继承类 tornado.Web.RequestHandler 来编写 Web 服务器端程序的,并编写 get()、post()业务方法,以实现对客户端指定 URL 的 GET 请求和 POST 请求的回应。然后启动框架中提供的服务器以等待客户端连接,处理相关数据并返回请求信息。下面的实例文件 app.py 演示了使用 Python 编写一个基本 Tornado 程序的过程。

源码路径:daima\1\1-2\app.py

```
import tornado.ioloop                #导入Tornado框架中的相关模块
```

```
import tornado.web                    #导入Tornado框架中的相关模块
#定义子类MainHandler
class MainHandler(tornado.web.RequestHandler):
    def get(self):                    #定义请求业务函数get()
        self.write("Hello, world")    #输出文本

def make_app():                       #定义应用配置函数
    return tornado.web.Application([
        (r"/", MainHandler),          #定义URL映射列表
    ])

if __name__ == "__main__":
    app = make_app()                  #调用配置函数
    app.listen(8888)                  #设置监听服务器8888端口
    tornado.ioloop.IOLoop.current().start()   #启动服务器
```

在上述实例代码中，首先导入了 Tornado 框架的相关模块，然后自定义 URL 的响应业务方法（GET、POST 等）。接下来，实例化 Tornado 模块中提供的 Application 类，并传 URL 映射列表及有关参数。最后启动服务器。在命令提示符下的对应子目录中执行：

```
python app.py
```

如果没有语法错误，服务器就已经启动并等待客户端连接了。在服务器运行以后，在浏览器地址栏中输入 http://localhost:8888，就可以访问服务器，看到默认主页页面了。在浏览器中的执行效果如图 1-2 所示。

图 1-2　执行效果

通过上述实例可以看出，使用 Tornado 框架编写的服务器端程序的结构是非常清晰的。其基本工作就是编写相关的业务处理类，并将它们和某一特定的 URL 映射起来，Tornado 框架服务器收到对应的请求后进行调用。一般来说，对于比较简单的网站项目，可以把所有的代码放入同一个模块之中。但为了维护方便，可以按照功能将其划分到不同的模块中，其一般模块结构（目录结构）如下。

```
proj\
    manage.py                    #服务器启动入口
    settings.py                  #服务器配置文件
    url.py                       #服务器URL配置文件
    handler\
            login.py             #相关URL业务请求处理类
    db\                          #数据库操作模块目录
    static\                      #静态文件存放目录
            js\                  #JavaScript文件存放目录
            css\                 #CSS文件目录
            img\                 #图片资源文件目录
    templates\                   #网页模板文件目录
```

1.2.2　获取请求参数

在 Python 程序中，客户端经常需要获取如下 3 类参数。
- URL 中的参数；
- GET 的参数；
- POST 中的参数。

1. 获取 URL 中的参数

在 Tornado 框架中，要获取 URL 中包含的参数，需要在 URL 定义中获取参数，并在对应的业务方法中给出相应的参数名进行获取。在 Tornado 框架的 URL 定义字符串中，使用正则表达式来匹配 URL 及 URL 中的参数，比如：

```
(r"uid/([0-9]+)",UserHdl)
```

上述形式的 URL 字符串定义可以接受形如"uid/"后跟一位或多位数字的客户端 URL 请求。针对上面的 URL 定义，可以如下方式定义 get()方法：

```
def get (self,uid):
    pass
```

此时，当发来匹配的 URL 请求时会截取与正则表达式匹配的部分，传递给 get()方法，这样可以把数据传递给 uid 变量，以在方法 get()中使用。

下面的实例文件 can.py 演示了在 GET 方法中获取 URL 参数的过程。

源码路径：daima\1\1-2\can.py

```
import tornado.ioloop                           #导入Tornado框架中的相关模块
import tornado.web                              #导入Tornado框架中的相关模块
class zi(tornado.web.RequestHandler):           #定义子类zi
    def get(self,uid):                          #获取URL参数
        self.write('你好，你的UID是%s!' % uid)   #显示UID，来源于下面的正则表达式

app = tornado.web.Application([                 #使用正则表达式获取参数
    (r'/([0-9]+)',zi),
    ],debug=True)

if __name__ == '__main__':
    app.listen(8888)                            #设置监听服务器的8888端口
    tornado.ioloop.IOLoop.instance().start()    #启动服务器
```

在上述实例代码中，使用正则表达式定义了 URL 字符串，使用 get()方法获取了 URL 参数中的 UID。在浏览器中的执行效果如图 1-3 所示。

2．获取 GET 和 POST 中的参数

在 Tornado 框架中，要获取 GET 或 POST 中的请求参数，只需要调用从类 RequestHandler 中继承来的 get_argument()方法即可。方法 get_argument()的原型如下。

图 1-3　执行效果

```
get_argument('name', default='',strip=False)
```

- name：请求中的参数名称。
- default：当没有获取参数时给定一个默认值。
- strip：指定是否去掉参数两侧的空格。

下面的实例文件 po.py 演示了获取 POST 参数的过程。

源码路径：daima\1\1-2\po.py

```
import tornado.ioloop                           #导入Tornado框架中的相关模块
import tornado.web                              #导入Tornado框架中的相关模块
html_txt = """                                  #初始化变量html_txt
<!DOCTYPE html>                                 #下面是一段普通的HTML代码
<html>
    <body>
        <h2>收到GET请求</h2>
        <form method='post'>
            <input type='text' name='name' placeholder='请输入你的姓名' />
            <input type='submit' value='发送POST请求' />
        </form>
    </body>
</html>
"""
class zi(tornado.web.RequestHandler):           #定义子类zi
    def get(self):                              #定义方法get()以处理get请求
        self.write(html_txt)                    #作为页面内容进行处理
    def post(self):                             #定义方法post()以处理post请求
        name = self.get_argument('name',default='匿名',strip=True)  #获取上面表单中的姓名
        self.write("你的姓名是:%s" % name)      #显示姓名
app = tornado.web.Application([                 #实例化Application对象
    (r'/get',zi),
    ],debug=True)

if __name__ == '__main__':
    app.listen(8888)                            #设置监听服务器的8888端口
    tornado.ioloop.IOLoop.instance().start()    #启动服务器
```

在上述实例代码中，当服务器收到 GET 请求时返回一个带有表单的页面内容；当用户填写自己的姓名并单击"发送 POST 请求"按钮时，将用户输入的姓名以 POST 参数形式发送到服务器端。最后在服务器端调用方法 get_argument()来获取输出请求。在浏览器中输入"http://localhost:8888/get"后的初始执行效果如图 1-4 所示。

在表单中输入姓名"浪潮软件"，然后单击"发送 POST 请求"按钮后的执行效果如图 1-5 所示。

图 1-4　初始执行效果

图 1-5　输入信息并单击按钮后的执行效果

1.2.3　使用 cookie

cookie（有时也用其复数形式 cookies）指某些网站为了辨别用户身份、进行会话跟踪而存储在用户本地终端上的数据（通常经过加密）。在现实应用中，服务器可以利用 cookie 包含信息的任意性来筛选并经常性维护这些信息，以判断在 HTTP 传输中的状态。cookie 最典型的应用是判定注册用户是否已经登录网站，用户可能会得到提示，是否在下一次进入此网站时保留用户信息以便简化登录手续，这些都是 cookie 的功用。另一个重要应用场合是"购物车"之类处理。用户可能会在一段时间内在同一家网站的不同页面中选择不同的商品，这些信息都会写入 cookie，以便在最后付款时提取信息。

在 Tornado 框架中提供了直接作用于一般 cookie 和安全 cookie 的方法。安全的 cookie 指存储在客户端的 cookie 是经过加密的，客户端只能查看到加密后的数据。在 Tornado 框架中，使用 cookie 和安全 cookie 的常用方法原型如下。

- set_cookie('name',value)：设置 cookie。
- get_cookie('name')：获取 cookie 值。
- set_secure_cookie ('name',value)：设置安全 cookie 值。
- get_secure_cookie('name')：获取安全 cookie 值。
- clear_cookie('name')：清除名为 name 的 cookie 值。
- clear_all_cookies()：清除所有 cookie。

下面的实例文件 co.py 演示了在不同页面中设置与获取 cookie 值的过程。

源码路径：daima\1\1-2\co.py

```
import tornado.ioloop              #导入Tornado框架中的相关模块
import tornado.web                 #导入Tornado框架中的相关模块
import tornado.escape              #导入Tornado框架中的相关模块
#定义处理类aaaa,用于设置cookie的值
class aaaa(tornado.web.RequestHandler):
    def get(self):                 #处理get请求
        #URL编码处理
        self.set_cookie('odn_cookie',tornado.escape.url_escape("未加密COOKIE串"))
        #设置普通cookie
        self.set_secure_cookie('scr_cookie',"加密SCURE_COOKIE串")
        #设置加密cookie
        self.write("<a href='/shcook'>查看设置的COOKIE</a>")
#定义处理类shcookHdl,用于获取cookie的值
class shcookHdl(tornado.web.RequestHandler):
    def get(self):                 #处理get请求
```

```
            #获取普通cookie
            odn_cookie = tornado.escape.url_unescape(self.get_cookie('odn_cookie'))
            #进行URL解码
            scr_cookie = self.get_secure_cookie('scr_cookie').decode('utf-8')
            #获取加密cookie
            self.write("普通COOKIE:%s,<br/>安全COOKIE:%s" % (odn_cookie,scr_cookie))
app = tornado.web.Application([
    (r'/sscook',aaaa),
    (r'/shcook',shcookHdl),
    ],cookie_secret='abcddddkdk##$$34323sdDsdfdsf#23')
if __name__ == '__main__':
    app.listen(8888)                                    #设置监听服务器的8888端口
    tornado.ioloop.IOLoop.instance().start()            #启动服务器
```

在上述实例代码中定义了两个类，分别用于设置 cookie 的值和获取 cookie 的值。当在浏览器中输入"http://localhost:8888/sscook"时开始设置 cookie，初始执行效果如图 1-6 所示。

当单击页面中的"查看设置的 COOKIE"链接时，会访问"shcook"，显示出刚刚设置的 cookie 值。执行效果如图 1-7 所示。

图 1-6 初始执行效果

图 1-7 单击"查看设置的 COOKIE"链接后的执行效果

1.2.4 URL 转向

所谓 URL 转向，是通过服务器的特殊设置，将访问当前域名的用户引导到用户指定的另一个 URL 页面。在 Tornado 框架中可以实现 URL 转向的功能，这需要借助如下两个方法实现 URL 转向功能。

- redirect(url)：在业务逻辑中转向 URL。
- RedirectHandler：实现某个 URL 的直接转向。

使用类 RedirectHandler 的语法格式如下。

```
(r'/aaa', tornado.Web.RedirectHandler, dict (url='/abc'))
```

下面的实例文件 zh.py 演示了实现两种 URL 转向功能的过程。

源码路径：daima\1\1-2\zh.py

```
import tornado.ioloop                    #导入Tornado框架中的相关模块
import tornado.web                       #导入Tornado框架中的相关模块
#定义类DistA，作为转向的目标URL请求处理程序
class DistA(tornado.web.RequestHandler):
    def get(self):                                      #获取get请求
        self.write("被转向的目标页面！")                  #显示一个字符串
#定义转向处理器类SrcA
class SrcA(tornado.web.RequestHandler):
    def get(self):                                      #获取get请求
        self.redirect('/dist')                          #业务逻辑转向，指向一个URL

app = tornado.web.Application([
    (r'/dist',DistA),                                   #指向DistA类
    (r'/src',SrcA),                                     #指向SrcA类
    (r'/rdrt',tornado.web.RedirectHandler,{'url':'/src'}) #定义一个直接转向URL
    ])

if __name__ == '__main__':
    app.listen(8888)                                    #设置监听服务器的8888端口
    tornado.ioloop.IOLoop.instance().start()            #启动服务器
```

在上述实例代码中定义了两个类，其中类 DistA 作为转向的目标 URL 请求处理程序，类

SrcA 是转向处理程序。当访问指向这个业务类时，会转向"/dist"网址。最后，在类 Application 中定义一个直接转向，只要访问"/rdrt"就会直接转向"/src"。在执行后如果试图访问"/rdrt"的 URL，会转向"/src"，再最终转向"/dist"。也就是说，无论是访问"/rdrt"，还是访问"/src"，最终的执行效果都如图 1-8 所示。

图 1-8　执行效果

1.2.5　使用静态资源文件

大多数 Web 应用程序都有一组对所有用户一视同仁的文件，这些文件在应用程序运行时是不会发生变化的。这些可以是用于网站装饰的媒体文件（如图片），用于描述如何在屏幕上绘制网页的 CSS，能够被浏览器下载和执行的 JavaScript 代码，不含动态内容的 HTML 页面等。这些不发生变化的文件称为静态资源文件。

Tornado 框架支持在网站页面中直接使用静态的资源文件，如图片、JavaScript 脚本、CSS 等。当需要用到静态文件资源时，需要在 Application 类初始化时提供"static_path"参数。下面的实例文件 tu.py 演示了使用图片静态资源文件的过程。

源码路径：daima\1\1-2\tu.py

```
import tornado.ioloop                              #导入Tornado框架中的相关模块
import tornado.web                                 #导入Tornado框架中的相关模块
#定义类AAA，用于访问输出静态图片文件
class AAA(tornado.web.RequestHandler):
    def get(self):                                 #获取get请求
        self.write("<img src='/static/ttt.jpg' />")    #使用一幅本地图片
app = tornado.web.Application([
    (r'/stt',AAA),                                 #参数"/stt"的请求
    ],static_path='./static')                      #调用本网站中的图片"static/ttt.jpg"
if __name__ == '__main__':
    app.listen(8888)                               #设置监听的服务器8888端口
    tornado.ioloop.IOLoop.instance().start()       #启动服务器
```

在上述实例代码中，通过参数"/stt"请求返回 HTML 代码中的一个 img 标签，并调用本网站中的图片"static/ttt.jpg"。在初始化类 Application 时提供了 static_path 参数，以指明静态资源的目录。最终的执行效果如图 1-9 所示。

图 1-9　执行效果

1.3　表单和模板操作

在 Tornado 框架中可以灵活地使用表单和模板技术，通过使用这些技术可以实现动态 Web 功能。

1.3.1　一个基本的注册表单

在下面的实例文件 001.py 中，首先实现一个让用户填写注册信息的 HTML 表单，然后显示表单处理结果。

源码路径：daima\1\1-3\001.py

```python
import os.path

import tornado.httpserver
import tornado.ioloop
import tornado.options
import tornado.web

from tornado.options import define, options
define("port", default=8001, help="运行在指定端口", type=int)

class IndexHandler(tornado.web.RequestHandler):
    def get(self):
        self.render('index.html')

class PoemPageHandler(tornado.web.RequestHandler):
    def post(self):
        noun1 = self.get_argument('noun1')
        noun2 = self.get_argument('noun2')
        verb = self.get_argument('verb')
        noun3 = self.get_argument('noun3')
        self.render('poem.html', roads=noun1, wood=noun2, made=verb,
                    difference=noun3)

if __name__ == '__main__':
    tornado.options.parse_command_line()
    app = tornado.web.Application(
        handlers=[(r'/', IndexHandler), (r'/poem', PoemPageHandler)],
        template_path=os.path.join(os.path.dirname(__file__), "templates")
    )
    http_server = tornado.httpserver.HTTPServer(app)
    http_server.listen(options.port)
    tornado.ioloop.IOLoop.instance().start()
```

为了突出 Web 程序界面的美观性，接下来我们将使用模板技术。框架 Tornado 自身提供了一个轻量级的模板模块 tornado.template，用于快速并且灵活地实现模板功能。我们将模板文件保存在 "templates" 文件夹中，其中文件 index.html 作为注册表单。具体实现代码如下。

源码路径：daima\1\1-3\templates\index.html

```html
<!DOCTYPE html>
<html>
    <head><title>会员登录</title></head>
    <body>
        <h1>输入注册信息.</h1>
        <form method="post" action="/poem">
        <p>用户名<br><input type="text" name="noun1"></p>
        <p>密码<br><input type="text" name="noun2"></p>
        <p>确认密码<br><input type="text" name="verb"></p>
        <p>性别<br><input type="text" name="noun3"></p>
        <input type="submit">
        </form>
    </body>
</html>
```

模板文件 poem.html 用于显示注册结果信息。具体实现代码如下。

源码路径：daima\1\1-3\templates\poem.html

```html
<!DOCTYPE html>
<html>
    <head><title>注册结果</title></head>
    <body>
        <h1>下面是你的注册信息</h1>
        <p>用户名：{{roads}}<br>密码：{{wood}}<br>确认密码：{{made}}<br>性别：
           {{difference}}.</p>
    </body>
</html>
```

开始调试本实例。首先运行前面的 Python 文件 001.py，然后在浏览器中输入 http://localhost:8001/，接下来会显示注册表单,这是由模板文件 index.html 实现的。执行效果如图 1-10 所示。在表单中输入注册信息,并单击"提交查询内容"按钮后显示注册结果,这是由模板文件 poem.html 实现的。执行效果如图 1-11 所示。

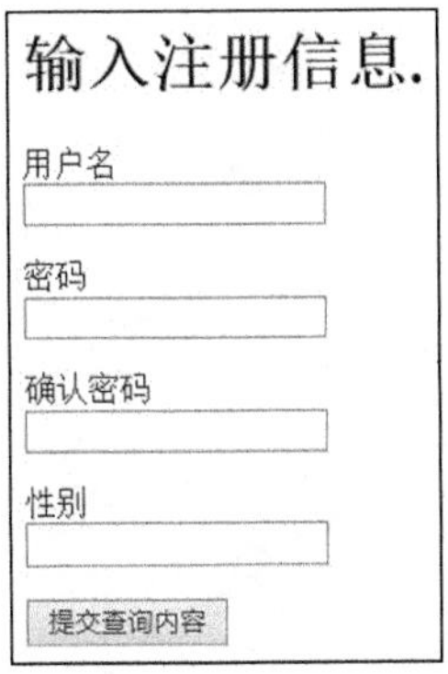

图 1-10　注册表单 　　　　　　图 1-11　注册结果

在上面的实例文件 001.py 中,定义了 RequestHandler 子类,并把它们传给 tornado.web.Application 对象。通过如下代码向 Application 对象中的 __init__()方法传递一个 template_path 参数。

```
template_path=os.path.join(os.path.dirname(__file__), "templates")
```

参数 template_path 的功能是告诉 Tornado 模板文件的具体位置,模板是一个允许你嵌入 Python 代码片段的 HTML 文件。通过上述代码告诉 Python,在 Tornado 应用文件相同目录下的 templates 文件夹中寻找模板文件。当告诉 Tornado 在哪里可以找到模板文件后,就可以使用类 RequestHandler 中的 render()函数告诉 Tornado 读入模板文件,插入其中的模板代码,并返回结果给浏览器。例如,在 IndexHandler 中通过如下代码告诉 Tornado 在文件夹"templates"下找到一个名为 index.html 的文件,读取其中的内容,并发送给浏览器。

```
self.render('index.html')
```

1.3.2　在模板中使用函数

在框架 Tornado 中,为模板功能提供了如下内置函数。
- escape(s):替换字符串 s 中的&、<、>为它们对应的 HTML 字符。
- url_escape(s):使用 urllib.quote_plus 替换字符串 s 中的字符为 URL 编码形式。
- json_encode(val):将 val 编码成 JSON 格式。(在系统底层,这是一个对 JSON 库的 dumps 函数的调用。)
- squeeze(s):过滤字符串 s,把连续的多个空白字符替换成一个空格。

在模板中可以使用一个自己编写的函数,这时只需要将函数名作为模板的参数传递即可,就像使用其他变量一样。例如:

```
>>> from tornado.template import Template
>>> def disemvowel(s):
...     return ''.join([x for x in s if x not in 'aeiou'])
...
>>> disemvowel("george")
'grg'
>>> print Template("my name is {{d('mortimer')}}").generate(d=disemvowel)
my name is mrtmr
```

再看下面的演示实例。首先在模板文件界面中提示用户输入两个文本:"源"文本和"替代"文本。单击"提交"按钮后会返回替代文本的一个副本,并将其中每个单词替换成源文本中首字母相同的某个单词。本实例包括 4 个文件:002.py(Tornado 程序)、style.css(CSS 文件)、index.html 和 munged.html(Tornado 模板)。

第 1 章　Tornado 框架

（1）在 Python 文件 002.py 中定义了两个请求处理类——IndexHandler 和 MungedPageHandler。具体实现代码如下。

源码路径：daima\1\1-3\moban2\002.py

```python
from tornado.options import define, options
define("port", default=8001, help="运行给定的端口", type=int)

class IndexHandler(tornado.web.RequestHandler):
    def get(self):
        self.render('index.html')

class MungedPageHandler(tornado.web.RequestHandler):
    def map_by_first_letter(self, text):
        mapped = dict()
        for line in text.split('\r\n'):
            for word in [x for x in line.split(' ') if len(x) > 0]:
                if word[0] not in mapped: mapped[word[0]] = []
                mapped[word[0]].append(word)
        return mapped

    def post(self):
        source_text = self.get_argument('source')
        text_to_change = self.get_argument('change')
        source_map = self.map_by_first_letter(source_text)
        change_lines = text_to_change.split('\r\n')
        self.render('munged.html', source_map=source_map, change_lines=change_lines,
                    choice=random.choice)

if __name__ == '__main__':
    tornado.options.parse_command_line()
    app = tornado.web.Application(
        handlers=[(r'/', IndexHandler), (r'/poem', MungedPageHandler)],
        template_path=os.path.join(os.path.dirname(__file__), "templates"),
        static_path=os.path.join(os.path.dirname(__file__), "static"),
        debug=True
    )
    http_server = tornado.httpserver.HTTPServer(app)
    http_server.listen(options.port)
    tornado.ioloop.IOLoop.instance().start()
```

- 类 IndexHandler——简单渲染了 index.html 中的模板，其中包括一个允许用户发送一个源文本（在 source 域中）和一个替换文本（在 change 域中）到 /poem 的表单。
- 类 MungedPageHandler——用于处理到 /poem 的 POST 请求。当一个请求到达时，它对传入的数据进行一些基本的处理，然后为浏览器渲染模板。map_by_first_letter 方法将传入的文本（从 source 域）分割成单词，然后创建一个字典，其中每个字母表中的字母对应文本中所有以指定字母开头的单词（我们将其放入一个叫作 source_map 的变量中）。再把这个字典和用户在替代文本（表单的 change 域）中指定的内容一起传给模板文件 munged.html。此外，我们还将 Python 标准库的 random.choice 函数传入模板，这个函数以一个列表作为输入，返回列表中的任一元素。
- 参数 static_path——指定了应用程序放置静态资源的目录，如图像、CSS 文件、JavaScript 文件等。
- 在编写 Web 应用程序时，总希望提供像样式表、JavaScript 文件和图像这样不需要为每个文件编写独立处理程序的"静态内容"。例如，可以通过向 Application 类的构造函数传递一个名为 static_path 的参数来告诉 Tornado 从文件系统的一个特定位置提供静态文件。下面是 Alpha Munger 中的相关代码片段。

```python
app = tornado.web.Application(
    handlers=[(r'/', IndexHandler), (r'/poem', MungedPageHandler)],
    template_path=os.path.join(os.path.dirname(__file__), "templates"),
    static_path=os.path.join(os.path.dirname(__file__), "static"),
```

```
        debug=True
)
```

这样设置了一个当前应用目录下名为 static 的子目录作为 static_path 的参数。现在应用将以读取 static 目录下的 filename.ext 来响应诸如/static/filename.ext 的请求，并在响应的主体中返回。

(2) 模板文件 index.html 用于显示主表单界面。具体实现代码如下。

源码路径：daima\1\1-3\moban2\templates\index.html

```
<!DOCTYPE html>
<html>
    <head>
        <link rel="stylesheet" href="{{ static_url("style.css") }}">
        <title>操作</title>
    </head>
    <body>
        <h1>替换操作</h1>
        <p>在下面输入两个文本，替换文本将用源文本中同一字母开头的单词替换单词。</p>
        <form method="post" action="/poem">
        <p>Source text<br>
            <textarea rows=4 cols=55 name="source"></textarea></p>
        <p>Text for replacement<br>
            <textarea rows=4 cols=55 name="change"></textarea></p>
        <input type="submit">
        </form>
    </body>
</html>
```

在 Tornado 模板中提供了 static_url()函数来生成 static 目录下文件的 URL。下面是在文件 index.html 中 static_url 调用的代码。

```
<link rel="stylesheet" href="{{ static_url("style.css") }}">
```

通过上述代码调用 static_url 生成了 URL 的值，并输出类似下面的代码。

```
<link rel="stylesheet" href="/static/style.css?v=ab12">
```

此处为什么使用 static_url 而不是在模板中使用硬编码方式呢？有如下两个原因。

- 函数 static_url()创建了一个基于文件内容的散列值，并将其添加到 URL 末尾（查询字符串的参数 v）。这个散列值确保浏览器总是加载一个文件的最新版而不是之前的缓存版本。无论是在应用的开发阶段，还是在部署到生产环境使用时，这个方式都非常有用，因为用户不必再为了看到静态内容而清除浏览器缓存了。
- 可以改变应用 URL 的结构，而不需要改变模板中的代码。例如，可以配置 Tornado 响应来自路径/s/filename.ext 的请求时提供静态内容，而不是默认的/static 路径。如果使用 static_url 而不是硬编码，代码不需要改变。比如，要把静态资源从我们刚才使用的/static 目录移到新的/s 目录，可以简单地改变静态路径 static 为 s，然后每个使用 static_url 包裹的引用都会自动更新。如果你在每个引用静态资源的文件中硬编码静态路径部分，将不得不手动修改每个模板。

(3) 模板文件 munged.html 用于显示替换结果。具体实现代码如下。

源码路径：daima\1\1-3\moban2\templates\munged.html

```
<!DOCTYPE html>
<html>
    <head>
        <link rel="stylesheet" href="{{ static_url("style.css") }}">
        <title>结果</title>
    </head>
    <body>
        <h1>现在的文本是</h1>
        <p>
{% for line in change_lines %}
    {% for word in line.split(' ') %}
        {% if len(word) > 0 and word[0] in source_map %}
            <span class="replaced"
                    title="{{word}}">{{ choice(source_map[word[0]]) }}</span>
```

```
                {% else %}
                    <span class="unchanged" title="unchanged">{{word}}</span>
                {% end %}
            {% end %}
                <br>
        {% end %}
            </p>
    </body>
</html>
```

在上述代码中迭代替代文本中的每行，再迭代每行中的每个单词。如果当前单词的第一个字母是 source_map 字典的一个键，则使用 random.choice 函数从字典的值中随机选择一个单词并展示它。如果字典的键中没有这个字母，则展示源文本中的原始单词。每个单词包括一个 span 标签，其中的 class 属性指定这个单词是替换后的（class="replaced"）还是原始的（class="unchanged"）。（我们还将原始单词放到了 span 标签的 title 属性中，以便于用户在鼠标指针经过单词时可以查看哪些单词被替换。下面看执行效果。假设在表单中分别输入 "from tornado.template import Template" 和 "print Template("my name is {{d('mortimer')}}").generate(d=disemvowel)"，如图 1-12 所示。

图 1-12　输入表单

单击"提交查询内容"按钮后执行替换操作，并显示替换后的结果，如图 1-13 所示。

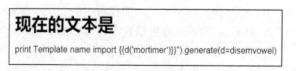

图 1-13　替换后的结果

1.4　数据库操作

在下面的内容中，假设已经在计算机上安装了 MongoDB，并通过 PyMongo 作为驱动来连接 MongoDB。本节将介绍在 Tornado 框架中实现数据库操作的知识。

1.4.1　实现持久化 Web 服务

假设我们需要编写一个只从 MongoDB 读取数据的 Web 服务，然后编写一个可以读写数据的服务。例如，将要创建的应用是一个基于 Web 的简单字典，在发送一个指定单词的请求后返回这个单词的定义。一个典型的交互如下。

```
$ curl http://localhost:8000/oarlock
{definition: "连接到一个设备",
"word": "oarlock"}
```

这个Web服务将从MongoDB数据库中取得数据。具体来说，将根据word属性查询文档。在查看Web应用本身的源码之前，先从Python解释器向数据库中添加一些单词。例如，通过如下文件001.py，可以向MongoDB数据库中添加指定的单词。

源码路径：daima\1\1-4\001.py

```
import pymongo
conn = pymongo.MongoClient("localhost", 27017)
db = conn.example
db.words.insert({"word": "oarlock", "definition": "A device attached to a rowboat to hold the oars in
    place"})
db.words.insert({"word": "seminomadic", "definition": "Only partially nomadic"})
db.words.insert({"word": "perturb", "definition": "Bother, unsettle, modify"})
```

通过如下命令开启MongoDB服务：

```
mongod --dbpath "h:\data"
```

在上述命令中，"h:\data"是一个保存MongoDB数据库数据的目录，读者可以随意在本地计算机硬盘中创建，并且还可以自定义目录名字。然后运行文件001.py，执行后会向MongoDB数据库中添加指定的单词。

为了验证上述添加的单词，我们编写如下所示的文件 definitions_readonly.py，在Tornado框架中实现对MongoDB数据库的访问。

源码路径：daima\1\1-4\definitions_readonly.py

```
import tornado.httpserver
import tornado.ioloop
import tornado.options
import tornado.web

import pymongo

from tornado.options import define, options
define("port", default=8000, help="run on the given port", type=int)

class Application(tornado.web.Application):
    def __init__(self):
        handlers = [(r"/(\w+)", WordHandler)]
        conn = pymongo.MongoClient("localhost", 27017)
        self.db = conn["example"]
        tornado.web.Application.__init__(self, handlers, debug=True)

class WordHandler(tornado.web.RequestHandler):
    def get(self, word):
        coll = self.application.db.words
        word_doc = coll.find_one({"word": word})
        if word_doc:
            del word_doc["_id"]
            self.write(word_doc)
        else:
            self.set_status(404)
            self.write({"error": "word not found"})

if __name__ == "__main__":
    tornado.options.parse_command_line()
    http_server = tornado.httpserver.HTTPServer(Application())
    http_server.listen(options.port)
    tornado.ioloop.IOLoop.instance().start()
```

运行上述实例文件definitions_readonly.py，然后在浏览器中输入"http://localhost:8000/perturb"后会显示：

```
{"word": "perturb", "definition": "Bother, unsettle, modify"}
```

这说明在Tornado框架中实现了对MongoDB数据库数据的访问功能。如果在浏览器中请求一个数据库中没有添加的单词，会得到一个404错误以及一条错误消息。

```
{"error": "word not found"}
```

那么，这个程序是如何工作的呢？让我们看看这个程序的主线。开始，我们在程序的最上面导入了 pymongo 库。然后在 TornadoApplication 对象的 __init__ 方法中实例化了一个 pymongo 连接对象。我们在 Application 对象中创建了一个 db 属性，指向 MongoDB 的 example 数据库。下面是相关的代码。

```
conn = pymongo.MongoClient ("localhost", 27017)
self.db = conn["example"]
```

一旦我们在 Application 对象中添加了 db 属性，就可以在任何 RequestHandler 对象中使用 self.application.db 访问它。其实这正是我们为了取出 pymongo 的 words 集合对象而在 WordHandler 的 get 方法中所做的事情。

```
def get(self, word):
    coll = self.application.db.words
    word_doc = coll.find_one({"word": word})
    if word_doc:
        del word_doc["_id"]
        self.write(word_doc)
    else:
        self.set_status(404)
        self.write({"error": "word not found"})
```

在我们将集合对象指定给变量 coll 后，我们使用用户在 HTTP 路径中请求的单词调用 find_one 方法。如果我们发现这个单词，则从字典中删除 _id 键（以便 Python 的 JSON 库可以将其序列化），然后将其传递给 RequestHandler 的 write 方法。write 方法将会自动序列化字典为 JSON 格式。

如果 find_one 方法没有匹配任何对象，则返回 None。这时将响应状态设置为 404，并且写一个简短的 JSON 来提示用户这个单词在数据库中没有找到。

在上述实例中，虽然可以很简单地在字典中查询单词，但是在交互解释器中添加单词的过程会非常麻烦。其实我们完全可以使 HTTP 在请求网站服务时创建和修改单词：首先发出一个特定单词的 POST 请求，然后根据请求中给出的定义修改已经存在的定义。如果这个单词并不存在，则创建它。例如，通过如下过程创建一个新的单词"pants"。

```
http://localhost:8000/pants
{"definition": "a leg shirt", "word": "pants"}
```

下面的实例文件 definitions_readwrite.py 演示了实现一个可读写 Web 服务的过程。

源码路径：daima\1\1-4\definitions_readwrite.py

```
define("port", default=8000, help="run on the given port", type=int)

class Application(tornado.web.Application):
    def __init__(self):
        handlers = [(r"/(\w+)", WordHandler)]
        conn = pymongo.MongoClient("localhost", 27017)
        self.db = conn["definitions"]
        tornado.web.Application.__init__(self, handlers, debug=True)

class WordHandler(tornado.web.RequestHandler):
    def get(self, word):
        coll = self.application.db.words
        word_doc = coll.find_one({"word": word})
        if word_doc:
            del word_doc["_id"]
            self.write(word_doc)
        else:
            self.set_status(404)
    def post(self, word):
        definition = self.get_argument("definition")
        coll = self.application.db.words
        word_doc = coll.find_one({"word": word})
        if word_doc:
            word_doc['definition'] = definition
            coll.save(word_doc)
        else:
```

```
                word_doc = {'word': word, 'definition': definition}
                coll.insert(word_doc)
            del word_doc["_id"]
            self.write(word_doc)

if __name__ == "__main__":
    tornado.options.parse_command_line()
    http_server = tornado.httpserver.HTTPServer(Application())
    http_server.listen(options.port)
    tornado.ioloop.IOLoop.instance().start()
```

在上述代码中，使用 get_argument()函数获取了 POST 请求中传递的 definition 参数，然后使用 find_one()函数从数据库中加载给定单词的文档。如果发现这个单词的文档，将 definition 条目的值设置为从 POST 参数中取得的值，然后调用集合对象的 save()函数将改变写到数据库中。如果没有发现文档则创建一个新的，并使用 insert()函数将文档保存到数据库中。无论上述哪种情况，都要在数据库操作执行之后在响应中写文档（注意，首先要删掉_id 属性）。

1.4.2　图书管理系统

接下来，通过一个图书管理系统的实现过程，介绍在 Tornado 框架中使用 MongoDB 数据库实现动态 Web 的过程。

源码路径：daima\1\1-4\BookManger

（1）在 MongoDB 服务器中创建一个数据库和集合，并用图书内容进行填充。例如，下面的演示过程。

```
>>> import pymongo
>>> conn = pymongo.MongoClient ()
>>> db = conn["bookstore"]
>>> db.books.insert({
...     "title":"Python开发从入门到精通",
...     "subtitle": "Python",
...     "image":"123.gif",
...     "author": "浪潮",
...     "date_added":20171231,
...     "date_released": "August 2007",
...     "isbn":"978-7-596-52932-1",
...     "description":"<p>[...]</p>"
... })
ObjectId('4eb6f1a6136fc42171000000')
>>> db.books.insert({
...     "title":"PHP从入门到精通",
...     "subtitle": "Web服务",
...     "image":"345.gif",
...     "author": "学习PHP",
...     "date_added":20171231,
...     "date_released": "May 2007",
...     "isbn":"978-7-534-52926-0",
...     "description":"<p>[...]/p>"
... })
ObjectId('4eb6f1cb136fc42171000001')
```

（2）编写 Python 程序文件 burts_books_db.py。首先在程序中添加一个 db 属性来连接 MongoDB 服务器，然后使用连接的 find()函数从数据库中取得图书文档的列表，并在渲染 recommended.html 时将这个列表传递给 RecommendedHandler 的 get 方法。文件 burts_books_db.py 的具体实现代码如下。

```
#!/usr/bin/env python
import os.path

import tornado.auth
import tornado.escape
import tornado.httpserver
import tornado.ioloop
import tornado.options
import tornado.web
```

```python
from tornado.options import define, options

import pymongo

define("port", default=8001, help="请运行在给定的端口", type=int)
class Application(tornado.web.Application):
    def __init__(self):
        handlers = [
            (r"/", MainHandler),
            (r"/recommended/", RecommendedHandler),
        ]
        settings = dict(
            template_path=os.path.join(os.path.dirname(__file__), "templates"),
            static_path=os.path.join(os.path.dirname(__file__), "static"),
            ui_modules={"Book": BookModule},
            debug=True,
        )
        conn = pymongo.MongoClient("localhost", 27017)
        self.db = conn["bookstore"]
        tornado.web.Application.__init__(self, handlers, **settings)


class MainHandler(tornado.web.RequestHandler):
    def get(self):

        self.render(
            "index.html",
            page_title = "图书管理| 主页",
            header_text = "欢迎使用图书管理系统!",
        )

class RecommendedHandler(tornado.web.RequestHandler):
    def get(self):
        coll = self.application.db.books
        books = coll.find()
        self.render(
            "recommended.html",
            page_title = "图书系统 | 图书信息",
            header_text = "图书信息",
            books = books
        )

class BookModule(tornado.web.UIModule):
    def render(self, book):
        return self.render_string(
            "modules/book.html",
            book=book,
        )

    def css_files(self):
        return "css/recommended.css"

    def javascript_files(self):
        return "js/recommended.js"


def main():
    tornado.options.parse_command_line()
    http_server = tornado.httpserver.HTTPServer(Application())
    http_server.listen(options.port)
    tornado.ioloop.IOLoop.instance().start()


if __name__ == "__main__":
    main()
```

如果此时在浏览器中输入 http://localhost:8001/recommended/，会读取并显示数据库中的图书信息。执行效果如图 1-14 所示。

图 1-14 执行效果

（3）编写 Python 文件 burts_books_rwdb.py 实现图书添加和修改两个功能。具体实现代码如下。

```python
define("port", default=8001, help="请运行在给定的端口", type=int)

class Application(tornado.web.Application):
    def __init__(self):
        handlers = [
            (r"/", MainHandler),
            (r"/recommended/", RecommendedHandler),
            (r"/edit/([0-9Xx\-]+)", BookEditHandler),
            (r"/add", BookEditHandler)
        ]
        settings = dict(
            template_path=os.path.join(os.path.dirname(__file__), "templates"),
            static_path=os.path.join(os.path.dirname(__file__), "static"),
            ui_modules={"Book": BookModule},
            debug=True,
        )
        conn = pymongo.MongoClient("localhost", 27017)
        self.db = conn["bookstore"]
        tornado.web.Application.__init__(self, handlers, **settings)


class MainHandler(tornado.web.RequestHandler):
    def get(self):

        self.render(
            "index.html",
            page_title = "图书管理 | 主页",
            header_text = "欢迎使用图书管理系统！",
        )

class BookEditHandler(tornado.web.RequestHandler):
    def get(self, isbn=None):
        book = dict()
        if isbn:
            coll = self.application.db.books
            book = coll.find_one({"isbn": isbn})
        self.render("book_edit.html",
            page_title="Burt's Books",
            header_text="Edit book",
            book=book)

    def post(self, isbn=None):
        import time
```

```python
                    book_fields = ['isbn', 'title', 'subtitle', 'image', 'author',
                        'date_released', 'description']
                    coll = self.application.db.books
                    book = dict()
                    if isbn:
                            book = coll.find_one({"isbn": isbn})
                    for key in book_fields:
                            book[key] = self.get_argument(key, None)

                    if isbn:
                            coll.save(book)
                    else:
                            book['date_added'] = int(time.time())
                            coll.insert(book)
                    self.redirect("/recommended/")

class RecommendedHandler(tornado.web.RequestHandler):
        def get(self):
                coll = self.application.db.books
                books = coll.find()
                self.render(
                        "recommended.html",
                        page_title = "Burt's Books | Recommended Reading",
                        header_text = "Recommended Reading",
                        books = books
                )

class BookModule(tornado.web.UIModule):
        def render(self, book):
                return self.render_string(
                        "modules/book.html",
                        book=book,
                )

        def css_files(self):
                return "css/recommended.css"

        def javascript_files(self):
                return "js/recommended.js"

def main():
        tornado.options.parse_command_line()
        http_server = tornado.httpserver.HTTPServer(Application())
        http_server.listen(options.port)
        tornado.ioloop.IOLoop.instance().start()

if __name__ == "__main__":
        main()
```

在上述代码中，BookEditHandler 主要完成如下两个功能。

- GET 请求渲染显示一个已存在图书数据的 HTML 表单（在模板 book_edit.html 中）。
- POST 请求从表单中取得数据，更新数据库中已存在的图书记录或根据提供的数据添加一本新的图书。

BookEditHandler 实现了两个不同路径模式的请求：其中一个是实现图书添加功能的/add，用于提供不存在信息的编辑表单，因此你可以向数据库中添加一本新的图书；另一个是实现图书修改功能的/edit/([0-9Xx\-]+)，用于根据图书的 ISBN 参数修改已存在图书的信息。

函数 get()的功能是从数据库中取出图书信息，如果该函数作为/add 请求的结果被调用，Tornado 将调用一个没有第二个参数的 get 方法（因为路径中没有正则表达式的匹配组）。在这种情况下，默认将一个空的 book 字典传递给 book_edit.html 模板。如果该方法作为/edit/0-123-456 请求的结果被调用，那么 isbn 参数被设置为 0-123-456。在这种情况下，我们从程序实例中取得 books 集合，并用它查询 isbn 匹配的图书，然后传递 book 字典给模板。

函数 post()的功能是将表单中的数据保存到数据库中,具体来说有两个功能:处理修改已存在图书信息的请求以及添加新图书信息的请求。如果有 isbn 参数(即路径的请求类似于/edit/0-123-4567),则编辑给定 isbn。如果这个参数没有提供,则添加新图书信息。先设置一个空的字典变量 book,如果正在编辑已存在的图书信息,则使用 book 集合的 find_one()函数从数据库中加载和传入与 isbn 值对应的文档。无论是哪一种情况,book_fields 列表都指定哪些域应该出现在图书文档中,再迭代这个列表,使用 RequestHandler 对象的 get_argument 方法从 POST请求中抓取对应的值。

图书修改功能的模板文件是 book_edit.html。具体实现代码如下所示。

```
{% extends "main.html" %}
{% autoescape None %}

{% block body %}
<form method="POST">
    ISBN <input type="text" name="isbn"
        value="{{ book.get('isbn', '') }}"><br>
    书名  <input type="text" name="title"
        value="{{ book.get('title', '') }}"><br>
    标题  <input type="text" name="subtitle"
        value="{{ book.get('subtitle', '') }}"><br>
    图片  <input type="text" name="image"
        value="{{ book.get('image', '') }}"><br>
    作者  <input type="text" name="author"
        value="{{ book.get('author', '') }}"><br>
    出版时间 <input type="text" name="date_released"
        value="{{ book.get('date_released', '') }}"><br>
    内容简介<br>
    <textarea name="description" rows="5"
        cols="40">{% raw book.get('description', '')%}</textarea><br>
    <input type="submit" value="Save">
</form>
{% end %}
```

上述代码实现了一个基本的 HTML 表单,如果请求处理函数传进来了 book 字典,那么将用它预填充带有已存在图书数据的表单中。如果键不在字典中,则使用 Python 字典对象的 get方法为其提供默认值,标签 input 中的 name 属性被设置为 book 字典的对应键。因为 form 标签没有 action 属性,所以表单中的 POST 将会定向到当前 URL。如果页面以/edit/0-123-4567 进行加载,POST 请求将转向/edit/0-123-4567;如果页面以/add 进行加载,则 POST 将转向/add。

添加新图书的界面如图 1-15 所示。

单击图 1-14 中的"编辑"链接后会弹出图书修改界面,在此界面中显示修改此图书信息的界面。执行效果如图 1-16 所示。

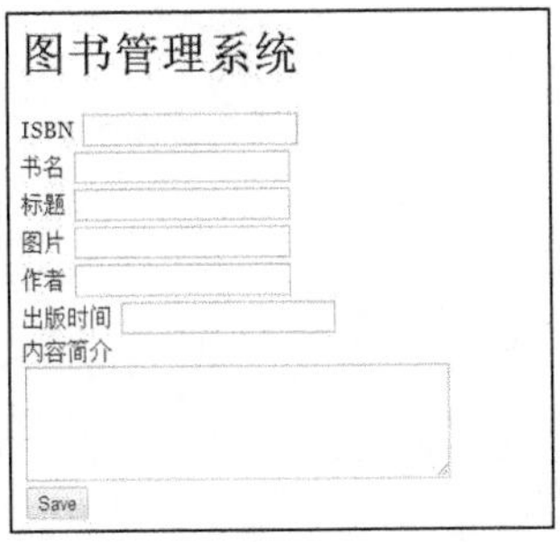

图 1-15　添加新图书的界面 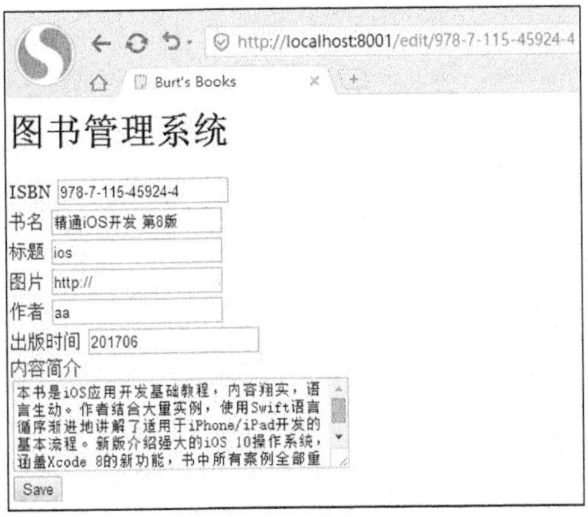　　　　　图 1-16　图书修改界面

第 2 章

使用 Django

　　Django 是一个开放源代码的 Web 应用框架，由 Python 写成。Django 遵守 BSD 版权，初次发布于 2005 年 7 月，并于 2008 年 9 月发布了第一个正式版本 Django 1.0。Django 采用了模型-视图-控制器 MVC（Model-View-Controller，MVC）的软件设计模式。本章将详细讲解使用 Django 框架开发动态 Web 程序的核心知识。

2.1 Django 简介

Django 自称是"能够很好地应对应用上线期限的 Web 框架"。其由劳伦斯日报负责在线业务的 Web 开发者创建。

在安装 Django 之前，必须先安装 Python。Apache 是 Web 服务器中使用量最多的，因此大多数部署都会使用这款服务器。Django 团队建议使用 mod_wdgi 这个 Apache 模块，并提供了安装指南和完整的开发文档，具体内容见 Django 官方网站。

在使用 Django 框架时需要使用数据库，当前的标准版 Django 只能够运行基于 SQL 的关系数据库管理系统（RDBMS）。开发者主要使用 4 种数据库，分别是 PostgreSQL、MySQL、Oracle 和 SQLite。其中最容易设置的是 SQLite。另外，SQLite 是这 4 个数据库中唯一一个无须部署数据库服务器的工具，所以使用起来也是最简单的。当然，简单并不代表功能弱，SQLite 的功能和另外 3 个一样强大。另外还需要提醒读者，在生产环境的服务器中并不是一定要使用 Apache，还可以有其他选择，其中有些服务器的内存占用量更少，速度更快。可以在 Django 官网中查找符合要求的 Web 服务器。

为什么 SQLite 很容易设置？因为 SQLite 数据库适配器是所有 Python 版本中自带的（从 2.5 版本开始）。注意，这里说的是适配器。有些 Python 发行版自带了 SQLite 本身，有些会使用系统上安装的 SQLite。

Django 支持众多关系数据库，SQLite 只是其中一种，所以如果不喜欢 SQLite，可以使用其他数据库。最近还有快速发展的非关系数据库（NoSQL）。这种类型的数据库提供了额外的可扩展性，能面对不断增长的数据量。如果处理像 Facebook、Twitter 那样的海量数据，关系数据库需要手动分区（切分）。如果需要使用 NoSQL 类数据库，如 MongoDB 或 Google App Engine 的原生数据库，建议尝试 Django-nonrel，这样用户就可以选择使用关系或非关系数据库。

2.2 Django 开发基础

本节将详细讲解 Django 开发的基础知识，包括如何搭建开发环境和常用命令等内容。

2.2.1 搭建 Django 环境

在当今技术环境下，有多种安装 Django 框架的方法。下面对这些安装方法按难易程度进行排序，其中越靠前的越简单：

- Python 包管理器；
- 操作系统包管理器；
- 官方发布的压缩包；
- 源码库。

最简单的下载和安装方式是使用 Python 包管理工具，建议读者使用这种安装方式。例如，可以使用 Setuptools 中的 easy_install 或 pip。目前在所有的操作系统平台上都可以使用这两个工具。对于 Windows 用户来说，在使用 Setuptools 时需要将 easy_install.exe 文件放在 Python 安装目录下的 Scripts 文件夹中。此时只要在 DOS 命令行窗口中使用一条命令就可以安装 Django。可以使用 easy_install 命令进行安装：

```
easy_install django
```

也可以使用 pip 命令进行安装：

```
pip install django
```
本书使用的 Django 版本是 1.10.4，控制台安装界面如图 2-1 所示。

图 2-1　控制台中安装 Django 的界面

2.2.2　常用的 Django 命令

接下来将要讲解 Django 框架中常用的命令。读者需要打开 Linux 或 Mac OS 的终端，直接在终端中输入这些命令（不是 Python 的 shell 中）。如果读者使用的是 Windows 系统，则在命令行窗口中输入操作命令。

1. 新建一个 Django 项目

要新建 Django 项目，可使用以下命令。

```
django-admin.py startproject project-name
```

其中，"project-name"表示项目名称。在 Windows 系统中需要使用如下命令创建项目：

```
django-admin startproject project-name
```

2. 新建应用

要新建应用，可使用以下命令。

```
python manage.py startapp app-name
```

或：

```
django-admin.py startapp app-name
```

通常一个项目有多个应用。当然，通用的应用也可以在多个项目中使用。

3. 同步数据库

要同步数据库，可使用以下命令。

```
python manage.py syncdb
```

读者需要注意，在 Django 1.7.1 及以上的版本中需要用以下命令。

```
python manage.py makemigrations
python manage.py migrate
```

这种方法可以创建表。当在 models.py 中新增类时，运行以上命令就可以自动在数据库中创建表，不需要手动创建。

4. 使用开发服务器

开发服务器在开发时使用,在修改代码后会自动重启,这会方便程序的调试和开发。但是由于性能问题,建议开发服务器只用来测试,不要用于生产环境。

```
python manage.py runserver
# 当提示端口被占用的时候,可以用其他端口
python manage.py runserver 8001
python manage.py runserver 9999
(当然,也可以终止占用端口的进程)

#监听所有可用IP(计算机可能有一个或多个内网IP地址,一个或多个外网IP地址,即有多个IP地址)
python manage.py runserver 0.0.0.0:8000
#访问对应的地址,比如 http://172.16.20.2:8000
```

5. 清空数据库

要清空数据库,可使用以下命令。

```
python manage.py flush
```

此命令会询问是 yes 还是 no,选择 yes 会把数据全部清空掉,只留下空表。

6. 创建超级管理员

要创建超级管理员,可使用以下命令。

```
python manage.py createsuperuser
# 按照提示输入用户名和对应的密码即可,邮箱可以为空,用户名和密码必填
# 修改用户密码
python manage.py changepassword username
```

7. 导出数据与导入数据

要导出数据,可使用以下命令。

```
python manage.py dumpdata appname > appname.json
```

要导入数据,可使用以下命令。

```
python manage.py loaddata appname.json
```

8. 启动 Django 项目的终端并打开命令行操作界面

要启动 Django 项目的终端并打开命令行操作界面,可使用以下命令。

```
python manage.py shell
```

如果安装了 bpython 或 ipython,会自动调用它们的界面,推荐安装 bpython。这个命令和直接运行 Python 或 bpython 进入 shell 的区别是:可以在这个 shell 里面调用当前项目的 models.py 中的 API。

9. 显示数据库的基本信息

要显示数据库的基本信息,可使用以下命令。

```
python manage.py dbshell
```

Django 会自动进入 settings.py 中设置的数据库,如果是 MySQL 或 postgreSQL,会要求输入用户名和密码。在这个终端可以执行数据库的 SQL 语句。如果用户对 SQL 比较熟悉,可能喜欢这种方式。

2.2.3 第一个 Django 项目

下面的实例代码演示了创建并运行第一个 Django 项目的过程。

源码路径:daima\2\2-2\mysite

(1) 在命令行窗口中定位到 H 盘,然后通过如下命令创建一个 "mysite" 目录。

```
django-admin startproject mysite
```

创建成功后会看到如下所示的目录样式。

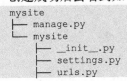

```
└── wsgi.py
```

也就是说,在 H 盘中新建了一个 mysite 目录,其中还有一个 mysite 子目录,这个子目录 mysite 中是一些项目的设置文件 settings.py、总的 urls 配置文件 urls.py 以及部署服务器时用到的 wsgi.py 文件。文件 __init__.py 是 Python 包的目录结构必需的项,与调用有关。

- mysite:项目的容器,保存整个项目。
- manage.py:一个实用的命令行工具,可让你以各种方式与该 Django 项目进行交互。
- mysite/__init__.py:一个空文件,告诉 Python 该目录是一个 Python 包。
- mysite/settings.py:该 Django 项目的设置/配置。
- mysite/urls.py:该 Django 项目的 URL 声明,一份由 Django 驱动的网站"目录"。
- mysite/wsgi.py:一个 WSGI 兼容的 Web 服务器的入口,以便运行你的项目。

(2)在命令行窗口中定位到 mysite 目录下(注意,不是 mysite 中的 mysite 目录),然后通过如下命令新建一个应用,名称为 learn。

```
H:\mysite>python manage.py startapp learn
```

此时可以看到在主 mysite 目录中多出了一个 learn 文件夹,在里面有如下所示的文件。

```
learn/
├── __init__.py
├── admin.py
├── apps.py
├── models.py
├── tests.py
└── views.py
```

(3)为了将新创建的应用添加到 settings.py 文件的 INSTALL_APPS 中,需要对文件 mysite/mysite/settings.py 进行如下修改。

```
INSTALLED_APPS = [
    'django.contrib.admin',
    'django.contrib.auth',
    'django.contrib.contenttypes',
    'django.contrib.sessions',
    'django.contrib.messages',
    'django.contrib.staticfiles',
    'learn',
]
```

这一步的目的是将新建的应用"learn"添加到 INSTALL_APPS 中,如果不这样做,Django 就不能自动找到应用中的模板文件(app-name/templates/下的文件)和静态文件(app-name/static/中的文件)。

(4)定义视图函数,用于显示访问页面时的内容。在 learn 目录中打开文件 views.py,然后进行如下所示的修改。

```
#编码:utf-8
from django.http import HttpResponse
def index(request):
    return HttpResponse(u"欢迎光临,浪潮软件欢迎您!")
```

对上述代码的具体说明如下。

- 第 1 行:声明编码为 utf-8,因为我们在代码中用到了中文,如果不声明就会报错。
- 第 2 行:引入 HttpResponse,用来向网页返回内容,就像 Python 中的 print 函数一样,只不过 HttpResponse 是把内容显示到网页上。
- 第 3~4 行:定义一个 index()函数,第一个参数必须是 request,与网页发来的请求有关。

在 request 变量里面包含 get/post 的内容、浏览器和系统等信息。函数 index()返回了一个 HttpResponse 对象,经过一些处理,最终可以在网页上显示几个字。

现在问题来了:用户应该访问什么网址才能看到刚才写的这个函数呢?怎么让网址和函数关联起来呢?接下来需要定义和视图函数相关的 URL。

(5)开始定义和视图函数相关的 URL,对文件 mysite/mysite/urls.py 进行如下修改。

```
from django.conf.urls import url
from django.contrib import admin
from learn import views as learn_views # new

urlpatterns = [
    url(r'^$', learn_views.index), # new
    url(r'^admin/', admin.site.urls),
]
```

(6）在终端运行如下命令进行测试。

```
python manage.py runserver
```

测试成功后显示图 2-2 所示的执行效果。

在浏览器中的执行效果如图 2-3 所示。

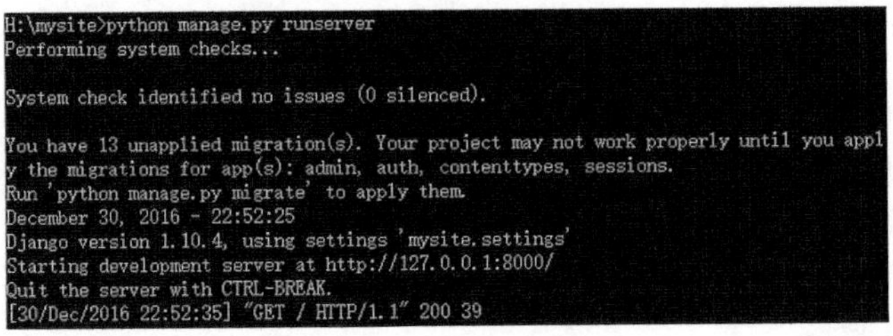

图 2-2　控制台执行效果 　　　　　　图 2-3　执行效果

2.2.4　在 URL 中传递参数

和前面学习的 Tornado 框架一样，使用 Django 框架也可以实现对 URL 参数的处理。下面的实例代码演示了使用 Django 框架实现参数相加功能的过程。

源码路径：daima\2\2-2\zqxt_views

（1）在命令行窗口中定位到 H 盘，然后通过如下命令创建一个"zqxt_views"目录。

```
django-admin startproject zqxt_views
```

也就是说，在 H 盘中新建了一个 zqxt_views 目录，其中还有一个 zqxt_views 子目录。这个子目录 mysite 中是一些项目的设置文件 settings.py，总的 urls 配置文件 urls.py，以及部署服务器时用到的 wsgi.py 文件。文件 __init__.py 是 Python 包的目录结构必需的项，与调用有关。

（2）在命令行窗口中定位到 zqxt_views 目录下（注意，不是 zqxt_views 中的 zqxt_views 目录），然后通过如下命令新建一个应用，名称为 calc。

```
cd zqxt_views
python manage.py startapp calc
```

此时自动生成的目录结构大致如下。

```
zqxt_views/
├── calc
│   ├── __init__.py
│   ├── admin.py
│   ├── apps.py
│   ├── models.py
│   ├── tests.py
│   └── views.py
├── manage.py
└── zqxt_views
    ├── __init__.py
    ├── settings.py
    ├── urls.py
    └── wsgi.py
```

（3）为了将新定义的应用添加到 settings.py 文件的 INSTALLED_APPS 中，需要对文件 zqxt_views/zqxt_views/settings.py 进行如下修改。

```
INSTALLED_APPS = [
    'django.contrib.admin',
    'django.contrib.auth',
```

```
        'django.contrib.contenttypes',
        'django.contrib.sessions',
        'django.contrib.messages',
        'django.contrib.staticfiles',
        'calc',
]
```

这一步的目的是将新建的应用"calc"添加到 INSTALLED_APPS 中，如果不这样做，Django 就不能自动找到应用中的模板文件（app-name/templates/下的文件）和静态文件（app-name/static/中的文件）。

（4）定义视图函数，用于显示访问页面时的内容。对文件 calc/views.py 的代码进行如下修改。

```
from django.shortcuts import render
from django.http import HttpResponse

def add(request):
    a = request.GET['a']
    b = request.GET['b']
    c = int(a)+int(b)
    return HttpResponse(str(c))
```

在上述代码中，request.GET 类似于一个字典，当没有传递 a 的值时，a 的默认值为 0。

（5）开始定义视图函数相关的 URL，添加一个网址来对应刚才新建的视图函数。对文件 zqxt_views/zqxt_views/urls.py 进行如下修改。

```
from django.conf.urls import url
from django.contrib import admin
from learn import views as learn_views  # new

urlpatterns = [
    url(r'^$', learn_views.index),  # new
    url(r'^admin/', admin.site.urls),
]
```

（6）在终端运行如下命令进行测试。

```
python manage.py runserver
```

在浏览器中输入"http://localhost:8000/add/"后的执行效果如图 2-4 所示。

图 2-4　执行效果

如果在 URL 中输入数字参数，例如，在浏览器地址栏中输入"http://localhost:8000/add/ ?a=4&b=5"，执行后会显示这两个数字（4 与 5）的和，执行效果如图 2-5 所示。

在 Python 程序中，也可以采用"/add/3/4/"这样的方式对 URL 中的参数进行求和处理。

图 2-5　执行效果

这时需要修改文件 calc/views.py 的代码，在里面新定义一个求和函数 add2()。具体代码如下所示。

```
def add2(request, a, b):
    c = int(a) + int(b)
    return HttpResponse(str(c))
```

接着修改文件 zqxt_views/urls.py 的代码，再添加一个新的 URL。具体代码如下所示。

```
url(r'^add/(\d+)/(\d+)/$', calc_views.add2, name='add2'),
```

此时可以看到网址中多了"\d+"，正则表达式中的"\d"代表一个数字，"+"代表一个或多个前面的字符，写在一起"\d+"就表示是一个或多个数字，用括号括起来的意思是另存为一个子组，每一个子组将作为一个参数，被文件 views.py 中的对应视图函数接收。此时输入如下网址，执行后就可以看到和图 2-1 一样的执行效果。

```
http://localhost:8000/add/?add/4/5/
```

2.2.5 使用模板

在 Tornado 框架中，模板是一个文本，用于分离文档的表现形式和具体内容。为了方便开发者进行开发，Tornado 框架提供了很多模板标签，具体说明如下。

- autoescape：控制当前自动转义的行为，有 on 和 off 两个选项，例如以下代码。

```
{% autoescape on %}
    {{ body }}
{% endautoescape %}
```

- block：定义一个子模板可以覆盖的块。
- comment：注释，例如{% comment %} 和 {% endcomment %}之间的内容表示注释。
- crsf_token：一个防止跨站点请求伪造（CSRF）攻击的标签。
- cycle：循环给出的字符串或者变量，可以混用，下面给出一段示例代码。

```
{% for o in some_list %}
    <tr class="{% cycle 'row1' rowvalue2 'row3' %}">
        ...
    </tr>
{% endfor %}
```

值得注意的是，这里的变量值默认不是自动转义的，下面给出一段示例代码。

```
{% for o in some_list %}
    <tr class="{% filter force_escape %}{% cycle rowvalue1 rowvalue2 %}{% endfilter %}">
        ...
    </tr>
{% endfor %}
```

在某些情况下，可能想在循环外部引用循环中的下一个值，这时需要用 as 给 cycle 标签设置一个名字，这个名字代表的是当前循环的值。但是在 cycle 标签里面可以用这个变量来获得循环中的下一个值。下面给出一段示例代码。

```
<tr>
    <td class = "{%cycle 'row1' 'row2' as rowcolors %}">...</td>
    <td class="{% cycle 'row1' 'row2' as rowcolors %}">...</td>
    <td class="{{ rowcolors }}">...</td>
</tr>
<tr>
    <td class="{% cycle rowcolors %}">...</td>
    <td class="{{ rowcolors }}">...</td>
</tr>
```

对应的渲染结果是：

```
<tr>
    <td class="row1">...</td>
    <td class="row1">...</td>
</tr>
<tr>
    <td class="row2">...</td>
    <td class="row2">...</td>
</tr>
```

但是一旦定义了 cycle 标签，默认就会使用循环中的第一个值。当你只是想定义一个循环而不想输出循环的值时（比如，在父模板中定义变量以方便继承），可以用 cycle 的 silent 参数

（必须保证 silent 是 cycle 的最后一个参数），并且 silent 也具有继承的特点。尽管 < td class = "{% cycle 'row1' 'row2' as rowcolors %}" > ... < /td > 中的 cycle 没有 silent 参数，但是因为 rowcoclors 是前面定义的且包含 silent 参数，所以第 2 个 cycle 也具有 silent 循环的特点。

```
{% cycle 'row1' 'row2' as rowcolors silent %}
{% cycle rowcolors %}
```

- debug：输出所有的调试信息，包括当前上下文和导入的模块。
- extends：表示当前模板继承了一个父模板，接受一个包含父模板名字的变量或者字符串常量。
- filter：通过可用的过滤器过滤内容，过滤器之间还可以相互调用。例如：

```
{% filter force_escape|lower %}
    This text will be HTML-escaped, and will appear in all lowercase.
{% endfilter %}
```

- firstof：返回列表中第一个可用（非 False）的变量或者字符串，注意，firstof 中的变量不是自动转义的。例如：

```
{% firstof var1 var2 var3 "fallback value" %}
```

- for：for 循环，可以在后面加入 reversed 参数来遍历逆序的列表。例如：

```
{% for obj in list reversed %}
```

还可以根据列表的数据来写 for 语句，例如，下面是对应字典类型数据的 for 循环。

```
{% for key, value in data.items %}
    {{ key }}: {{ value }}
{% endfor %}
```

另外，在 for 循环中还有一系列有用的变量，具体说明如表 2-1 所示。

表 2-1　　　　　　　　　　　　　　for 循环中的变量

变量	描述
forloop.counter	当前循环的索引，从 1 开始
forloop.counter0	当前循环的索引，从 0 开始
forloop.revcounter	当前循环的索引（从后面算起），从 1 开始
forloop.revcounter0	当前循环的索引（从后面算起），从 0 开始
forloop.first	如果这是第一次循环，返回真
forloop.last	如果这是最后一次循环，返回真
forloop.parentloop	如果是嵌套循环，指的是外一层循环

- for...empty：如果 for 循环中的参数列表为空，则执行 empty 里面的内容。例如：

```
<ul>
{% for athlete in athlete_list %}
    <li>{{ athlete.name }}</li>
{% empty %}
    <li>Sorry, no athlete in this list!</li>
{% endfor %}
<ul>
```

- if：这是一个条件语句。例如：

```
{% if athlete_list %}
    Number of athletes: {{ athlete_list|length }}
{% elif athlete_in_locker_room_list %}
    Athletes should be out of the locker room soon!
{% else %}
    No athletes.
{% endif %}
```

- 布尔操作符：在 if 标签中只可以使用 and、or 和 not 这 3 个布尔操作符，还有==、!=、<、>、<=、>=、in、not in 等操作符。在 if 标签里面，通过这些操作符可以创建复杂的表达式。

- ifchange：检测一个值在循环的最后有没有改变，这个标签是在循环里面使用的，有如下两个用法。

① 当没有接受参数时，比较的是 ifchange 标签里面的内容相比以前是否有变化，如果有变化，生效。

② 当接受一个或一个以上的参数的时候，如果有一个或者一个以上的参数发生变化，生效。

在 ifchange 中可以有 else 标签。例如：

```
{% for match in matches %}
    <div style="background-color:
        {% ifchanged match.ballot_id %}
            {% cycle "red" "blue" %}
        {% else %}
            grey
        {% endifchanged %}
    ">{{ match }}</div>
{% endfor %}
```

- ifequal：仅当两个参数相等的时候输出块中的内容，可以配合 else 输出。例如：

```
{% ifequal user.username "adrian" %}
...
{% endifequal %}
```

- ifnotequal：功能和用法与 ifequal 标签类似。
- include：用于加载一个模板并用当前上下文（include 该模板的上下文）渲染它，接受一个变量或者字符串参数，也可以在使用 include 的时候传递一些参数进来。例如：

```
{% include "name_snippet.html" with person="Jane" greeting="Hello" %}
```

如果只想接受传递的参数，不接受当前模板的上下文，可以使用 only 参数。例如：

```
{% include "name_snippet.html" with greeting="Hi" only %}
```

- load：加载一个自定义的模板标签集合。
- now：显示当前的时间，接受格式化的字符串参数。例如：

```
It is {% now "jS F Y H:i" %}
```

在现实中已经定义好了一些格式化的字符串参数。例如：

- DATE_FORMAT（月日年）；
- DATETIME_FORMAT（月日年时）；
- SHORT_DATE_FORMAT（月/日/年）；
- SHORT_DATETIME_FORMAT（月/日/年/时）；
- regroup：通过共同的属性对一个列表的相似对象重新分组，假如存在如下一个城市列表。

```
cities = [
    {'name': 'Mumbai', 'population': '19,000,000', 'country': 'India'},
    {'name': 'Calcutta', 'population': '15,000,000', 'country': 'India'},
    {'name': 'New York', 'population': '20,000,000', 'country': 'USA'},
    {'name': 'Chicago', 'population': '7,000,000', 'country': 'USA'},
    {'name': 'Tokyo', 'population': '33,000,000', 'country': 'Japan'},
]
```

如果想按照 country 属性来重新分组，得到每个国家（地区）中不同城市的人口，则可以通过如下代码实现分组功能。

```
{% regroup cities by country as country_list %}
<ul>
{% for country in country_list %}
    <li>{{ country.grouper }}
    <ul>
        {% for item in country.list %}
            <li>{{ item.name }}: {{ item.population }}</li>
        {% endfor %}
    </ul>
```

```
        </li>
    {% endfor %}
</ul>
```

值得注意的是，regroup 并不会重新排序，所以必须确保 city 在 regroup 之前已经按照 country 排好序，否则将得不到预期的结果。如果不确定 city 在 regroup 之前已经按照 country 排好序，可以用 dictsort 进行过滤器排序。例如：

```
{% regroup cities|dictsort:"country" by country as country_list %}
```

- spaceless：移除 html 标签之间的空格，需要注意的是，标签之间的空格、标签与内容之间的空格不会被删除。例如：

```
{% spaceless %}
    <p>
        <a href="foo/">Foo</a>
    </p>
{% endspaceless %}
```

运行结果是：

```
<p><a href="foo/">Foo</a></p>
```

- ssi：在页面上输出给定文件的内容。例如：

```
{% ssi /home/html/ljworld.com/includes/right_generic.html %}
```

使用参数 parsed 可以使得输入的内容作为一个模板，从而可以使用当前模板的上下文。例如：

```
{% ssi /home/html/ljworld.com/includes/right_generic.html parsed %}
```

- url：返回一个绝对路径的引用（没有域名的 URL），接受的第一个参数是一个视图函数的名字，然后从 urls 配置文件里面找到那个视图函数对应的 URL。
- widthratio：计算给定值与最大值的比率，然后把这个比率与一个常数相乘，返回最终的结果。例如：

```
<img src="bar.gif" height="10" width="{% widthratio this_value max_value 100 %}" />
```

- with：用更简单的变量名缓存复杂的变量名。例如：

```
{% with total=business.employees.count %}
    {{ total }} employee{{ total|pluralize }}
{% endwith %}
```

下面的实例代码演示了在 Django 框架中使用模板的过程。

源码路径：daima\2\2-2\zqxt_tmpl

（1）分别创建一个名为"zqxt_tmpl"的项目和一个名为"learn"的应用。
（2）将"learn"应用添加到 settings.INSTALLED_APPS 中。具体实现代码如下。

```
INSTALLED_APPS = (
    'django.contrib.admin',
    'django.contrib.auth',
    'django.contrib.contenttypes',
    'django.contrib.sessions',
    'django.contrib.messages',
    'django.contrib.staticfiles',
    'learn',
)
```

（3）打开文件 learn/views.py 编写一个首页的视图。具体实现代码如下。

```
from django.shortcuts import import render
def home(request):
    return render(request, 'home.html')
```

（4）在"learn"目录下新建一个"templates"文件夹用于保存模板文件，然后在里面新建一个 home.html 文件作为模板。文件 home.html 的具体实现代码如下。

```
<!DOCTYPE html>
<html>
<head>
    <title>欢迎光临</title>
</head>
<body>
欢迎选择浪潮产品！
```

```
</body>
</html>
```

（5）为了将视图函数对应到网址，对文件 zqxt_tmpl/urls.py 的代码进行如下修改。

```
from django.conf.urls import include, url
from django.contrib import admin
from learn import views as learn_views
urlpatterns = [
    url(r'^$', learn_views.home, name='home'),
    url(r'^admin/', admin.site.urls),
]
```

（6）输入如下命令启动服务器。

```
python manage.py runserver
```

执行后将显示模板的内容，执行效果如图 2-6 所示。

图 2-6　执行效果

2.2.6　使用表单

在动态 Web 应用中，表单是实现动态网页效果的核心。下面的实例代码演示了在 Django 框架中使用表单计算数字和的过程。

源码路径：daima\2\2-2\zqxt_form2

（1）新建一个名为"zqxt_form2"的项目，然后进入"zqxt_form2"文件夹并新建一个名为"tools"的应用。

```
django-admin startproject zqxt_form2
python manage.py startapp tools
```

（2）在"tools"文件夹中新建文件 forms.py。具体实现代码如下。

```
from django import forms
class AddForm(forms.Form):
    a = forms.IntegerField()
    b = forms.IntegerField()
```

（3）编写视图文件 views.py，求两个数字的和。具体实现代码如下。

```
# coding:utf-8
from django.shortcuts import render
from django.http import HttpResponse

#导入我们创建的表单类
from .forms import AddForm

def index(request):
    if request.method == 'POST':#当提交表单时

        form = AddForm(request.POST)  # form包含提交的数据

        if form.is_valid():#如果提交的数据合法
            a = form.cleaned_data['a']
            b = form.cleaned_data['b']
            return HttpResponse(str(int(a) + int(b)))

    else:#当正常访问时
        form = AddForm()
    return render(request, 'index.html', {'form': form})
```

（4）编写模板文件 index.html，实现一个简单的表单。具体实现代码如下。

```
<form method='post'>
{% csrf_token %}
{{ form }}
```

```
<input type="submit" value="提交">
</form>
```

(5) 在文件 urls.py 中将视图函数对应到网址。具体实现代码如下。

```
from django.conf.urls import include, url
from django.contrib import admin
from tools import views as tools_views
urlpatterns = [
    url(r'^$', tools_views.index, name='home'),
    url(r'^admin/', admin.site.urls),
]
```

在浏览器中运行后会显示一个表单，在表单中输入两个数字，如图 2-7 所示。

单击"提交"按钮后会计算这两个数字的和，并显示求和结果。执行效果如图 2-8 所示。

图 2-7　在表单中输入数字

图 2-8　显示求和结果

2.2.7　实现基本的数据库操作

在动态 Web 应用中，数据库技术永远是核心技术。Django 模型是与数据库相关的，与数据库相关的代码一般保存在文件 models.py 中。Django 框架支持 SQLite3、MySQL 和 PostgreSQL 等数据库工具，开发者只需要在文件 settings.py 中进行配置，不用修改文件 models.py 中的代码。下面的实例代码演示了在 Django 框架中实现数据库操作的过程。

源码路径：daima\2\2-2\learn_models

（1）新建一个名为"learn_models"的项目，然后进入"learn_models"文件夹中并新建一个名为"people"的应用。

```
django-admin startproject learn_models # 新建一个项目
cd learn_models # 进入该项目的文件夹
django-admin startapp people # 新建一个 people 应用
```

（2）将新建的应用（people）添加到文件 settings.py 中的 INSTALLED_APPS 中，也就是告诉 Django 有这么一个应用。

```
INSTALLED_APPS = (
    'django.contrib.admin',
    'django.contrib.auth',
    'django.contrib.contenttypes',
    'django.contrib.sessions',
    'django.contrib.messages',
    'django.contrib.staticfiles',
    'people',
)
```

（3）打开文件 people/models.py，新建一个继承自类 models.Model 的子类 Person，此类中有姓名和年龄这两个字段。具体实现代码如下。

```
from django.db import models
class Person(models.Model):
    name = models.CharField(max_length=30)
    age = models.IntegerField()
    def __str__(self):
        return self.name
```

在上述代码中，name 和 age 这两个字段中不能有双下划线"＿＿"，这是因为它在 Django QuerySetAPI 中有特殊含义（表示关系、包含、不区分大小写、以什么指定字符或结尾、日期的大于或小于、正则等）。另外，在代码中也不能有 Python 中的关键字、所以 name 是合法的，student_name 也是合法的，但是 student＿＿name 不合法，try、class 和 continue 也不合法，因为它们是 Python 的关键字。

（4）开始同步数据库操作，在此使用默认数据库 SQLite3，无须进行额外配置，具体命令如下。

```
# 进入 manage.py 所在的那个文件夹并输入这个命令
python manage.py makemigrations
python manage.py migrate
```

通过上述命令可以创建一个数据库表，当在前面的文件 models.py 中新增类 people 时，运行上述命令后就可以自动在数据库中创建对应数据库表，不用开发者手动创建。命令行运行后会发现 Django 生成了一系列的表，也生成了上面刚刚新建的表 people_person。命令运行界面中显示的内容如图 2-9 所示。

```
mac:learn_models tu$ python manage.py syncdb
Creating tables ...
Creating table django_admin_log
Creating table auth_permission
Creating table auth_group_permissions
Creating table auth_group
Creating table auth_user_groups
Creating table auth_user_user_permissions
Creating table auth_user
Creating table django_content_type
Creating table django_session
Creating table people_person
```

图 2-9　命令行运行界面中显示的内容

（5）输入命令进行测试。整个测试过程如下。

```
$ python manage.py shell
>>> from people.models import Person
>>> Person.objects.create(name="haoren", age=24)
<Person: haoren>
>>> Person.objects.get(name="haoren")
<Person: haoren>
```

2.3　使用 Django 后台系统开发一个博客系统

在动态 Web 应用中，后台管理系统十分重要，网站管理员通过后台实现对整个网站的管理。Django 框架的功能十分强大，为开发者提供了现成的后台管理系统，程序员只需要编写很少的代码就可以实现功能强大的后台管理系统。下面的实例代码演示了使用 Django 框架开发一个博客系统的过程。

源码路径：daima\2\2-3\

（1）新建一个名为 "zqxt_admin" 的项目，然后进入 "zqxt_admin" 文件夹并新建一个名为 "blog" 的应用。

```
django-admin startproject zqxt_admin
cd zqxt_admin
# 创建 blog 应用
python manage.py startapp blog
```

（2）修改 "blog" 文件夹中的文件 models.py。具体实现代码如下。

```
# -*- coding: utf-8 -*-
from __future__ import unicode_literals

from django.db import models
from django.utils.encoding import python_2_unicode_compatible
@python_2_unicode_compatible
class Article(models.Model):
    title = models.CharField('标题', max_length=256)
    content = models.TextField('内容')
```

```python
    pub_date = models.DateTimeField('发表时间', auto_now_add=True, editable=True)
    update_time = models.DateTimeField('更新时间', auto_now=True, null=True)
    def __str__(self):
        return self.title
class Person(models.Model):
    first_name = models.CharField(max_length=50)
    last_name = models.CharField(max_length=50)
    def my_property(self):
        return self.first_name + ' ' + self.last_name
    my_property.short_description = "Full name of the person"
    full_name = property(my_property)
```

（3）将"blog"添加到 settings.py 文件中的 INSTALLED_APPS 中。具体实现代码如下。

```
INSTALLED_APPS = (
    'django.contrib.admin',
    'django.contrib.auth',
    'django.contrib.contenttypes',
    'django.contrib.sessions',
    'django.contrib.messages',
    'django.contrib.staticfiles',
    'blog',
)
```

（4）通过如下命令同步所有的数据库表。

```
# 进入包含有 manage.py 的文件夹
python manage.py makemigrations
python manage.py migrate
```

（5）进入文件夹"blog"，修改里面的文件 admin.py（如果没有此文件，则新建一个）。具体实现代码如下。

```python
from django.contrib import admin
from .models import Article, Person
class ArticleAdmin(admin.ModelAdmin):
    list_display = ('title', 'pub_date', 'update_time',)
class PersonAdmin(admin.ModelAdmin):
    list_display = ('full_name',)
admin.site.register(Article, ArticleAdmin)
admin.site.register(Person, PersonAdmin)
```

输入下面的命令启动服务器。

```
python manage.py runserver
```

然后在浏览器地址栏中输入"http://localhost:8000/admin"会显示一个用户登录界面，如图 2-10 所示。

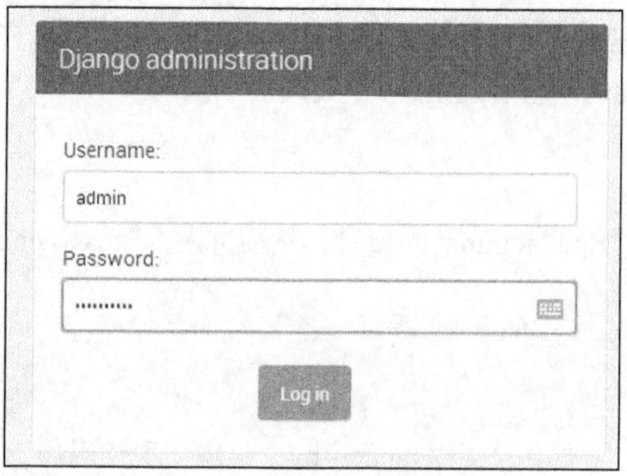

图 2-10 用户登录界面

我们可以创建一个超级管理员用户，使用命令进入包含 manage.py 的文件夹"zqxt_admin"。然后输入如下命令创建一个超级账号，根据提示分别输入账号、邮箱地址和密码。

```
python manage.py createsuperuser
```

此时可以使用超级账号登录后台管理系统。登录成功后的界面如图 2-11 所示。

2.3 使用 Django 后台系统开发一个博客系统

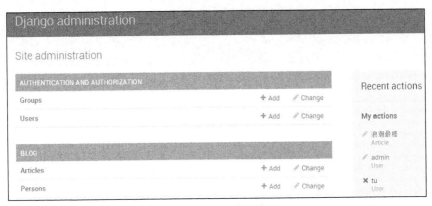

图 2-11 登录成功后的界面

管理员可以修改、删除或添加账号信息，如图 2-12 所示。

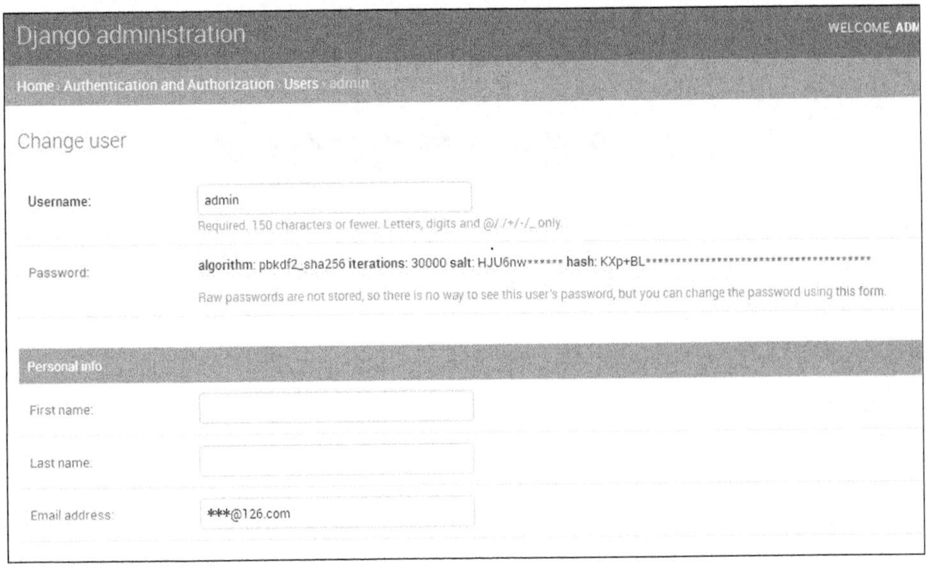

图 2-12 账号管理

也可以对系统内已经发布的博客信息进行管理维护，如图 2-13 所示。

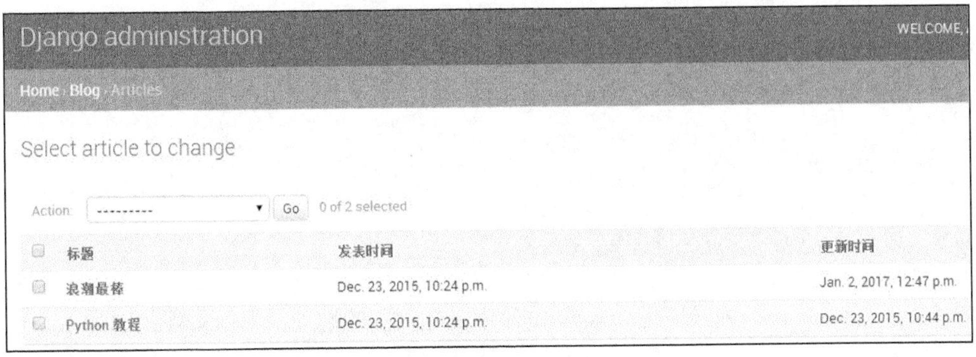

图 2-13 博客信息管理

还可以直接修改用户账号信息中的密码，如图 2-14 所示。

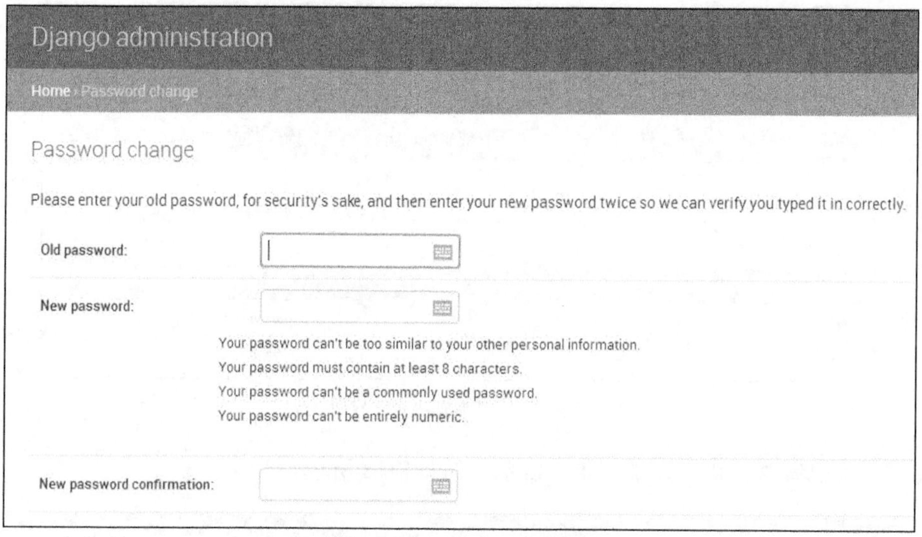

图 2-14 修改用户账号信息中的密码

2.4 开发一个新闻聚合系统

本实例使用聚合技术抓取某网站中的新闻信息。通过聚合技术，可以直接调用其他网站的内容，而无须自己创建或维护网站的内容。本实例的功能是使用知乎网提供的 API 抓取知乎当日新闻，并显示新闻列表。

源码路径：daima\2\2-4\

2.4.1 基本设置

通过以下步骤完成基本设置。

（1）新建一个名为"news"的项目，在文件 settings.py 中设置使用的模块和系统目录。涉及的变动代码如下。

```
INSTALLED_APPS = (
    'django.contrib.admin',
    'django.contrib.auth',
    'django.contrib.contenttypes',
    'django.contrib.sessions',
    'django.contrib.messages',
    'django.contrib.staticfiles',
    'news',
    'bs4',
)

STATIC_URL = '/static/'
STATIC_ROOT = 'static_files'
STATICFILES_DIRS = (
    os.path.join(BASE_DIR, "static"),
)
MAIN_SERVER = True
```

（2）在"main_app"目录下的 urls.py 文件中设置网页的地址。具体实现代码如下。

```
urlpatterns = [
    url('^$', home),
    url(r'^admin/', admin.site.urls),
    url(r'^news/', include('news.urls')),
]
```

（3）在"news"目录下的 urls.py 文件中设置地址的映射关系。一定不要忘记创建名为 app_name 的"news"，这将作为命名空间来使用。具体实现代码如下。

```python
urlpatterns = [
    url(r'^$', news.views.news_home, name='news_home'),
    url(r'^about/$', news.views.about, name='about'),
    url(r'^(?P<source>\w+)/$', news.views.NewsList.as_view(), name='story_list'),
    url(r'^detail/(?P<story_id>\d+)/$', news.views.StoryDetail.as_view(), name="story_detail"),
    url(r'^convertlist/(?P<source>\w+)/$', news.views.ConvertList.as_view(), name='convert_list'),
    url(r'^convertdetail/(?P<source>\w+)/(?P<id>\d+)/$', news.views.ConvertDetail.as_view(),
        name='convert_detail'),
]
app_name = 'news'
```

2.4.2 获取聚合信息

编写文件 fetcher.py 来获取知乎新闻客户端的新闻。具体实现流程如下。

（1）定义类 FetchError 实现错误处理功能，并输出对应的错误类型。对应代码如下。

```python
class FetchError(Exception):

    def __init__(self, errtype=''):
        self.errtype = errtype

    def __str__(self, *args, **kwargs):
        return self.__class__.__name__ + ':' + self.errtype


class PageNotFoundError(FetchError):
    pass
```

（2）定义类 NewsFetcher。首先设置解析目标网页的文件头信息，然后分别通过函数 fetch_json()和 fetch_html()获取目标 JSON 与 HTML 信息。具体实现代码如下。

```python
class NewsFetcher():
    headers = {"Accept": "text/html,application/xhtml+xml,application/xml;",
               "Accept-Encoding": "gzip",
               "Accept-Language": "zh-CN,zh;q=0.8",
               "Referer": "http://www.example.com/",
               "User-Agent": "Mozilla/5.0 (Windows NT 6.1; WOW64) AppleWebKit/537.36 "
                   "(KHTML, like Gecko) Chrome/42.0.2311.90 Safari/537.36"
               }

    def fetch_json(self, url, method='get'):
        r = requests.request(
            method=method,
            url=url,
            headers=self.headers
        )
        try:

            r.raise_for_status()
            if r.encoding.lower() == 'iso-8859-1':
                r.encoding = 'utf-8'
            json = r.json()
        except HTTPError as e:
            raise FetchError('invalid_url')
        return json

    def fetch_html(self, url, method='get'):
        r = requests.request(
            method=method,
            url=url,
            headers=self.headers
        )
        try:
            r.raise_for_status()
            if r.encoding.lower() == 'iso-8859-1':
```

```
            r.encoding = 'utf-8'
        html = r.text
    except HTTPError as e:
        raise FetchError('invalid_url')
    return html
```

（3）编写类 ZhihuDailyFetcher，通过函数 get_latest_news()解析当日知乎客户端的新闻；通过函数 get_before_news()解析过去某日知乎客户端的新闻；通过函数 get_story_detail()解析知乎客户端某具体 ID 的新闻，并显示这条新闻的详细内容。具体实现代码如下。

```
class ZhihuDailyFetcher(NewsFetcher):

    def get_latest_news(self):
        response_json = self.fetch_json(
            'http://news.at.知乎网站域名/api/4/news/latest')
        return response_json

    def get_before_news(self, date_str):
        response_json = self.fetch_json(
            'http://news.at.zhihu.com/api/4/news/before/' + date_str)
        return response_json

    def get_story_detail(self, story_id):
        response_json = self.fetch_json(
            'http://news.at.知乎网站域名/api/4/news/' + str(story_id))
        return response_json
```

（4）编写类 CBFetcher 实现具体的解析功能，这是整个实例的核心——聚合解析。通过函数 get_news_list()解析新闻列表；通过函数 get_story_comment()解析指定 ID 新闻的具体内容；通过函数 get_story_detail()解析某 ID 新闻的详细内容；通过函数 get_before_news()解析过去某日新闻的内容。类 CBFetcher 的具体实现代码如下。

```
class CBFetcher(NewsFetcher):

    def get_news_list(self, page_number=1):
        url = 'http://m.cnbeta.com/list_latest_' + str(page_number) + '.htm'
        html = self.fetch_html(url)

        soup = BeautifulSoup(html, "html.parser")
        news_list = []
        ul = soup.html.body.find('ul')
        if not ul:
            raise FetchError('parse_error')
        for li in ul.find_all('li'):
            a = li.div.a
            news_id = int(a.attrs['href'].split('.')[0].split('/')[2])
            news_list.append((news_id, a.text))
        if news_list:
            first_id = news_list[0][0]
            last_id = news_list[-1][0]
            return {'news_list': news_list, 'first_id': first_id, 'last_id': last_id}
        else:
            raise FetchError('parse_error')

    def get_story_comment(self, story_id):
        url = 'http://m.cnbeta.com/comments_' + str(story_id) + '.htm'
        html = self.fetch_html(url)
        soup = BeautifulSoup(html, "html.parser")
        J_commt_list = soup.html.body.find('ul', id="J_commt_list")
        if not J_commt_list:
            return [], 0, 0
        comments = str(J_commt_list)
        comment_count_list = soup.html.body.find(
            'span', class_="morComment").find_all('b')
        comment_count_all = int(comment_count_list[0].string)
        comment_count_show = int(comment_count_list[1].string)
        return comments, comment_count_all, comment_count_show
```

```python
def get_story_detail(self, story_id, update_comment=True):
    url = 'http://www.cnbeta.com/articles/' + str(story_id) + '.htm'
    html = self.fetch_html(url)
    soup = BeautifulSoup(html, "html.parser")
    introduction = soup.html.body.find('div', class_="introduction")
    if not introduction:
        raise PageNotFoundError()
    theme = introduction.find('img')
    theme_id = introduction.find('a')['href'].split('/')[2].split('.')[0]
    theme_text = theme['title']
    theme_img = theme['src']
    summary = introduction.p
    title = soup.html.body.find('h2', id="news_title").string
    title_bar = soup.html.body.find('div', class_="title_bar")
    time = datetime.datetime.strptime(
        title_bar.find('span', class_="date").string, "%Y-%m-%d %H:%M:%S")
    where = title_bar.find('span', class_="where").string
    body = soup.html.body.find('div', class_="content")
    author = soup.html.body.find('span', class_="author")
    if update_comment:
        comments, comment_count_all, comment_count_show = self.get_story_comment(
            story_id)
    else:
        comments, comment_count_all, comment_count_show = '', 0, 0
    result = {'body': str(body),
              'share_url': url,
              'image': str(theme_img),
              'section': {'thumbnail': str(theme_img),
                          'id': str(theme_id),
                          'name': str(theme_text),
                          },
              'title': str(title),
              'id': story_id,

              'author': str(author),
              'summary': (summary),
              'time': time,
              'where': str(where),
              'comments': comments,
              'comment_count_all': comment_count_all,
              'comment_count_show': comment_count_show,
              }
    return result

def get_before_news(self, date_str, ratio=3):

    date = datetime.datetime.strptime(date_str, '%Y-%m-%d').date()
    page_number = 1
    a_page = self.get_news_list(page_number)
    b_page = self.get_news_list(page_number * ratio)
    top_id = int(a_page['news_list'][0][0])
    bottom_id = int(b_page['news_list'][-1][0])
    top_time = self.get_story_detail(top_id, update_comment=False)['time']
    bottom_time = self.get_story_detail(
        bottom_id, update_comment=False)['time']
    print(top_time)
    print(bottom_time)
    while bottom_time.date() > date:
        page_number *= ratio
        a_page = b_page
        top_id = int(a_page['news_list'][0][0])
        top_time = self.get_story_detail(
            top_id, update_comment=False)['time']
        try:
            b_page = self.get_news_list(page_number * ratio)
        except FetchError:
```

```
                    break
                bottom_id = int(b_page['news_list'][-1][0])
                bottom_time = self.get_story_detail(
                    bottom_id, update_comment=False)['time']
                print(top_time)
                print(bottom_time)
    print('finish')
    a_number = page_number
    b_number = page_number * ratio
    while b_number - a_number > 1:
        print(str(a_number) + ' ' + str(b_number))
        mid_number = int(math.ceil((a_number + b_number) / 2))
        try:
            mid_page = self.get_news_list(mid_number)
        except FetchError:
            b_number = mid_number
            continue
        mid_page_top_id = int(mid_page['news_list'][0][0])
        mid_page_top_time = self.get_story_detail(
            mid_page_top_id, update_comment=False)['time']
        if mid_page_top_time.date() > date:
            a_number = mid_number
        else:
            b_number = mid_number
    while True:
        news_list = self.get_news_list(a_number)['news_list']
        for news in news_list:
            news_id = int(news[0])
            news_dict = self.get_story_detail(news_id, False)
            print(news_dict['title'])

        a_number += 1
```

2.4.3 视图处理

编写文件 views.py 获取知乎新闻客户端的新闻信息。具体实现流程如下。

(1) 编写函数 news_home()设置系统主页。具体实现代码如下。

```
def news_home(request):
    return HttpResponseRedirect(reverse('news:story_list', kwargs={'source': 'zhihudaily'}))
```

(2) 编写函数 about()设置"关于我们"页面。具体实现代码如下。

```
@gzip_page
def about(request):
    return render_to_response('about.html', {'nav_item': 'about'})
```

(3) 定义类 NewsViewBase 实现基本的视图功能，通过函数 get()获取指定页面的视图；通过函数 media_display()设置是否显示新闻中的图片信息，通过函数 get_date()获取新闻的时间；通过函数 hide_media()隐藏新闻中的图片信息。具体实现代码如下。

```
class NewsViewBase(TemplateView):

    @method_decorator(gzip_page)
    def get(self, request, *args, **kwargs):
        return TemplateView.get(self, request, *args, **kwargs)

    def media_display(self):
        image_show = self.request.GET.get("image", None)
        if image_show:
            self.request.session['image'] = image_show
        else:
            image_show = self.request.session.get('image', 'hide')
        if image_show == 'show':
            return True
        else:
            return False
```

```python
    def get_date(self):
        date_str = self.request.GET.get("date", None)
        if date_str:
            date = datetime.datetime.strptime(date_str, '%Y-%m-%d').date()
        else:
            date = datetime.datetime.now().date()
        date_str = date.strftime('%Y-%m-%d')
        last_date = date - datetime.timedelta(days=1)
        last_date_str = last_date.strftime('%Y-%m-%d')
        next_date = date + datetime.timedelta(days=1)
        next_date_str = next_date.strftime('%Y-%m-%d')
        if date == datetime.datetime.now().date():
            next_date_str = ''
        return date, last_date_str, date_str, next_date_str

    def hide_media(self, raw_html):
        soup = BeautifulSoup(raw_html, "html.parser")
        for img in soup.find_all('img'):
            tag = soup.new_tag("a")
            tag['href'] = img['src']
            tag.string = img.get('title', '图片')
            img.insert_before(tag)
            img.decompose()

        for embed in soup.find_all('embed'):
            tag = soup.new_tag("a")
            tag['href'] = embed['src']
            tag.string = '视频' + embed['src']
            embed.insert_before(tag)
            embed.decompose()

        for script in soup.find_all('script'):
            script.decompose()

        return str(soup)

    def get_context_data(self, request, *args, **kwargs):
        context = {}
        return context
```

(4) 定义类 NewsList 显示新闻列表。具体实现代码如下。

```python
class NewsList(NewsViewBase):
    template_name = 'newslist.html'

    def get_context_data(self, source, **kwargs):
        if not self.media_display():
            self.template_name = 'newslist_no_pic.html'

        date, last_date_str, date_str, next_date_str = self.get_date()
        dailydate = DailyDate.objects.update_daily_date_with_date(
            date, source=source)
        story_qs = dailydate.get_daily_stories()
        context = {}
        context['date_str'] = date_str
        context['story_qs'] = story_qs
        context['last_date_str'] = last_date_str
        context['next_date_str'] = next_date_str
        context['nav_item'] = 'zhihu'
        return context
```

(5) 定义类 StoryDetail 显示新闻的详细内容。具体实现代码如下。

```python
class StoryDetail(NewsViewBase):
    template_name = 'story_detail.html'

    def get_context_data(self, story_id, **kwargs):
        story = Story.objects.get(story_id=story_id)
        story.update()
```

```python
        if not self.media_display():
            story.body = self.hide_media(story.body)
        context = {}
        context['story'] = story
        context['nav_item'] = 'zhihu'
        return context
```

（6）定义类 ConvertList 实现"上一页"和"下一页"功能。具体实现代码如下。

```python
class ConvertList(NewsViewBase):
    template_name = 'convert_list.html'

    def get_context_data(self, source, **kwargs):
        page_number = int(self.request.GET.get('page', 1))
        if source == 'cb':
            result = CBFetcher().get_news_list(page_number)
            news_list = result['news_list']
            context = {}
            context['page_number'] = page_number
            if page_number > 1:
                context['previous_page'] = page_number - 1
            context['next_page'] = page_number + 1

            context['news_list'] = news_list
            context['nav_item'] = 'cnbeta'
            return context
        else:
            raise Exception('wrong source name')
```

2.4.4 模板文件

"关于我们"页面对应的模板文件是 about.html。具体实现代码如下。

```html
{% extends 'bootstrap_base.html' %}

{% block title %}
    关于
{% endblock %}

{% block content %}
    <div class="row">
        <h1>啥</h1>
          <h3>都没有</h3>
    </div><!-- /.row -->
{% endblock %}
```

新闻列表页面的模板文件是 newslist.html。具体实现代码如下。

```html
{% extends 'bootstrap_base.html' %}

{% block title %}
    知乎日报 {{date_str}}
{% endblock %}

{% block extra_js %}

{% endblock %}

{% block content %}

    <script type="text/javascript">
        function showImg( url ) {
        var frameid = 'frameimg' + Math.random();
        window.img = '<img id="img" jump_url class="img-thumbnail"
            src=\''+url+'?'+Math.random()+'\' /><script>window.onload = function()
            { parent.document.getElementById(\''+frameid+'\').height =
            document.getElementById(\'img\').height+\'px\'; }<'+'/script>';
        document.write('<iframe id="'+frameid+'" src="javascript:parent.img;" frameBorder="0"
            scrolling="no" width="100%"></iframe>');
        }
    </script>
```

```html
        <div class="row">
            {% for story in story_qs %}
                <div class="col-xs-6 col-sm-4 col-md-3">
                    <a href="{% url 'news:story_detail' story_id=story.story_id %}">
                        <!--<img class="img-thumbnail" src="
                                //v4.bootcss.com/examples/screenshots/starter-template.
                                jpg" alt="" />-->
                                <div id="hotlinking"><script
                                    type="text/javascript">showImg("{{story.cover_picture_fi
                                    rst}}");</script></div>
                    </a>
                    <div style= "height:50px;"><a href="{% url 'news:story_detail'
                        story_id=story.story_id %}">{{story.title}}</a></div>
                </div>
            {% endfor %}

    </div><!-- /.row -->
        <nav>
            <ul class="pager">
                <li><a href="?date={{last_date_str}}">Previous</a></li>

                {% if next_date_str %}
                <li><a href="?date={{next_date_str}}">Next</a></li>
                {% endif %}

            </ul>

            <form method="get" action="" >
                <input type="date" name="date" value="{{date_str}}">
                <button type="submit">跳转</button>
            </form>

        </nav>
{% endblock %}
```

隐藏图片的新闻列表页面的模板文件是 newslist_no_pic.html。具体实现代码如下。

```
{% extends 'bootstrap_base.html' %}
{% block title %} 知乎日报 {{date_str}}
{% endblock %} {% block extra_js %} {% endblock %}

{% block content %}
<table class='table'>
    {% for story in story_qs %}
    <tr>
        <td><a href={% url 'news:story_detail' story_id=story.story_id%}>{{story.title}}</a></td>
    </tr>
    {% endfor %}
</table>


<nav>
    <ul class="pager">
        <li><a href="?date={{last_date_str}}">Previous</a></li>
        {% if next_date_str %}
            <li><a href="?date={{next_date_str}}">Next</a></li>
        {% endif %}

    </ul>

    <form method="get" action="">
        <input type="date" name="date" value="{{date_str}}">
        <button type="submit">跳转</button>
    </form>

</nav>
{% endblock %}
```

无图版系统主页如图 2-15 所示。

图 2-15　无图版系统主页

有图版系统主页如图 2-16 所示。

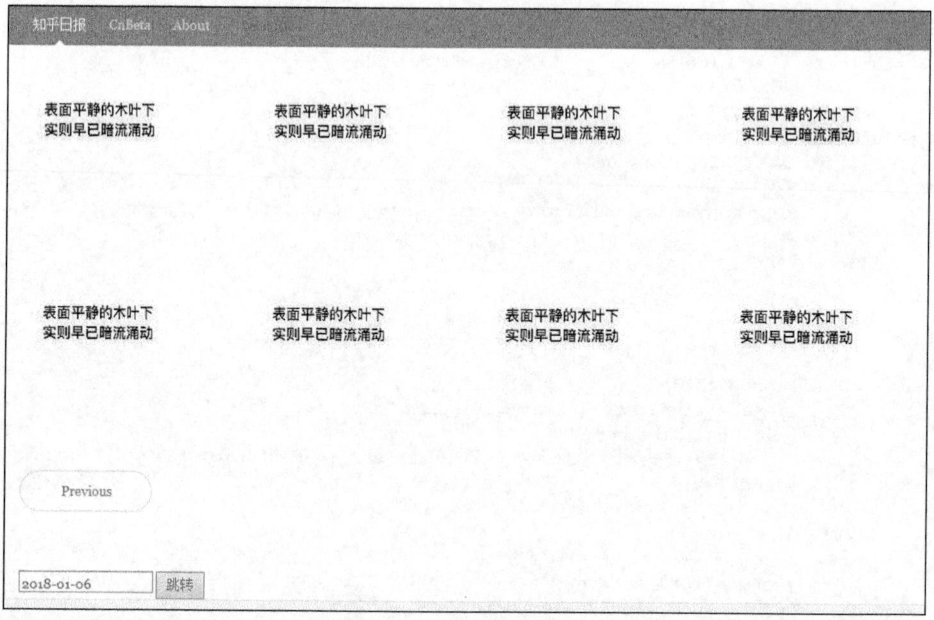

图 2-16　有图版系统主页

无图版新闻详情页面如图 2-17 所示。
有图版新闻详情页面如图 2-18 所示。

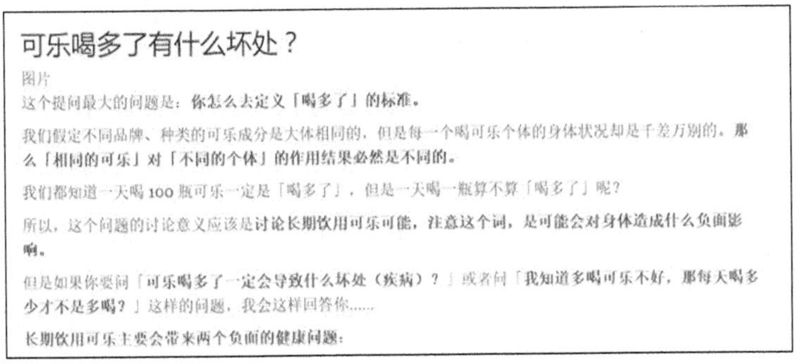

图 2-17　无图版新闻详情页面

图 2-18　有图版新闻详情页面

2.5　开发一个在线商城系统

本实例的功能是使用 Django 框架开发一个在线商城系统，在该系统中不但可以实现后台商品数据的添加和修改操作，而且能实现对商品分类的添加、修改和删除操作，还实现了商城系统的大量核心功能——订单处理和购物车处理。

源码路径：daima\2\2-5\

2.5.1　系统设置

在 "myshop" 子目录下的文件 settings.py 中实现系统设置功能，分别添加 shop、cart 和 orders 三大模块，设置数据库信息，设置 URL 链接路径。主要实现代码如下。

源码路径：daima\2\2-5\myshop\myshop\settings.py

```
INSTALLED_APPS = (
    'django.contrib.admin',
    'django.contrib.auth',
    'django.contrib.contenttypes',
    'django.contrib.sessions',
    'django.contrib.messages',
    'django.contrib.staticfiles',
    'shop',
    'cart',
    'orders',
)

DATABASES = {
    'default': {
        'ENGINE': 'django.db.backends.sqlite3',
        'NAME': os.path.join(BASE_DIR, 'db.sqlite3'),
    }
}

STATIC_URL = '/static/'
```

```
MEDIA_URL = '/media/'
MEDIA_ROOT = os.path.join(BASE_DIR, 'media/')

CART_SESSION_ID = 'cart'

EMAIL_BACKEND = 'django.core.mail.backends.console.EmailBackend'

cwd = os.getcwd()
if cwd == '/app' or cwd[:4] == '/tmp':
    import dj_database_url
    DATABASES = {
        'default': dj_database_url.config(default='postgres://localhost')
    }

    SECURE_PROXY_SSL_HEADER = ('HTTP_X_FORWARDED_PROTO', 'https')

    ALLOWED_HOSTS = ['*']
    DEBUG = False

    BASE_DIR = os.path.dirname(os.path.abspath(__file__))
    STATIC_ROOT = 'staticfiles'
    STATICFILES_DIRS = (
        os.path.join(BASE_DIR, 'static'),
    )
```

在"myshop"子目录下的文件 urls.py 中实现系统所有功能模块页面的布局功能。整个系统分为四大模块——前台商城展示、后台管理、订单处理和购物车处理。文件 urls.py 的具体实现代码如下。

源码路径：daima\2\2-5\myshop\myshop\urls.py

```
urlpatterns = [
    url(r'^admin/', admin.site.urls),
    url(r'^cart/', include('cart.urls', namespace='cart')),
    url(r'^orders/', include('orders.urls', namespace='orders')),
    url(r'^', include('shop.urls', namespace='shop')),
]

if settings.DEBUG:
    urlpatterns += static(settings.MEDIA_URL,
                          document_root=settings.MEDIA_ROOT)
```

2.5.2 前台商城展示模块

在前台商城展示模块中显示系统内所有商品的分类信息，并展示各个分类商品的信息，单击某个商品后可以展示这个商品的详细信息。本模块主要由以下 3 个文件实现：

- models.py；
- urls.py；
- views.py。

在文件 models.py 中编写两个类。其中，类 Category 实现商品分类展示功能，类 Product 实现产品展示功能。文件 models.py 的具体实现代码如下。

源码路径：daima\2\2-5\myshop\shop\models.py

```
class Category(models.Model):
    name = models.CharField(max_length=200, db_index=True)
    slug = models.SlugField(max_length=200, db_index=True, unique=True)

    class Meta:
        ordering = ('name',)
        verbose_name = 'category'
        verbose_name_plural = 'categories'

    def __str__(self):
        return self.name
```

```python
    def get_absolute_url(self):
        return reverse('shop:product_list_by_category', args=[self.slug])


class Product(models.Model):
    category = models.ForeignKey(Category, related_name='products',on_delete=models.CASCADE)
    name = models.CharField(max_length=200, db_index=True)
    slug = models.SlugField(max_length=200, db_index=True)
    image = models.ImageField(upload_to='products/%Y/%m/%d', blank=True)
    description = models.TextField(blank=True)
    price = models.DecimalField(max_digits=10, decimal_places=2)
    stock = models.PositiveIntegerField()
    available = models.BooleanField(default=True)
    created = models.DateTimeField(auto_now_add=True)
    updated = models.DateTimeField(auto_now=True)

    class Meta:
        ordering = ('-created',)
        index_together = (('id', 'slug'),)

    def __str__(self):
        return self.name

    def get_absolute_url(self):
        return reverse('shop:product_detail', args=[self.id, self.slug])
```

文件 urls.py 的功能是实现前台页面的 URL 处理，并分别实现"分类展示"和"产品详情展示"。具体实现代码如下。

源码路径：daima\2\2-5\myshop\shop\urls.py

```python
urlpatterns = [
    url(r'^$', views.product_list, name='product_list'),
    url(r'^(?P<category_slug>[-\w]+)/$', views.product_list, name='product_list_by_category'),
    url(r'^(?P<id>\d+)/(?P<slug>[-\w]+)/$', views.product_detail, name='product_detail'),
]
app_name = 'myshop'
```

在文件 views.py 中实现视图展示功能，通过函数 product_list()实现商品列表展示功能，通过函数 product_detail()实现商品详情展示功能。文件 views.py 的具体实现代码如下。

源码路径：daima\2\2-5\myshop\shop\views.py

```python
def product_list(request, category_slug=None):
    category = None
    categories = Category.objects.all()
    products = Product.objects.filter(available=True)
    if category_slug:
        category = get_object_or_404(Category, slug=category_slug)
        products = products.filter(category=category)
    return render(request, 'shop/product/list.html', {'category': category,
                                                       'categories': categories,
                                                       'products': products})


def product_detail(request, id, slug):
    product = get_object_or_404(Product, id=id, slug=slug, available=True)
    cart_product_form = CartAddProductForm()
    return render(request,
                  'shop/product/detail.html',
                  {'product': product,
                   'cart_product_form': cart_product_form
                   })
```

前台商品展示主页的模板文件是 base.html。具体实现代码如下。

```html
{% load static %}
<!DOCTYPE html>
<html>
<head>
    <meta charset="utf-8" />
    <title>{% block title %}My shop 我的商店{% endblock %}</title>
```

```
            <link href="{% static "css/base.css" %}" rel="stylesheet">
    </head>
    <body>
        <div id="header">
            <a href="/" class="logo">My shop 我的商店</a>
        </div>
        <div id="subheader">
            <div class="cart">
                {% with total_items=cart|length %}
                    {% if cart|length > 0 %}
                        Your cart:
                        <a href="{% url "cart:cart_detail" %}">
                            {{ total_items }} item{{ total_items|pluralize }}, ${{ cart.
                            get_total_price }}
                        </a>
                    {% else %}
                        Your cart is empty.
                    {% endif %}
                {% endwith %}
            </div>
        </div>
        <div id="content">
            {% block content %}
            {% endblock %}
        </div>
    </body>
</html>
```

商品列表展示功能的模板文件是 list.html。具体实现代码如下。

```
{% extends "shop/base.html" %}
{% load static %}

{% block title %}
    {% if category %}{{ category.name }}{% else %}Products{% endif %}
{% endblock %}

{% block content %}
    <div id="sidebar">
        <h3>Categories</h3>
        <ul>
            <li {% if not category %}class="selected"{% endif %}>
                <a href="{% url "shop:product_list" %}">All</a>
            </li>
            {% for c in categories %}
                <li {% if category.slug == c.slug %}class="selected"{% endif %}>
                    <a href="{{ c.get_absolute_url }}">{{ c.name }}</a>
                </li>
            {% endfor %}
        </ul>
    </div>
    <div id="main" class="product-list">
        <h1>{% if category %}{{ category.name }}{% else %}Products{% endif %}</h1>
        {% for product in products %}
            <div class="item">
                <a href="{{ product.get_absolute_url }}">
                    <img src="{% if product.image %}{{ product.image.url }}{% else %}{% static
                        "img/no_image.png" %}{% endif %}">
                </a>
                <a href="{{ product.get_absolute_url }}">{{ product.name }}</a><br>
                ${{ product.price }}
            </div>
        {% endfor %}
    </div>
{% endblock %}
```

某个商品详情展示功能的模板文件是 detail.html。具体实现代码如下。

```
{% extends "shop/base.html" %}
{% load static %}

{% block title %}
```

```
        {{ product.name }}
{% endblock %}

{% block content %}
    <div class="product-detail">
        <img src="{% if product.image %}{{ product.image.url }}{% else %}{% static
        "img/no_image.png" %}{% endif %}">
        <h1>{{ product.name }}</h1>
        <h2><a href="{{ product.category.get_absolute_url }}">{{ product.category }}</a></h2>
        <p class="price">${{ product.price }}</p>
        <form action="{% url "cart:cart_add" product.id %}" method="post">
            {{ cart_product_form }}
            {% csrf_token %}
            <input type="submit" value="Add to cart">
        </form>

        {{ product.description|linebreaks }}
    </div>
{% endblock %}
```

前台商品展示主页如图 2-19 所示。单击左侧分类链接后，会只显示这个分类下的所有商品信息。

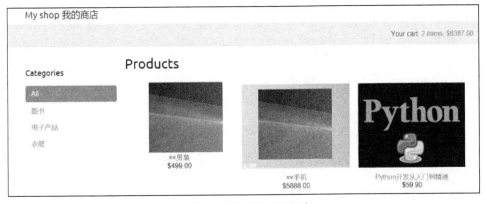

图 2-19 前台商品展示主页

某商品详情展示页面如图 2-20 所示。

图 2-20 某商品详情展示页面

2.5.3 购物车模块

购物车是在线商城系统的核心功能之一。该模块主要由如下 4 个文件实现：

- cart.py；
- urls.py；
- forms.py；
- views.py。

文件 urls.py 的功能是实现购物车页面的 URL 处理，以及实现购物车详情展示、商品的添加和商品的删除。具体实现代码如下。

源码路径：daima\2\2-5\myshop\cart\urls.py

```python
urlpatterns = [
    url(r'^$', views.cart_detail, name='cart_detail'),
    url(r'^add/(?P<product_id>\d+)/$', views.cart_add, name='cart_add'),
    url(r'^remove/(?P<product_id>\d+)/$', views.cart_remove, name='cart_remove'),
]
app_name = 'myshop'
```

在文件 cart.py 中实现和购物车相关的操作：通过函数 __init__() 获取登录用户的账号信息，通过函数 __len__() 统计当前账户购物车中的商品数量，通过函数 __iter__() 展示购物车内各个商品的信息，通过函数 add() 向购物车内添加新的商品信息，通过函数 remove() 删除购物车内的某个商品信息，通过函数 save() 保存当前购物车内的商品信息。文件 cart.py 的具体实现代码如下。

源码路径：daima\2\2-5\myshop\cart\cart.py

```python
class Cart(object):
    def __init__(self, request):
        """
        Initialize the cart.
        """
        self.session = request.session
        cart = self.session.get(settings.CART_SESSION_ID)
        if not cart:
            cart = self.session[settings.CART_SESSION_ID] = {}
        self.cart = cart

    def __len__(self):
        """
        Count all items in the cart.
        """
        return sum(item['quantity'] for item in self.cart.values())

    def __iter__(self):
        """
        Iterate over the items in the cart and get the products from the database.
        """
        product_ids = self.cart.keys()
        # get the product objects and add them to the cart
        products = Product.objects.filter(id__in=product_ids)
        for product in products:
            self.cart[str(product.id)]['product'] = product

        for item in self.cart.values():
            item['price'] = Decimal(item['price'])
            item['total_price'] = item['price'] * item['quantity']
            yield item

    def add(self, product, quantity=1, update_quantity=False):
        """
        Add a product to the cart or update its quantity.
        """
        product_id = str(product.id)
        if product_id not in self.cart:
            self.cart[product_id] = {'quantity': 0,
                                     'price': str(product.price)}
        if update_quantity:
            self.cart[product_id]['quantity'] = quantity
        else:
            self.cart[product_id]['quantity'] += quantity
        self.save()

    def remove(self, product):
        """
        Remove a product from the cart.
```

```python
            """
            product_id = str(product.id)
            if product_id in self.cart:
                del self.cart[product_id]
                self.save()

    def save(self):
        self.session[settings.CART_SESSION_ID] = self.cart
        self.session.modified = True

    def clear(self):
        self.session[settings.CART_SESSION_ID] = {}
        self.session.modified = True

    def get_total_price(self):
        return sum(Decimal(item['price']) * item['quantity'] for item in self.cart.values())
```

文件 forms.py 的功能是处理购物车表单中商品的变动信息。具体实现代码如下。

源码路径：daima\2\2-5\myshop\cart\forms.py

```python
PRODUCT_QUANTITY_CHOICES = [(i, str(i)) for i in range(1, 21)]

class CartAddProductForm(forms.Form):
    quantity = forms.TypedChoiceField(choices=PRODUCT_QUANTITY_CHOICES,
                                      coerce=int)
    update = forms.BooleanField(required=False,
                                initial=False,
                                widget=forms.HiddenInput)
```

在文件 views.py 中实现视图展示功能，通过函数 cart_add()向购物车中添加商品，通过函数 cart_remove()从购物车中删除商品，通过函数 cart_detail()实现购物车详情展示功能。文件 views.py 的具体实现代码如下。

源码路径：daima\2\2-5\myshop\cart\views.py

```python
def cart_add(request, product_id):
    cart = Cart(request)
    product = get_object_or_404(Product, id=product_id)
    form = CartAddProductForm(request.POST)
    if form.is_valid():
        cd = form.cleaned_data
        cart.add(product=product,
                 quantity=cd['quantity'],
                 update_quantity=cd['update'])
    return redirect('cart:cart_detail')

def cart_remove(request, product_id):
    cart = Cart(request)
    product = get_object_or_404(Product, id=product_id)
    cart.remove(product)
    return redirect('cart:cart_detail')

def cart_detail(request):
    cart = Cart(request)
    for item in cart:
        item['update_quantity_form'] = CartAddProductForm(initial={'quantity': item['quantity'],
                                                                   'update': True})
    return render(request, 'cart/detail.html', {'cart': cart})
```

购物车模块的模板文件是 detail.html。具体实现代码如下。

```django
{% extends "shop/base.html" %}
{% load static %}

{% block title %}
    Your shopping cart
{% endblock %}

{% block content %}
    <h1>Your shopping cart</h1>
```

```html
<table class="cart">
    <thead>
        <tr>
            <th>Image</th>
            <th>Product</th>
            <th>Quantity</th>
            <th>Remove</th>
            <th>Unit price</th>
            <th>Price</th>
        </tr>
    </thead>
    <tbody>
    {% for item in cart %}
        {% with product=item.product %}
        <tr>
            <td>
                <a href="{{ product.get_absolute_url }}">
                    <img src="{% if product.image %}{{ product.image.url }}{% else %}{% static "img/no_image.png" %}{% endif %}">
                </a>
            </td>
            <td>{{ product.name }}</td>
            <td>
                <form action="{% url "cart:cart_add" product.id %}" method="post">
                    {{ item.update_quantity_form.quantity }}
                    {{ item.update_quantity_form.update }}
                    <input type="submit" value="Update">
                    {% csrf_token %}
                </form>
            </td>
            <td><a href="{% url "cart:cart_remove" product.id %}">Remove</a></td>
            <td class="num">${{ item.price }}</td>
            <td class="num">${{ item.total_price }}</td>
        </tr>
        {% endwith %}
    {% endfor %}
        <tr class="total">
            <td>Total</td>
            <td colspan="4"></td>
            <td class="num">${{ cart.get_total_price }}</td>
        </tr>
    </tbody>
</table>
<p class="text-right">
    <a href="{% url "shop:product_list" %}" class="button light">Continue shopping</a>
    <a href="{% url 'orders:order_create' %}" class="button">Checkout</a>
</p>
```

购物车界面的执行效果如图 2-21 所示。我们可以灵活地增加或删除里面的商品，也可以修改里面的商品数量。

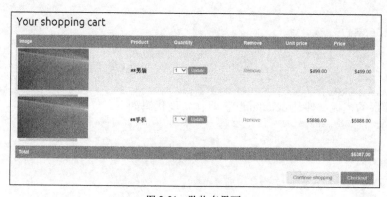

图 2-21 购物车界面

2.5.4 订单模块

订单处理也是在线商城系统的核心功能之一。本模块主要由如下文件实现。

(1) 文件 urls.py 的功能是实现订单页面的 URL 处理，并创建订单。具体实现代码如下。

源码路径：daima\2\2-5\myshop\orders\urls.py

```python
urlpatterns = [
    url(r'^create/$', views.order_create, name='order_create'),
]
app_name = 'myshop'
{% endblock %}
```

(2) 文件 admin.py 的功能是实现订单展示和订单管理。具体实现代码如下。

源码路径：daima\2\2-5\myshop\orders\admin.py

```python
class OrderItemInline(admin.TabularInline):
    model = OrderItem
    raw_id_fields = ['product']


class OrderAdmin(admin.ModelAdmin):
    list_display = ['id', 'first_name', 'last_name', 'email', 'address', 'postal_code',
                    'city', 'paid', 'created', 'updated']
    list_filter = ['paid', 'created', 'updated']
    inlines = [OrderItemInline]

admin.site.register(Order, OrderAdmin)
```

(3) 文件 forms.py 的功能是创建订单列表。具体实现代码如下。

源码路径：daima\2\2-5\myshop\orders\forms.py

```python
class OrderCreateForm(forms.ModelForm):
    class Meta:
        model = Order
        fields = ['first_name', 'last_name', 'email', 'address', 'postal_code', 'city']
```

(4) 文件 models.py 的功能是实现订单和订单列表的处理。具体实现代码如下。

源码路径：daima\2\2-5\myshop\orders\models.py

```python
class Order(models.Model):
    first_name = models.CharField(max_length=50)
    last_name = models.CharField(max_length=50)
    email = models.EmailField()
    address = models.CharField(max_length=250)
    postal_code = models.CharField(max_length=20)
    city = models.CharField(max_length=100)
    created = models.DateTimeField(auto_now_add=True)
    updated = models.DateTimeField(auto_now=True)
    paid = models.BooleanField(default=False)

    class Meta:
        ordering = ('-created',)

    def __str__(self):
        return 'Order {}'.format(self.id)

    def get_total_cost(self):
        return sum(item.get_cost() for item in self.items.all())


class OrderItem(models.Model):
    order = models.ForeignKey(Order, related_name='items', on_delete=models.CASCADE)
    product = models.ForeignKey(Product, related_name='order_items', on_delete=models.CASCADE)
    price = models.DecimalField(max_digits=10, decimal_places=2)
    quantity = models.PositiveIntegerField(default=1)

    def __str__(self):
        return '{}'.format(self.id)
```

```python
    def get_cost(self):
        return self.price * self.quantity
```

（5）文件 tasks.py 的功能是创建新的订单。具体实现代码如下。

源码路径：daima\2\2-5\myshop\orders\tasks.py

```python
def order_created(order_id):
    """
    Task to send an e-mail notification when an order is successfully created.
    """
    order = Order.objects.get(id=order_id)
    subject = 'Order nr. {}'.format(order.id)
    message = 'Dear {},\n\nYou have successfully placed an order. Your order id is
        {}.'.format(order.first_name,
                                                                                  order.id)
    mail_sent = send_mail(subject, message, 'admin@myshop.com', [order.email])
    return mail_sent
```

（6）文件 views.py 的功能是创建订单视图。具体实现代码如下。

源码路径：daima\2\2-5\myshop\orders\views.py

```python
def order_create(request):
    cart = Cart(request)
    if request.method == 'POST':
        form = OrderCreateForm(request.POST)
        if form.is_valid():
            order = form.save()
            for item in cart:
                OrderItem.objects.create(order=order,
                                         product=item['product'],
                                         price=item['price'],
                                         quantity=item['quantity'])
            cart.clear()
            # order_created.delay(order.id)
            return render(request, 'orders/order/created.html', {'order': order})
    else:
        form = OrderCreateForm()
    return render(request, 'orders/order/create.html', {'cart': cart,
                                                                              'form': form})
```

创建订单功能的模板文件是 create.html。具体实现代码如下。

```html
{% extends "shop/base.html" %}

{% block title %}
    Checkout
{% endblock %}

{% block content %}
    <h1>Checkout</h1>

    <div class="order-info">
        <h3>Your order</h3>
        <ul>
            {% for item in cart %}
                <li>{{ item.quantity }}x {{ item.product.name }}
                    <span>${{ item.total_price }}</span></li>
            {% endfor %}
        </ul>
        <p>Total: ${{ cart.get_total_price }}</p>
    </div>

    <form action="." method="post" class="order-form">
        {{ form.as_p }}
        <p><input type="submit" value="Place order"></p>
        {% csrf_token %}
    </form>
{% endblock %}
```

创建订单成功的页面的模板文件是 created.html。具体实现代码如下。

```html
{% extends "shop/base.html" %}
```

```
{% block title %}
    Thank you
{% endblock %}

{% block content %}
    <h1>Thank you</h1>
    <p>Your order has been successfully completed. Your order number is
    <strong>{{ order.id }}</strong>.</p>
{% endblock %}
```

在购物车界面中单击"Checkout"按钮后会弹出创建订单界面,如图 2-22 所示。填写配送信息完毕,并单击"Place order"按钮,会成功创建订单。

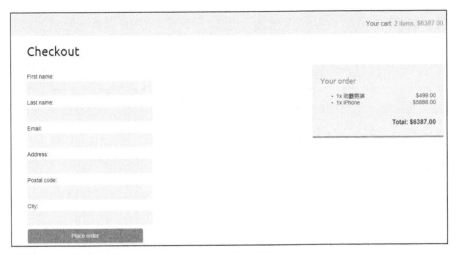

图 2-22　创建订单界面

整个实例全部介绍完毕,后台系统是通过 Django 框架自动实现的。订单管理界面如图 2-23 所示。

图 2-23　订单管理界面

商品分类管理界面如图 2-24 所示。

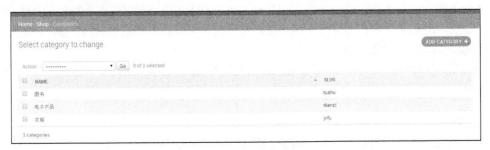

图 2-24　商品分类管理界面

商品管理界面如图 2-25 所示。

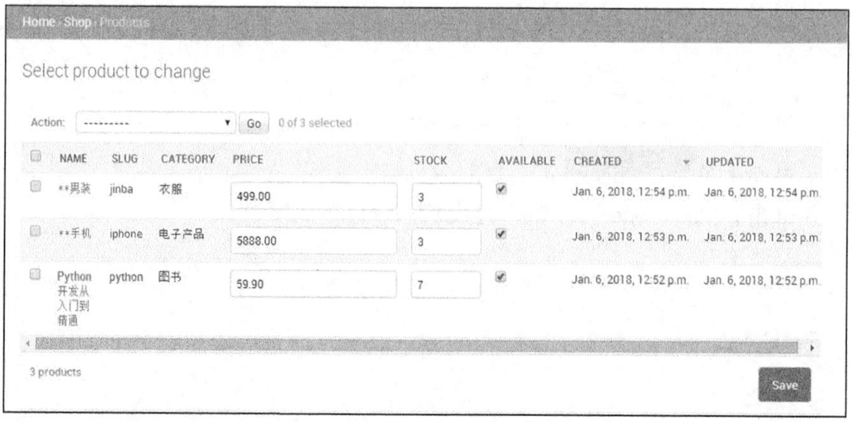

图 2-25　商品管理界面

添加商品的界面如图 2-26 所示。

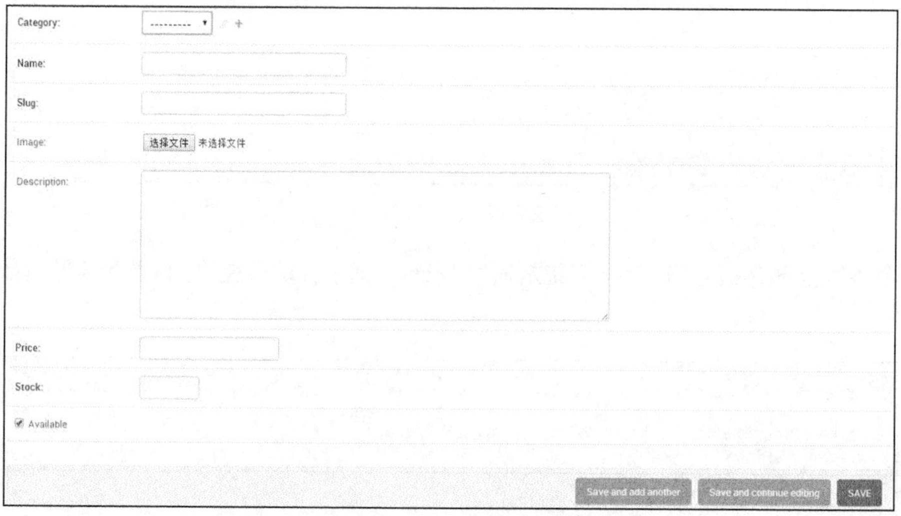

图 2-26　添加商品的界面

2.6　使用 Mezzanine 库

Mezzanine 是一款开源的、基于 Django 的 CMS (Content Management System)框架。其实任何一个网站都可以被看作一个特定的内容管理系统，只不过每个网站发布和管理的具体内容不一样。例如，携程网发布的是航班、酒店的信息，而淘宝网发布的是商品的信息。下面将详细介绍框架 Mezzanine 的使用知识。

在安装 Mezzanine 之前需要先确保已经安装了 Django，然后使用如下命令安装 Mezzanine。

```
pip install mezzanine
```

接下来便可以使用 Mezzanine 快速创建一个 CMS。具体实现流程如下。

源码路径：daima\2\2-6\testing

（1）使用如下命令创建一个 Mezzanine 项目。项目名为"testing"。

```
mezzanine-project testing
```

(2)使用如下命令进入项目目录。

```
cd testing
```

(3)使用如下命令初始化创建一个数据库。

```
python manage.py createdb
```

这个过程需要填写如下基本信息。

- 域名和端口：默认为 http://127.0.0.1:8000/。
- 默认的超级管理员账号和密码。
- 默认主页。

(4)使用如下命令启动这个项目。

```
python manage.py runserver
```

当显示如下信息时，说明成功运行了新建的 Mezzanine 项目"testing"。

```
             .....
         _d^^^^^^^^^b_
      .d''           ``b.
    .p'                `q.
   .d'                   `b.
  .d'                     `b.   * Mezzanine 4.2.3
  ::                       ::   * Django 1.10.8
  ::    M E Z Z A N I N E  ::   * Python 3.6.0
  ::                       ::   * SQLite 3.2.2
  `p.                     .q'   * Windows 10
   `p.                   .q'
    `b.                 .d'
      `q..           ..p'
        ^q........p^
            ''''

Performing system checks...

System check identified no issues (0 silenced).
April 17, 2018 - 14:07:33
Django version 1.10.8, using settings 'testing.settings'
Starting development server at http://127.0.0.1:8000/
```

在浏览器中输入"http://127.0.0.1:8000/"后进入前台主页，如图 2-27 所示。

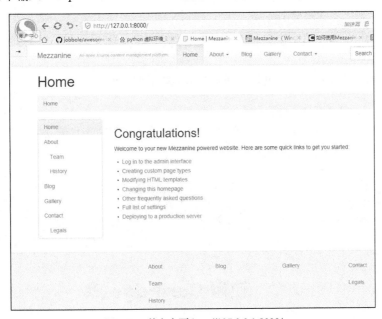

图 2-27　前台主页 http://127.0.0.1:8000/

(5)后台管理首页是 http://127.0.0.1:8000/admin，如图 2-28 所示。在登录后台管理页面时，使用在创建数据库时设置的管理员账号登录。

图 2-28　后台管理首页 http://127.0.0.1:8000/admin

后台管理系统的主要功能如下。

- 在 Dashboard 中，选择 Content→Pages，可配置导航、页脚信息。
- 在 Dashboard 中，选择 Content→Blog posts，可添加分类、发布文章。
- 在 Dashboard 中，选择 Site→Settings，可配置网站的 Title、Tagline。

（6）系统主页默认显示 Home 页面，如果想以 Blog 的列表作为主页，只须将文件 url.py 中的代码行 un-comment 修改为：

```
url("^$", "mezzanine.blog.views.blog_post_list", name="home")
```

也就是将文件 url.py 中的如下代码注释掉。

```
#url("^$", direct_to_template, {"template": "index.html"}, name="home")
```

在 url.py 中把如下代码取消注释。

```
url("^$", "mezzanine.blog.views.blog_post_list", name="home")
```

（7）如果想去掉导航栏中的 Search 输入框，需要添加如下配置项。

```
SEARCH_MODEL_CHOICES = []
```

如果想去掉左侧导航链接和页脚，则需要添加如下配置项。

```
PAGE_MENU_TEMPLATES = ( (1, "Top navigation bar", "pages/menus/dropdown.html"), )
```

（8）Mezzanine 默认支持 4 种数据库，分别是 postgresql_psycopg2、MySQL、SQLite3 和 Oracle，在默认情况下使用 SQLite3。我们可以在文件 local_settings.py 中的如下代码段中进行修改设置。

```
DATABASES = {
    "default": {
        "ENGINE": "django.db.backends.sqlite3",
        "NAME": "dev.db",
        "USER": "",
        "PASSWORD": "",
        "HOST": "",
        "PORT": "",
    }
}
```

2.7　使用 Cartridge 库

Cartridge 库是一个基于 Mezzanine 构建的购物车应用框架，通过一个基于 Mezzanine 构建的购物车应用可以快速实现电子商务应用中的购物车程序。下面将详细讲解 Cartridge 库的使用方法。

2.7 使用 Cartridge 库

在安装 Cartridge 库之前需要先确保已经安装了 Mezzanine。使用如下命令安装 Cartridge。

```
pip install Cartridge
```

接下来便可以使用 Cartridge 库快速创建一个购物车应用程序系统，具体实现流程如下。

源码路径：daima\2\2-7\car

（1）使用如下命令创建一个 Cartridge 项目，项目名称是"car"。

```
mezzanine-project -a cartridge car
```

（2）使用如下命令进入项目目录。

```
cd car
```

（3）使用如下命令初始化创建一个数据库，默认数据库类型是 SQLite。

```
python manage.py createdb --noinput
```

在这个过程中需要注意系统默认的管理员账号信息，其中用户名默认为 admin，密码默认为 default。

（4）使用如下命令启动这个项目。

```
python manage.py runserver
```

当显示如下所示的信息时，说明成功运行了新建的 Cartridge 项目"car"。

```
               .....
           .d^^^^^^^^^b_
         .d''           ``b.
       .p'                `q.
      .d'                   `b.
     .d'                     `b.   * Mezzanine 4.2.3
     ::                       ::   * Django 1.10.8
     ::    M E Z Z A N I N E   ::   * Python 3.6.0
     ::                       ::   * SQLite 3.2.2
     `p.                     .q'   * Windows 10
      `p.                   .q'
       `b.                 .d'
         `q..           ..p'
           ^q.........p^
               ''''

Performing system checks...

System check identified no issues (0 silenced).
April 17, 2018 - 21:05:02
Django version 1.10.8, using settings 'car.settings'
Starting development server at http://127.0.0.1:8000/
Quit the server with CTRL-BREAK.
```

在浏览器地址栏中输入"http://127.0.0.1:8000/"后，进入前台主页，如图 2-29 所示。

图 2-29　前台主页 http://127.0.0.1:8000/

（5）后台管理首页地址是 http://127.0.0.1:8000/admin，如图 2-30 所示。在登录后台管理页面时，使用在创建数据库时提供的默认账号信息。

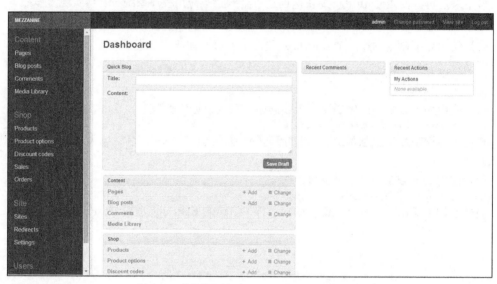

图 2-30　后台管理首页 http://127.0.0.1:8000/admin

与传统的 Django 和 Mezzanine 项目相比，Cartridge 提供了和电子商务功能密切相关的模块，具体说明如下。

- Products：实现商品管理功能。
- Product options：设置商品规格信息，包括颜色、尺寸和其他规格信息。
- Discount codes：设置系统商品折扣信息。
- Sales：设置销售信息。
- Orders：实现订单管理功能。

（6）系统主页默认显示 Home 页面，如果想以 Blog 的列表作为主页，只须将文件 url.py 中的代码行 un-comment 修改为如下内容即可。

```
url("^$", "mezzanine.blog.views.blog_post_list", name="home")
```

也就是将文件 url.py 中的如下代码注释掉。

```
#url("^$", direct_to_template, {"template": "index.html"}, name="home")
```

然后在 url.py 中对如下代码取消注释。

```
url("^$", "mezzanine.blog.views.blog_post_list", name="home")
```

（7）如果想去掉导航栏中的 Search 输入框，需要添加如下配置项。

```
SEARCH_MODEL_CHOICES = []
```

如果想去掉左侧导航链接和页脚，则需要添加如下配置项。

```
PAGE_MENU_TEMPLATES = ( ( 1, "Top navigation bar", "pages/menus/dropdown.html"), )
```

（8）Mezzanine 默认支持 4 种数据库，分别是 postgresql_psycopg2、MySQL、SQLite3 和 Oracle，在默认情况下使用 SQLite3。我们可以在文件 local_settings.py 中的如下代码段中进行修改设置。

```
DATABASES = {
    "default": {
        "ENGINE": "django.db.backends.sqlite3",
        "NAME": "dev.db",
        "USER": "",
        "PASSWORD": "",
        "HOST": "",
        "PORT": "",
    }
}
```

2.8 使用 django-oscar 库

django-oscar 库是一个基于 Django 的、开源的电子商务框架，其官网是 oscarcommerce 网站，其源码参见 GitHub 网站。在安装 django-oscar 库之前需要先确保已经安装了 Django 库。使用如下命令安装 django-oscar 库。

```
pip install django-oscar
```

虽然安装 django-oscar 库很简单，但是 django-oscar 库的一些依赖项不支持 Windows 系统，并且安装起来十分困难，可能会遇到一些错误从而阻止安装成功。所以，建议读者在 Linux 系统下安装 django-oscar 库，或者使用虚拟环境进行安装。

在浏览器地址栏中输入"http://127.0.0.1:8000/"后，进入前台主页，如图 2-31 所示。可见，这是一个在线商城风格的页面。

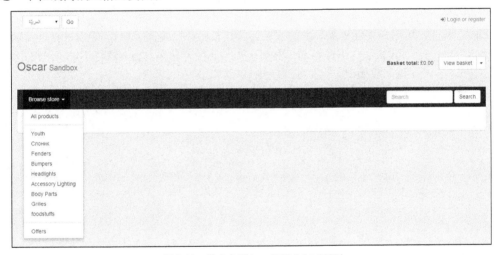

图 2-31　前台主页 http://127.0.0.1:8000/

商品展示界面如图 2-32 所示。

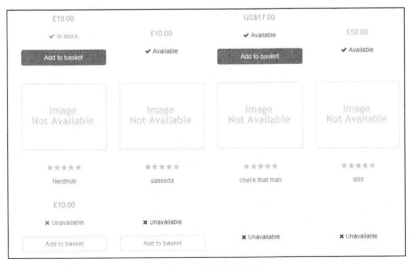

图 2-32　商品展示界面

第 3 章

Flask 框架

Flask 是一个免费的 Web 框架，它的文档齐全，社区活跃度高。Flask 的设计目标是实现一个 WSGI 的微框架，其核心代码保持简单和可扩展性，很容易学习。本章将详细讲解使用 Flask 框架开发动态 Web 程序的知识。

3.1　Flask 开发基础

因为 Flask 框架并不是 Python 的标准库，所以在使用它之前必须先进行安装。本节将详细讲解 Flask 开发的基础知识。

3.1.1　Flask 框架简介

Flask 一般用于开发轻量级的 Web 应用程序。Flask 框架的基本结构如图 3-1 所示。

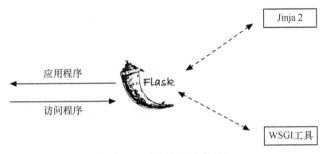

图 3-1　Flask 框架的基本结构

Flask 框架依赖两个外部库——Werkzeug 和 Jinja2。Werkzeug 是一个 WSGI（在 Web 应用和多种服务器之间的标准 Python 接口）工具集，Jinja2 负责渲染模板。为了理解 Flask 框架如何抽象出 Web 开发中的共同部分，我们先来看看 Web 应用程序的一般流程。对于 Web 应用来说，当要获取动态资源时，客户端就会发起一个 HTTP 请求（比如，用浏览器访问一个 URL），Web 应用程序会在后台进行相应的业务处理，取出用户需要的数据，生成相应的 HTTP 响应。如果访问静态资源，则直接返回资源，不需要进行业务处理。整个 Web 应用程序的处理过程如图 3-2 所示。

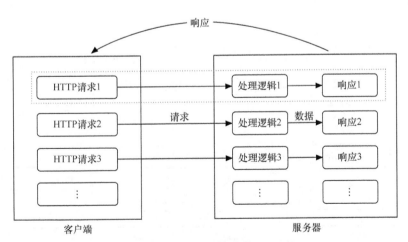

图 3-2　整个 Web 应用程序的处理过程

在实际应用中，不同的请求可能会调用相同的处理逻辑。例如，在论坛站点中，所有获取 topic 内容的请求都可以用 topic/<topic_id>/ 这类 URL 来表示。这里的 topic_id 用来区分不同的 topic。接着在后台定义一个 get_topic(topic_id)函数，用来获取 topic 相应的数据，此外，还需要建立 URL 和函数之间的一一对应关系。这就是 Web 开发中所谓的路由分发，如

图 3-3 所示。

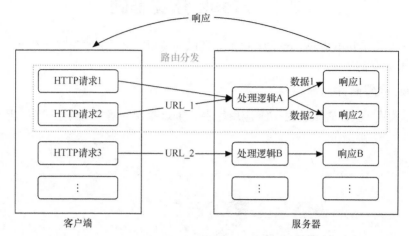

图 3-3　Web 开发中的路由分发

Flask 底层使用 Werkzeug 实现路由分发，代码十分简单，例如下面的代码。

```
@app.route('/topic/<int:topic_id>/')
defget_topic(topic_id):
```

通过业务逻辑函数获取数据后，需要根据这些数据生成 HTTP 响应（对于 Web 应用来说，HTTP 响应一般是一个 HTML 文件）。Web 开发中的一般做法是提供一个 HTML 模板文件，然后将数据传入模板，经过渲染后得到最终需要的 HTML 响应文件。

一种比较常见的场景是：请求虽然不同，但响应中数据的展示方式是相同的。仍以论坛为例。对不同 topic 而言，其具体 topic 内容虽然不同，但页面展示的方式是一样的，都有标题栏、内容栏等。也就是说，对于 topic 来说，我们只需提供一个 HTML 模板，然后传入不同 topic 数据，即可得到不同的 HTTP 响应。这就是所谓的模板渲染，如图 3-4 所示。

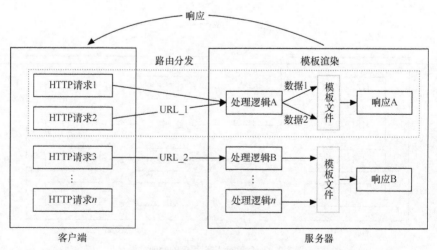

图 3-4　模板渲染

Flask 使用 Jinja2 模板渲染引擎来实现模板渲染功能，例如下面的演示代码。

```
@app.route('/topic/<int:topic_id>/')
defget_topic(topic_id):
    returnrender_template('path/to/template.html', data_needed)
```

综上所述，Flask 处理一个访问请求的流程如下。

（1）根据 URL 决定由哪个函数来处理。
（2）在函数中进行操作，取得所需的数据。
（3）将数据传给相应的模板文件，由 Jinja2 负责渲染并得到 HTTP 响应内容，然后由 Flask 返回响应内容。

3.1.2 Django 和 Flask 的对比

现在市面中上可选的基于 Python 的 Web 框架有很多，如 Django、Flask、Pyramid、Tornado、Bottle、Diesel、Pecan、Falcon 等。其中最受开发者欢迎的是 Flask 和 Django，这两个框架涵盖了从小微项目到企业级的 Web 服务。

Flask 是一个面向简单需求的小型应用的"微框架"（microframework），而 Django 是面向大型应用的，但是有不同的拓展性和灵活性。Django 的目的是囊括 Web 应用的所有内容，所以开发者只需要开箱即用。

Django 包括一个开箱即用的 ORM，而 Pyramid 和 Flask 让开发者自己选择如何存储或者是否存储他们的数据。到目前为止，对于非 Django 的 Web 应用来说，最流行的 ORM 是 SQLAlchemy，同时还有多种其他选择，从 DynamoDB 和 MongoDB 到简单本地存储的 LevelDB 或朴实的 SQLite。Pyramid 可使用任何数据持久层。

Django 的 "batteries included" 特性让开发者不需要提前为他们的应用程序基础设施做决定，因为他们知道 Python 已经深入到了 Web 应用当中。Django 已经内建了模板、表单、路由、认证、基本数据库管理等内容。

Flask 框架比较年轻，诞生于 2010 年。2006 年 Django 发布了第一个版本，它是一个非常成熟的框架，积累了众多插件和扩展以满足巨大的需求。

虽然 Flask 历史相对更短，但它能够学习之前出现的框架优点，并且把注意力放在了微小项目上。它大多数情况下使用在一些只有一两个功能的小型项目上。例如，httpbin，一个简单的（但很强大的）调试和测试 HTTP 库的项目。

最具活力的社区当属 Django，其有 80 000 个 StackOverflow 问题和一系列来自开发者和优秀用户的良好博客。Flask 社区并没有那么大，但它们的社区在邮件列表和 IRC 上相当活跃。StackOverflow 上仅有 5 000 个相关的标签，Flask 比 Django 小 1/15。在 GitHub 上，它们的星级近乎相当，Django 有 11 300 个，Flask 有 10 900 个。这两个框架使用的都是 BSD 衍生的协议。Flask 和 Django 的协议是 BSD 3 条款。

Django 内建了 Bootstrapping 工具，而 Flask 没有包含类似的工具，因为 Flask 的目标用户不是那些试图构建大型 MVC 应用的人。

3.1.3 安装 Flask

建议使用 pip 命令快速安装 Flask，因为 pip 会自动帮用户安装 Flask 依赖的第三方库。在命令行窗口中使用如下命令进行安装。

```
pip install flask
```

成功安装时的界面如图 3-5 所示。

安装完 Flask 框架后，可以在交互式环境下使用 import flask 语句进行验证。如果没有错误提示，则说明成功安装了 Flask 框架。另外，也可以通过手动下载的方式进行手动安装。要手动安装，必须先下载安装 Flask 依赖的两个外部库，即 Werkzeug 和 Jinja2，分别解压后进入对应的目录，在命令行窗口中使用 python setup.py install 来安装它们。然后可以在 Python 官网中下载 Flask，下载后再使用 python setup.py install 命令来安装它。

注意：Flask 的依赖外部库可以在 GitHub 中得到。

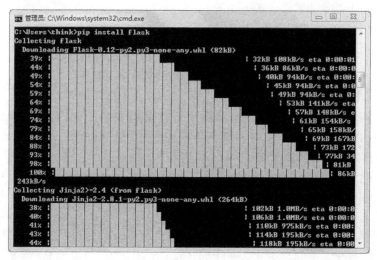

图 3-5　成功安装时的界面

3.1.4　第一个 Flask Web 程序

下面的实例文件 flask1.py 演示了使用 Flask 框架开发一个简单 Web 程序的过程。

源码路径：daima\3\3-1\flask1.py

```
import flask                          #导入Flask模块
app = flask.Flask(__name__)           #实例化类Flask
@app.route('/')                       #装饰器操作,用于关联URL hello()
def hello():                          #定义业务处理函数hello()
    return '你好,这是第一个Flask程序!'
if __name__ == '__main__':
    app.run()                         #运行程序
```

上述实例代码中导入了 Flask 框架，然后实例化主类并自定义只返回一串字符的函数 hello()，然后使用@app.route('/')修饰器将 URL 和函数 hello()联系起来，使得服务器收到对应的 URL 请求时调用这个函数，返回这个函数生成的数据。

执行后会显示一行提醒语句，如图 3-6 所示。这表示 Web 服务器已经正常启动运行了，它的默认服务器端口为 5000，IP 地址为 127.0.0.1。

在浏览器地址栏中输入网址 "http://127.0.0.1:5000/" 后便可以测试上述 Web 程序，执行效果如图 3-7 所示。通过按下键盘中的 Ctrl+C 组合键可以退出当前的服务器。

图 3-6　表示服务器正常运行的语句　　　　　图 3-7　执行效果

当浏览器访问发出的请求被服务器收到后，服务器还会显示出相关信息，如图 3-8 所示，表示访问该服务器的客户端地址、访问的时间、请求的方法以及访问结果的状态码。

图 3-8　服务器显示的相关信息

在上述实例代码中，方法 run()的功能是启动一个服务器，在调用时可以通过参数来设置服务器。常用的主要参数如下。

- host：服务器的 IP 地址，默认为 None。
- port：服务器的端口，默认为 None。

- debug：表示是否开启调试模式，默认为 None。

在现实开发应用中，建议读者使用集成开发工具 PyCharm 来开发 Flask 程序。具体流程如下。

(1) 打开 PyCharm，单击"Create New Project"按钮，如图 3-9 所示。

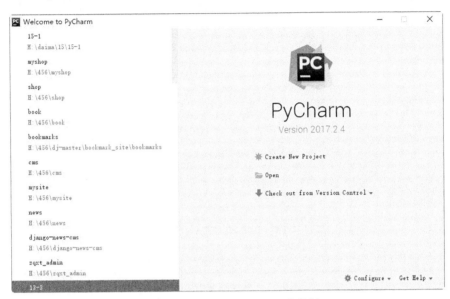

图 3-9　单击"Create New Project"按钮

(2) 弹出"New Project"对话框，在左侧列表中选择"Flask"选项，在"Location"中设置项目的保存路径，如图 3-10 所示。

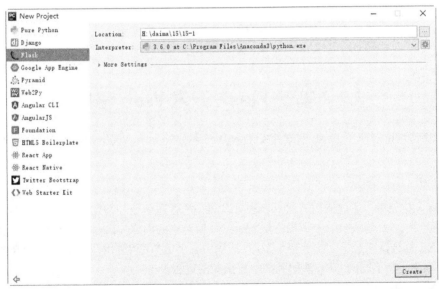

图 3-10　"New Project"对话框

(3) 单击"Create"按钮后会创建一个 Flask 项目，并自动创建保存模板文件和静态文件的文件夹。

(4) 在项目中可以新建一个 Python 文件，其代码可以和前面的实例文件 flask1.py 完全一样，如图 3-11 所示。

图 3-11　创建实例文件 flask1.py

（5）可以直接在 PyCharm 中调试与运行实例文件 flask1.py，运行后在 PyCharm 中会显示图 3-12 所示的效果。

图 3-12　PyCharm 中的执行效果

单击链接"http://127.0.0.1:5000/"会显示具体的执行效果，如图 3-13 所示。

图 3-13　实例文件 flask1.py 的执行效果

3.2　基于 Flask 开发 Web 程序

本节将详细讲解使用 Flask 框架开发基本 Web 程序的知识，为读者学习本书后面的知识打下基础。

3.2.1　传递 URL 参数

在 Flask 框架中，通过使用方法 route() 可以将一个普通函数与特定的 URL 关联起来。当服务器收到这个 URL 请求时，会调用方法 route() 返回对应的内容。Flask 框架中的一个函数可以由多个 URL 修饰器来装饰，实现多个 URL 请求由一个函数产生的内容响应。下面的实例文件 flask2.py 演示了将不同的 URL 映射到同一个函数的过程。

源码路径：daima\3\3-2\flask2.py

```
import flask                          #导入Flask模块
app = flask.Flask(__name__)           #实例化类
@app.route('/')                       #修饰器操作，实现第1个URL的映射
@app.route('/aaa')                    #修饰器操作，实现第2个URL的映射
def hello():
    return '你好，这是一个Flask程序！'
if __name__ == '__main__':
    app.run()                         #运行程序
```

执行本实例后，无论是在浏览器地址栏中输入"http:// 127.0.0.1:5000/"，还是输入"http://127.0.0.1:5000/aaa"，都会在服务器端将这两个 URL 请求映射到同一个函数 hello()，所以输入两个 URL 后的效果一样。执行效果如图 3-14 所示。

图 3-14 执行效果

在现实应用中，传递 HTTP 请求最常用的两种方法是 GET 和 POST。在 Flask 框架中，URL 装饰器的默认方法为 GET，通过使用 Flask 中 URL 装饰器的参数"方法类型"，可以让同一个 URL 的两种请求方法都映射在同一个函数上。

在默认情况下，当通过浏览器传递参数或相关数据时，都是通过 GET 或 POST 请求中包含的参数来实现的。其实通过 URL 也可以传递参数，此时直接将数据放入 URL 中，然后在服务器端获取传递的数据。

在 Flask 框架中，获取的 URL 参数需要在 URL 装饰器和业务函数中分别进行定义或处理。有如下两种形式的 URL 变量规则（URL 装饰器中的 URL 字符串写法）。

```
/hello/<name>        #例如，获取URL"/hello/wang"中的参数"wang"并赋予name变量
/hello/<int: id>     #例如，获取URL"/hello/5"中的参数"5"，自动转换为整数5并赋予id变量
```

要获取和处理 URL 中传递来的参数，需要在对应业务函数的参数列表中列出变量名，具体语法格式如下。

```
@app.route("/hello/<name>")
    def get_url_param (name):
        pass
```

这样在列表中列出变量名后，就可以在业务函数 get_url_param()中引用这个变量，并可以进一步使用从 URL 中传递过来的参数。

下面的实例文件 flask3.py 演示了使用 get 请求获取 URL 参数的过程。

源码路径：daima\3\3-2\flask3.py

```
import flask                        #导入Flask模块
html_txt = """                      #初始化变量html_txt，作为GET请求的页面
<!DOCTYPE html>
<html>
    <body>
        <h2>如果收到了GET请求</h2>
        <form method='post'>        #设置请求方法是post
            <input type='submit' value='按下我发送POST请求' />
        </form>
    </body>
</html>
"""
app = flask.Flask(__name__)         #实例化类Flask
#不管是GET方法还是POST方法，请求都映射到hello()函数
@app.route('/aaa',methods=['GET','POST'])
def hello():                        #定义业务处理函数hello()
    if flask.request.method == 'GET':   #如果接收到的请求是GET
        return html_txt             #返回html_txt的页面内容
    else:                           #否则，接收到的请求是POST
        return '我已经收到POST请求！'
if __name__ == '__main__':
    app.run()                       #运行程序
```

本实例演示了使用参数"方法类型"的 URL 装饰器实例的过程。在上述实例代码中，预先定义了 GET 请求要返回的页面内容字符串 html_txt，在函数 hello()的装饰器中提供了参数 methods 为 GET 和 POST 的字符串列表，表示不管是 GET 方法还是 POST 方法，URL 为"/aaa"的请求都被映射到 hello()函数。在函数 hello()内部使用 flask.request.method 来判断收到的请求方法是 GET 还是 POST，然后分别返回不同的内容。

执行本实例，在浏览器地址栏中输入"http://127.0.0.1:5000/aaa"后的效果如图 3-15 所示。单击"按下我发送 POST 请求"按钮后的效果如图 3-16 所示。

图 3-15 执行效果

图 3-16 单击"按下我发送 POST 请求"按钮后的效果

3.2.2 使用会话和 cookie

通过本书前面的内容可知，使用 cookie 和会话可以存储客户端与服务器端的交互状态。其中 cookie 能够运行在客户端并存储交互状态，而会话能够在服务器端存储交互状态。在 Flask 框架中提供了上述两种常用交互状态的存储方式，其中会话存储方式与其他 Web 框架有一些不同。Flask 框架中的会话使用了密钥签名的方式进行加密。也就是说，虽然用户可以查看你的 cookie，但是如果没有密钥就无法修改它，并且 cookie 只保存在客户端。

在 Flask 框架中，可以通过如下代码获取 cookie。

```
flask.request.cookies.get('name')
```

在 Flask 框架中，可以使用 make_response 对象设置 cookie，例如下面的代码。

```
resp = make_response (content)     #content返回页面内容
resp.set_cookie ('username', 'the username')      #设置名为username的cookie
```

下面的实例文件 flask4.py 演示了使用 cookie 跟踪用户的过程。

源码路径：daima\3\3-2\flask4.py

```python
import flask                        #导入Flask模块
html_txt = """                      #初始化变量html_txt，作为GET请求的页面
<!DOCTYPE html>
<html>
    <body>
        <h2>可以收到GET请求</h2>
        <a href='/get_xinxi'>单击我获取Cookie信息</a>
    </body>
</html>
"""
app = flask.Flask(__name__)                     #实例化类Flask
@app.route('/set_xinxi/<name>')                 #把URL映射到指定目录中的文件
def set_cks(name):                              #函数set_cks()用于从URL中获取参数并将其存入cookie中
    name = name if name else 'anonymous'
    resp = flask.make_response(html_txt)        #构造响应对象
    resp.set_cookie('name',name)                #设置cookie
    return resp
@app.route('/get_xinxi')
def get_cks():                                  #函数get_cks()用于从cookie中读取数据并显示在页面中
    name = flask.request.cookies.get('name')    #获取cookie信息
    return '获取的cookie信息是:' + name          #显示获取到的cookie信息
if __name__ == '__main__':
    app.run(debug=True)
```

在上述实例代码中，首先定义了两个函数：第一个函数用于从 URL 中获取参数并将其存入 cookie 中；第二个函数的功能是从 cookie 中读取数据并显示在页面中。

当在浏览器地址栏中使用"http://127.0.0.1:5000/set_xinxi/langchao"浏览时，表示设置了名为 langchao 的 cookie 信息，执行效果如图 3-17 所示。单击"单击我获取 cookie 信息"链接后时，会在新页面中显示在 cookie 中保存的名为"langchao"的信息，效果如图 3-18 所示。

图 3-17 执行效果

图 3-18 单击"单击我获取 cookie 信息"链接后的效果

3.2.3 使用 Flask-Script 扩展

Flask 被设计为可扩展形式，因此它没有提供一些重要的功能，如数据库和用户认证，于是开发者可以自由选择最适合程序的包，或者按需求自行开发。社区成员开发了大量不同用途的扩展，如果这不能满足需求，开发者还可使用所有 Python 标准包或代码库。

虽然用 Flask 框架开发的 Web 服务器支持很多启动设置选项，但是只能在脚本中作为参数传递给 app.run()函数。这种方式不太方便，传递设置选项的最佳方式是使用命令行参数。Flask-Script 是一个著名的 Flask 扩展，为 Flask 程序添加了一个命令行解析器。Flask-Script 自带了一组常用选项，而且还支持自定义命令。可以通过如下 pip 命令安装 Flask-Script 扩展。

```
pip install flask-script
```

下面的实例文件 hello.py 演示了使用 Flask-Script 扩展增强程序功能的过程。实例文件 hello.py 的具体实现代码如下。

源码路径：daima\3\3-2\kuo\hello.py

```python
from flask import Flask
from flask_script import Manager

app = Flask(__name__)

manager = Manager(app)

@app.route('/')
def index():
    return '<h1>Hello World!</h1>'

@app.route('/user/<name>')
def user(name):
    return '<h1>Hello, %s!</h1>' % name

if __name__ == '__main__':
    manager.run()
```

在上述代码中，Flask-Script 输出了一个名为 Manager 的类，这可以从 flask_script 中导入。这个扩展的初始化方法也适用于其他很多扩展。例如，把程序实例作为参数传给构造函数，初始化主类的实例。创建的对象可以在各个扩展中使用，上述代码中的服务器由 manager.run()启动，启动后就能解析命令行了。在 PyCharm 中运行上述代码后会输出下面的结果。

```
usage: hello.py [-?] {shell,runserver} ...

positional arguments:
  {shell,runserver}
    shell              Runs a Python shell inside Flask application context.
    runserver          Runs the Flask development server i.e. app.run()

optional arguments:
  -?, --help           show this help message and exit
```

shell 命令用于在程序的上下文中启动 Python shell 会话。你可以在这个会话中运行维护任务或测试，还可调试异常。顾名思义，runserver 命令用来启动 Web 服务器。运行 python hello.py runserver 后会以调试模式启动 Web 服务器。在实际应用中，还有如下的可用选项。

```
$ python hello.py runserver --help
usage: hello.py runserver [-h] [-t HOST] [-p PORT] [--threaded]
[--processes
```

使用 app.run()运行 Flask 开发服务器后，输出以下信息。

```
optional arguments:
    -h, --help      显示帮助信息并退出
    -t HOST, --host HOST
    -p PORT, --port PORT
    --threaded
    --processes PROCESSES
    --passthrough-errors
```

```
        -d, --no-debug
        -r, --no-reload
```

在上述输出中，参数"--host"是一个很有用的选项，它告诉 Web 服务器在哪个网络接口上监听来自客户端的连接。在默认情况下，Flask 开发 Web 服务器监听 localhost 上的连接，所以只接受服务器所在计算机发起的连接。通过下面的命令可以让 Web 服务器监听公共网络接口上的连接，允许同一个网络中的其他计算机连接服务器。

```
$ python hello.py runserver --host 127.0.0.1
* Running on http://127.0.0.1:5000/
* Restarting with reloader
```

现在，Web 服务器可使用 http://*a.b.c.d*:5000/ 网络中的任一台计算机进行访问，其中"*a.b.c.d*"是服务器所在计算机的外网 IP 地址。

3.2.4 使用模板

模板是一个包含响应文本的文件，其中包含用占位变量表示的动态部分，其具体值只在请求的上下文中才能知道。使用真实值替换变量，再返回最终得到的响应字符串，这一过程称为渲染。为了渲染模板，Flask 使用了一个名为 Jinja2 的强大模板引擎。形式最简单的 Jinja2 模板就是一个包含响应文本的文件。例如，下面就是一个 Jinja2 模板的实现代码，它和前面实例文件 flask2.py 中的 hello()视图函数的响应一样。

```
<h1>Hello World!</h1>
```

下面的实例演示了使用上述模板文件的过程。

（1）假设定义的两个模板文件保存在 templates 文件夹中，这两个模板文件分别命名为 index.html 和 user.html。其中模板文件 index.html 只有一行代码，具体实现代码如下。

源码路径：daima\3\3-2\moban\templates\index.html

```
<h1>Hello World!</h1>
```

模板文件 user.html 也只有一行代码，具体实现代码如下。

源码路径：daima\3\3-2\moban\templates\user.html

```
<h1>Hello, {{ name }}!</h1>
```

（2）在默认情况下，Flask 在程序文件夹中的 templates 子文件夹中寻找模板。接下来可以在 Python 程序中通过视图函数处理上面的模板文件，以便渲染这些模板。实例文件 moban.py 的内容如下。

源码路径：daima\3\3-2\moban\moban.py

```
import flask
from flask import Flask, render_template
app = flask.Flask(__name__)

@app.route('/')
def index():
    return render_template('index.html')
@app.route('/user/<name>')
def user(name):
    return render_template('user.html', name=name)

if __name__ == '__main__':
    app.run()
```

在上述代码中，Flask 提供的 render_template 函数把 Jinja2 模板引擎集成到了程序中。render_template 函数的第一个参数是模板的文件名。随后的参数都是键值对，表示模板中变量对应的真实值。在上述代码中，第二个模板收到一个名为 name 的变量。代码中的 name=name 是关键字参数。这类关键字参数很常见，但如果你不熟悉它们，可能会觉得迷惑且难以理解。左边的"name"表示参数名，就是模板中使用的占位符；右边的"name"是当前作用域中的变量，表示同名参数的值。

在浏览器地址栏中输入"http://127.0.0.1:5000/"后，会调用模板文件 index.html，执行效果如图 3-19 所示。

执行"http://127.0.0.1:5000/user/guanguan"后会调用模板文件 user.html，表示用户名为"guanguan"，执行效果如图 3-20 所示。

图 3-19　调用模板文件 index.html

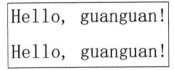

图 3-20　用户名为"guanguan"

1. 变量

在上面的模板文件 user.html 中，{{ name }}代码部分表示一个变量，这是一种特殊的占位符，目的是告诉模板引擎这个位置的值从渲染模板时使用的数据中获取。Jinja2 能识别所有类型的变量，甚至是一些复杂的类型，如列表、字典和对象等。下面是一些在模板中使用变量的演示代码。

```
<p>A value from a dictionary: {{ mydict['key'] }}.</p>
<p>A value from a list: {{ mylist[3] }}.</p>
<p>A value from a list, with a variable index: {{ mylist[myintvar] }}.</p>
<p>A value from an object's method: {{ myobj.somemethod() }}.</p>
```

开发者可以使用过滤器修改变量，将过滤器名添加在变量名后面，中间使用竖线分隔。在下面的演示代码中，模板以首字母大写形式显示变量 name 的值。

```
Hello, {{ name|capitalize }}
```

下面列出了 Jinja2 提供的几种常用过滤器。
- safe：在渲染值时不转义。
- capitalize：把值的首字母转换成大写形式，把其余的字母转换成小写形式。
- lower：把值转换成小写形式。
- upper：把值转换成大写形式。
- title：把值中每个单词的首字母都转换成大写形式。
- trim：把值的首尾空格去掉。
- striptags：渲染之前把值中所有的 HTML 标签都删掉。

其中，关于 safe 过滤器需要特别提醒一下。在默认情况下，出于安全方面的考虑，Jinja2 会转义所有变量。如果一个变量的值为'<h1>Hello</h1>'，则 Jinja2 会将其渲染成：

```
&lt;h1&gt;Hello&lt;/h1&gt;
```

浏览器能显示这个 h1 元素，但不会解释它。在很多情况下，如果需要显示变量中存储的 HTML 代码，就可以使用 safe 过滤器。

2. 控制结构

Jinja2 提供了多种控制结构，可用来改变模板的渲染流程。下面的演示代码展示了在模板中使用条件控制语句的过程。

```
{% if user %}
Hello, {{ user }}!
{% else %}
Hello, Stranger!
{% endif %}
```

另一种常见需求是在模板中渲染一组元素。下面的演示代码展示了使用 for 循环渲染一组元素的过程。

```
<ul>
{% for comment in comments %}
<li>{{ comment }}</li>
{% endfor %}
</ul>
```

Jinja2 还支持宏功能，宏类似于 Python 代码中的函数。例如下面的演示代码。

```
{% macro render_comment(comment) %}
<li>{{ comment }}</li>
{% endmacro %}
<ul>
{% for comment in comments %}
{{ render_comment(comment) }}
{% endfor %}
</ul>
```

为了可以重复使用宏功能，可以将其保存在单独的文件中，然后在需要使用的模板中导入。例如下面的演示代码。

```
{% import 'macros.html' as macros %}
<ul>
{% for comment in comments %}
{{ macros.render_comment(comment) }}
{% endfor %}
</ul>
```

把需要多处重复使用的模板代码片段写入单独的文件，然后再包含在所有模板中，这样可以避免重复。例如下面的演示代码。

```
{% include 'common.html' %}
```

另一种重复使用代码的强大方式是模板继承，它类似于 Python 代码中的类继承。首先，创建一个名为 base.html 的基模板。

```
<html>
<head>
{% block head %}
<title>{% block title %}{% endblock %} - My Application</title>
{% endblock %}
</head>
<body>
{% block body %}
{% endblock %}
</body>
</html>
```

然后，使用标签 block 定义的元素在衍生模板中修改，假如定义了名为 head、title 和 body 的块，其中 title 包含在 head 中。下面的演示代码是基模板的衍生模板。

```
{% extends "base.html" %}
{% block title %}Index{% endblock %}
{% block head %}
{{ super() }}
<style>
</style>
{% endblock %}
{% block body %}
<h1>Hello, World!</h1>
{% endblock %}
```

指令 extends 的功能是声明这个模板衍生自 base.html。在 extends 指令之后，基模板中的 3 个块被重新定义，模板引擎会将其插入适当的位置。注意新定义的 head 块，在基模板中其内容不是空的，所以使用 super() 获取原来的内容。

3.2.5 使用 Flask-Bootstrap 扩展

Bootstrap 是 Twitter 开发的一个开源框架，它提供的用户界面组件可用于创建整洁且具有吸引力的网页，而且这些网页还能兼容所有的现代 Web 浏览器。因为 Bootstrap 是客户端框架，所以不会直接涉及服务器。服务器需要做的只是提供引用了 Bootstrap 层叠样式表（CSS）和 JavaScript 文件的 HTML 响应，并在 HTML、CSS 和 JavaScript 代码中实例化所需组件，这些操作最理想的执行场所就是模板。

要在 Python 程序中集成 Bootstrap，要对模板做必要的改动。不过，更简单的方法是使用一个名为 Flask-Bootstrap 的 Flask 扩展。Flask-Bootstrap 使用如下 pip 命令进行安装。

3.2 基于 Flask 开发 Web 程序

```
pip install flask-bootstrap
```

Flask 扩展一般都在创建程序实例时进行初始化。下面是初始化 Flask-Bootstrap 的演示代码。

```
from flask.ext.bootstrap import Bootstrap
# ...
bootstrap = Bootstrap(app)
```

Flask-Bootstrap 是从 flask.ext 命名空间中导入的,然后把程序实例传入构造方法进行初始化。在初始化 Flask-Bootstrap 之后,就可以在程序中使用一个包含所有 Bootstrap 文件的基模板。这个模板利用 Jinja2 的模板继承机制,让程序扩展一个具有基本页面结构的基模板,其中就有用来导入 Bootstrap 的元素。

下面的实例演示了在 Flask 中使用 Flask-Bootstrap 扩展的过程。

(1) 看 Python 文件 untitled.py。通过代码 "bootstrap=Bootstrap(app)" 为 Flask 扩展 Bootstrap 实现实例初始化,这行代码是 Flask-Bootstrap 的初始化方法。具体实现代码如下。

源码路径:daima\3\3-2\untitled\untitled.py

```
from flask import Flask,render_template
from flask_bootstrap import Bootstrap
app=Flask(__name__)
bootstrap=Bootstrap(app)
@app.route('/')
def index():
    return render_template('index.html')
if __name__=="__main__":
    app.run(debug=True)
```

(2) 看模板文件 base.html。通过 Jinja2 中的 extends 指令从 Flask-Bootstrap 中导入 bootstrap/base.html,从而实现模板继承。Flask-Bootstrap 中的基模板提供了一个网页框架,导入了 Bootstrap 中的所有 CSS 和 JavaScript 文件。基模板中定义了可在衍生模板中重定义的块。block 和 endblock 指令定义的块中的内容可添加到基模板中。文件 base.html 的具体实现代码如下。

源码路径:daima\3\3-2\untitled\base.html

```
{% extends "bootstrap/base.html" %} <!-- base.html模板继承自bootstrap/base.html -->
{% block title %}Flask{% endblock %}
{% block navbar %}
<div class="navbar navbar-inverse" role="navigation">
 <div class="container">
 <div class="navbar-header">
 <button type="button" class="navbar-toggle" data-toggle="collapse" data-taget=".navbar-collapse">
 <span class="sr-only">Toggle navigation</span>
 <sapn class="icon-bar"></sapn>
 <span class="icon-bar"></span>
 <span class="icon-bar"></span>
 </button>
 <a class="navbar-brand" href="/">Flasky</a>
 </div>
 <div class="navbar=collapse collapse">
 <ul class="nav navbar-nav">
 <li>
 <a href="/">Home</a>
 </li>
 </ul>
 </div>
 </div>
</div>
{% endblock %}


{% block content %}
<div class="container">
 {% block page_content %}{% endblock %}
</div>
{% endblock %}
```

在上述模板文件 base.html 中定义了 3 个块,分别为 title、navbar 和 content。这些块都是基模板提供的,可以在衍生模板中重新定义。title 块的作用很明显,其中的内容会出现在渲染后的 HTML 文档头部,放在<title> 标签中。navbar 和 content 这两个块分别表示页面中的导航条与主体内容。在这个模板中,navbar 块使用 Bootstrap 组件定义了一个简单的导航条。content 块中有个<div> 容器,其中包含一个页面头部。之前版本的模板的欢迎信息,现在就放在这个页面头部。

(3)在模板文件 index.html 中继承上面的 base.html。index.html 的具体实现代码如下。

源码路径:daima\3\3-2\untitled\index.html

```
{% extends "base.html" %}
{% block title %}首页{% endblock %}
{% block page_content %}
<h2>这里是首页,welcome</h2>
Technorati Tags: flask
{% endblock %}
```

在浏览器地址栏中输入 "http://127.0.0.1:5000/" 后会显示指定的模板样式,并实现一个导航效果,如图 3-21 所示。

在 Flask-Bootstrap 的 base.html 模板中定义了很多其他块,这些块都可以在衍生模板中使用。下面列出了所有可用的块。

- doc:整个 HTML 文档。
- html_attribs:<html> 标签的属性。
- html:<html> 标签中的内容。
- head:<head> 标签中的内容。
- title:<title> 标签中的内容。
- metas:一组<meta> 标签。
- styles:层叠样式表的定义。
- body_attribs:<body> 标签的属性。
- body:<body> 标签中的内容。
- navbar:用户定义的导航条。
- content:用户定义的页面内容。
- scripts:文档底部的 JavaScript 声明。

图 3-21 导航效果

上面的很多块都是 Flask-Bootstrap 自用的,如果直接重定义可能会导致一些问题。例如,Bootstrap 所需的文件在 styles 和 scripts 块中声明。如果程序需要向已经有内容的块中添加新内容,必须使用 Jinja2 提供的 super() 函数。如果要在衍生模板中添加新的 JavaScript 文件,需要像下面这样定义 scripts 块。

```
{% block scripts %}
{{ super() }}
<script type="text/javascript" src="my-script.js"></script>
{% endblock %}
```

3.2.6 使用 Flask-Moment 扩展本地化日期和时间

Moment.js 是一个简单易用的轻量级 JavaScript 日期处理类库,提供了日期格式化和日期解析等功能。Moment.js 支持在浏览器和 Node.js 两种环境中运行,此类库能够将给定的任意日期转换成多种不同的格式,具有强大的日期计算功能,同时也内置了能显示多样的日期形式的函数。另外,Moment.js 也支持多种语言,开发者可以任意新增一种新的语言包。Flask-Moment 是一个集成 moment.js 到 Jinja2 模板的 Flask 扩展。下面是初始化 Flask-Moment

的演示代码。

```
from flask.ext.moment import Moment
moment = Moment(app)
```

安装 Flask-Moment 的指令如下。

```
pip install flask-moment
```

Flask-Moment 依赖于 moment.js 和 jquery.js，需要直接包含在 HTML 文档中。例如通过如下代码在 base.html 模版中的 head 标签中导入 moment.js 和 jquery.js。

```
<html>
    <head>

        {{ moment.include_jquery() }}
        {{ moment.include_moment() }}

        <!--使用中文,默认是英语的-->
        {{ moment.lang("zh-CN") }}

</head> <body> ... </body> </html>
```

如果使用了 bootstrap，可以不用导入 jquery.js，因为在 bootstrap 中包含了 jquery.js。

下面的实例演示了使用 Flask-Moment 扩展的过程。

（1）编写程序文件 hello.py，为了处理时间戳，Flask-Moment 向模板开放了 moment 类，把变量 current_time 传入模板进行渲染。具体实现代码如下。

源码路径：daima\3\3-2\flasky3e\hello.py

```
from datetime import datetime
from flask import Flask, render_template
from flask_script import Manager
from flask_bootstrap import Bootstrap
from flask_moment import Moment

app = Flask(__name__)

manager = Manager(app)
bootstrap = Bootstrap(app)
moment = Moment(app)
@app.errorhandler(404)
def page_not_found(e):
    return render_template('404.html'), 404


@app.errorhandler(500)
def internal_server_error(e):
    return render_template('500.html'), 500


@app.route('/')
def index():
    return render_template('index.html',
                           current_time=datetime.utcnow())


@app.route('/user/<name>')
def user(name):
    return render_template('user.html', name=name)


if __name__ == '__main__':
    manager.run()
```

（2）在模板文件 index.html 中使用 Flask-Moment 渲染时间戳。

源码路径：daima\3\3-2\flasky3e\templates\index.html

```
{% extends "base.html" %}

{% block title %}Flask教程{% endblock %}
```

```
{% block page_content %}
<div class="page-header">
    <h1>Hello World!</h1>
</div>
<p>当前时间是：{{ moment(current_time).format('LLL') }}.</p>
<p>这是{{ moment(current_time).fromNow(refresh=True) }}.</p>
{% endblock %}
```

在上述代码中，format('LLL') 会根据客户端计算机中的时区和区域设置渲染日期和时间。参数决定了渲染的方式，'L' 到'LLLL' 分别对应不同的复杂度。format()函数还可接受自定义的格式说明符。上述代码中的 fromNow()函数用于渲染相对时间戳，而且会随着时间的推移自动刷新显示的时间。这个时间戳最开始显示为"a few seconds ago"，但指定 refresh 参数后，其内容会随着时间的推移而更新。如果一直停留在这个页面上，几分钟后，会看到显示的文本变成"这是 123 minute ago"之类的提示文本。

在浏览器地址栏中输入"http://127.0.0.1:5000/"后，显示的本地化时间如图 3-22 所示。

图 3-22　显示本地化时间

3.3　表单操作

表单是动态 Web 开发的最核心模块之一，绝大多数的 Web 动态交互功能是通过表单实现的。本节将详细讲解在 Flask 框架中实现表单操作的知识。

3.3.1　使用 Flask-WTF 扩展

虽然 Flask 请求对象提供的信息足够用于处理 Web 表单，但是有一些任务很单调，而且要重复操作，例如，生成表单的 HTML 代码和验证提交的表单数据。通过使用 Flask-WTF 扩展，对独立的 WTForms 包进行包装，可以方便集成到 Flask 程序中。

可以使用如下所示的 pip 命令来安装 Flask-WTF 及其依赖。

```
pip install flask-wtf
```

在默认情况下，Flask-WTF 能使所有表单免受跨站请求伪造（Cross-Site Request Forgery，CSRF）攻击。当恶意网站把请求发送到攻击者已登录的其他网站时就会引发 CSRF 攻击。为了防止受到 CSRF 攻击，Flask-WTF 需要程序设置一个密钥。Flask-WTF 使用这个密钥生成加密令牌，再用令牌验证请求中表单数据的真伪。下面是一段设置密钥的演示代码。

```
app = Flask(__name__)
app.config['SECRET_KEY'] = 'aaa bbb ccc'
```

在 Flask 框架中，使用 app.config 字典来存储框架、扩展程序的配置变量。使用标准的字典句法就能把配置值添加到 app.config 对象中。这个对象还提供了一些方法，用于从文件或环境中导入配置值。SECRET_KEY 配置变量是通用密钥，可以在 Flask 和多个第三方扩展中使用。加密的强度取决于变量值的机密程度，不同的程序要使用不同的密钥，而且要保证其他人不知道你所用的字符串。

在使用 Flask-WTF 时，每个 Web 表单都由一个继承自 Form 的类表示。这个类定义表单中的一组字段，每个字段都用对象表示。字段对象可附属一个或多个验证函数，通过验证函数可以验证用户提交的输入值是否符合要求。下面的实例演示了使用 Flask-WTF 实现表单处理的过程。

（1）编写程序文件 hello.py 创建一个简单的 Web 表单，该表单包含一个文本字段和一个提交按钮。将表单中的字段都定义为类变量，类变量的值是相应字段类型的对象。其中在 NameForm 表单中有一个名为 name 的文本字段和一个名为 submit 的提交按钮。类 StringField

表示属性为 type="text" 的<input> 元素，类 SubmitField 表示属性为 type="submit" 的<input>元素。字段构造函数 NameForm()的第一个参数是把表单渲染成 HTML 时使用的标号。文件 hello.py 的具体实现代码如下。

源码路径：daima\3\3-3\biaodan01\hello.py

```python
from flask import Flask, render_template
from flask_script import Manager
from flask_bootstrap import Bootstrap
from flask_moment import Moment
from flask_wtf import FlaskForm
from wtforms import StringField, SubmitField
from wtforms.validators import Required

app = Flask(__name__)
app.config['SECRET_KEY'] = 'aaa bbb ccc'

manager = Manager(app)
bootstrap = Bootstrap(app)
moment = Moment(app)


class NameForm(FlaskForm):
    name = StringField('你叫什么名字?', validators=[Required()])
    submit = SubmitField('提交')
@app.route('/', methods=['GET', 'POST'])
def index():
    name = None
    form = NameForm()
    if form.validate_on_submit():
        name = form.name.data
        form.name.data = ''
    return render_template('index.html', form=form, name=name)
```

在上述代码中，视图函数 index()不仅要渲染表单，还要接收表单中的数据。app.route 使用修饰器中添加的 methods 参数告诉 Flask，在 URL 映射中把这个视图函数注册为 GET 和 POST 请求的处理程序。如果没指定 methods 参数，则只把视图函数注册为 GET 请求的处理程序。很有必要把 POST 加入方法列表中，因为将提交表单作为 POST 请求进行处理更加便利。表单也可作为 GET 请求提交，不过 GET 请求没有主体，提交的数据以查询字符串的形式附加到 URL 中，这可以在浏览器的地址栏中看到。基于这个以及其他多个原因，提交的表单大多作为 POST 请求进行处理。局部变量 name 用于存放表单中输入的有效名字，如果没有输入，其值为 None。例如，在上述代码中，在视图函数中创建一个 NameForm 类实例用于表示表单。提交表单后，如果数据能被所有验证函数接受，那么 validate_on_submit()函数的返回值为 True；否则，返回 False。

WTForms 可以支持如下 HTML 标准字段。

- StringField：文本字段。
- TextAreaField：多行文本字段。
- PasswordField：密码文本字段。
- HiddenField：隐藏文本字段。
- DateField：文本字段，值为 datetime.date 格式。
- DateTimeField：文本字段，值为 datetime.datetime 格式。
- IntegerField：文本字段，值为整数。
- DecimalField：文本字段，值为 decimal.Decimal。
- FloatField：文本字段，值为浮点数。
- BooleanField：复选框，值为 True 和 False。

- RadioField：一组单选框。
- SelectField：下拉列表。
- SelectMultipleField：下拉列表，可选择多个值。
- FileField：文件上传字段。
- SubmitField：表单提交按钮。
- FormField：把表单作为字段嵌入另一个表单。
- FieldList：一组指定类型的字段。

在 WTForms 中包含如下内置验证函数。

- Email：验证电子邮件地址。
- EqualTo：比较两个字段的值，常用于要求对两次输入的密码进行确认的情况。
- IPAddress：验证 IPv4 网络地址。
- Length：验证输入字符串的长度。
- NumberRange：验证输入的值在数字范围内。
- Optional：当无输入值时跳过其他验证函数。
- Required：确保字段中有数据。
- Regexp：使用正则表达式验证输入值。
- URL：验证 URL。
- AnyOf：确保输入值在可选值列表中。
- NoneOf：确保输入值不在可选值列表中。

（2）在模板文件 index.html 中使用 Flask-WTF 和 Flask-Bootstrap 来渲染表单。具体实现代码如下。

源码路径：daima\3\3-3\biaodan01\templates\index.html

```
{% extends "base.html" %}
{% import "bootstrap/wtf.html" as wtf %}

{% block title %}Flasky{% endblock %}

{% block page_content %}
<div class="page-header">
    <h1>Hello, {% if name %}{{ name }}{% else %}Stranger{% endif %}!</h1>
</div>
{{ wtf.quick_form(form) }}
{% endblock %}
```

上述模板代码的内容区有两部分。第一部分是页面头部，用于显示欢迎消息，在这里用到了一个模板条件语句。Jinja2 中的条件语句格式为{% if condition %}...{% else %}...{% endif %}。如果条件的判断结果为 True，那么渲染 if 和 else 指令之间的值。如果条件的判断结果为 False，则渲染 else 和 endif 指令之间的值。在上述代码中，如果没有定义模板变量 name，则会渲染字符串"Hello, Stranger!"。内容区的第二部分使用 wtf.quick_form() 函数渲染 NameForm 对象。

在浏览器地址栏中输入"http://127.0.0.1:5000/"后的初始执行效果如图 3-23 所示。

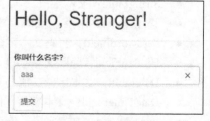

图 3-23 初始执行效果

在表单中随便输入一个用户名，例如，输入"aaa"并单击"提交"按钮后会在表单上面显示对用户"aaa"的欢迎信息，如图 3-24 所示。

如果在表单为空时单击"提交"按钮，会显示"这是必填字段"的提示，如图 3-25 所示。

图3-24 显示对用户名"aaa"的欢迎信息 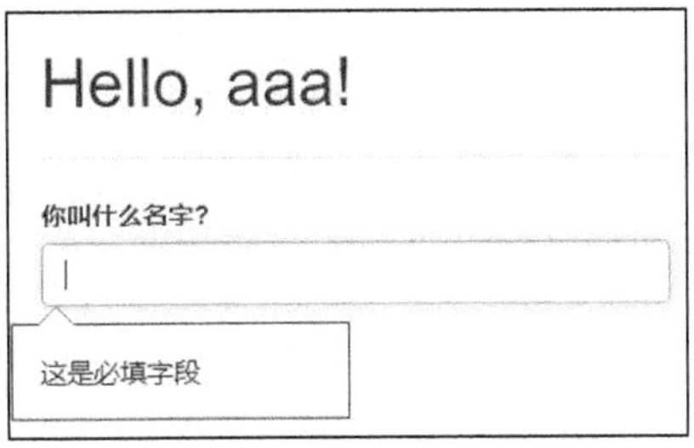　　　图3-25 在表单为空时的提示

3.3.2 文件上传

在Flask框架中实现文件上传的方法非常简单，与传递GET或POST参数十分相似。在Flask框架实现文件上传的基本流程如下。

(1) 将在客户端上传的文件保存在flask.request.files对象中。
(2) 使用flask.request.files对象获取上传的文件名和文件对象。
(3) 调用文件对象中的方法save()将文件保存到指定的目录中。

下面的实例文件flask5.py演示了在Flask框架中实现文件上传系统的过程。

源码路径：daima\3\3-3\up\flask5.py

```python
import flask                              #导入Flask模块
app = flask.Flask(__name__)               #实例化类Flask
#URL映射操作,用于处理GET请求和POST请求
@app.route('/upload',methods=['GET','POST'])
def upload():                             #定义文件上传函数upload()
    if flask.request.method == 'GET':     #如果是GET请求
        return flask.render_template('upload.html')   #返回上传页面
    else:                                 #如果是POST请求
        file = flask.request.files['file']            #获取文件对象
        if file:                          #如果文件不为空
            file.save(file.filename)      #保存上传的文件
            return '上传成功！'            #显示提示信息
if __name__ == '__main__':
    app.run(debug=True)
```

在上述实例代码中，只定义了一个实现文件上传功能的函数upload()，该函数能够同时处理GET请求和POST请求。当获得GET请求时，返回上传页面；当获得POST请求时，获取上传文件，并保存到当前的目录下。

模板文件upload.html的具体实现代码如下。

源码路径：daima\3\3-3\up\templates\upload.html

```html
<!DOCTYPE html>
<html>
    <body>
        <h2>亲,你可以选择一个将要上传的文件</h2>
        <form method='post' enctype='multipart/form-data'>
            <input type='file' name='file' />
            <input type = 'submit' value='单击我上传'/>
        </form>
    </body>
</html>
```

当在浏览器中使用"http://127.0.0.1:5000/upload"运行时，会显示一个文件上传界面，效果如图3-26所示。单击"浏览"按钮可以选择一个要上传的文件，单击"上传"按钮会上传这个文件，并显示上传成功的提示。执行效果如图3-27所示。

图 3-26　执行效果

图 3-27　上传成功的提示

3.4　使用数据库

在数据库中，先按照一定规则保存数据，之后再发起查询取回所需的数据。Web 程序最常用基于关系模型的数据库，这种数据库也称为 SQL 数据库，因为它们使用结构化查询语言。不过最近几年，文档数据库和键值对数据库成了替代选择，这两种数据库合称 NoSQL 数据库。本节将详细讲解在 Flask 框架中使用数据库技术的基本知识。

3.4.1　Python 数据库框架

大多数的数据库引擎都有对应的 Python 包，包括开源包和商业包。因为 Flask 并不会限制开发者使用何种类型的数据库包，所以用户可以根据自己的喜好选择使用 MySQL、Postgres、SQLite、Redis、MongoDB 或 CouchDB。如果这些都无法满足用户的需求，还有一些数据库抽象层代码包可供选择，如 SQLAlchemy 和 MongoEngine。开发者可以使用这些抽象包直接处理高级的 Python 对象，而不用处理如表、文档或查询语言等数据库实体。在选择数据库框架时需要考虑很多因素，这些因素的具体说明如下。

- 易用性。如果直接比较数据库引擎和数据库抽象层，显然后者更易于使用。抽象层也称为对象关系映射（Object-Relational Mapper，ORM）或对象文档映射（Object-Document Mapper，ODM），在用户不察觉的情况下把高层的面向对象操作转换成低层的数据库指令。
- 性能。ORM 和 ODM 在把对象业务转换成数据库业务时会有一定的性能损耗。大多数情况下，这种性能的降低微不足道，但也不一定都是如此。一般情况下，ORM 和 ODM 对生产率的提升远远超过了这微不足道的性能降低，所以性能降低这个理由不足以说服用户完全放弃 ORM 和 ODM。真正的关键点在于如何选择一个能直接操作低层数据库的抽象层，以防特定的操作需要直接使用数据库原生指令进行优化。
- 可移植性。在选择数据库时，必须考虑其是否能在你的开发平台和生产平台中使用。例如，如果打算利用云平台托管程序，就要知道这个云服务提供了哪些数据库。
- Flask 集成度。在选择框架时，不一定非得选择已经集成了 Flask 的框架，但选择这些框架可以节省编写集成代码的时间。使用集成了 Flask 的框架可以简化配置和操作，所以专门为 Flask 开发的扩展是首选。例如，本节后面介绍的数据库框架是 Flask-SQLAlchemy，这个 Flask 扩展包装了 SQLAlchemy。

3.4.2　会员注册和登录

在下面的实例中将实现一个简单的会员注册和登录系统，将会员的注册信息保存到 SQLite3 数据库中。在表单中输入登录信息后，会将输入信息和数据库中保存的信息进行对比。如果一致，则成功登录；否则，提示"登录失败"。

实例文件 flask6.py 的具体实现代码如下。

源码路径：daima\3\3-4\user\flask6.py

```python
DBNAME = 'test.db'

app = flask.Flask(__name__)
app.secret_key = 'dfadff#$#5dgfddgssgfgsfgr4$T^%^'

@app.before_request
def before_request():
    g.db = connect(DBNAME)

@app.teardown_request
def teardown_request(e):
    db = getattr(g,'db',None)
    if db:
        db.close()
    g.db.close()

@app.route('/')
def index():
    if 'username' in session:
        return "你好, " + session['username'] + '<p><a href="/logout">注销</a></p>'
    else:
        return '<a href="/login">登录</a>,<a href="/signup">注册</a>'

@app.route('/signup',methods=['GET','POST'])
def signup():
    if request.method == 'GET':
        return render_template('signup.html')
    else:
        name = 'name' in request.form and request.form['name']
        passwd = 'passwd' in request.form and request.form['passwd']
        if name and passwd:
            cur = g.db.cursor()
            cur.execute('insert into user (name,passwd) values (?,?)',(name,passwd))
            cur.connection.commit()
            cur.close()
            session['username'] = name
            return redirect(url_for('index'))
        else:
            return redirect(url_for('signup'))

@app.route('/login',methods=['GET','POST'])
def login():
    if request.method == 'GET':
        return render_template('login.html')
    else:
        name = 'name' in request.form and request.form['name']
        passwd = 'passwd' in request.form and request.form['passwd']
        if name and passwd:
            cur = g.db.cursor()
            ur.execute('select * from user where name=?',(name,))
            res = cur.fetchone()
            if res and res[1] == passwd:
                session['username'] = name
                return redirect(url_for('index'))
            else:
                return '登录失败!'
        else:
            return '参数不全!'

@app.route('/logout')
def logout():
    session.pop('username',None)
    return redirect(url_for('index'))

def init_db():
    if not os.path.exists(DBNAME):
        cur = connect(DBNAME).cursor()
        cur.execute('create table user (name text,passwd text)')
```

```
                cur.connection.commit()
                print('数据库初始化完成!')
if __name__ == '__main__':
    init_db()
    app.run(debug=True)
```

本实例用到了模板技术。其中用户注册功能的实现模板是 signup.html。具体实现代码如下。

源码路径：daima\3\3-4\user\templates\signup.html

```
<!DOCTYPE html>
<html>
    <body>
        <form method='post'>
        <input type='text' name='name' placeholder='用户名' />
        <input type='password' name='passwd' placeholder='密码' />
        <input type='submit' value='注册' />
        </form>
    </body>
</html>
```

用户登录功能的实现模板是 login.html。具体实现代码如下。

源码路径：daima\3\3-4\user\templates\login.html

```
<!DOCTYPE html>
<html>
    <body>
        <form method='post'>
        <input type='text' name='name' placeholder='用户名' />
        <input type='password' name='passwd' placeholder='密码' />
        <input type='submit' value='登录' />
        </form>
    </body>
</html>
```

执行后将显示"注册"和"登录"链接，如图3-28所示。

单击"注册"链接后，弹出注册界面，如图3-29所示。

图3-28 "注册"和"登录"链接

图3-29 注册界面

单击"登录"链接后，弹出登录界面，如图3-30所示。登录成功后显示"你好，aaa"的提示信息，并显示"注销"链接。执行效果如图3-31所示。

图3-30 登录界面

图3-31 登录成功的界面

3.4.3 使用 Flask-SQLAlchemy 管理数据库

Flask-SQLAlchemy 是一个常用的 Flask 扩展，简化了在 Flask 程序中使用 SQLAlchemy 的操作。SQLAlchemy 是一个很强大的关系型数据库框架，支持多种数据库后台。SQLAlchemy 提供了高层 ORM，也提供了使用数据库原生 SQL 的低层功能。可以使用如下 pip 命令安装 Flask-SQLAlchemy。

```
pip install flask-sqlalchemy
```

在 Flask-SQLAlchemy 中，数据库使用 URL 指定。其中最流行的数据库引擎采用的数据库 URL 格式如下。

- MySQL 的 URL 格式是 mysql://username:password@hostname/database。
- Postgres 的 URL 格式是 postgresql://username:password@hostname/database。
- SQLite（UNIX）的 URL 格式是 sqlite:////absolute/path/to/database。
- SQLite（Windows）的 URL 格式是 sqlite:///c:/absolute/path/to/database。

在上述 URL 中，hostname 表示 MySQL 服务所在的主机，可以是本地主机（localhost），也可以是远程服务器。数据库服务器上可以托管多个数据库，因此 database 表示要使用的数据库名。如果数据库需要进行认证，username 和 password 表示数据库用户名与密码。

在程序中使用的数据库 URL 必须保存到 Flask 配置对象的 SQLALCHEMY_DATABASE_URI 键中。在配置对象中还有一个很有用的选项，即 SQLALCHEMY_COMMIT_ON_TEARDOWN 键，当将其设为 True 时，每次请求结束后都会自动提交数据库中的变动。其他配置选项的作用请参阅 Flask-SQLAlchemy 的官方文档。

下面的实例演示了使用 SQLAlchemy 扩展库的基本知识。

首先看程序文件 hello.py。具体实现流程如下。

源码路径：daima\3\3-4\sql\hello.py

（1）配置数据库，其中对象 db 是 SQLAlchemy 类的实例，表示程序使用的数据库，同时还获得了 Flask-SQLAlchemy 提供的所有功能，对应代码如下。

```
basedir = os.path.abspath(os.path.dirname(__file__))

app = Flask(__name__)
app.config['SECRET_KEY'] = 'hard to guess string'
app.config['SQLALCHEMY_DATABASE_URI'] =\
    'sqlite:///' + os.path.join(basedir, 'data.sqlite')
app.config['SQLALCHEMY_COMMIT_ON_TEARDOWN'] = True
app.config['SQLALCHEMY_TRACK_MODIFICATIONS'] = False

manager = Manager(app)
bootstrap = Bootstrap(app)
moment = Moment(app)
db = SQLAlchemy(app)
```

（2）定义 Role 和 User 模型，SQLAlchemy 创建的数据库实例为模型提供了一个基类以及一系列辅助类和辅助函数，可用于定义模型的结构。本实例中的数据库表 roles 与 users 可以分别定义为模型 Role 和 User，对应代码如下。

```
class Role(db.Model):
    __tablename__ = 'roles'
    id = db.Column(db.Integer, primary_key=True)
    name = db.Column(db.String(64), unique=True)
    users = db.relationship('User', backref='role', lazy='dynamic')

    def __repr__(self):
        return '<Role %r>' % self.name


class User(db.Model):
    __tablename__ = 'users'
    id = db.Column(db.Integer, primary_key=True)
    username = db.Column(db.String(64), unique=True, index=True)
    role_id = db.Column(db.Integer, db.ForeignKey('roles.id'))

    def __repr__(self):
        return '<User %r>' % self.username
```

在上述代码中，类变量 __tablename__ 定义在数据库中使用的表名。如果没有定义

__tablename__，SQLAlchemy 会使用一个默认名字，但默认的表名没有遵守使用复数形式进行命名的约定，所以最好由用户自己来指定表名。其余的类变量都是该模型的属性，被定义为 db.Column 类的实例。类 db.Column 的构造函数的第一个参数是数据库列和模型属性的类型。下面列出了一些常用的列类型以及在模型中使用的 Python 数据类型。

- Integer：int 类型，普通整数，一般是 32 位。
- SmallInteger：int 类型，取值范围小的整数，一般是 16 位。
- BigInteger：int 或 long 类型，不限制精度的整数。
- Float：float 类型，表示浮点数。
- Numeric：decimal.Decimal 类型，表示定点数。
- String：str 类型，表示变长字符串。
- Text：str 类型，表示变长字符串，对较长或不限长度的字符串进行了优化。
- Unicode：unicode 类型，表示变长 Unicode 字符串。
- UnicodeText：unicode 类型，表示变长 Unicode 字符串，对较长或不限长度的字符串进行了优化。
- Boolean：bool 类型，表示布尔值。
- Date：datetime.date 类型，表示日期。
- Time：datetime.time 类型，表示时间。
- DateTime：datetime.datetime 类型，表示日期和时间。
- Interval：datetime.timedelta 类型，表示时间间隔。
- Enum：str 类型，表示一组字符串。
- PickleType：任何 Python 对象类型，表示自动使用 Pickle 序列化。
- LargeBinary：str 类型，表示二进制文件。

在 db.Column 中还包含了其余的参数，功能是指定属性的配置选项。下面列出了一些常用的选项。

- unique：如果设为 True，此列不允许出现重复的值。
- index：如果设为 True，为此列创建索引，提升查询效率。
- nullable：如果设为 True，此列允许使用空值；如果设为 False，此列不允许使用空值。
- default：为此列定义默认值。

在上述代码中，关系使用 users 表中的外键连接了两行。添加到 User 模型中的 role_id 列被定义为外键，就是这个外键建立起了关系。传递给 db.ForeignKey() 的参数'roles.id'表明，此列的值是 roles 表中行的 id 值。添加到 Role 模型中的 users 属性代表这个关系的面向对象视角。对于一个 Role 类的实例，其 users 属性将返回与角色相关联的用户组成的列表。db.relationship() 的第一个参数表明这个关系的另一端是哪个模型。如果模型类尚未定义，可使用字符串形式指定。db.relationship() 中的 backref 参数向 User 模型中添加一个 role 属性，从而定义反向关系。这一属性可替代 role_id 访问 Role 模型，此时获取的是模型对象，而不是外键的值。在大多数情况下，db.relationship() 都能自行找到关系中的外键，但有时无法确定以哪一列作为外键。例如，如果 User 模型中有两个或两个以上的列定义为 Role 模型的外键，SQLAlchemy 就不知道该使用哪一列。如果无法确定外键，则需要为 db.relationship()提供额外参数，从而确定所用外键。

模板文件 index.html 非常简单。具体实现代码如下。

源码路径：daima\3\3-4\sql\templates\index.html

```
{% extends "base.html" %}
{% import "bootstrap/wtf.html" as wtf %}
```

```
{% block title %}Flasky{% endblock %}

{% block page_content %}
<div class="page-header">
    <h1>Hello, {% if name %}{{ name }}{% else %}Stranger{% endif %}!</h1>
</div>
{{ wtf.quick_form(form) }}
{% endblock %}
```

在浏览器地址栏中输入"http://127.0.0.1:5000/"后的初始执行效果如图 3-32 所示。

在表单中随便输入一个名字，例如，输入"aaa"并单击"提交"按钮后会在表单上面显示对用户"aaa"的欢迎信息，如图 3-33 所示。

如果在表单中输入另外一个名字，例如，输入"bbb"，单击"提交"按钮后会显示对用户"bbb"的欢迎信息，并在上方显示"看来你改了名字"的文本提示，如图 3-34 所示。

图 3-32　初始执行效果

图 3-33　对用户"aaa"的欢迎信息

图 3-34　修改名字后的提示信息

3.5　收发电子邮件

在现实应用中，很多应用程序都需要在特定事件发生时提醒用户，常用的通信方法是电子邮件。虽然 Python 标准库中的 smtplib 包可用在 Flask 程序中发送电子邮件，但包装了 smtplib 的 Flask-Mail 扩展可以更好地和 Flask 集成。本节将详细讲解在 Flask 框架中实现邮件收发的核心知识。

3.5.1　使用 Flask-Mail 扩展

Flask-Mail 能够连接到简单邮件传输协议（Simple Mail Transfer Protocol，SMTP）服务器，并由这个服务器发送邮件。如果不进行配置，Flask-Mail 会连接本地主机上的端口 25，无须验证即可发送电子邮件。

可以使用如下 pip 命令来安装 Flask-Mail。

```
$ pip install flask-mail
```

下面列出了可用来设置 SMTP 服务器的配置。
- MAIL_SERVER：默认值是 localhost，表示电子邮件服务器的主机名或 IP 地址。
- MAIL_PORT：默认值是 25，表示电子邮件服务器的端口。
- MAIL_USE_TLS：默认值是 False，表示启用传输层安全（Transport Layer Security，TLS）协议。
- MAIL_USE_SSL：默认值是 False，表示启用安全套接层（Secure Socket Layer，SSL）协议。

- MAIL_USERNAME：默认值是 None，表示邮件账户的用户名。
- MAIL_PASSWORD：默认值是 None，表示邮件账户的密码。

下面的实例文件 123.py 演示了使用 Flask-Mail 扩展发送带附件的邮件的过程。实例文件 123.py 的具体实现代码如下。

源码路径：daima\3\3-5\123.py

```python
from flask import Flask
from flask_mail import Mail, Message
import os

app = Flask(__name__)
app.config.update(
    DEBUG = True,
    MAIL_SERVER='smtp.qq.com',
    MAIL_PROT=25,
    MAIL_USE_TLS = True,
    MAIL_USE_SSL = False,
    MAIL_USERNAME = '输入发件人邮箱',
    MAIL_PASSWORD = '这里输入授权码',
    MAIL_DEBUG = True
)

mail = Mail(app)

@app.route('/')
def index():
# sender 表示发件人，recipients表示邮件收件人列表
    msg = Message("Hi!This is a test ",sender='输入发件人邮箱', recipients=['输入收件人邮箱'])
# msg.body表示邮件正文
    msg.body = "This is a first email"
# msg.attach表示邮件的附件
# msg.attach("文件名", "类型", 读取文件)
    with app.open_resource("123.jpg", 'rb') as fp:
        msg.attach("image.jpg", "image/jpg", fp.read())

    mail.send(msg)
    print("Mail sent")
    return "Sent"

if __name__ == "__main__":
    app.run()
```

在上述代码中，利用 QQ 邮箱实现了邮件发送功能。在设置时一定要登录 QQ 邮件中心进行设置，开启 POP3/SMTP 服务，如图 3-35 所示。

图 3-35　开启 POP3/SMTP 服务

读者一定要注意，在 MAIL_PASSWORD 中输入的不是 QQ 邮箱的登录密码，而是在开启 POP3/SMTP 服务时得到的授权密码。通过如下命令运行上述程序。

```
python 123.py runserver
```

然后，在浏览器地址栏中输入 "http://127.0.0.1:5000/"。接下来，会得到一个简单的网页，并成功实现发送邮件功能，如图 3-36 所示。

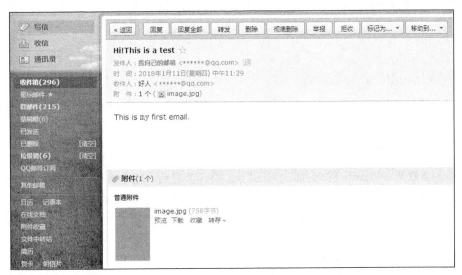

图 3-36　成功收到含有附件的邮件

配置 Hotmail 邮箱服务器的基本参数格式如下。

```
MAIL_SERVER = 'smtp.live.com',
MAIL_PROT = 25,
MAIL_USE_TLS = True,
MAIL_USE_SSL = False,
MAIL_USERNAME = "",
MAIL_PASSWORD = "",
MAIL_DEBUG = True
```

配置 126 邮箱服务器的基本参数格式如下。

```
MAIL_SERVER = 'smtp.126.com',
MAIL_PROT = 25,
MAIL_USE_TLS = True,
MAIL_USE_SSL = False,
MAIL_USERNAME = "",
MAIL_PASSWORD = "",
MAIL_DEBUG = True
```

下面是一段使用 163 邮箱发送邮件的演示代码。

```
from flask import Flask
from flask_mail import Mail, Message

app = Flask(__name__)

app.config.update(
    #EMAIL SETTINGS
    MAIL_SERVER='smtp.163.com',
    MAIL_PORT=465,
    MAIL_USE_SSL=True,
    # 下面写的是你的账号
    MAIL_USERNAME = 'm15913955859@163.com',
    # 下面写的是授权码，不是你的密码
    MAIL_PASSWORD = '你的授权码'
    )

mail = Mail(app)

@app.route("/")
def index():
    msg = Message(subject="这里不能写英文的你好",
                  # 这个账号和上面的那个MAIL_USERNAME 一样
                  sender='m15913955859@163.com',
                  # 这个收件人
                  recipients=['你的qq@qq.com'])
    msg.html = "<b>testing 这都不行？</b> html"
```

```
            mail.send(msg)
            return '<h1>Sent</h1>'
    if __name__ == '__main__':
        app.run(debug=True)
```

3.5.2 使用 SendGrid 发送邮件

SendGrid 是一个电子邮件服务平台，可以帮助市场营销人员跟踪他们的电子邮件统计数据。SendGrid 致力于帮助公司管理事务性邮件，包括航运通知、简报和注册确认等。在 Python 程序中，可以使用 sendgrid 库实现邮件发送功能。通过如下 pip 命令可以安装 sendgrid。

```
pip install sendgrid
```

然后需要登录 SendGrid 的官方网站申请自己的 API 密钥。接下来就可以使用这个 API 密钥开发自己的邮件发送程序了。下面的实例演示了在 Flask 程序中使用 SendGrid 发送邮件的过程。

（1）编写程序文件 mailPage.py 以获取表单中的参数值，根据在表单中输入的邮件信息来发送邮件。文件 mailPage.py 的具体实现代码如下。

源码路径：daima\3\3-5\youjian03\mailPage.py

```python
from flask import Flask,render_template,request
import send2

app = Flask(__name__)

@app.route("/")
def my_form(name=None):
    return render_template("mailForm.html",name=name)

@app.route("/",methods=['POST'])
def my_form_post():
    if request.method=='POST':
        myMailId = request.form.get("myMailId")
        otherMailIds = request.form.get("otherMailIds").strip().split(',')
        sub = request.form.get("sub")
        body = request.form.get("body")
        send2.mailSend(myMailId,otherMailIds,sub,body)
        return "mail sent!!!"

if __name__=="__main__":
    app.run()
```

（2）在程序文件 send2.py 中调用申请的 API 密钥，并根据从表单中获取的信息实现邮件发送功能。文件 send2.py 的具体实现代码如下。

源码路径：daima\3\3-5\youjian03\send2.py

```python
import sendgrid
import os

def mailSend(mailId,emailList,sub,body):

    sg = sendgrid.SendGridAPIClient(apikey='这里写你的API密钥')
    for toMailId in emailList:
        data = {
          "personalizations": [
            {
              "to": [
                {
                  "email": toMailId
                }
              ],
              "subject": sub
            }
          ],
          "from": {
            "email": mailId
          },
          "content": [
```

```
                    {
                        "type": "text/plain",
                        "value": body
                    }
                ]
            }
response = sg.client.mail.send.post(request_body=data)
print(response.status_code)
print(response.body)
print(response.headers)
```

(3) 在模板文件 mailForm.html 中创建一个邮件发送表单，要求分别输入发件人邮箱地址、收件人邮箱地址、邮件主题和邮件内容。文件 mailForm.html 的具体实现代码如下。

源码路径：daima\3\3-5\youjian03\templates\mailForm.html

```
<!doctype html>
<head>
    <title>The super awesome mail page!!</title>
</head>
<body>
    <link rel="stylesheet" type="text/css" href='../static/style.css'>
    <div class="top">
        <br>
        <h1>MAIL EXPRESS</h1>
        <br>.....sending e-mails has never been so easy :)
    </div>
    <div class="form1">
        <form method="POST" align=center>
            <br>
            <label>enter your email address</label>
            <br>
            <input type="text" name="myMailId">
            <p>
            <label>send to: (input mail-ids of receivers seperated by ',')</label>
            <br>
            <input type="text" name="otherMailIds">
            <p>
            <label>Subject:</label>
            <br>
            <input type="text" name="sub">
            <p>
            <label>Enter your message here-</label>
            <br>
            <textarea name="body" rows='3' cols='50'> </textarea>
            <p>
            <input class="button" type="submit" value="send">
        </form>
    </div>
</body>
```

执行后会显示一个邮件发送表单（见图 3-37），分别输入发件人邮箱地址、收件人邮箱地址、邮件主题和邮件内容，单击"send"按钮即可发送邮件。

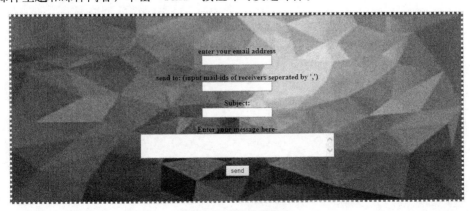

图 3-37 邮件发送表单

3.6 Flask+MySQL+ SqlAlchemy 信息发布系统

本节将通过一个信息发布系统的具体实现过程，详细讲解使用 Flask 技术实现动态 Web 项目的方法。读者可以在此系统的基础上进行升级，把它改造为 BBS 论坛系统或新闻系统。

3.6.1 使用 Virtualenv 创建虚拟环境

本项目是一个小型的信息发布系统，是 BBS 论坛的缩小版。本实例采用 Python + Flask + MySQL 进行开发，前端用的是 Bootstrap 框架，界面比较简单。用户可以发布问题，也可以在发布的问题后面留言。本项目主要的功能包括：新用户注册、用户登录、显示首页、发布问题、查询问题、发表评论和显示头像等功能。

为了避免本项目污染 Python 环境，建议读者在 Virtualenv 创建虚拟环境下运行本项目。具体流程如下。

（1）使用如下所示的命令安装 Virtualenv。

```
pip install virtualenv
```

使用 CD 命令进入一个希望创建虚拟 Python 环境的文件夹下面，例如，本地 D 盘的"virtualenv"目录。

```
D:>cd virtualenv
D:\virtualenv>virtualenv venv
```

注意：使用 Virtualenv 创建的虚拟环境与主机的 Python 环境完全无关，在主机中配置的库不能在 Virtualenv 中直接使用，需要在虚拟环境中利用 pip install 命令再次安装配置后才能使用。

（2）本实例所需要的框架安装信息保存在文件中，通过命令定位到上面刚刚创建的 Virtualenv 虚拟环境中，然后运行如下所示的命令安装本实例所需要的各种库。

```
pip install -r requirements.txt
```

（3）为了在 PyCharm 中使用配置好的 Virtualenv 环境，首先打开"setting"面板，添加本地 Python 环境。在"Project Interpreter"中选择上面刚刚创建的虚拟环境，然后单击"OK"按钮，如图 3-38 所示。注意，随着时间的推移，很多库可能出现了较新的版本，可以直接在此界面中单击向上的箭头图标进行升级。

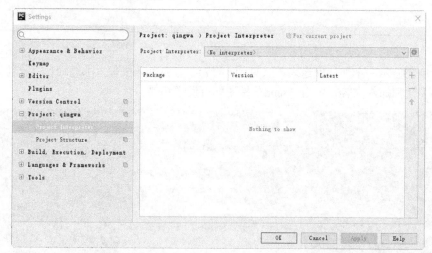

图 3-38 在"Project Interpreter"中选择虚拟环境

3.6.2 使用 Flask 实现数据库迁移

使用 Flask 中强大的 SqlAlchemy 可以轻松实现数据库迁移，具体流程如下。

3.6 Flask+MySQL+ SqlAlchemy 信息发布系统

(1) 在安装好 MySQL 数据库后创建一个 databases，在 MySQL 命令行中输入使用如下命令。

```
create database db_demo8 charset urf-8;
```

通过上述命令创建了一个名为 "db_demo8" 的数据库。

(2) 使用 Flask 中的 SqlAlchemy 实现数据库的迁移，具体命令如下。

```
python manage.py db init         # 创建迁移的仓库
Python manage.py db migrate      # 创建迁移的脚本
python manage.py db upgrade      # 更新数据库
```

如果出现 "alembic.util.exc.CommandError: Directory migrations already exists" 的提示错误信息，只需要执行上面的第 3 条命令 "python manage.py db upgrade" 更新数据库即可。数据迁移成功后会在 MySQL 数据库中成功创建指定的表，如图 3-39 所示。

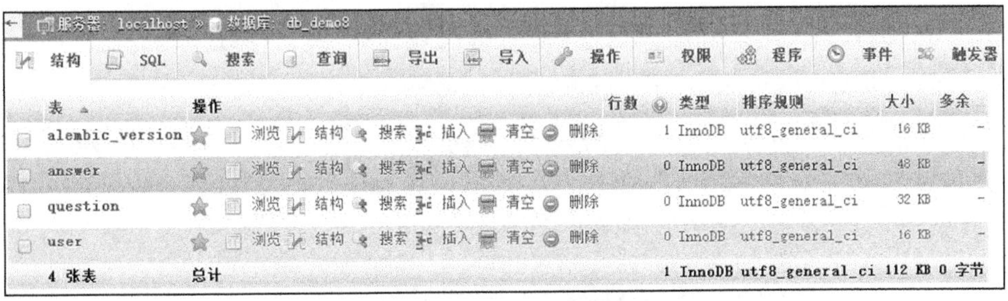

图 3-39 迁移后成功创建的表

实现数据迁移功能的代码保存在 "migrations" 目录下，其主要功能是实现数据表的创建。例如，文件 6f25708c588d_.py 的功能是创建 user 和 question 两个表，主要实现代码如下。

```python
def upgrade():
    # ### commands auto generated by Alembic - please adjust! ###
    op.create_table('user',
    sa.Column('id', sa.Integer(), nullable=False),
    sa.Column('telephone', sa.String(length=11), nullable=False),
    sa.Column('username', sa.String(length=50), nullable=False),
    sa.Column('password', sa.String(length=10), nullable=False),
    sa.PrimaryKeyConstraint('id')
    )
    op.create_table('question',
    sa.Column('id', sa.Integer(), nullable=False),
    sa.Column('title', sa.String(length=100), nullable=False),
    sa.Column('content', sa.Text(), nullable=False),
    sa.Column('create_time', sa.DateTime(), nullable=True),
    sa.Column('author_id', sa.Integer(), nullable=True),
    sa.ForeignKeyConstraint(['author_id'], ['user.id'], ),
    sa.PrimaryKeyConstraint('id')
```

3.6.3 具体实现

信息发布系统的具体实现方式如下。

(1) 在文件 config.py 中配置数据库连接参数，实现和指定数据库的连接。具体实现代码如下。

```python
import os

DEBUG = True

SECRET_KEY = os.urandom(24)

SQLALCHEMY_DATABASE_URI = 'mysql+pymysql://root:66688888@localhost/db_demo8'
SQLALCHEMY_TRACK_MODIFICATIONS = True
```

在上述代码中，"root" 表示登录 MySQL 数据库的用户名，"66688888" 表示登录密码，"localhost" 表示服务器地址，"db_demo8" 表示数据库名。

(2) 在文件 platform.py 中实现本项目的各个功能模块，具体实现流程如下。

① 通过函数 index()加载系统主页模板，将发布的问题按照发布时间进行排序。对应代码如下。

```python
@app.route('/')
def index():
    context = {
        'questions': Question.query.order_by('-create_time').all()
    }
    return render_template('index.html', **context)
```

② 通过函数 login()加载用户登录页面模板，获取用户在表单中输入的电话和密码，通过验证后将登录数据存储在会话中。对应代码如下。

```python
@app.route('/login/', methods=['GET', 'POST'])
def login():
    if request.method == 'GET':
        return render_template('login.html')
    else:
        telephone = request.form.get('telephone')
        password = request.form.get('password')
        user = User.query.filter(User.telephone == telephone, User.password ==
                                 password).first()
        if user:
            session['user_id'] = user.id
            #如果想在31天内都不需要登录
            session.permanent = True
            return redirect(url_for('index'))
        else:
            return u'手机号码或者密码错误，请确认好再登录'
```

③ 通过函数 regist()加载新用户注册页面模板，获取用户在表单中输入的注册信息，如果手机号码已经注册过则提示更换手机信息，并验证两次输入的密码一致，注册成功后将表单中的信息添加到数据库中。对应代码如下。

```python
@app.route('/regist/', methods=['GET', 'POST'])
def regist():
    if request.method == 'GET':
        return render_template('regist.html')
    else:
        telephone = request.form.get('telephone')
        username = request.form.get('username')
        password1 = request.form.get('password1')
        password2 = request.form.get('password2')

        #验证手机号码，如果手机号码已被注册了就不能用了
        user = User.query.filter(User.telephone == telephone).first()
        if user:
            return u'该手机号码已被注册，请更换手机号码'
        else:
            # password1 要和password2相同才可以
            if password1 != password2:
                return u'两次密码不相同，请核实后再填写'
            else:
                user = User(telephone=telephone, username=username, password=password1)
                db.session.add(user)
                db.session.commit()
                # 如果注册成功，就跳转到登录的页面
                return redirect(url_for('login'))
```

④ 通过函数 logout()注销用户。首先判断用户是否登录，如果已经登录，则从会话清空登录数据即可。对应代码如下。

```python
@app.route('/logout/')
def logout():
    # session.pop('user_id')
    # del session('user_id')
    session.clear()
    return redirect(url_for('login'))
```

⑤ 通过函数 question()加载问题发布页面模板，分别获取在表单中输入的问题标题、内容、发布者信息，将获取的信息添加到数据库中。对应代码如下。

```python
@app.route('/question/', methods=['GET', 'POST'])
@login_required
def question():
    if request.method == 'GET':
        return render_template('question.html')
    else:
        title = request.form.get('title')
        content = request.form.get('content')
        question = Question(title=title, content=content)
        user_id = session.get('user_id')
        user = User.query.filter(User.id == user_id).first()
        question.author = user
        db.session.add(question)
        db.session.commit()
        return redirect(url_for('index'))
```

⑥ 通过函数 detail()显示发布的某个问题的详情。对应代码如下。

```python
@app.route('/detail/<question_id>/')
def detail(question_id):
    question_model = Question.query.filter(Question.id == question_id).first()
    return render_template('detail.html', question=question_model)
```

⑦通过函数 add_answer()向数据库中添加针对某个问题发布的评论。对应代码如下。

```python
@app.route('/add_answer/', methods=['POST'])
@login_required
def add_answer():
    content = request.form.get('answer_content')
    question_id = request.form.get('question_id')
    answer = Answer(content=content)
    user_id = session['user_id']
    user = User.query.filter(User.id == user_id).first()
    answer.author = user
    question = Question.query.filter(Question.id == question_id).first()
    answer.question = question
    db.session.add(answer)
    db.session.commit()
    return redirect(url_for('detail', question_id=question_id))
```

⑧ 通过函数 search()快速查找当前系统中某个关键字的信息。对应代码如下。

```python
@app.route('/search/')
def search():
    q = request.args.get('q')
    # title, content
    #通过标题或内容来查找
    # questions = Question.query.filter(or_(Question.title.contains(q),
    #
    Question.content.constraints(q))).order_by('-create_time')
    #只能通过标题来查找
    questions = Question.query.filter(Question.title.contains(q), Question.content.contains(q))
    return render_template('index.html', questions=questions)
```

⑨ 定义钩子函数 my_context_processor()。对应代码如下。

```python
# 钩子函数(注销)
@app.context_processor
def my_context_processor():
    user_id = session.get('user_id')
    if user_id:
        user = User.query.filter(User.id == user_id).first()
        if user:
            return {'user': user}
    return {}
```

(3) 模板文件 base.html 用于实现 Web 导航功能。具体实现代码如下。

```html
<head>
    <meta charset="UTF-8">
    <title>{% block title %}{% endblock %}-问答平台</title>
    <link rel="stylesheet" href="https://cdn.bootcss.com/bootstrap/3.3.7/css/bootstrap.min.css" integrity=
        "sha384-BVYiiSIFeK1dGmJRAkycuHAHRg32OmUcww7on3RYdg4Va+PmSTsz/K68vbdEjh4u"
    crossorigin="anonymous">
```

```html
        <script> src="https://cdn.bootcss.com/jquery/3.2.1/jquery.min.js"></script>

        <script src="https://cdn.bootcss.com/bootstrap/3.3.7/js/bootstrap.min.js" integrity=
            "sha384-Tc5IQib027qvyjSMfHjOMaLkfuWVxZxUPnCJA7l2mCWNIpG9mGCD8wGNIcPD7Txa"
            crossorigin="anonymous"></script>

        <link rel="stylesheet" href="{{ url_for('static', filename='css/base.css') }}">

        {% block head %}{% endblock %}
</head>
<body>
    <nav class="navbar navbar-default">
            <div class="container">
                <!-- Brand and toggle get grouped for better mobile display -->
                <div class="navbar-header">
                    <button type="button" class="navbar-toggle collapsed" data-toggle="collapse"
                        data-target="#bs-example-navbar-collapse-1" aria-expanded="false">
                        <span class="sr-only">Toggle navigation</span>
                        <span class="icon-bar"></span>
                        <span class="icon-bar"></span>
                        <span class="icon-bar"></span>
                    </button>
                    <a class="navbar-brand" href="#">
                        <img class="logo" src="{{ url_for('static', filename=
                            'images/qingwalogo.jpg') }}" alt="">
                    </a>
                </div>

                <!-- Collect the nav links, forms, and other content for toggling -->
                <div class="collapse navbar-collapse" id="bs-example-navbar-collapse-1">
                    <ul class="nav navbar-nav">
                        <li class="active"><a href="/">首页 <span
                            class="sr-only">(current)</span></a></li>
                        <li><a href="{{ url_for('question') }}">发布问题</a></li>
                    </ul>
                    <form class="navbar-form navbar-left" action="{{ url_for('search') }}"
                        method="get">
                        <div class="form-group">
                            <input name="q" type="text" class="form-control" placeholder="
                                请输入关键字">
                        </div>
                        <button type="submit" class="btn btn-default">查找</button>
                    </form>
                    <ul class="nav navbar-nav navbar-right">
                        {% if user %}
                            <li><a href="#">{{ user.username }}</a></li>
                            <li><a href="{{ url_for('logout') }}">注销</a></li>
                        {% else %}
                            <li><a href="{{ url_for('login') }}">登录</a></li>
                            <li><a href="{{ url_for('regist') }}">注册</a></li>
                        {% endif %}
                    </ul>
                </div><!-- /.navbar-collapse -->
            </div><!-- /.container-fluid -->
    </nav>

            <div class='main'>
                {% block main %}{% endblock %}
            </div>
```

（4）模板文件 index.html 用于实现系统首页。具体实现代码如下。

```
{% extends 'base.html' %}

{% block title %}首页{% endblock %}

{% block head %}
    <link rel="stylesheet" href="{{ url_for('static', filename='css/index.css') }}">
{% endblock %}
```

3.6 Flask+MySQL+SqlAlchemy 信息发布系统

```
{% block main %}
    <ul>
        {% for question in questions %}
            <li>
                <div class="avatar-group">
                    <img src="{{ url_for('static', filename='images/qingwalogo.jpg') }}"
                        alt="" class="avatar">
                </div>
                <div class="question-group">
                    <p class="question-title"><a href="{{ url_for('detail',question
                    _id=question.id)
                    }}">{{ question.title }}</a></p>
                    <p class="question-content">{{ question.content }}</p>
                    <div class="question-info">
                        <span class="question-author">{{ question.author.username }}</span>
                        <span class="question-time">{{ question.create_time }}</span>
                    </div>
                </li>
        {% endfor %}
    </ul>
{% endblock %}
```

系统首页如图 3-40 所示。

图 3-40 系统主页

（5）模板文件 detail.html 用于显示某个问题的详情。具体实现代码如下。

```
{% extends 'base.html' %}

{% block title %}详情-{% endblock %}

{% block head %}
    <link rel="stylesheet" href="{{ url_for('static', filename='css/detail.css') }}">

{% endblock %}

{% block main %}
    <h3 class="page-title">{{ question.title }}</h3>
    <p class="question-info">
        <sqan>作者： {{ question.author.username }}</sqan>
        <span>时间： {{ question.create_time }}</span>
    </p>
    <hr>
    <p class="question-content">{{ question.content }}</p>
    <hr>
    <h4>评论： （0）</h4>
    <form action="{{ url_for('add_answer') }}" method="post">
        <div class="form-group">
            <input name="answer_content" type="text" class="from-control" placeholder="请填写评论">
            <input type="hidden" name="question_id" value="{{ question.id }}">
        </div>
        <div class="form-group" style="text-align: right;">
            <button class="btn btn-primary">立即评论</button>
        </div>
    </form>
```

```html
            <ul class="answer-list">
                {% for answer in question.answers %}
                    <li>
                        <div class="user-info">
                            <img src="{{ url_for('static', filename='images/qingwalogo.jpg') }}" alt=""
                                class="avatar">
                            <span class="username">{{ answer.author.username }}</span>
                            <span class="create-time">{{ answer.create_time }}</span>
                        </div>
                        <p class="answer-content">{{ answer.content }}</p>
                    </li>
                {% endfor %}
            </ul>
{% endblock %}
```

在首页中单击某个问题后会显示这个问题的详情页面，执行效果如图 3-41 所示。

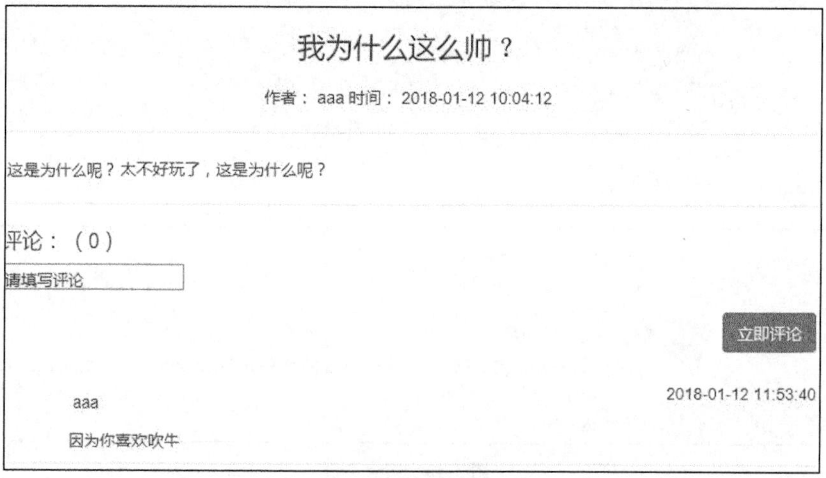

图 3-41　问题详情页面

（6）模板文件 question.html 用于发布问题界面。具体实现代码如下。

```html
{% extends 'base.html' %}

{% block title %}发布问题{% endblock %}

{% block head %}
    <link rel="stylesheet" href="{{ url_for('static', filename='css/question.css') }}">
{% endblock %}

{% block main %}
    <h3 class="page-title">发布问答</h3>
    <form action="" method="post">
        <div class="form-container">
            <div class="form-group">
                <input type="text" placeholder="请输入标题" name="title"
                    class="form-control">
            </div>
            <div class="form-group">
                <textarea name="content" rows="5" placeholder="请输入内容"
                    class="form-control"></textarea>
            </div>
            <div class="form-group">
                <button class="btn btn-primary">立即发布</button>
            </div>
        </div>
    </form>
{% endblock %}
```

问题发布界面如图 3-42 所示。

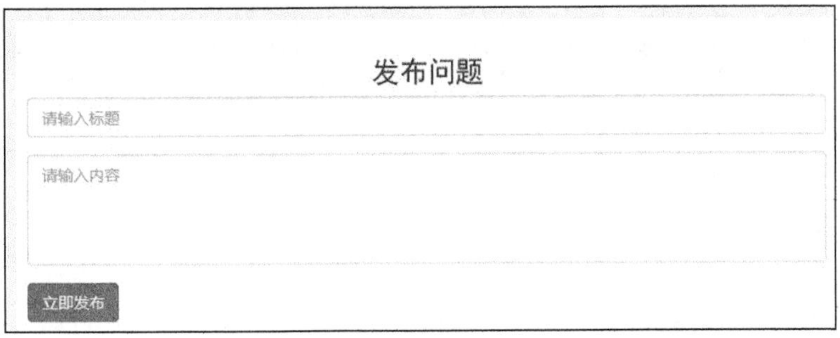

图 3-42　问题发布界面

（7）模板文件 login.html 用于实现用户登录界面。具体实现代码如下。

```
{% extends 'base.html' %}

{% block head %}
    <link rel="stylesheet" href="{{ url_for('static', filename='css/login_regist.css') }}">
{% endblock %}

{% block main %}
    <h3 class="page-title">登录</h3>
    <form action="" method="POST">
        <div class='form-container'>
            <div class='form-group'>
                <input type="text" class="form-control" placeholder="手机号码"
                       name="telephone">
            </div>
            <div class='form-group'>
                <input type="password" class="form-control" placeholder="密码"
                       name="password">
            </div>
            <div class="form-group">
                <button class="btn btn-primary btn-block">登录</button>
            </div>
        </div>
    </form>
{% endblock %}
```

用户登录界面如图 3-43 所示。

（8）模板文件 regist.html 用于实现会员用户注册界面。具体实现代码如下。

```
{% extends 'base.html' %}

{% block title %}注册{% endblock %}

{% block head %}
    <link rel="stylesheet" href="{{ url_for('static', filename='css/login_regist.css') }}">
{% endblock %}

{% block main %}
    <h3 class="page-title">注册</h3>
    <form action="" method="POST">
        <div class='form-container'>
            <div class='form-group'>
                <input type="text" class="form-control" placeholder="手机号码"
                       name="telephone">
            </div>
            <div class='form-group'>
                <input type="text" class="form-control" placeholder="用户名"
                       name="username">
            </div>
            <div class='form-group'>
                <input type="password" class="form-control" placeholder="密码"
                       name="password1">
```

```html
            </div>
            <div class='form-group'>
                <input type="password" class="form-control" placeholder="确认密码"
                    name="password2">
            </div>
            <div class="form-group">
                <button class="btn btn-primary btn-block">立即注册</button>
            </div>
        </div>
    </form>
```
{% endblock %}

会员用户注册界面如图 3-44 所示。

图 3-43　用户登录界面

图 3-44　用户注册界面

3.7　图书借阅管理系统

本节将通过一个图书借阅管理系统的实现过程，详细讲解使用 Flask+SQLite3 开发动态 Web 项目的过程。这是一个典型的管理项目，读者可以在此基础上开发出自己需要的管理类系统。

3.7.1　数据库设置

本项目使用的是 SQLite3 数据库，在程序文件 book.py 中通过如下代码设置数据库。

```python
DATABASE = 'book.db'
DEBUG = True
SECRET_KEY = 'development key'
def get_db():
    top = _app_ctx_stack.top
    if not hasattr(top, 'sqlite_db'):
        top.sqlite_db = sqlite3.connect(app.config['DATABASE'])
        top.sqlite_db.row_factory = sqlite3.Row
    return top.sqlite_db

@app.teardown_appcontext
def close_database(exception):
    top = _app_ctx_stack.top
    if hasattr(top, 'sqlite_db'):
        top.sqlite_db.close()

def init_db():
    with app.app_context():
        db = get_db()
        with app.open_resource('book.sql', mode='r') as f:
            db.cursor().executescript(f.read())
        db.commit()

def query_db(query, args=(), one=False):
    cur = get_db().execute(query, args)
    rv = cur.fetchall()
    return (rv[0] if rv else None) if one else rv
```

3.7.2 登录验证与管理

要验证用户输入的用户名和密码是否正确，在程序文件 book.py 中输入如下代码。如果正确则通过会话存储用户信息，将此用户设置为登录状态。

```python
def get_user_id(username):
    rv = query_db('select user_id from users where user_name = ?',
                  [username], one=True)
    return rv[0] if rv else None


@app.before_request
def before_request():
    g.user = None
    if 'user_id' in session:
        g.user = session['user_id']
```

通过函数 manager_login() 判断是否是管理员登录了。只要输入的用户名和密码与 app.config 中设置的相同，则说明管理员登录了系统。在程序文件 book.py 中通过如下代码实现上述功能。

```python
@app.route('/manager_login', methods=['GET', 'POST'])
def manager_login():
    error = None
    if request.method == 'POST':
        if request.form['username'] != app.config['MANAGER_NAME']:
            error = 'Invalid username'
        elif request.form['password'] != app.config['MANAGER_PWD']:
            error = 'Invalid password'
        else:
            session['user_id'] = app.config['MANAGER_NAME']
            return redirect(url_for('manager'))
    return render_template('manager_login.html', error = error)4
```

通过函数 reader_login() 判断输入的登录信息是否合法。如果非法，则显示提示信息。在程序文件 book.py 中通过如下代码实现上述功能。

```python
@app.route('/reader_login', methods=['GET', 'POST'])
def reader_login():
    error = None
    if request.method == 'POST':
        user = query_db('''select * from users where user_name = ?''',
                        [request.form['username']], one=True)
        if user is None:
            error = 'Invalid username'
        elif not check_password_hash(user['pwd'], request.form['password']):
            error = 'Invalid password'
        else:
            session['user_id'] = user['user_name']
            return redirect(url_for('reader'))
    return render_template('reader_login.html', error = error)
```

通过函数 register() 实现注册功能。首先判断用户是否在表单中输入了合法的用户名和密码。如果合法，则将表单中的数据插入数据库中，然后进行注册。在程序文件 book.py 中通过如下代码实现上述功能。

```python
@app.route('/register', methods=['GET', 'POST'])
def register():
    error = None
    if request.method == 'POST':
        if not request.form['username']:
            error = 'You have to enter a username'
        elif not request.form['password']:
            error = 'You have to enter a password'
        elif request.form['password'] != request.form['password2']:
            error = 'The two passwords do not match'
        elif get_user_id(request.form['username']) is not None:
            error = 'The username is already taken'
        else:
            db = get_db()
            db.execute('''insert into users (user_name, pwd, college, num, email) \
```

```
                            values (?, ?, ?, ?, ?) ''', [request.form['username'], generate
                            _password_hash(
                            request.form['password']), request.form['college'], request.form['number'],
                                    request.form['email']])
                db.commit()
                return redirect(url_for('reader_login'))
    return render_template('register.html', error = error)

@app.route('/logout')
```

通过函数 logout() 实现注销功能。在程序文件 book.py 中通过如下代码实现上述功能。

```
def logout():
    session.pop('user_id', None)
    return redirect(url_for('index'))
```

3.7.3 安全检查与页面跳转管理

通过函数 manager_judge() 实现安全检查。在程序文件 book.py 中通过如下代码实现上述功能。

```
#添加简单的安全性检查
def manager_judge():
    if not session['user_id']:
        error = 'Invalid manager, please login'
        return render_template('manager_login.html', error = error)

def reader_judge():
    if not session['user_id']:
        error = 'Invalid reader, please login'
        return render_template('reader_login.html', error = error)
```

通过函数 manager_books() 获取系统内的所有图书信息,并使页面跳转到模板文件 manager.html, 通过函数 manager() 使页面跳转到模板文件 manager.html, 通过函数 reader() 使页面跳转到模板文件 reader.html。在程序文件 book.py 中通过如下代码实现上述功能。

```
@app.route('/manager/books')
def manager_books():
    manager_judge()
    return render_template('manager_books.html',
            books = query_db('select * from books', []))

@app.route('/manager')
def manager():
    manager_judge()
    return render_template('manager.html')

@app.route('/reader')
def reader():
    reader_judge()
    return render_template('reader.html')
```

3.7.4 后台用户管理

通过函数 manager_users() 获取系统数据库中的所有用户信息。在程序文件 book.py 中通过如下代码实现上述功能。

```
def manager_users():
    manager_judge()
    users = query_db('''select * from users''', [])
    return render_template('manager_users.html', users = users)
```

通过函数 manger_user_modify() 修改系统数据库中某个指定 id 的用户信息。在程序文件 book.py 中通过如下代码实现上述功能。

```
@app.route('/manager/user/modify/<id>', methods=['GET', 'POST'])
def manger_user_modify(id):

    error = None
    user = query_db('''select * from users where user_id = ?''', [id], one=True)
    if request.method == 'POST':
        if not request.form['username']:
            error = 'You have to input your name'
        elif not request.form['password']:
```

```
                    db = get_db()
                    db.execute('''update users set user_name=?, college=?, num=? \
                        , email=? where user_id=? ''', [request.form['username'],
                        request.form['college'], request.form['number'],
                        request.form['email'], id])
                    db.commit()
                    return redirect(url_for('manager_user', id = id))
            else:
                    db = get_db()
                    db.execute('''update users set user_name=?, pwd=?, college=?, num=? \
                        , email=? where user_id=? ''', [request.form['username'],
                            generate_password_hash(request.form['password']),
                        request.form['college'], request.form['number'],
                        request.form['email'], id])
                    db.commit()
                    return redirect(url_for('manager_user', id = id))
    return render_template('manager_user_modify.html', user=user, error = error)
```

通过函数 manger_user_delete()删除系统数据库中某个指定 id 的用户信息。在程序文件 book.py 中通过如下代码实现上述功能。

```
@app.route('/manager/user/deleter/<id>', methods=['GET', 'POST'])
def manger_user_delete(id):
    manager_judge()
    db = get_db()
    db.execute('''delete from users where user_id=? ''', [id])
    db.commit()
    return redirect(url_for('manager_users'))
```

3.7.5 图书管理

通过函数 manager_books_add()向数据库中添加新的图书信息。在程序文件 book.py 中通过如下代码实现上述功能。

```
@app.route('/manager/books/add', methods=['GET', 'POST'])
def manager_books_add():
    manager_judge()
    error = None
    if request.method == 'POST':
        if not request.form['id']:
            error = 'You have to input the book ISBN'
        elif not request.form['name']:
            error = 'You have to input the book name'
        elif not request.form['author']:
            error = 'You have to input the book author'
        elif not request.form['company']:
            error = 'You have to input the publish company'
        elif not request.form['date']:
            error = 'You have to input the publish date'
        else:
            db = get_db()
            db.execute('''insert into books (book_id, book_name, author, publish_com,
                publish_date) values (?, ?, ?, ?, ?) ''', [request.form['id'],
                    request.form['name'], request.form['author'], request.form
                    ['company'],
                    request.form['date']])
            db.commit()
            return redirect(url_for('manager_books'))
    return render_template('manager_books_add.html', error = error)
```

通过函数 manager_books_delete()在数据库中删除指定 id 的图书信息。在程序文件 book.py 中通过如下代码实现上述功能。

```
@app.route('/manager/books/delete', methods=['GET', 'POST'])
def manager_books_delete():
    manager_judge()
    error = None
    if request.method == 'POST':
        if not request.form['id']:
            error = 'You have to input the book name'
        else:
            book = query_db('''select * from books where book_id = ?''',
```

```
                            [request.form['id']], one=True)
        if book is None:
            error = 'Invalid book id'
        else:
            db = get_db()
            db.execute('''delete from books where book_id=? ''', [request.form['id']])
            db.commit()
            return redirect(url_for('manager_books'))
    return render_template('manager_books_delete.html', error = error)
```

通过函数 manager_book() 在数据库中查询指定 id 的图书信息，并查询这本图书是否处于借出状态。在程序文件 book.py 中通过如下代码实现上述功能。

```
@app.route('/manager/book/<id>', methods=['GET', 'POST'])
def manager_book(id):
    manager_judge()
    book = query_db('''select * from books where book_id = ?''', [id], one=True)
    reader = query_db('''select * from borrows where book_id = ?''', [id], one=True)
    name = query_db('''select user_name from borrows where book_id = ?''', [id], one=True)

    current_time = time.strftime('%Y-%m-%d',time.localtime(time.time()))
    if request.method == 'POST':
        db = get_db()
        db.execute('''update histroys set status = ?, date_return = ?  where book_id=?
            and user_name=? and status=? ''',
                ['retruned', current_time, id, name[0], 'not return'])
        db.execute('''delete from borrows where book_id = ? ''' , [id])
        db.commit()
        return redirect(url_for('manager_book', id = id))
    return render_template('manager_book.html', book = book, reader = reader)
```

通过函数 manager_modify() 在数据库中修改指定 id 的图书信息。在程序文件 book.py 中通过如下代码实现上述功能。

```
@app.route('/manager/modify/<id>', methods=['GET', 'POST'])
def manager_modify(id):
    manager_judge()
    error = None
    book = query_db('''select * from books where book_id = ?''', [id], one=True)
    if request.method == 'POST':
        if not request.form['name']:
            error = 'You have to input the book name'
        elif not request.form['author']:
            error = 'You have to input the book author'
        elif not request.form['company']:
            error = 'You have to input the publish company'
        elif not request.form['date']:
            error = 'You have to input the publish date'
        else:
            db = get_db()
            db.execute('''update books set book_name=?, author=?, publish_com=?,
                publish_date=? where book_id=? ''', [request.form['name'], request.
                form['author'], request.form['company'], request.form['date'], id])
            db.commit()
            return redirect(url_for('manager_book', id = id))
    return render_template('manager_modify.html', book = book, error = error)
```

3.7.6 前台用户管理

通过函数 reader_query() 在系统数据库中快速查询指定关键字的图书信息，分别通过书名和图书作者两种 SQL 语句进行查询。在程序文件 book.py 中通过如下代码实现上述功能。

```
@app.route('/reader/query', methods=['GET', 'POST'])
def reader_query():
    reader_judge()
    error = None
    books = None
    if request.method == 'POST':
        if request.form['item'] == 'name':
            if not request.form['query']:
                error = 'You have to input the book name'
            else:
```

3.7 图书借阅管理系统

```
                    books = query_db('''select * from books where book_name = ?''',
                                [request.form['query']])
                    if not books:
                        error = 'Invalid book name'
            else:
                if not request.form['query']:
                    error = 'You have to input the book author'
                else:
                    books = query_db('''select * from books where author = ?''',
                                [request.form['query']])
                    if not books:
                        error = 'Invalid book author'
    return render_template('reader_query.html', books = books, error = error)
```

通过函数 reader_book() 在前台向用户展示某本图书的详细信息，分别用 SQL 图书查询语句、SQL 图书借阅语句和 SQL 统计语句进行查询。在程序文件 book.py 中通过如下代码实现上述功能。

```
@app.route('/reader/book/<id>', methods=['GET', 'POST'])
def reader_book(id):
    reader_judge()
    error = None
    book = query_db('''select * from books where book_id = ?''', [id], one=True)
    reader = query_db('''select * from borrows where book_id = ?''', [id], one=True)
    count = query_db('''select count(book_id) from borrows where user_name = ? ''',
                [g.user], one = True)

    current_time = time.strftime('%Y-%m-%d',time.localtime(time.time()))
    return_time = time.strftime('%Y-%m-%d',time.localtime(time.time() + 2600000))
    if request.method == 'POST':
        if reader:
            error = 'The book has already borrowed.'
        else:
            if count[0] == 3:
                error = 'You can\'t borrow more than three books.'
            else:
                db = get_db()
                db.execute('''insert into borrows (user_name, book_id, date_borrow, \
                    date_return) values (?, ?, ?, ?) ''', [g.user, id,
                                                current_time, return_time])
                db.execute('''insert into histroys (user_name, book_id, date_borrow, \
                    status) values (?, ?, ?, ?) ''', [g.user, id,
                                                current_time, 'not return'])
                db.commit()
                return redirect(url_for('reader_book', id = id))
    return render_template('reader_book.html', book = book, reader = reader, error = error)
```

通过函数 reader_histroy() 展示当前用户的借阅图书历史记录信息。在程序文件 book.py 中通过如下代码实现上述功能。

```
@app.route('/reader/histroy', methods=['GET', 'POST'])
def reader_histroy():
    reader_judge()
    histroys = query_db('''select * from histroys, books where histroys.book_id =
        books.book_id and histroys.user_name=? ''', [g.user], one = False)

    return render_template('reader_histroy.html', histroys = histroys)
```

读者登录界面如图 3-45 所示，图书详情页面如图 3-46 所示。

图 3-45　读者登录界面　　　　　　　　图 3-46　图书详情页面

图书查询页面如图 3-47 所示。

图 3-47　图书查询页面

后台图书管理页面如图 3-48 所示。

图 3-48　后台图书管理页面

第 4 章

数据库存储框架

在 Python 程序中，经常需要实现和数据库操作相关的功能。数据库工具有多种，如 MySQL 和 SQLite 等。本章将详细讲解在 Python 程序中可以使用的其他数据库工具，为读者学习本书后面的知识打下基础。

4.1 安装与使用 pickleDB

pickleDB 是一个轻量级、简单的数据库，本节将详细讲解在 Python 程序中使用 pickleDB 的基本知识。

4.1.1 安装 pickleDB

在实际应用中，使用如下命令即可安装 pickleDB。

```
$ pip install pickledb
```

下面的实例文件 pickleDB01.py 演示了创建、添加并读取 pickleDB 数据库中数据的过程。

源码路径：daima\4\4-1\pickleDB01.py

```python
import pickledb
db = pickledb.load('example.db', False)
db.set('key', 'value')
print(db.get('key'))
```

在上述实例代码中，首先创建了一个名为 example.db 的 pickleDB 数据库，然后向里面添加"key，value"或对的数据，最后通过 print() 函数输出 key 对应的值。执行后会输出：

```
value
```

4.1.2 使用 pickleDB

下面的实例文件 pickleDB02.py 综合演示了创建、添加并读取 pickleDB 数据库中数据的过程。

源码路径：daima\4\4-1\pickleDB02.py

```python
import pickledb
db = pickledb.load('test.db', False)

testVariable = 'STRING STRING STRIN@!#$G$%SFSDF'
testList = ['STRING STRING RWDFWRADF@##@$^Y$#^&@$! ','STRING AGAIN!!!!!']
testDict = { 'STRING 1': 'STRING CHEESE', 'STRING #2*!': 'STRING KEYS ON CRACK!'}

db.set('testVariable', testVariable)
db.set('testList', testList)
db.set('testJson', testDict)

print(db.get('testVariable'))
print(db.get('testList')[0])
print(db.get('testList')[1])
print(db.get('testJson')['STRING 1'])
print(db.get('testJson')['STRING #2*!'])
print(db.get('testJson'))
```

执行后会输出：

```
STRING STRING STRIN@!#$G$%SFSDF
STRING STRING RWDFWRADF@##@$^Y$#^&@$!
STRING AGAIN!!!!!
STRING CHEESE
STRING KEYS ON CRACK!
{'STRING 1': 'STRING CHEESE', 'STRING #2*!': 'STRING KEYS ON CRACK!'}
```

4.2 安装与使用 TinyDB

TinyDB 是使用纯 Python 编写的 NoSQL 数据库，和 SQLite 数据库对应。SQLite 是小型、嵌入式的关系型数据库，而 TinyDB 是小型、嵌入式的 NoSQL 数据库，它不需要外部服务器，也没有任何依赖，使用 JSON 文件存储数据。本节将详细讲解在 Python 程序中使用 TinyDB 的基本知识。

4.2.1 安装 TinyDB

关于 TinyDB 的文档参见其官方网站。TinyDB 源代码参见 Github 网站。在现实应用中，使用如下命令安装 TinyDB。

```
pip install tinydb
```

下面的实例文件 TinyDB01.py 演示了创建、插入、查询、删除和更新 TinyDB 数据库中数据的过程。

源码路径：daima\4\4-2\TinyDB01.py

```python
from tinydb import TinyDB, Query, where
db = TinyDB('db.json')
#插入两条记录
db.insert({'name': 'John', 'age': 22})
db.insert({'name': 'apple', 'age': 7})
#输出所有记录
print(db.all())
# [{u'age': 22, u'name': u'John'}, {u'age': 7, u'name': u'apple'}]
#查询
User = Query()
print(db.search(User.name == 'apple'))
# [{u'age': 7, u'name': u'apple'}]
#查询
print(db.search(where('name') == 'apple'))
#更新记录
db.update({'age': 10}, where('name') == 'apple')
# [{u'age': 10, u'name': u'apple'}]
#删除age大于20的记录
db.remove(where('age') > 20)
#清空数据库
db.purge()
```

执行后会输出：

```
[{'name': 'John', 'age': 22}, {'name': 'apple', 'age': 7}]
[{'name': 'apple', 'age': 7}]
[{'name': 'apple', 'age': 7}]
```

4.2.2 使用 TinyDB

下面的实例综合演示了插入并读取 TinyDB 数据库中数据的过程。实例文件 test.py 的具体实现代码如下。

源码路径：daima\4\4-2\tinyDB\test.py

```python
from tinydb import TinyDB, where
from tinydb.storages import MemoryStorage


def testBasicOperation():
    def addone(x):
        x['int'] += 1
    default_db = TinyDB('test/default.json')
    real_table = default_db.table("real")

    print("{a}打开了数据库{tablename}{a}".format(
        a="*" * 20, tablename=default_db.name))

    default_db.insert({'int': 1, 'char': 'a'})
    default_db.insert({'int': 2, 'char': 'b'})
    default_db.insert({'int': 3, 'char': 'c'})
    default_db.insert({'int': 4, 'char': 'd'})
    real_table.insert({'int': 5, 'char': 'e'})
    real_table.insert({'int': 6, 'char': 'f'})
    real_table.insert({'int': 7, 'char': 'g'})
    real_table.insert({'int': 8, 'char': 'h'})

    print('对每一个元素进行打印的操作：')
    default_db.process_elements(lambda data, doc_id: print(data[doc_id]))
```

```python
        print("default_db中每一个int字段加1")
        default_db.update(addone)

        print('对每一个元素进行打印的操作:')
        default_db.process_elements(lambda data, doc_id: print(data[doc_id]))

        print("default_db中有的所有表段为: ", default_db.tables())
        print("default_db中所有的数据为: ", default_db.all())

        default_db.purge_tables()
        print("{a}清除了所有表{a}".format(a="*" * 20))
        print("db中所有的表段为: ", default_db.tables())
        print("default_db中所有的数据为: ", default_db.all())

        print("{a}关闭了表{tablename}{a}".format(
            a="*" * 20, tablename=default_db.name))
        default_db.close()

def testMemoryStorage():
    db = TinyDB(storage=MemoryStorage)
    db.insert({'data': 5})
    print(db.search(where('data') == 5))

testMemoryStorage()
```

执行后会输出:

```
platform win32 -- Python 3.6.2, pytest-3.3.1, py-1.5.2, pluggy-0.6.0
rootdir: C:\123\Python专题\Python第三方库开发实战\4\4-2\tinyDB, inifile:
collected 2 items
test.py .********************打开了数据库_default********************
对每一个元素进行打印的操作:
{'int': 1, 'char': 'a'}
{'int': 2, 'char': 'b'}
{'int': 3, 'char': 'c'}
{'int': 4, 'char': 'd'}
default_db中每一个int字段加1
对每一个元素进行打印的操作:
{'int': 2, 'char': 'a'}
{'int': 3, 'char': 'b'}
{'int': 4, 'char': 'c'}
{'int': 5, 'char': 'd'}
default_db中所有的表段为:   {'real', '_default'}
default_db中所有的数据为: [{'int': 2, 'char': 'a'}, {'int': 3, 'char': 'b'}, {'int': 4, 'char': 'c'}, {'int': 5, 'char': 'd'}]
********************清除了所有表********************
db中所有的表段为:   set()
default_db中所有的数据为:   []
********************关闭了表_default********************
.                                                          [100%][{'data': 5}]
```

4.3 如何使用 ZODB

关系数据库并不是 Python 程序员唯一可用的解决方案,通常对象数据库可能更适合解决某些问题。ZODB 是一个可扩展和冗余的对象数据库,其专注于存储可扩展的对象,而没有天生"对象—关系"不匹配情况:当尝试在面向对象的语言与关系查询系统的映射对象之间建立关系时,可能会出现这种情况。本节将详细讲解在 Python 程序中使用 ZODB 的基本知识。

4.3.1 安装并使用 ZODB

在实际应用中,使用如下命令安装 ZODB。

```
pip install ZODB
```

4.3 如何使用 ZODB

1. 准备使用 ZODB

下面的实例文件 myzodb.py 演示了创建 ZODB 对象的过程。

源码路径：daima\4\4-3\myzodb.py

```
#-*-coding: UTF-8 -*-
from ZODB import FileStorage, DB
import transaction

class MyZODB(object):
    def __init__ (self, path):
        self.storage = FileStorage.FileStorage(path)
        self.db = DB(self.storage)
        self.connection = self.db.open()
        self.dbroot = self.connection.root()

    def close(self):
        self.connection.close()
        self.db.close()
        self.storage.close()
```

在上述代码中，通过函数 __init__() 实现和指定 ZODB 数据库的连接，通过函数 close() 关闭当前数据库的连接。

2. 存储简单 Python 数据

ZODB 数据库可以存储所有类型的 Python 对象。下面的实例文件 ZODB01.py 演示了创建一个指定 ZODB 数据库的过程。

源码路径：daima\4\4-3\ZODB01.py

```
from myzodb import MyZODB, transaction
db = MyZODB('./Data.fs')
dbroot = db.dbroot
dbroot['a_number'] = 3
dbroot['a_string'] = 'Gift'
dbroot['a_list'] = [1, 2, 3, 5, 7, 12]
dbroot['a_dictionary'] = { 1918: 'Red Sox', 1919: 'Reds' }
dbroot['deeply_nested'] = {
 1918: [ ('Red Sox', 4), ('Cubs', 2) ],
 1919: [ ('Reds', 5), ('White Sox', 3) ],
}
transaction.commit()
db.close()
```

运行后会创建一个名为"Data.fs"的数据库，并向里面添加了指定的数据。

3. 获取数据库中的数据

下面的实例文件 ZODB02.py 演示了获取指定 ZODB 数据库"Data.fs"中数据的过程。

源码路径：daima\4\4-3\ZODB02.py

```
from myzodb import MyZODB
db = MyZODB('./Data.fs')
dbroot = db.dbroot
for key in dbroot.keys():
    print(key + ':', dbroot[key])
db.close()
```

执行后会输出：

```
a_number: 3
a_string: Gift
a_list: [1, 2, 3, 5, 7, 12]
a_dictionary: {1918: 'Red Sox', 1919: 'Reds'}
deeply_nested: {1918: [('Red Sox', 4), ('Cubs', 2)], 1919: [('Reds', 5), ('White Sox', 3)]}
```

4. 更改数据

当将某个键设置为新值时，ZODB 始终能够了解这一点。因此，对上面的数据库进行如下更改时将会自动识别和持久化。

```
db = MyZODB('./Data.fs')
    dbroot = db.dbroot
```

```
dbroot['a_string'] = 'Something Else'
transaction.commit()
db.close()
```

但是需要显式地将对列表或字典的更改告诉 ZODB，因为 ZODB 无法了解所做的更改。例如，下面的代码不会导致 ZODB 看到更改。

```
a_dictionary = dbroot['a_dictionary']
a_dictionary[1920] = 'Indians'
transaction.commit()
db.close()
```

如果打算更改而不是完全替换，则需要设置数据库根的属性 _p_changed，以通知它需要重新存储其下的属性，例如，下面的演示代码。

```
a_dictionary = dbroot['a_dictionary']
a_dictionary[1920] = 'Indians'
db._p_changed = 1
transaction.commit()
db.close()
```

5. 删除数据

删除 ZODB 数据库中数据的方法十分简单，通过 del 语句即可实现，例如，下面的演示代码。

```
del dbroot['a_number']
transaction.commit()
db.close()
```

4.3.2 模拟银行存取款系统

下面的实例演示了使用 ZODB 数据库模拟在银行存款并取款的过程。

（1）编写实例文件 customer.py，定义一个继承于 Persistent 的 Account 类。具体实现代码如下。

源码路径：daima\4\4-3\customer.py

```python
import persistent
class OutOfFunds(Exception):
    pass
class Account(persistent.Persistent):
    def __init__(self,name,start_balance=0):
        self.name = name
        self.balance = start_balance
    def __str__(self):
        return "Account: %s, balance: %s" %(self.name,self.balance)
    def __repr__(self):
        return "Account: %s, balance: %s" %(self.name,self.balance)

    def deposit(self,amount):
        """save amount into balance"""
        self.balance += amount

    def withdraw(self,amount):
        """withdraw from balance"""
        if amount > self.balance:
            raise OutOfFunds
        self.balance -= amount
        return self.balance
```

（2）编写实例文件 zodb_customer_app.py 模拟银行的存款和取款。具体实现代码如下。

源码路径：daima\4\4-3\zodb_customer_app.py

```python
import ZODB
import ZODB.FileStorage as ZFS
import transaction
import customer


class ZODBUtils:

    conn = None
    filestorage = None
```

```python
        def openConnection(self,file_name):
            self.filestorage = ZFS.FileStorage(file_name)
            db = ZODB.DB(self.filestorage)
            self.conn = db.open()
            return self.conn

        def closeConnection(self):
            self.conn.close()
            self.filestorage.close()


def init_balance():
    zodbutils = ZODBUtils()
    conn = zodbutils.openConnection('zodb_filestorage.db')
    root = conn.root()

    noah = customer.Account('noah',1000)
    print(noah)
    root['noah'] = noah

    jermy = customer.Account('jermy',2000)
    print(jermy)
    root['jermy'] = jermy

    transaction.commit()
    zodbutils.closeConnection()


def app():
    zodbutils = ZODBUtils()
    conn = zodbutils.openConnection('zodb_filestorage.db')
    root = conn.root()
    noah = root['noah']
    print("Before Deposit Or Withdraw")
    print("=" * 30)
    print(noah)
    jermy = root['jermy']
    print(jermy)
    print('-' * 30)

    transaction.begin()
    noah.deposit(300)
    jermy.withdraw(300)
    transaction.commit()

    print("After Deposit Or Withdraw")
    print("=" * 30)
    print(noah)
    print(jermy)
    print("-" * 30)

    zodbutils.closeConnection()


if __name__ == '__main__':
    init_balance()
    app()
```

执行后会输出：

```
Account: noah, balance: 1000
Account: jermy, balance: 2000
Before Deposit Or Withdraw
==============================
Account: noah, balance: 1000
Account: jermy, balance: 2000
------------------------------
After Deposit Or Withdraw
==============================
```

```
Account: noah, balance: 1300
Account: jermy, balance: 1700
------------------------------
```

4.4 个人日志系统（使用 Flask 与 TinyDB 实现）

本节将通过一个综合实例的实现过程，介绍联合使用 Flask+TinyDB 实现一个个人日志系统的过程。

4.4.1 系统设置

为了便于系统维护，在系统设置文件中实现一些常用的功能。具体来说，本项目主要涉及如下的系统设置文件。

(1) 初始化文件 __init__.py，功能是设置用户登录模块，导入指定模块的视图文件，通过 app.config 命令设置系统加密信息和邮件服务器信息。具体实现代码如下。

源码路径：daima\4\4-4\chronoflask\app__init__.py

```python
app = Flask(__name__)

from app.admin import admin
app.register_blueprint(admin, url_prefix='/admin')
from app.auth import auth
app.register_blueprint(auth, url_prefix='/admin')
from app.main import main
app.register_blueprint(main, url_prefix='/')

from app.main.views import *
from app.auth.views import *
from app.main.errors import *

app.config['SECRET_KEY'] = os.environ.get('you-will-never-guess')
app.config['MAIL_SERVER'] = os.environ.get('MAIL_SERVER')
app.config['MAIL_PORT'] = 587
app.config['MAIL_USE_TLS'] = True
app.config['MAIL_USERNAME'] = os.environ.get('MAIL_USERNAME')
app.config['MAIL_PASSWORD'] = os.environ.get('MAIL_PASSWORD')
app.config['MAIL_SUBJECT_PREFIX'] = '[Chronoflask]'
app.config['MAIL_SENDER'] = 'Chronoflask <admin@chronoflask.com>'
app.config['DEFAULT_NAME'] = 'Chronoflask'
app.config['DEFAULT_AUTHOR'] = 'Chronologist'

mail = Mail(app)

bootstrap = Bootstrap(app)
```

(2) 编写文件 db.py 实现 TinyDB 数据库的操作，包括查询、检索、添加和更新功能。具体实现代码如下。

源码路径：daima\4\4-4\chronoflask\app\db.py

```python
from tinydb import TinyDB, Query
import ujson

def get_db():
    db = TinyDB('db.json')
    return db

def get_table(table_name):
    table = get_db().table(table_name)
    return table

def get_record(table_name, query):
    result = get_table(table_name).get(query)
    return result
```

4.4 个人日志系统（使用 Flask 与 TinyDB 实现）

```python
def search_records(table_name, query):
    results = get_table(table_name).search(query)
    return results

def insert_record(table_name, record):
    element_id = get_table(table_name).insert(record)
    return element_id

def update_record(table_name, field, query):
    get_table(table_name).update(field, query)
    return True

def get_element_id(table_name, query):
    element = get_record(table_name, query)
    element_id = element.eid
    return element_id
```

(3) 编写文件 pagination.py 实现日志信息分页显示功能。具体实现代码如下。

源码路径：daima\4\4-4\chronoflask\app\pagination.py

```python
def update_pagination():
    ''' Creates a table of entries organized by page
    for use when browsing entries. '''
    results = search_records('entries', \
        Query().creator_id == session.get('user_id'))
    all_entries = results[::-1]
    limit = len(all_entries)
    total = 0
    count = 0
    page = 1
    p_entries = list()
    get_table('pagination').purge() # clear the old pagination table
    for entry in all_entries:
        total += 1
        count += 1
        p_entries.append(entry)
        if count == 10 or total == limit:
            insert_record('pagination', {'page': page, 'entries': p_entries})
            del p_entries[:]
            page += 1
            count = 0
    return True

def get_entries_for_page(page):
    results = get_record('pagination', Query().page == page)
    if not results:
        return abort(404)
    else:
        return results['entries']

def check_next_page(page):
    next_page = page + 1
    if not get_record('pagination', Query().page == next_page):
        return None
    else:
        return next_page
```

(4) 编写解析文件 parse.py，功能是根据不同的节点显示日志信息。例如：可以根据"tags"参数来发布新信息，根据"days"参数来浏览某一天的日志信息，根据"raw_entry"参数显示不同的日志信息界面或管理界面。具体实现代码如下。

源码路径：daima\4\4-4\chronoflask\app\parse.py

```python
def parse_input(raw_entry, current_time):
    '''Parse input and either create a new entry using the input
    or call a function (that may take part of the input as an argument).'''
    if raw_entry == 'browse all':
        return redirect(url_for('main.browse_all_entries'))
    elif raw_entry[:3] == 't: ':
```

```python
                return redirect(url_for('main.view_single_entry', \
                                        timestamp=raw_entry[3:]))
        elif raw_entry[:5] == 'tag: ':
            return redirect(url_for('main.view_entries_for_tag', \
                                    tag=raw_entry[5:]))
        elif raw_entry[:5] == 'day: ':
            return redirect(url_for('view_entries_for_day', day=raw_entry[5:]))
        elif raw_entry == 'login':
            return redirect(url_for('auth.login'))
        elif raw_entry == 'logout':
            return redirect(url_for('auth.logout'))
        elif raw_entry == 'change email':
            return redirect(url_for('auth.change_email'))
        elif raw_entry == 'change password':
            return redirect(url_for('auth.change_password'))
        elif raw_entry == 'about':
            return redirect(url_for('admin.get_details'))
        elif raw_entry == 'rename chrono':
            return redirect(url_for('admin.rename_chronofile'))
        elif raw_entry == 'rename author':
            return redirect(url_for('admin.rename_author'))
        else:
            return process_entry(raw_entry, current_time)

def process_entry(raw_entry, current_time):
    '''Take the user's input and UTC datetime and return a clean, formatted
    entry, a timestamp as a string, and a list of tags'''
    raw_tags = find_and_process_tags(raw_entry)
    clean_entry = clean_up_entry(raw_entry, raw_tags)
    clean_tags = clean_up_tags(raw_tags)
    timestamp = create_timestamp(current_time)
    return create_new_entry(clean_entry, timestamp, clean_tags)

def clean_up_entry(raw_entry, raw_tags):
    '''Strip tags from end of entry.'''
    bag_of_words = raw_entry.split()
    if bag_of_words == raw_tags:
        clean_entry = bag_of_words[0]
    else:
        while bag_of_words[-1] in raw_tags:
            bag_of_words.pop()
        stripped_entry = ' '.join(bag_of_words)
        clean_entry = stripped_entry[0].upper() + stripped_entry[1:]
    return clean_entry

def create_timestamp(current_time):
    '''TinyDB can't handle datetime objects; convert datetime to string.'''
    timestamp = datetime.strftime(current_time, '%Y-%m-%d %H:%M:%S')
    return timestamp
```

（5）因为本项目用到了信息加密和认证功能，所以需要编写独立文件 config.py 来保存 secret_key 信息。在创建 Flask Web 项目的时候，如果没有独立设置 secret_key，则会显示"Must provide secret_key to use csrf"提醒信息。不能将 SECRET_KEY 写在程序代码中，需要单独将其放在文件 config.py 中。具体实现代码如下。

源码路径：daima\4\4-4\chronoflask\config.py

```
CSRF_ENABLED = True
SECRET_KEY = 'you-will-never-guess'
```

（6）编写文件 run.py 作为 Fask 项目的启动文件，在里面调用了文件 config.py 中的 SECRET_KEY 信息。具体实现代码如下。

源码路径：daima\4\4-4\chronoflask\run.py

```
app.config.from_object('config')
```

4.4 个人日志系统（使用 Flask 与 TinyDB 实现）

```
if __name__ == '__main__':
    app.run(debug=app.config['DEBUG'])
```

（7）编写文件 mail.py 实现邮件发送，邮件服务器的设置在文件 "app__init__.py" 中实现。具体实现代码如下。

源码路径：daima\4\4-4\chronoflask\app\mail.py

```
mail = Mail(app)

def send_async_email(app, msg):1
    with app.app_context():
        mail.send(msg)

def send_email(to, subject, template, **kwargs):
    msg = Message(app.config['MAIL_SUBJECT_PREFIX'] + ' ' + subject,
                  sender=app.config['MAIL_SENDER'], recipients=[to])
    msg.body = render_template(template + '.txt', **kwargs)
    msg.html = render_template(template + '.html', **kwargs)
    thr = Thread(target=send_async_email, args=[app, msg])
    thr.start()
    return thr
```

4.4.2 后台管理

在 "admin" 目录中保存了和后台管理相关的程序文件，接下来详细介绍这部分程序文件的具体实现。具体流程如下。

（1）编写程序文件 forms.py，功能是修改系统名和系统作者名。具体实现代码如下。

源码路径：daima\4\4-4\chronoflask\app\admin\forms.py

```
class RenameChronofileForm(Form):
    new_name = StringField('Enter new name for chronofile:', \
                           validators=[Required()])
    submit = SubmitField('Rename chronofile')


class RenameAuthorForm(Form):
    new_name = StringField('Enter new author name:', \
                           validators=[Required()])
    submit = SubmitField('Rename author')
```

（2）编写程序文件 views.py，功能是根据用户操作跳转到指定的后台管理页面，实现修改系统名和系统作者名的功能。具体实现代码如下。

源码路径：daima\4\4-4\chronoflask\app\admin\views.py

```
@admin.route('/')
@login_required
def view_admin():
    '''Display name of site, author, etc. as well as links to edit
    those details and change email, password, etc.'''
    details = get_details()
    return render_template('admin.html', details=details)


@admin.route('/rename_chronofile', methods=['GET', 'POST'])
@login_required
def rename_chronofile():
    details = get_details()
    form = RenameChronofileForm()
    if form.validate_on_submit():
        update_record('admin', {'chronofile_name': form.new_name.data}, \
                      Query().creator_id == session.get('user_id'))
        flash('Chronfile name updated.')
        return redirect(url_for('admin.view_admin'))
    return render_template('rename_chronofile.html', \
                           form=form, details=details)


@admin.route('/rename_author', methods=['GET', 'POST'])
```

```python
@login_required
def rename_author():
    details = get_details()
    form = RenameAuthorForm()
    if form.validate_on_submit():
        test=update_record('admin', {'author_name': form.new_name.data}, \
                 Query().creator_id == session.get('user_id'))
        flash('Author name updated.')
        return redirect(url_for('admin.view_admin'))
    return render_template('rename_author.html', form=form, details=details)
```

4.4.3 登录认证管理

在"auth"目录中保存了和用户登录认证相关的程序文件,接下来将详细介绍这部分程序文件的具体实现。具体流程如下。

(1)编写程序文件 forms.py,功能是实现账号检测、邮箱检测、登录验证、注册信息验证、登录表单信息处理、邮箱设置、重设邮箱、修改密码和重置密码等功能。具体实现代码如下。

源码路径:daima\4\4-4\chronoflask\app\auth\forms.py

```python
# 基本验证
def account_exists(form, field):
    user = get_record('auth', Query().email == field.data)
    if not user:
        raise ValidationError('Create an account first.')


def email_exists(form, field):
    user = get_record('auth', Query().email == field.data)
    if not user:
        raise ValidationError('Please verify that you typed your email \
            correctly.')


def authorized(form, field):
    '''Verify user through password.'''
    user = get_table('auth').get(eid=session.get('user_id'))
    if not pwd_context.verify(field.data, user['password_hash']):
        raise ValidationError('Invalid login credentials. Please try again.')


def has_digits(form, field):
    if not bool(re.search(r'\d', field.data)):
        raise ValidationError('Your password must contain at least one \
            number.')


def has_special_char(form, field):
    if not bool(re.search(r'[^\w\*]', field.data)):
        raise ValidationError('Your password must contain at least one \
            special character.')


class PasswordCorrect(object):
    '''Verify email/password combo before validating form.'''
    def __init__(self, fieldname):
        self.fieldname = fieldname

    def __call__(self, form, field):
        try:
            email = form[self.fieldname]
        except KeyError:
            raise ValidationError(field.gettext("Invalid field name '%s'.") \
                    % self.fieldname)
        user = get_record('auth', Query().email == email.data)
        if not pwd_context.verify(field.data, user['password_hash']):
            raise ValidationError('Invalid password. Please try again.')
```

4.4 个人日志系统（使用 Flask 与 TinyDB 实现）

```python
class LoginForm(Form):
    email = StringField('Email:', validators=[Required(), Email(), \
        account_exists])
    password = PasswordField('Password:', validators=[Required(), \
        PasswordCorrect('email')])
    submit = SubmitField('Log in')


class RegistrationForm(Form):
    email = StringField('Enter email address:', \
        validators=[Required(), Email()])
    password = PasswordField('Enter password: ' +\
        '(min 12 char., must incl. number and special character)', \
        validators=[Required(), Length(min=12), has_digits, has_special_char])
    submit = SubmitField('Create account')


class ChangeEmailForm(Form):
    password = PasswordField('Enter your password:', validators=[Required(), \
        authorized])
    new_email = StringField('New email address:', \
        validators=[Required(), Email(), EqualTo('verify_email', \
        message='Emails must match')])
    verify_email = StringField('Re-enter new email address:', \
        validators=[Required(), Email()])
    submit = SubmitField('Change email')


class ChangePasswordForm(Form):
    current_password = PasswordField('Your current password:', \
        validators=[Required(), authorized])
    new_password = PasswordField('New password: ' +\
        '(min 12 char., must incl. number and special character)', \
        validators=[Required(), Length(min=12), EqualTo('verify_password', \
        message='New passwords must match.'), has_digits, has_special_char])
    verify_password = PasswordField('Re-enter new password:', \
        validators=[Required(), Length(min=12)])
    submit = SubmitField('Change password')


class ResetPasswordForm(Form):
    email = StringField('Your registered email address:',
        validators=[Required(), Email(), email_exists])
    submit = SubmitField('Request password reset link')


class SetNewPasswordForm(Form):
    new_password = PasswordField('New password: ' +\
        '(min 12 char., must incl. number and special character)', \
        validators=[Required(), Length(min=12), EqualTo('verify_password', \
        message='New passwords must match.'), has_digits, has_special_char])
    verify_password = PasswordField('Re-enter new password:', \
        validators=[Required(), Length(min=12)])
    submit = SubmitField('Set new password')
```

(2) 编写程序文件 views.py，功能是根据用户操作跳转到指定的认证页面，实现登录、注销、密码重置和邮箱修改等功能。具体实现代码如下。

源码路径：daima\4\4-4\chronoflask\app\auth\views.py

```python
@auth.route('/login', methods=['GET', 'POST'])
def login():
    details = get_details()
    if not get_record('auth', Query().email.exists()):
        flash('You need to register first.')
        return redirect(url_for('auth.register'))
    if session.get('logged_in'):
        return redirect(url_for('main.browse_all_entries'))
    form = LoginForm()
```

```python
        if form.validate_on_submit():
            session['logged_in'] = True
            user_id = get_element_id('auth', Query().email == form.email.data)
            session['user_id'] = user_id
            if request.args.get('next'):
                return redirect(request.args.get('next'))
            else:
                return redirect(url_for('main.browse_all_entries'))
        return render_template('login.html', form=form, details=details)


@auth.route('/logout')
@login_required
def logout():
    session['logged_in'] = None
    flash('You have been logged out.')
    return redirect(url_for('main.browse_all_entries'))


@auth.route('/register', methods=['GET', 'POST'])
def register():
    ''' Register user and create pagination table with one page
    and no entries.'''
    details = get_details()
    if details:
        flash('A user is already registered. Log in.')
        return redirect(url_for('auth.login'))
    details = {'chronofile_name': current_app.config['DEFAULT_NAME'], \
               'author_name': current_app.config['DEFAULT_AUTHOR']}
    register = True
    form = RegistrationForm()
    if form.validate_on_submit():
        password_hash = pwd_context.hash(form.password.data)
        creator_id = insert_record('auth', {'email': form.email.data, \
                                   'password_hash': password_hash})
        insert_record('admin', {'chronofile_name': \
                                    current_app.config['DEFAULT_NAME'], \
                                'author_name': \
                                    current_app.config['DEFAULT_AUTHOR'], \
                                'creator_id': creator_id})
        insert_record('pagination', {'page': 1, 'entries': None})
        flash('Registration successful. You can login now.')
        return redirect(url_for('auth.login'))
    return render_template('register.html', form=form, details=details, \
                                   register=register)


@auth.route('/reset_password', methods=['GET', 'POST'])
def request_reset():
    details = get_details()
    if not details:
        return abort(404)
    form = ResetPasswordForm()
    if form.validate_on_submit():
        email = form.email.data
        user_id = get_element_id('auth', Query().email == email)
        token = generate_confirmation_token(user_id)
        send_email(email, 'Link to reset your password',
                   'email/reset_password', token=token)
        flash('Your password reset token has been sent.')
        return redirect(url_for('auth.login'))
    return render_template('reset_password.html', form=form, details=details)


@auth.route('/reset_password/<token>', methods=['GET', 'POST'])
def confirm_password_reset(token):
    details = get_details()
    if not details:
        return abort(404)
```

4.4 个人日志系统（使用 Flask 与 TinyDB 实现）

```python
        s = Serializer(current_app.config['SECRET_KEY'])
        try:
                data = s.loads(token)
        except:
                flash('The password reset link is invalid or has expired.')
                return redirect(url_for('auth.request_reset'))
        if not data.get('confirm'):
                flash('The password reset link is invalid or has expired.')
                return redirect(url_for('auth.request_reset'))
        user_id = data.get('confirm')
        form = SetNewPasswordForm()
        if form.validate_on_submit():
                new_password_hash = pwd_context.hash(form.new_password.data)
                get_table('auth').update({'password_hash': new_password_hash}, \
                                        eids=[user_id])
                flash('Password updated—you can now log in.')
                return redirect(url_for('auth.login'))
        return render_template('set_new_password.html', form=form, token=token, \
                                        details=details)


@auth.route('/change_email', methods=['GET', 'POST'])
@login_required
def change_email():
        details = get_details()
        form = ChangeEmailForm()
        if form.validate_on_submit():
                new_email = form.new_email.data
                user_id = session.get('user_id')
                get_table('auth').update({'email': new_email}, eids=user_id)
                flash('Your email address has been updated.')
                return redirect(url_for('admin.view_admin'))
        return render_template('change_email.html', form=form, details=details)


@auth.route('/change_password', methods=['GET', 'POST'])
@login_required
def change_password():
        details = get_details()
        form = ChangePasswordForm()
        if form.validate_on_submit():
                new_password_hash = pwd_context.hash(form.new_password.data)
                get_table('auth').update({'password_hash': new_password_hash}, \
                                        eids=[session.get('user_id')])
                flash('Your password has been updated.')
                return redirect(url_for('admin.view_admin'))
        return render_template('change_password.html', form=form, details=details)


def generate_confirmation_token(user_id, expiration=3600):
        serial = Serializer(current_app.config['SECRET_KEY'], expiration)
        return serial.dumps({'confirm': user_id})
```

4.4.4 前台日志展示

在"main"目录中保存了和用户登录认证相关的程序文件，接下来详细介绍这部分程序文件的具体实现。具体流程如下。

（1）编写程序文件 forms.py，功能是实现日志信息发布和修改功能。具体实现代码如下。

源码路径：daima\4\4-4\chronoflask\app\main\forms.py

```python
class RawEntryForm(Form):
        raw_entry = StringField('New entry in chronofile:', validators=[Required()])
        submit = SubmitField('Post entry')

def no_hashtags(form, field):
        if bool(re.search(r'#', field.data)):
                raise ValidationError(Markup("Don't start tags with \
                        <code>#</code> here."))
```

```
def use_commas(form, field):
    tags = field.data.split(' ')
    items = len(tags)
    print(items)
    if items != 1:
        count = 1
        for i in tags:
            print(i)
            print(i[-1])
            if i[-1] == ',':
                print(i[-1])
                count += 1
        print(count)
        if count < items:
            raise ValidationError(Markup('Separate tags \
                with commas like this: <code>tag1, tag2</code>.'))


class EditEntryForm(Form):
    new_entry = StringField('Edit entry text:', validators=[Required()])
    new_tags = StringField("Edit tags: (separate with commas, don't use #)", \
        validators=[no_hashtags, use_commas])
    submit = SubmitField('Save edited entry')
```

(2) 编写程序文件 views.py，功能是根据用户操作跳转到指定的前台展示页面，实现发布新日志信息、浏览全部日志信息和浏览某日的日志信息等功能。具体实现代码如下。

源码路径：daima\4\4-4\chronoflask\app\main\views.py

```
@main.route('/', methods=['GET', 'POST'])
def browse_all_entries():
    '''Returns all entries (most recent entry at the top of the page).'''
    details = get_details()
    if not session.get('logged_in'):
        if details:
            register = False
        else:
            register = True
            details = {'chronofile_name': current_app.config['DEFAULT_NAME'], \
                'author_name': current_app.config['DEFAULT_AUTHOR']}
        return render_template('welcome.html', details=details, \
            register=register)
    form = RawEntryForm()
    if form.validate_on_submit():
        return parse_input(form.raw_entry.data, datetime.utcnow())
    page = 1
    entries_for_page = get_entries_for_page(page)
    next_page = check_next_page(page)
    return render_template('home.html', entries_for_page=entries_for_page, \
        form=form, details=details, next_page=next_page)


@main.route('page/<page>', methods=['GET', 'POST'])
@login_required
def view_entries_for_page(page):
    '''Returns entries for given page in reverse chronological order.'''
    try:
        int(page)
    except:
        TypeError
        return abort(404)
    page = int(page)
    if page == 1:
        return redirect(url_for('main.browse_all_entries'))
    details = get_details()
    form = RawEntryForm()
    if form.validate_on_submit():
        return parse_input(form.raw_entry.data, datetime.utcnow())
    entries_for_page = get_entries_for_page(page)
```

4.4 个人日志系统（使用 Flask 与 TinyDB 实现）

```python
        next_page = check_next_page(page)
        prev_page = page - 1
        return render_template('page.html', form=form, \
            entries_for_page=entries_for_page, details=details, \
            page=page, next_page=next_page, prev_page=prev_page)


@main.route('day/<day>', methods=['GET', 'POST'])
@login_required
def view_entries_for_day(day):
    '''Returns entries for given day in chronological order.'''
    details = get_details()
    form = RawEntryForm()
    if form.validate_on_submit():
        return parse_input(form.raw_entry.data, datetime.utcnow())
    entries_for_day = search_records('entries', Query().timestamp.all([day]))
    if not entries_for_day:
        return abort(404)
    return render_template('day.html', form=form, day=day, \
                            entries_for_day=entries_for_day, details=details)


@main.route('timestamp/<timestamp>', methods=['GET', 'POST'])
@login_required
def view_single_entry(timestamp):
    '''Return a single entry based on given timestamp.'''
    entry = get_record('entries', Query().timestamp == timestamp)
    if not entry:
        return abort(404)
    form = RawEntryForm()
    if form.validate_on_submit():
        return parse_input(form.raw_entry.data, datetime.utcnow())
    details = get_details()
    return render_template('entry.html', form=form, timestamp=timestamp, \
                            entry=entry, details=details)


@main.route('timestamp/<timestamp>/edit', methods=['GET', 'POST'])
@login_required
def edit_entry(timestamp):
    '''Edit an entry and return a view of the edited entry'''
    entry = get_record('entries', Query().timestamp == timestamp)
    if not entry:
        return abort(404)
    form = EditEntryForm()
    if form.validate_on_submit():
        tags = form.new_tags.data.split(", ")
        update_record('entries', {'entry': form.new_entry.data, \
            'tags': tags}, (Query().creator_id == 1) & \
            (Query().timestamp == timestamp))
        flash('Entry updated.')
        update_pagination()
        return redirect(url_for('main.view_single_entry', timestamp=timestamp))
    form.new_entry.default = entry['entry']
    form.new_tags.default = ', '.join(entry['tags'])
    form.process()
    details = get_details()
    return render_template('edit_entry.html', form=form, timestamp=timestamp, \
        details=details)


@main.route('tags', methods=['GET', 'POST'])
@login_required
def view_all_tags():
    details = get_details()
    form = RawEntryForm()
    if form.validate_on_submit():
        return parse_input(form.raw_entry.data, datetime.utcnow())
    all_entries = search_records('entries', \
```

```python
            all_tags = list()
            for entry in all_entries:
                for tag in entry['tags']:
                    if tag not in all_tags:
                        all_tags.append(tag)
            all_tags.sort()
            return render_template('tags.html', all_tags=all_tags, form=form, \
                                   details=details)


@main.route('days', methods=['GET', 'POST'])
@login_required
def view_all_days():
    details = get_details()
    form = RawEntryForm()
    if form.validate_on_submit():
        return parse_input(form.raw_entry.data, datetime.utcnow())
    all_entries = search_records('entries', \
                                  Query().creator_id == session.get('user_id'))
    all_days = list()
    for entry in all_entries:
        if entry['timestamp'][:10] not in all_days:
            all_days.append(entry['timestamp'][:10])
    return render_template('days.html', all_days=all_days, form=form, \
                           details=details)


@main.route('tags/<tag>', methods=['GET', 'POST'])
@login_required
def view_entries_for_tag(tag):
    '''Return entries for given tag in chronological order.'''
    details = get_details()
    form = RawEntryForm()
    if form.validate_on_submit():
        return parse_input(form.raw_entry.data, datetime.utcnow())
    entries_for_tag = search_records('entries', Query().tags.all([tag]))
    if not entries_for_tag:
        return abort(404)
    return render_template('tag.html', form=form, tag=tag, \
                           entries_for_tag=entries_for_tag, details=details)
```

(3) 编写程序文件 errors.py，用于在获取数据库信息出错时跳转到指定的 HTML 页面。具体实现代码如下。

源码路径：daima\4\4-4\chronoflask\app\main\errors.py

```python
@main.app_errorhandler(404)
def page_not_found(e):
    details = get_details()
    return render_template('404.html', details=details), 404


@main.app_errorhandler(500)
def internal_server_error(e):
    details = get_details()
    return render_template('500.html', details=details), 500
```

4.4.5 系统模板

在"templates"目录中保存了 Flask 系统模板文件，接下来详细介绍这部分程序文件的具体实现。具体流程如下。

(1) 编写程序文件 welcome.html，功能是显示欢迎信息，提供登录链接。具体实现代码如下。

源码路径：daima\4\4-4\chronoflask\app\templates\welcome.html

```
{% extends "base.html" %}
{% block page_content %}
```

4.4 个人日志系统（使用 Flask 与 TinyDB 实现）

```html
<div class="page-header">
    <h1>Hello</h1>
</div>

<div>
    {% if register == True %}
    <p>Chronoflask is a minimalist diary/journal application using Python 3, Flask,
        and TinyDB inspired by Warren Ellis's Chronofile Minimal and Buckminster
        Fuller's Dymaxion Chronofile.</p></br>
    <p>Add new entries (with or witout hashtags) in a single input field. Each
        entry is stored with a UTC timestamp.</p></br>
    <p>View recent entries, view all entries for a single day or date-range (chronologically),
        view a single entry, view all entries associated with a tag, and view a
        list of tags.</p></br>
    </p>Private by default. Please <a href="{{ url_for('auth.register') }}">register</a>
        to begin using Chronoflask.</p>
    {% else %}
    <h2>Please <a href="{{ url_for('auth.login') }}">log in</a>.</h2>
    {% endif %}
</div>
```

执行后显示的欢迎信息和登录链接如图 4-1 所示。

图 4-1 欢迎信息和登录链接

（2）编写程序文件 login.html，功能是提供登录界面，并提供找回密码的链接。具体实现代码如下。

源码路径：daima\4\4-4\chronoflask\app\templates\login.html

```html
{% extends "base.html" %}
{% block page_content %}

    <div class="page-header">
        <h1>Login</h1>
    </div>

    <div>
        {{ wtf.quick_form(form) }}
        <p>Forgot your password? Click <a href="{{ url_for('auth.request_reset') }}">
            here</a> to reset it.</p>
    </div>

{% endblock %}
```

执行后显示的用户登录如图 4-2 所示。

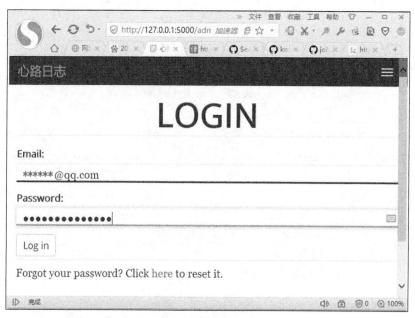

图 4-2 用户登录界面

（3）编写程序文件 home.html，功能是以分页的样式展示系统数据库内所有的日志信息。具体实现代码如下。

源码路径：daima\4\4-4\chronoflask\app\templates\home.html

```
{% extends "base.html" %}
{% block page_content %}

    <div class="page-header">
        <h1>Stream</h1>
    </div>

    <div>
        {{ wtf.quick_form(form) }}
    </div>

    <div>
    {% if not entries_for_page %}
        <p>Nothing in your chronofile yet.</p>
    {% else %}
        <ul>
        {% for entry in entries_for_page %}
            <li><p><a href="{{ url_for('main.view_entries_for_day', day=entry['timestamp'][:10],
                _external=True) }}">{{ entry['timestamp'][:10] }}</a> at <a href="{{ url_for
                ('main.view_single_entry', timestamp=entry['timestamp'], _external=True) }}">
                {{ entry['timestamp'][11:] }}</a>:</br>{{entry['entry']}}</br>
                Tags:
                    {% if entry['tags']|length != 0 %}
                        {% for tag in entry['tags'] %}
                            <a href="{{ url_for('main.view_entries_for_tag', tag=tag,
                                _external=True) }}">#{{ tag }}</a>
                        {% endfor %}
                    {% else %}
                        none
                    {% endif %}
                </p></li>
        {% endfor %}
        </ul>
    {% endif %}
    </div>
```

```
        </br>
        <div>
            {% if next_page %}
            <p><a class="btn btn-default" href="{{ url_for('main.view_entries_for_page',
                page=next_page, _external=True) }}">Older entries</a></p>
            {% endif %}
        </div>

{% endblock %}
```

执行后显示的日志信息如图 4-3 所示。

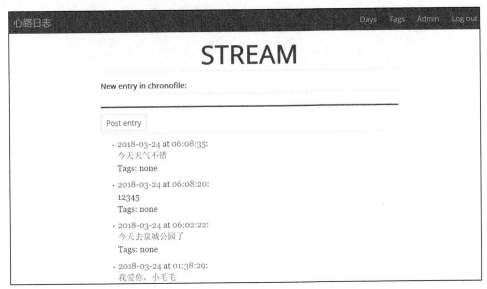

图 4-3 分页显示的日志信息

（4）编写程序文件 days.html，功能是以"日"为单位来查看系统数据库内的日志信息。具体实现代码如下。

源码路径：daima\4\4-4\chronoflask\app\templates\days.html

```
{% extends "base.html" %}
{% block page_content %}

    <div class="page-header">
        <h1>Days</h1>
    </div>

    <div>
        {{ wtf.quick_form(form) }}
    </div>

    <div>
    {% if all_days|length == 0 %}
        <p>No entries yet.</p>
    {% else %}
        <ul>
        {% for day in all_days %}
            <li><p>
                <a href="{{ url_for('main.view_entries_for_day', day=day,
                    _external=True) }}">{{ day }}</a>
            </p></li>
        {% endfor %}
        </ul>
    {% endif %}
    </div>

{% endblock %}
```

执行后按日显示日志信息的界面如图 4-4 所示。

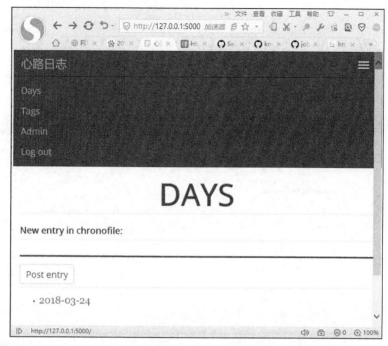

图 4-4　按日显示日志信息的界面

（5）编写程序文件 day.html，功能是显示系统数据库内某日的日志信息。具体实现代码如下。

源码路径：daima\4\4-4\chronoflask\app\templates\day.html

```
{% extends "base.html" %}
{% block page_content %}

<div class="page-header">
    {% if entries_for_day|length == 0 %}
        <h1>No Entries</h1>
    {% else %}
        <h1>{{ day[2:] }}</h1>
    {% endif %}
</div>

<div>
    {{ wtf.quick_form(form) }}
</div>

<div>
    <ul>
    {% for entry in entries_for_day %}
        <li><p><a href="{{ url_for('main.view_single_entry', timestamp=entry['timestamp'],
            _external=True) }}">{{ entry['timestamp'][11:] }}</a>: {{entry['entry']}}</br>
        Tags:
            {% if entry['tags']|length != 0 %}
                {% for tag in entry['tags'] %}
                    <a href="{{ url_for('main.view_entries_for_tag', tag=tag,
                        _external=True) }}">#{{ tag }}</a>
                {% endfor %}
            {% else %}
                none
            {% endif %}
        </p></li>
    {% endfor %}
```

```
        </ul>
    </div>
{% endblock %}
```

执行后显示的日志信息如图 4-5 所示。

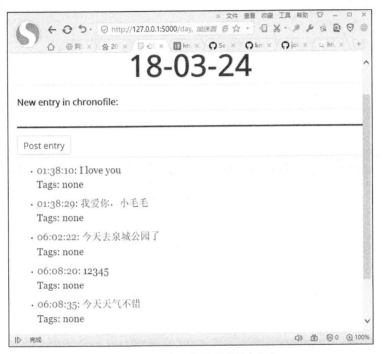

图 4-5　2018 年 3 月 24 日的日志信息

(6) 编写程序文件 admin.html，功能是显示系统管理主页，提供如下按钮和链接。
- Chronofile name：用于修改系统名。
- Author name：用于修改系统作者名。
- Change email address：用于修改邮箱地址。
- Change password：用于修改密码。

文件 admin.html 的具体实现代码如下。

源码路径：daima\4\4-4\chronoflask\app\templates\admin.html

```
{% extends "base.html" %}
{% block page_content %}

    <div class="page-header">
        <h1>Admin</h1>
    </div>

    <div>
        <p>Chronofile name:</p>
        <p>{{ details['chronofile_name'] }} <a class="btn btn-default"
            href="{{ url_for('admin.rename_chronofile') }}">Edit</a></p>
    </div>
    </br>
    <div>
        <p>Author name:</p>
        <p>{{ details['author_name'] }} <a class="btn btn-default"
            href="{{ url_for('admin.rename_author') }}">Edit</a></p>
    </div>
    </br>
    <div>
```

```
            <p>Change email address:</p>
            <p><a class="btn btn-default" href="{{ url_for('auth.change_email') }}">Click here</a></p>
        </div>
        </br>
        <div>
            <p>Change password:</p>
            <p><a class="btn btn-default" href="{{ url_for('auth.change_password') }}">Click here</a></p>
        </div>
{% endblock %}
```

执行后显示的系统管理主页如图 4-6 所示。

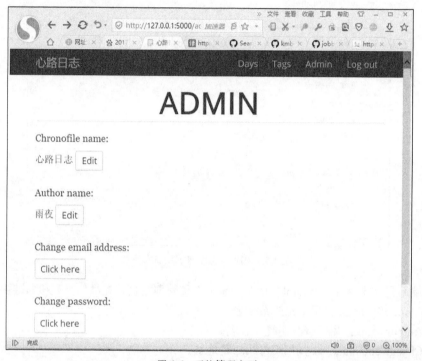

图 4-6　系统管理主页

为节省本书篇幅，本项目中的其他模板文件将不再详细讲解。相关内容请查看本书配套源码。

第 5 章

数据库驱动框架

在 Python 程序中，经常需要实现和数据库操作相关的功能。在连接不同的数据库时，需要使用不同的连接驱动。本章将详细讲解在 Python 程序中常用的数据库驱动框架。

5.1 连接 MySQL 数据库

在 Python 3.x 版本中，使用内置库 PyMySQL 来连接 MySQL 数据库服务器，Python 2 版本中使用 mysqldb 库。PyMySQL 完全遵循 Python 数据库 API v2.0 规范，并包含了纯 Python 的 MySQL 客户端库。本节将详细讲解在 Python 程序中连接 MySQL 数据库的基本知识。

5.1.1 使用 mysqlclient

mysqlclient 是 mysql-python 的分支，完全支持 Python 3。在使用 mysqlclient 之前需要先使用如下命令安装 mysqlclient。

```
$ pip install mysqlclient
```

如果是 Windows 系统，也可以登录美国国立卫生研究院设在加利福尼亚大学生物医学荧光光谱学研究中心的网站（后面简称"LFD"）来下载对应版本的 mysqlclient，然后使用 pip install 命令进行安装。

如果是 Linux 系统，需要先通过如下命令安装 mysql-devel。

```
yum install mysql-devel
```

如果在当前的 Linux 系统中已经安装了 mysql-devel，通过如下命令安装 mysql-client。

```
$ pip install mysqlclient
```

下面的实例演示了创建 MySQL 数据库表并添加、修改和删除指定数据的过程。

（1）在 MySQL 数据库中创建一个名为"1234"的数据库。

（2）编写程序文件 mysqlclient01.py，实现对数据库"1234"的操作。具体代码如下。

源码路径：daima\5\5-1\mysqlclient01.py

```python
import MySQLdb

#connect() 方法用于创建数据库的连接，里面可以指定参数：用户名、密码、主机等信息。
#这只连接到了数据库，要操作数据库，需要创建游标
conn= MySQLdb.connect(
        host='localhost',
        user='root',
        passwd='66688888',
        db ='1234',
        )

#通过获取到的数据库连接conn下的cursor()方法来创建游标。
cur = conn.cursor()

#创建数据表,通过游标cur 操作execute()方法可以写入纯sql语句。通过在execute()方法中写入sql语句对数据进行操作
cur.execute("create table student(id int ,name varchar(20),class varchar(30),age varchar(10))")

#插入一条数据
cur.execute("insert into student values('2','Tom','3 year 2 class','9')")

#修改查询条件的数据
cur.execute("update student set class='3 year 1 class' where name = 'Tom'")

#删除查询条件的数据
cur.execute("delete from student where age='9'")

#cur.close() 关闭游标
cur.close()

#conn.commit()方法在提交事物或在向数据库插入一条数据时必须要有，否则数据不会被真正地插入。
conn.commit()

#conn.close()关闭数据库连接
conn.close()
```

5.1 连接 MySQL 数据库

执行后会发现,在数据库"1234"中出现了一个名为"student"的表。要在数据库表"student"中插入新的数据,可以通过如下实例文件 mysqlclient02.py 实现。

源码路径:daima\5\5-1\mysqlclient02.py

```python
import MySQLdb

conn= MySQLdb.connect(
        host='localhost',
        user='root',
        passwd='66688888',
        db ='1234',
        )
cur = conn.cursor()

#插入一条记录
sqli="insert into student values(%s,%s,%s,%s)"
cur.execute(sqli,('3','Huhu','2 year 1 class','7'))

cur.close()
conn.commit()
conn.close()
```

执行后会发现在 MySQL 数据库中添加的这一条记录,如图 5-1 所示。

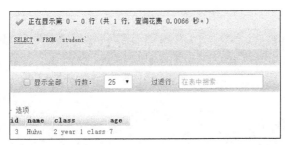

图 5-1 插入的记录

如果同时向数据库表"1234"中插入多条记录,可以通过如下实例文件 mysqlclient03.py 实现。

源码路径:daima\5\5-1\mysqlclient03.py

```python
import MySQLdb

conn= MySQLdb.connect(
        host='localhost',
        user='root',
        passwd='66688888',
        db ='1234',
        )
cur = conn.cursor()

#同时插入多条记录
sqli="insert into student values(%s,%s,%s,%s)"
cur.executemany(sqli,[
    ('3','Tom','1 year 1 class','6'),
    ('3','Jack','2 year 1 class','7'),
    ('3','Yaheng','2 year 2 class','7'),
    ])

cur.close()
conn.commit()
conn.close()
```

由此可见,方法 executemany()可以同时插入多条记录,执行单条 sql 语句,但是重复执行参数列表里的参数,返回值为受影响的行数。执行后会在 MySQL 数据库中发现添加的这 3 条记录,如图 5-2 所示。

第 5 章　数据库驱动框架

图 5-2　插入的 3 条记录

5.1.2　使用 PyMySQL

在 Python 3.x 版本中，使用内置库 PyMySQL 来连接 MySQL 数据库服务器。

1．安装 PyMySQL

在使用 PyMySQL 之前，必须先确保已经安装了 PyMySQL，读者可以在 GitHub 网站下载 PyMySQL。如果还没有安装，可以使用如下命令安装最新版的 PyMySQL。

```
pip install PyMySQL
```

安装成功后的界面如图 5-3 所示。

图 5-3　CMD 界面

如果当前系统不支持 pip 命令，可以使用如下两种方式进行安装。

（1）使用 git 命令下载安装包并安装。

```
$ git clone https://GitHub官方域名.com/PyMySQL/PyMySQL
$ cd PyMySQL/
$ python3 setup.py install
```

（2）如果需要指定版本号，可以使用 curl 命令进行安装。

```
$ # X.X 为 PyMySQL 的版本号
$ curl -L https://GitHub官方域名.com/PyMySQL/PyMySQL/tarball/pymysql-X.X | tar xz
$ cd PyMySQL*
$ python3 setup.py install
$ # 现在可以删除 PyMySQL* 目录
```

注意：你必须确保拥有 root 权限才可以安装上述模块。另外，在安装的过程中可能会出现 "ImportError: No module named setuptools" 提示信息，这个提示信息的意思是没有安装 setuptools，可以从 Python 官方网站的 setuptools 中找到各个系统的安装方法。在 Linux 系统中的安装实例是：

```
$ wget https://......ez_setup.py
$ python3 ez_setup.py
```

下面的实例文件 mysql.py 显示 PyMySQL 数据库的版本号。

源码路径：daima\5\5-1\mysql.py

```
import pymysql
#打开数据库连接
db = pymysql.connect("localhost","root","66688888","TESTDB" )
#使用cursor()方法创建一个游标对象cursor
```

```
cursor = db.cursor()
#使用 execute()方法执行SQL查询
cursor.execute("SELECT VERSION()")
#使用 fetchone() 方法获取单条记录
data = cursor.fetchone()
print ("Database version : %s " % data)
#关闭数据库连接
db.close()
```

执行后会输出：

```
Database version : 5.7.17-log
```

2. 创建数据库表

在 Python 程序中，可以使用方法 execute()在数据库中创建一个新表。下面的实例文件 new.py 演示了，在 PyMySQL 数据库中创建新表 EMPLOYEE 的过程。

源码路径：daima\5\5-1\new.py

```
import pymysql
#打开数据库连接
db = pymysql.connect("localhost","root","66688888","TESTDB" )
#使用cursor()方法创建一个游标对象 cursor
cursor = db.cursor()
#使用 execute() 方法执行sql语句，如果表存在则删除
cursor.execute("DROP TABLE IF EXISTS EMPLOYEE")
#使用预处理语句创建表
sql = """CREATE TABLE EMPLOYEE (
         FIRST_NAME  CHAR(20) NOT NULL,
         LAST_NAME  CHAR(20),
         AGE INT,
         SEX CHAR(1),
         INCOME FLOAT )"""
cursor.execute(sql)
#关闭数据库连接
db.close()
```

执行上述代码后，将在 MySQL 数据库中创建一个名为 "EMPLOYEE" 的新表，执行后的效果如图 5-4 所示。

3. 数据库插入操作

在 Python 程序中，可以使用 SQL 语句向数据库中插入新的数据信息。下面的实例文件 cha.py 演示了使用 INSERT 语句向表 EMPLOYEE 中插入数据信息的过程。

源码路径：daima\5\5-1\cha.py

图 5-4 执行效果

```
import pymysql
#打开数据库连接
db = pymysql.connect("localhost","root","66688888","TESTDB" )
#使用cursor()方法获取操作游标
cursor = db.cursor()
# SQL插入语句
sql = """INSERT INTO EMPLOYEE(FIRST_NAME,
         LAST_NAME, AGE, SEX, INCOME)
         VALUES ('Mac', 'Mohan', 20, 'M', 2000)"""
try:
   #执行SQL语句
   cursor.execute(sql)
   #提交到数据库执行
   db.commit()
except:
   #如果发生错误则回滚
   db.rollback()
# 关闭数据库连接
db.close()
```

执行上述代码后，打开 MySQL 数据库中的表 "EMPLOYEE"，会发现在里面插入了一条新的记录。执行后的效果如图 5-5 所示。

第5章 数据库驱动框架

图 5-5 执行效果

4. 数据库查询操作

在 Python 程序中，可以使用 fetchone() 方法获取 MySQL 数据库中的单条记录，使用 fetchall() 方法获取 MySQL 数据库中的多条数据。

下面的实例文件 fi.py 演示了如何查询并显示表 EMPLOYEE 中 INCOME（工资）大于 1000 的所有员工。

源码路径：daima\5\5-1\fi.py

```python
import pymysql
#打开数据库连接
db = pymysql.connect("localhost","root","66688888","TESTDB" )
#使用cursor()方法获取操作游标
cursor = db.cursor()
# sql查询语句
sql = "SELECT * FROM EMPLOYEE \
       WHERE INCOME > '%d'" % (1000)
try:
    #执行sql语句
    cursor.execute(sql)
    #获取所有记录列表
    results = cursor.fetchall()
    for row in results:
        fname = row[0]
        lname = row[1]
        age = row[2]
        sex = row[3]
        income = row[4]
        # 显示结果
        print ("fname=%s,lname=%s,age=%d,sex=%s,income=%d" % \
               (fname, lname, age, sex, income ))
except:
    print ("Error: unable to fetch data")
#关闭数据库连接
db.close()
```

执行后会输出：

```
fname=Mac,lname=Mohan,age=20,sex=M,income=2000
```

5. 数据库更新操作

在 Python 程序中，可以使用 UPDATE 语句更新数据库中的数据信息。在下面的实例文件 xiu.py 中，将数据库表中"SEX"字段为"M"的"AGE"字段递增 1。

源码路径：daima\5\5-1\xiu.py

```python
import pymysql
#打开数据库连接
db = pymysql.connect("localhost","root","66688888","TESTDB" )
#使用cursor()方法获取操作游标
cursor = db.cursor()
# sql 更新语句
sql = "UPDATE EMPLOYEE SET AGE = AGE + 1 WHERE SEX = '%c'" % ('M')
try:
    #执行SQL语句
    cursor.execute(sql)
```

```
        #提交到数据库
        db.commit()
    except:
        #当发生错误时回滚
        db.rollback()
#关闭数据库连接
db.close()
```

执行后的效果如图 5-6 所示。

(a) 修改前　　　　　　　　　　　　　(b) 修改后

图 5-6　执行效果

6. 数据库删除操作

在 Python 程序中，可以使用 DELETE 语句删除数据库中的数据。在下面的实例文件 del.py 中，删除了表 EMPLOYEE 中所有 AGE 大于 20 的数据。

源码路径：daima\5\5-1\del.py

```
import pymysql
#打开数据库连接
db = pymysql.connect("localhost","root","66688888","TESTDB" )
#使用cursor()方法获取操作游标
cursor = db.cursor()
# sql删除语句
sql = "DELETE FROM EMPLOYEE WHERE AGE > '%d'" % (20)
try:
    #执行sql语句
    cursor.execute(sql)
    #提交修改
    db.commit()
except:
    #当发生错误时回滚
    db.rollback()

#关闭连接
db.close()
```

执行后将删除表 EMPLOYEE 中所有 AGE 大于 20 的数据。执行后的效果如图 5-7 所示。

7. 执行事务

在 Python 程序中，使用事务机制可以确保数据一致性。通常来说，事务应该具有 4 个属性——原子性、一致性、隔离性、持久性。这 4 个属性通常称为 ACID 特性。

图 5-7　表 EMPLOYEE 中的数据已经为空

- 原子性：一个事务是一个不可分割的工作单元，事务中包括的诸操作要么都做，要么都不做。
- 一致性：事务必须是使数据库从一个一致性状态变成另一个一致性状态，一致性与原子性是密切相关的。
- 隔离性：一个事务的执行不能被其他事务干扰，即一个事务内部的操作及使用的数据对并发的其他事务是隔离的，并发执行的各个事务之间不能互相干扰。
- 持久性：持续性也称永久性，指一个事务一旦提交，它对数据库中数据的改变就应该是永久性的，接下来的其他操作或故障不应该对它有任何影响。

在 Python DB API 2.0 的事务机制中提供了两个处理方法，分别是 commit() 和 rollback()。在下面的实例文件 shi.py 中，通过执行事务的方式删除了表 EMPLOYEE 中所有 AGE 大于 19 的员工。

源码路径：daima\5\5-1\shi.py

```python
import pymysql
#打开数据库连接
db = pymysql.connect("localhost","root","66688888","TESTDB" )
#使用cursor()方法获取操作游标
cursor = db.cursor()
# SQL删除记录语句
sql = "DELETE FROM EMPLOYEE WHERE AGE > '%d'" % (19)
try:
    #执行SQL语句
    cursor.execute(sql)
    #向数据库提交
    db.commit()
except:
    #发生错误时回滚
    db.rollback()
```

执行后将删除表 EMPLOYEE 中所有 AGE 大于 19 的员工。

8. 综合 MySQL 数据库操作系统

在下面的实例中，模拟实现了足球游戏中的球员管理系统功能。这个实例综合演练了在 Python 中操作 MySQL 数据库的过程。

（1）创建一个新的 MySQL 数据库，如"TESTDB"，然后将 league.sql 文件导入 MySQL 数据库中以生成数据表。

（2）在实例文件 main.py 中编写代码，通过菜单提示可以实现对数据库的操作处理。文件 main.py 的具体实现代码如下。

源码路径：daima\5\5-1\main.py

```python
import CRUD

option = 0

print("\n欢迎光临，请选择数据库操作！")

while option != 6:
    print("\n---------------------------")
    print("选择操作： \n"
          "0．读取并显示所有数据\n"
          "1．创建数据库\n"
          "2．插入新的数据\n"
          "3．注册新球员\n"
          "4．预约\n"
          "5．删除所有的表\n"
          "6．退出")
    print("---------------------------")

    option = int(input())

    if option == 0:

        CRUD.readAll()

    elif option == 1:

        CRUD.createDB()

    elif option == 2:

        CRUD.insertRows()
```

```python
        elif option == 3:
            CRUD.insertRow(0)

        elif option == 4:
            print("\nConsultas: \n"
                  "0. 查看目前的冠军球队?\n"
                  "1. 游戏角色的比较?\n"
                  "2. 玩家ID的特点?\n"
                  "3. 其中最有特点的球员?\n"
                  "4. 你感觉最多球员的球队? \n"
                  "5. 让所有的球员开始比赛\n"
                  "6. 退出")
            readOpt = int(input())

            if readOpt is 0:
                CRUD.readTables(0)
            elif readOpt == 1:
                CRUD.readTables(1)
            elif readOpt == 2:
                CRUD.readTables(2)
            elif readOpt == 3:
                CRUD.readTables(3)
            elif readOpt == 4:
                CRUD.readTables(4)
            elif readOpt == 5:
                CRUD.readTables(5)

        elif option == 5:
            CRUD.deleteTables()

        elif option == 6:
            quit()
```

（3）编写实例文件 CRUD.py，功能是根据用户选择的菜单项执行对应的操作。具体实现流程如下。

① 建立和指定数据库的连接，填写正确的连接参数。对应代码如下。

```
con = mdb.connect('localhost', 'root', '66688888', 'TESTDB')
```

② 当用户在界面中输入"0"指令后执行函数 readAll()，功能是获取数据库内所有表的数据。对应代码如下。

```python
def readAll():
    with con:
        cur = con.cursor()

        cur.execute("SHOW TABLES;")
        lolbd = cur.fetchall()
        print("联赛表:")
        for table in lolbd:
            print(table[0])

        for table in lolbd:
            cur.execute("SELECT * FROM "+table[0])
            print("\nTabela - "+table[0])
            for row in cur.fetchall():
                print(row)
```

③ 当用户在界面中输入"1"指令后执行函数 createDB()，功能是重新创建数据库表。对应代码如下。

```python
def createDB():
    try:
        con = mdb.connect('localhost', 'root', '66688888', '123');

        cur = con.cursor()

        # cur.execute("DROP TABLE IF EXISTS jogador")
        cur.execute("CREATE TABLE `jogador` ( \
```

```python
                `jogador_id` int(11) NOT NULL,\
                `time_id` int(11) DEFAULT NULL,\
                `nome` varchar(45) DEFAULT NULL,\
                `abates_totais` int(11) DEFAULT NULL,\
                `mortes_totais` int(11) DEFAULT NULL\
              ) ENGINE=MyISAM DEFAULT CHARSET=latin1;")

# cur.execute("DROP TABLE IF EXISTS partida")
cur.execute("CREATE TABLE `partida` (\
                `partida_id` int(11) NOT NULL,\
                `timeA_id` int(11) DEFAULT NULL,\
                `timeB_id` int(11) DEFAULT NULL,\
                `abates_timeA` int(11) DEFAULT NULL,\
                `abates_timeB` int(11) DEFAULT NULL,\
                `vencedor_id` int(11) DEFAULT NULL,\
                `torneio_id` varchar(45) DEFAULT NULL\
              ) ENGINE=MyISAM DEFAULT CHARSET=latin1;")

# cur.execute("DROP TABLE IF EXISTS personagem")
cur.execute("CREATE TABLE `personagem` (\
                `personagem_id` int(11) NOT NULL,\
                `personagem_nome` varchar(45) DEFAULT NULL,\
                `personagem_preco` int(11) DEFAULT NULL\
              ) ENGINE=MyISAM DEFAULT CHARSET=latin1;")

# cur.execute("DROP TABLE IF EXISTS personagem_comprado")
cur.execute("CREATE TABLE `personagem_comprado` (\
                `personagem_id` int(11) NOT NULL,\
                `jogador_id` int(11) DEFAULT NULL\
              ) ENGINE=MyISAM DEFAULT CHARSET=latin1;")

# cur.execute("DROP TABLE IF EXISTS time")
cur.execute("CREATE TABLE `time` (\
                `time_id` int(11) NOT NULL,\
                `nome_time` varchar(45) DEFAULT NULL,\
                `abates` int(11) DEFAULT NULL,\
                `mortes` int(11) DEFAULT NULL\
              ) ENGINE=MyISAM DEFAULT CHARSET=latin1;")

# cur.execute("DROP TABLE IF EXISTS torneio")
cur.execute("CREATE TABLE `torneio` (\
                `torneio_id` int(11) NOT NULL,\
                `regiao_torneio` varchar(45) DEFAULT NULL\
              ) ENGINE=MyISAM DEFAULT CHARSET=latin1;")

cur.execute("ALTER TABLE `jogador`\
              ADD PRIMARY KEY (`jogador_id`);")

cur.execute("ALTER TABLE `partida`\
              ADD PRIMARY KEY (`partida_id`);")

cur.execute("ALTER TABLE `personagem`\
              ADD PRIMARY KEY (`personagem_id`);")

cur.execute("ALTER TABLE `personagem_comprado`\
              ADD KEY `jogador_id` (`jogador_id`),\
              ADD KEY `jogador_id_2` (`jogador_id`);")

cur.execute("ALTER TABLE `time`\
              ADD PRIMARY KEY (`time_id`);")

cur.execute("ALTER TABLE `torneio`\
              ADD PRIMARY KEY (`torneio_id`);")

cur.execute("ALTER TABLE `jogador`\
               MODIFY `jogador_id` int(11) NOT NULL AUTO_INCREMENT,
                 AUTO_INCREMENT=19;")
```

```python
            cur.execute("ALTER TABLE `partida`\
                        MODIFY `partida_id` int(11) NOT NULL AUTO_INCREMENT,\
                        AUTO_INCREMENT=5;")

            cur.execute("ALTER TABLE `personagem`\
                        MODIFY `personagem_id` int(11) NOT NULL AUTO_INCREMENT,\
                        AUTO_INCREMENT=13;")

            cur.execute("ALTER TABLE `time`\
                        MODIFY `time_id` int(11) NOT NULL AUTO_INCREMENT,\
                        AUTO_INCREMENT=4;")
    except mdb.Error as e:

        print("Error %d: %s" % (e.args[0], e.args[1]))
        sys.exit(1)

    finally:

        if con:
            con.close()
```

④ 编写函数 readTables()，功能是读取指定数据库表中的数据。根据用户在界面中输入的指令执行指定的查询操作。对应实现代码如下。

```python
def readTables(option: int):
    try:
        con = mdb.connect('localhost', 'root', '66688888', 'TESTDB');

        cur = con.cursor()

        cur.execute("SET sql_mode=(SELECT
            REPLACE(@@sql_mode,'ONLY_FULL_GROUP_BY',''));")

        if option == 0:
            cur.execute("SELECT time.nome_time AS Vencedor FROM partida\
                        JOIN time\
                        ON time.time_id = partida.vencedor_id\
                        GROUP BY time.time_id\
                        ORDER BY COUNT(*)DESC, time.abates DESC,\
                        time.mortes ASC LIMIT 1")
        if option == 1:
            cur.execute("SELECT personagem.personagem_nome AS Personagem,
                COUNT(personagem_comprado.personagem_id) AS Quantia FROM personagem_comprado\
                        JOIN personagem\
                        ON personagem.personagem_id = personagem_comprado.personagem_id\
                        GROUP BY personagem_comprado.personagem_id\
                        ORDER BY Quantia DESC LIMIT 1 ")
        if option == 2:
            cur.execute("SELECT personagem.personagem_nome FROM personagem\
                        JOIN personagem_comprado\
                        ON personagem_comprado.personagem_id =
                        personagem.personagem_id\
                        WHERE personagem_comprado.jogador_id = 1")
        if option == 3:
            cur.execute("SELECT jogador.nome, COUNT(personagem_comprado.jogador_id) AS
                Quantidade FROM personagem_comprado\
                        JOIN jogador\
                        ON jogador.jogador_id = personagem_comprado.jogador_id\
                        GROUP BY personagem_comprado.jogador_id\
                        ORDER BY Quantidade DESC LIMIT 1\
                        ")
        if option == 4:
            cur.execute("SELECT f.nome, time.nome_time, f.abates_totais\
                        FROM (\
                            SELECT jogador.nome, jogador.time_id, max(jogador.abates
                            _totais) AS minprice\
                            FROM jogador GROUP BY jogador.time_id\
                        ) AS x INNER JOIN jogador AS f ON f.time_id = x.time_id AND
                        f.abates_totais = x.minprice\
```

```
                              JOIN time\
                              ON time.time_id = f.time_id")
        if option == 5:
            cur.execute("SELECT time.nome_time, jogador.nome, jogador.abates_totais AS Abates,
                jogador.mortes_totais AS Mortes FROM time\
                              JOIN jogador\
                              ON time.time_id = jogador.time_id\
                              WHERE time.nome_time = 'BepidPower' \
                              ")

        for row in cur.fetchall():
            print(row)

except mdb.Error as e:

    print("Error %d: %s" % (e.args[0], e.args[1]))
    sys.exit(1)

finally:

    if con:
        con.close()
```

⑤ 编写函数 insertRows()，功能是当用户在界面中输入指令 2 后插入新的数据。对应实现代码如下。

```
def insertRows():
    with con:

        cur = con.cursor()

        cur.execute("INSERT INTO 'jogador' ('jogador_id', 'time_id', 'nome', 'abates_totais',
            `mortes_totais`) VALUES\
                        (1, 1, 'Italus', 5, 2),\
                        (2, 1, 'Thiago', 10, 7),\
                        (3, 1, 'Gabriela', 1, 3),\
                        (4, 1, 'Macabeus', 0, 3),\
                        (5, 1, 'Trabson', 14, 0),\
                        (6, 2, 'Douglas', 8, 8),\
                        (7, 2, 'Estela', 2, 2),\
                        (8, 2, 'Elias', 25, 15),\
                        (9, 2, 'Designer', 3, 1),\
                        (10, 2, 'Pistela', 2, 9),\
                        (12, 3, 'Moskito', 3, 5),\
                        (13, 3, 'Kira', 1, 9),\
                        (16, 3, 'Alsaher', 7, 1),\
                        (17, 3, 'Rodrix', 2, 20),\
                        (18, 3, 'Lalitax', 7, 5);")

        cur.execute("INSERT INTO 'partida' ('partida_id', 'timeA_id', 'timeB_id',
            'abates_timeA', 'abates_timeB', 'vencedor_id', 'torneio_id') VALUES\
                        (1, 1, 2, 20, 10, 1, '1'),\
                        (2, 2, 3, 30, 15, 2, '1'),\
                        (3, 1, 3, 10, 5, 1, '1');")

        cur.execute("INSERT INTO 'personagem' (`personagem_id`, 'personagem_nome',
            'personagem_preco') VALUES\
                        (1, 'Ashe', 1200),\
                        (2, 'Vayne', 2000),\
                        (3, 'Tryndamere', 1200),\
                        (4, 'Blitzcrank', 2000),\
                        (5, 'Amumu', 2000),\
                        (6, 'Fiora', 4500),\
                        (7, 'Sona', 6300),\
                        (10, 'Alistar', 900),\
                        (8, 'Morgana', 6300),\
                        (9, 'Kayle', 4500);")
```

5.1 连接 MySQL 数据库

```
        cur.execute("INSERT INTO 'personagem_comprado' ('personagem_id', 'jogador_id') VALUES\
                    (1, 1),\
                    (2, 1),\
                    (3, 1),\
                    (4, 1),\
                    (5, 1),\
                    (1, 2),\
                    (2, 2),\
                    (1, 3),\
                    (4, 3);")

        cur.execute("INSERT INTO 'time' ('time_id', 'nome_time', 'abates', 'mortes') VALUES\
                    (1, 'BepidPower', 30, 15),\
                    (2, 'AlbusNox', 40, 35),\
                    (3, 'Invocados', 20, 40);")

        cur.execute("INSERT INTO 'torneio' ('torneio_id', 'regiao_torneio') VALUES\
                    (1, 'Brasil');")
```

⑥ 当用户在界面中输入指令 3 后插入指定新数据。对应实现代码如下。

```
def insertRow(option: int):
    print("球员名字:")
    nome = str(input())
    while valueExistsInTable(nome, "nome", "jogador"):
        print("已注册的球员,要修改这一名字吗?(s/n)")
        nome = str(input())
        if nome == 's':
            print("球员名字:")
            nome = str(input())
        else:
            return

    print("Quantidade de Abates:")
    abates = str(input())
    print("Quantidade de Mortes")
    mortes = str(input())

    print("Time:")
    time = str(input())

    while valueExistsInTable(time, "nome_time", "time") == 0:
        print("时间是不存在的,需要修改?(s/n)")
        opt = str(input())
        if opt == 's':
            insertNewTeamNamed(time, abates, mortes)
            print("新的团队注册!")
        else:
            break

    team_id = getTeamID(time)

    insertNewPlayerNamed(nome,abates,mortes,team_id)
```

⑦ 编写函数 insertNewPlayerNamed(),功能是插入指定名字的新球员信息。对应实现代码如下。

```
def insertNewPlayerNamed(name: str, kills: str, deaths: str, teamID: str):
    with con:
        cur = con.cursor()

        cur.execute("INSERT INTO 'jogador' ('jogador_id', 'time_id', 'nome',
            'abates_totais', `mortes_totais`) VALUES\
                    ("+str(generateID("jogador", "jogador_id"))+", "+teamID+",
                    '"+name+"', "+kills+", "+deaths+")")
        cur.execute("SELECT * FROM jogador;")

        rows = cur.fetchall()
```

```
        for row in rows:
            print(row)
```

⑧ 编写函数 insertNewTeamNamed()，功能是插入指定名字的新球队信息。对应实现代码如下。

```
def insertNewTeamNamed(teamName: str, kills: str, deaths: str):
    with con:
        cur = con.cursor()

        cur.execute("INSERT INTO `time` (`time_id`, `nome_time`, `abates`, `mortes`) VALUES\
                    ("+str(generateID("time", "time_id"))+", '"+teamName+"', "+kills+", "
                    "+deaths+")")
        cur.execute("SELECT * FROM time;")

        rows = cur.fetchall()

        for row in rows:
            print(row)
```

⑨ 编写函数 getTeamID()，功能是插入球队 ID 信息。对应实现代码如下。

```
def getTeamID(teamName: str):
    with con:

        cur = con.cursor()

        cur.execute("SELECT time_id, nome_time FROM time")

        rows = cur.fetchall()

        for row in rows:
            if str(row[1]) == teamName:
                return str(row[0])
```

⑩ 编写函数 valueExistsInTable()，功能是统计数据表的信息。对应实现代码如下。

```
def valueExistsInTable(value: any, column: str, table: str):

    with con:
        cur = con.cursor()

        cur.execute("SELECT "+column+" FROM "+table)
        rows = cur.fetchall()

        for row in rows:
            if str(value) == str(row[0]):
                return 1
        return 0

def generateID(entity: str, column: str):
    try:
        con = mdb.connect('localhost', 'user', 'goodPassword', 'league');
        cur = con.cursor()
        getBiggestId = "SELECT "+column+" FROM " + entity
        cur.execute(getBiggestId)
        arrayIDs = cur.fetchall()
        return int(arrayIDs[len(arrayIDs)-1][0]) + 1

    except mdb.Error as e:

        print("Error %d: %s" % (e.args[0], e.args[1]))
        sys.exit(1)

    finally:
        if con:
            con.close()
```

⑪ 编写函数 deleteTables()，功能是当用户在界面中输入指令 5 后删除数据库中的数据。对

应实现代码如下。

```python
def deleteTables():
    try:
        con = mdb.connect('localhost', 'root', '66688888', 'TESTDB');
        
        cur = con.cursor()
        
        cur.execute("DROP TABLE jogador")
        cur.execute("DROP TABLE partida")
        cur.execute("DROP TABLE personagem")
        cur.execute("DROP TABLE personagem_comprado")
        cur.execute("DROP TABLE time")
        cur.execute("DROP TABLE torneio")
        
        cur.execute("SHOW TABLES")
        rows = cur.fetchall()
        
        for row in rows:
            print(row)
    
    except mdb.Error as e:
        
        print("Error %d: %s" % (e.args[0], e.args[1]))
        sys.exit(1)
    
    finally:
        
        if con:
            con.close()
```

当用户输入不同的指令后会执行不同的操作，例如，在作者的计算机中执行上述代码后会输出：

```
欢迎光临，请选择数据库操作！

---------------------------
选择操作：
0．读取并显示所有数据
1．创建数据库
2．插入新的数据
3．注册新球员
4．预约
5．删除所有的表
6．退出
---------------------------
0
联赛表：
employee
jogador
partida
personagem
personagem_comprado
testtable
time
torneio

Tabela - employee
('Mac', 'Mohan', 21, 'M', 2000.0)

Tabela - jogador
(1, 1, 'Italus', 5, 2)
(2, 1, 'Thiago', 10, 7)
(3, 1, 'Gabriela', 1, 3)
(4, 1, 'Macabeus', 0, 3)
(5, 1, 'Trabson', 14, 0)
(6, 2, 'Douglas', 8, 8)
(7, 2, 'Estela', 2, 2)
(8, 2, 'Elias', 25, 15)
(9, 2, 'Designer', 3, 1)
(10, 2, 'Pistela', 2, 9)
(12, 3, 'Moskito', 3, 5)
```

```
(13, 3, 'Kira', 1, 9)
(16, 3, 'Alsaher', 7, 1)
(17, 3, 'Rodrix', 2, 20)
(18, 3, 'Lalitax', 7, 5)

Tabela - partida
(1, 1, 2, 20, 10, 1, '1')
(2, 2, 3, 30, 15, 2, '1')
(3, 1, 3, 10, 5, 1, '1')

Tabela - personagem
(1, 'Ashe', 1200)
(2, 'Vayne', 2000)
(3, 'Tryndamere', 1200)
(4, 'Blitzcrank', 2000)
(5, 'Amumu', 2000)
(6, 'Fiora', 4500)
(7, 'Sona', 6300)
(10, 'Alistar', 900)
(8, 'Morgana', 6300)
(9, 'Kayle', 4500)

Tabela - personagem_comprado
(1, 1)
(2, 1)
(3, 1)
(4, 1)
(5, 1)
(1, 2)
(2, 2)
(1, 3)
(4, 3)

Tabela - testtable

Tabela - time
(1, 'BepidPower', 30, 15)
(2, 'AlbusNox', 40, 35)
(3, 'Invocados', 20, 40)

Tabela - torneio
(1, 'Brasil')

---------------------------
选择操作:
0. 读取并显示所有数据
1. 创建数据库
2. 插入新的数据
3. 注册新球员
4. 预约
5. 删除所有的表
6. 退出
---------------------------
```

5.2 连接 PostgreSQL 数据库

PostgreSQL 是一款开源、免费的数据库，其优势在于 SQL 标准的完备性、对于事务的支持、对于事务隔离级别的支持，以及数据类型、内置函数、索引的扩展性都很好。本节将详细讲解使用第三方库连接 PostgreSQL 数据库的知识。

5.2.1 下载并安装 PostgreSQL

在 Windows 系统下，下载并安装 PostgreSQL 的具体流程如下。

（1）在百度中搜索 PostgreSQL 官网，在官网首页单击"Downloads"链接进入下载界面，如图 5-8 所示。

5.2 连接 PostgreSQL 数据库

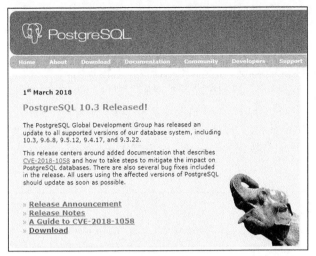

图 5-8 下载界面

（2）单击下方的"Windows"链接弹出 Windows 系统的下载界面，如图 5-9 所示。

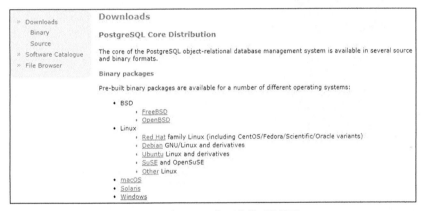

图 5-9 "Windows"系统的下载界面

（3）单击"Download the installer"链接，在弹出的页面中选择安装的 PostgreSQL 的版本和安装的操作系统版本，单击"DOWNLOAD NOW"按钮即可下载，如图 5-10 所示。

（4）下载完成后得到一个".exe"格式的运行文件，双击该文件后开始安装。注意，需要先安装微软的 Visual C++，如图 5-11 所示。

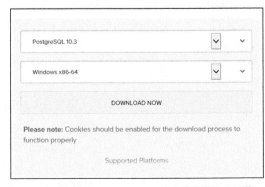

图 5-10 单击"DOWNLOAD NOW"按钮开始下载

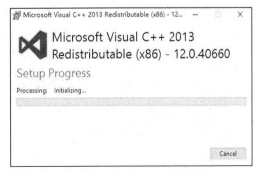

图 5-11 先安装 Visual C++

(5) 在弹出的 "Setup" 界面中，单击 "Next" 按钮。
(6) 在弹出的 "Installation Directory" 界面中选择安装路径，单击 "Next" 按钮。
(7) 在弹出的 "Select Components" 界面中选择要安装的组件，单击 "Next" 按钮。
(8) 在弹出的 "Data Directory" 界面中选择数据的保存路径，单击 "Next" 按钮。
(9) 在弹出的 "Password" 界面中设置数据库的密码，单击 "Next" 按钮。
(10) 在弹出的 "Port" 界面中设置数据库使用的端口，单击 "Next" 按钮。
(11) 在弹出的 "Advanced Options" 界面中设置本地语言为简体中文，单击 "Next" 按钮。
(12) 连续单击 "Next" 按钮，一直到弹出安装成功的界面，如图 5-12 所示。整个安装过程大约需要 2min。

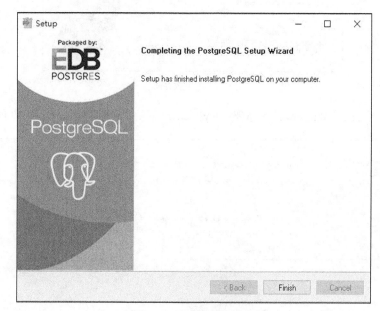

图 5-12　安装成功的界面

5.2.2　使用 psycopg2 模块

与前面介绍的 MySQL 数据库不同，Postgres 至少包含 3 种 Python 适配器驱动——psycopg、PyPgSQL 和 PyGreSQL。另外一种适配器 PoPy 目前已废弃，于 2003 年将其项目与 PyGreSQL 进行了合并。目前剩下的这 3 种适配器都有自己的特性和优缺点，建议开发者根据实际项目进行选择。注意，PyPgSQL 自 2006 年起就不再开发了，而 PyGreSQL 则是在 2009 年发布的最新版本（4.0）。这两种适配器不再活跃，使得 psycopg 成为 PostgreSQL 适配器的唯一引领者，因此本书的示例使用该适配器。psycopg 目前已进入到第二个版本 psycopg2，所以本书将以 psycopg2 模块为主来介绍连接 PostgreSQL 数据库的知识。

在使用 psycopg2 之前需要先安装，具体安装命令如下。

```
pip install python-psycopg2
```

在 Python 程序中使用 psycopg2 的基本流程如下。

(1) 导入 psycopg2。要使用 psycopg2，必须先用 import 语句导入该包。

```
import psycopg2
```

(2) 使用类 connection 进行连接。类 connection 表示数据库连接对象，由 psycopg2.connect() 方法创建。

(3) 创建 connection 对象。通过内置方法 psycopg2.connect() 创建一个新的数据库会话，并且返回一个连接对象。该方法各个参数的具体说明如下。
- dbname：数据库名。
- user：数据库角色名。
- password：数据库角色的密码。
- host：数据库地址。
- port：端口。

（4）创建 cursor 对象。在 psycopg2 中提供了一个 cursor 类，用来在数据库 Session 中执行 PostgreSQL 命令。cursor 对象由 connection.cursor()方法创建，例如：

```
cur = conn.cursor()
```

在 psycopg2 模块中主要包含如下内置方法。
- psycopg2.connect(database="testdb", user="postgres", password="cohondob", host="127.0.0.1", port="5432") 和指定的 PostgreSQL 数据库建立连接。如果成功建立连接，它返回一个连接对象。
- connection.cursor() 创建一个光标，将用于整个数据库中使用 Python 编程。
- cursor.execute(sql [, optional parameters]) 执行 SQL 语句,可使用参数化的 SQL 语句（即占位符，而不是 SQL 文字）。psycopg2 的模块支持占位符（用%s 标志），例如：

```
cursor.execute("insert into people values (%s, %s)", (who, age))
```

- curosr.executemany(sql, seq_of_parameters) 对于所有参数序列或序列中的 SQL 映射执行 SQL 命令。
- curosr.callproc(procname[, parameters]) 调用一个名为 procname 的存储过程。
- cursor.rowcount 这是一个只读属性，返回数据库中行的总数。
- connection.commit() 此方法提交当前事务。
- connection.rollback() 此方法会回滚自上次调用 commit()方法的任何更改。
- connection.close() 此方法会关闭数据库连接。请注意，这并不自动调用 commit()。如果只是关闭数据库连接而不调用 commit()方法，那么所有的更改将会丢失。
- cursor.fetchone() 用于提取查询结果集中的下一行，返回一个序列。
- cursor.fetchmany([size=cursor.arraysize]) 取出下一组中查询结果的行数，返回一个列表。如果没有找到记录，则返回空列表。
- cursor.fetchall() 用于获取所有查询结果（剩余）行，返回一个列表。

下面的实例文件 PostgreSQL01.py 演示了使用 Python 程序连接指定 PostgreSQL 数据库的过程。

源码路径：daima\5\5-2\PostgreSQL01.py

```
import psycopg2

def connectPostgreSQL():
    conn = psycopg2.connect(database="mydb", user="postgres", password="66688888",
        host="127.0.0.1", port="5432")
    print('connect successful!')

if __name__ == '__main__':
    connectPostgreSQL()
```

执行后会输出：

```
connect successful!
```

下面的实例文件 PostgreSQL02.py 演示了使用 Python 程序在 PostgreSQL 数据库中创建指定数据表的过程。

源码路径：daima\5\5-2\PostgreSQL02.py

```
import os
import sys
import psycopg2
```

```python
def connectPostgreSQL():
    conn = psycopg2.connect(database="mydb", user="postgres", password="66688888",
        host="127.0.0.1", port="5432")
    print('connect successful!')
    cursor = conn.cursor()
    cursor.execute('''
        create table public.member(
        id integer not null primary key,
        name varchar(32) not null,
        password varchar(32) not null,
        singal varchar(128)
    )''')
    conn.commit()
    conn.close()
    print('table public.member is created!')

if __name__ == '__main__':
    connectPostgreSQL()
```

执行后会输出：

```
connect successful!
table public.member is created!
```

下面的实例文件 PostgreSQL03.py 演示了使用 Python 程序创建 PostgreSQL 数据表并插入新数据的过程。

源码路径：daima\5\5-2\PostgreSQL03.py

```
import os
import sys
import psycopg2

def connectPostgreSQL():
    conn = psycopg2.connect(database="mydb", user="postgres", password="66688888",
        host="127.0.0.1", port="5432")
    print('connect successful!')
    cursor = conn.cursor()
    cursor.execute('''create table public.member(
id integer not null primary key,
name varchar(32) not null,
password varchar(32) not null,
singal varchar(128)
)''')
    conn.commit()
    conn.close()
    print('table public.member is created!')

def insertOperate():
    conn = psycopg2.connect(database="mydb", user="postgres", password="66688888",
        host="127.0.0.1", port="5432")
    cursor = conn.cursor()
    cursor.execute("insert into public.member(id,name,password,singal)\
        values(1,'member0','password0','singal0')")
    cursor.execute("insert into public.member(id,name,password,singal)\
        values(2,'member1','password1','singal1')")
    cursor.execute("insert into public.member(id,name,password,singal)\
        values(3,'member2','password2','singal2')")
    cursor.execute("insert into public.member(id,name,password,singal)\
        values(4,'member3','password3','singal3')")
    conn.commit()
    conn.close()

    print('insert records into public.member successfully')

if __name__ == '__main__':
    # connectPostgreSQL()
    insertOperate()
```

执行后会在数据表 public.member 中插入指定的数据，并输出如下提示。

```
insert records into public.member successfully
```

下面的实例文件 PostgreSQL04.py 演示了使用 Python 程序创建 PostgreSQL 数据表并插入新数据，然后查询显示指定数据库表中数据的过程。

源码路径：daima\5\5-2\PostgreSQL04.py

```python
import os
import sys
import psycopg2


def connectPostgreSQL():
    conn = psycopg2.connect(database="mydb", user="postgres", password="66688888",
        host="127.0.0.1", port="5432")
    print
    'connect successful!'
    cursor = conn.cursor()
    cursor.execute('''create table public.member(
id integer not null primary key,
name varchar(32) not null,
password varchar(32) not null,
singal varchar(128)
)''')
    conn.commit()
    conn.close()
    print('table public.member is created!')

def insertOperate():
    conn = psycopg2.connect(database="mydb", user="postgres", password="66688888",
        host="127.0.0.1", port="5432")
    cursor = conn.cursor()
    cursor.execute("insert into public.member(id,name,password,singal)\
values(1,'member0','password0','singal0')")
    cursor.execute("insert into public.member(id,name,password,singal)\
values(2,'member1','password1','singal1')")
    cursor.execute("insert into public.member(id,name,password,singal)\
values(3,'member2','password2','singal2')")
    cursor.execute("insert into public.member(id,name,password,singal)\
values(4,'member3','password3','singal3')")
    conn.commit()
    conn.close()

    print('insert records into public.member successfully')

def selectOperate():
    conn = psycopg2.connect(database="mydb", user="postgres", password="66688888",
        host="127.0.0.1", port="5432")
    cursor = conn.cursor()
    cursor.execute("select id,name,password,singal from public.member where id>2")
    rows = cursor.fetchall()
    for row in rows:
        print('id=', row[0], ',name=', row[1], ',pwd=', row[2], ',singal=', row[3], '\n')
    conn.close()

if __name__ == '__main__':
    # connectPostgreSQL()
    # insertOperate()
    selectOperate()
```

执行后会输出显示指定数据库表中的数据。

```
id= 3 ,name= member2 ,pwd= password2 ,singal= singal2

id= 4 ,name= member3 ,pwd= password3 ,singal= singal3
```

下面的实例文件 PostgreSQL05.py 演示了向指定 PostgreSQL 数据库中插入新数据并更新数据的过程。

源码路径：daima\5\5-2\PostgreSQL05.py

```python
def connectPostgreSQL():
    conn = psycopg2.connect(database="mydb", user="postgres", password="66688888",
```

```python
            host="127.0.0.1", port="5432")
        print('connect successful!')
        cursor = conn.cursor()
        cursor.execute('''create table public.member(
id integer not null primary key,
name varchar(32) not null,
password varchar(32) not null,
singal varchar(128)
)''')
        conn.commit()
        conn.close()
        print('table public.member is created!')

def insertOperate():
        conn = psycopg2.connect(database="mydb", user="postgres", password="66688888",
            host="127.0.0.1", port="5432")
        cursor = conn.cursor()
        cursor.execute("insert into public.member(id,name,password,singal)\
 values(1,'member0','password0','singal0')")
        cursor.execute("insert into public.member(id,name,password,singal)\
 values(2,'member1','password1','singal1')")
        cursor.execute("insert into public.member(id,name,password,singal)\
 values(3,'member2','password2','singal2')")
        cursor.execute("insert into public.member(id,name,password,singal)\
 values(4,'member3','password3','singal3')")
        conn.commit()
        conn.close()
        print('insert records into public.member successfully')

def selectOperate():
        conn = psycopg2.connect(database="mydb", user="postgres", password="66688888",
            host="127.0.0.1", port="5432")
        cursor = conn.cursor()
        cursor.execute("select id,name,password,singal from public.member where id>2")
        rows = cursor.fetchall()
        for row in rows:
            print('id=', row[0], ',name=', row[1], ',pwd=', row[2], ',singal=', row[3], '\n')
        conn.close()


def updateOperate():
        conn = psycopg2.connect(database="mydb", user="postgres", password="66688888",
            host="127.0.0.1", port="5432")
        cursor = conn.cursor()
        cursor.execute("update public.member set name='update ...' where id=2")
        conn.commit()
        print("Total number of rows updated :", cursor.rowcount)
        cursor.execute("select id,name,password,singal from public.member")
        rows = cursor.fetchall()
        for row in rows:
            print('id=', row[0], ',name=', row[1], ',pwd=', row[2], ',singal=', row[3], '\n')
        conn.close()

if __name__ == '__main__':
    # connectPostgreSQL()
    # insertOperate()
    # selectOperate()
    updateOperate()
```

执行后会显示更新数据表后的数据信息。

```
Total number of rows updated : 1
id= 1 ,name= member0 ,pwd= password0 ,singal= singal0

id= 3 ,name= member2 ,pwd= password2 ,singal= singal2

id= 4 ,name= member3 ,pwd= password3 ,singal= singal3

id= 2 ,name= update ... ,pwd= password1 ,singal= singal1
```

下面的实例文件 PostgreSQL06.py 演示了删除指定 PostgreSQL 数据库表中 id 为 2 的数

据的过程。

源码路径：daima\5\5-2\PostgreSQL06.py

```python
def connectPostgreSQL():
    conn = psycopg2.connect(database="mydb", user="postgres", password="66688888",
        host="127.0.0.1", port="5432")
    print('connect successful!')
    cursor = conn.cursor()
    cursor.execute('''
      create table public.member(
      id integer not null primary key,
      name varchar(32) not null,
      password varchar(32) not null,
      singal varchar(128)
    )''')
    conn.commit()
    conn.close()
    print( 'table public.member is created!')

def insertOperate():
    conn = psycopg2.connect(database="mydb", user="postgres", password="66688888",
        host="127.0.0.1", port="5432")
    cursor = conn.cursor()
    cursor.execute("insert into public.member(id,name,password,singal)\
        values(1,'member0','password0','singal0')")
    cursor.execute("insert into public.member(id,name,password,singal)\
        values(2,'member1','password1','singal1')")
    cursor.execute("insert into public.member(id,name,password,singal)\
        values(3,'member2','password2','singal2')")
    cursor.execute("insert into public.member(id,name,password,singal)\
        values(4,'member3','password3','singal3')")
    conn.commit()
    conn.close()
    print('insert records into public.member successfully')

def selectOperate():
    conn = psycopg2.connect(database="mydb", user="postgres", password="66688888",
        host="127.0.0.1", port="5432")
    cursor = conn.cursor()
    cursor.execute("select id,name,password,singal from public.member where id>2")
    rows = cursor.fetchall()
    for row in rows:
        print('id=', row[0], ',name=', row[1], ',pwd=', row[2], ',singal=', row[3], '\n')
    conn.close()

def updateOperate():
    conn = psycopg2.connect(database="mydb", user="postgres", password="66688888",
        host="127.0.0.1", port="5432")
    cursor = conn.cursor()
    cursor.execute("update public.member set name='update ...' where id=2")
    conn.commit()
    print("Total number of rows updated :", cursor.rowcount)
    cursor.execute("select id,name,password,singal from public.member")
    rows = cursor.fetchall()
    for row in rows:
        print('id=', row[0], ',name=', row[1], ',pwd=', row[2], ',singal=', row[3], '\n')
    conn.close()

def deleteOperate():
    conn = psycopg2.connect(database="mydb", user="postgres", password="66688888",
        host="127.0.0.1", port="5432")
    cursor = conn.cursor()

    cursor.execute("select id,name,password,singal from public.member")
    rows = cursor.fetchall()
    for row in rows:
        print('id=', row[0], ',name=', row[1], ',pwd=', row[2], ',singal=', row[3], '\n')
    print('begin delete')
```

```
        cursor.execute("delete from public.member where id=2")
        conn.commit()
        print('end delete')
        print("Total number of rows deleted :", cursor.rowcount)
        cursor.execute("select id,name,password,singal from public.member")
        rows = cursor.fetchall()
        for row in rows:
            print('id=', row[0], ',name=', row[1], ',pwd=', row[2], ',singal=', row[3], '\n')
        conn.close()

if __name__ == '__main__':
    # connectPostgreSQL()
    # insertOperate()
    # selectOperate()
    # updateOperate()
    deleteOperate()
```

执行后会显示删除前和删除后的数据库信息以进行对比。

```
id= 1 ,name= member0 ,pwd= password0 ,singal= singal0

id= 3 ,name= member2 ,pwd= password2 ,singal= singal2

id= 4 ,name= member3 ,pwd= password3 ,singal= singal3

id= 2 ,name= update ... ,pwd= password1 ,singal= singal1

begin delete
end delete
Total number of rows deleted : 1
id= 1 ,name= member0 ,pwd= password0 ,singal= singal0

id= 3 ,name= member2 ,pwd= password2 ,singal= singal2

id= 4 ,name= member3 ,pwd= password3 ,singal= singal3
```

下面的实例文件 PostgreSQL.py 演示了创建 PostgreSQL 数据表并分别实现插入、查询、删除和更新数据的过程。

源码路径：daima\5\5-2\PostgreSQL.py

```
import psycopg2

#连接数据库
conn = psycopg2.connect(dbname="mydb", user="postgres",
        password="6668888", host="127.0.0.1", port="5432")

#创建cursor以访问数据库
cur = conn.cursor()

#创建表
cur.execute(
        'CREATE TABLE stu ('
        'name    varchar(80),'
        'address varchar(80),'
        'age     int,'
        'date    date'
        ')'
    )

#插入数据
cur.execute("INSERT INTO stu "
        "VALUES('Gopher', 'China Beijing', 100, '2018-05-27')")

#查询数据
cur.execute("SELECT * FROM stu")
rows = cur.fetchall()
for row in rows:
    print('name=' + str(row[0]) + ' address=' + str(row[1]) +
        ' age=' + str(row[2]) + ' date=' + str(row[3]))
```

```
#更新数据
cur.execute("UPDATE stu SET age=12 WHERE name='Gopher'")

#删除数据
cur.execute("DELETE FROM stu WHERE name='Gopher'")

#提交事务
conn.commit()

#关闭连接
conn.close()
```

执行后会输出：

```
name=Gopher address=China Beijing age=100 date=2018-05-27
```

5.2.3 使用 queries 模块

queries 模块是对前面讲解的 psycopg2 库的封装，通过使用 queries 模块，Python 程序可以更加灵活地操作 Postgres 数据库。安装 queries 模块的具体命令如下。

```
pip install queries
```

下面的实例文件 simple-tornado.py 演示了使用 queries 模块查询指定 PostgreSQL 数据库中数据的过程。

源码路径：daima\5\5-2\simple-tornado.py

```python
import queries
with queries.Session('postgresql://postgres:66688888@127.0.0.1:5432/mydb') as session:
    for row in session.query('SELECT * FROM public.member'):
        print(row)
```

执行后会输出表 public.member 中的数据。

```
{'id': 1, 'name': 'member0', 'password': 'password0', 'singal': 'singal0'}
{'id': 3, 'name': 'member2', 'password': 'password2', 'singal': 'singal2'}
{'id': 4, 'name': 'member3', 'password': 'password3', 'singal': 'singal3'}
```

再看下面的实例文件 simple.py。这是一个 Tornado Web 程序，它演示了使用 queries 模块查询并显示指定 PostgreSQL 数据库中数据的过程。

源码路径：daima\5\5-2\simple.py

```python
import datetime
import logging

from queries import pool
import queries
from tornado import gen, ioloop, web


class ExampleHandler(web.RequestHandler):
    queries.Session('postgresql://postgres:66688888@127.0.0.1:5432/mydb')
    SQL = 'SELECT * FROM public.member'

    @gen.coroutine
    def get(self):
        try:
            result = yield self.application.session.query(self.SQL)
        except queries.OperationalError as error:
            logging.error('Error connecting to the database: %s', error)
            raise web.HTTPError(503)

        rows = []
        for row in result.items():
            row = dict([(k, v.isoformat()
                         if isinstance(v, datetime.datetime) else v)
                        for k, v in row.items()])
            rows.append(row)
        result.free()
        self.finish({'pg_stat_activity': rows})
```

```
class ReportHandler(web.RequestHandler):

    @gen.coroutine
    def get(self):
        self.finish(pool.PoolManager.report())


if __name__ == '__main__':
    logging.basicConfig(level=logging.DEBUG)
    application = web.Application([
        (r'/', ExampleHandler),
        (r'/report', ReportHandler)
    ], debug=True)
    application.session = queries.TornadoSession()
    application.listen(8000)
    ioloop.IOLoop.instance().start()
```

5.3 连接 SQLite3 数据库

在开发 Python 程序的过程中，可以使用 apsw 模块实现和 SQLite 数据库的连接和操作。apsw 模块是一个第三方库，实现对 SQLite 的封装。开发者可以通过如下命令安装 apsw。

```
pip install apsw
```

下面的实例文件 apsw01.py 演示了使用 apsw 模块创建并操作 SQLite 数据库数据的过程。

源码路径：daima\5\5-3\apsw01.py

```
import apsw
con=apsw.Connection(":memory:")
cur=con.cursor()
for row in cur.execute("create table foo(x,y,z);insert into foo values (?,?,?);"
                       "insert into foo values(?,?,?);select * from foo;drop table foo;"
                       "create table bar(x,y);insert into bar values(?,?);"
                       "insert into bar values(?,?);select * from bar;",
                       (1,2,3,4,5,6,7,8,9,10)):
    print(row)
```

执行后会输出添加到 SQLite 数据库中的数据。

```
(1, 2, 3)
(4, 5, 6)
(7, 8)
(9, 10)
```

下面的实例文件 apsw02.py 演示了使用 apsw 模块在 SQLite 数据库中同时批处理上千条数据的过程。

源码路径：daima\5\5-3\apsw02.py

```
import threading, apsw
import queue
import sys
class TestThr(threading.Thread):
    def __init__(self):
        threading.Thread.__init__(self)
        self.IQ = queue.Queue()
        self.OQ = queue.Queue()

    def run(self):
        try:
            print("*THREAD: Thread started")
            while self.IQ.empty(): pass
            self.IQ.get()
            print("*THREAD: <<< Prepare database")
            con = apsw.Connection('test.db')
            c = con.cursor()
            try:
                c.execute('create table a(a integer)')
                c.execute('end')
            except:
```

```
                pass
                c.execute('begin')
                c.execute('delete from a')
                c.execute('end')
                print("*THREAD: >>> Prepare database")

                self.OQ.put(1)
                while self.IQ.empty(): pass
                self.IQ.get()
                print("*THREAD: <<< Fillup 1000 values")
                c.execute('begin')
                print("*THREAD: Trans. started")
                for i in range(1000):
                        c.execute('insert into a values(%d)' % i)
                print("*THREAD: >>> Fillup 1000 values")
                self.OQ.put(1)
                while self.IQ.empty(): pass
                self.IQ.get()
                c.execute('end')
                print("*THREAD: Trans. finished")
                self.OQ.put(1)
                while self.IQ.empty(): pass
                self.IQ.get()
                print("*THREAD: <<< Fillup 1000 values")
                c.execute('begin')
                print("Trans. started")
                for i in range(1000, 2000):
                        c.execute('insert into a values(%d)' % i)
                print("*THREAD: >>> Fillup 1000 values")
                c.execute('end')
                print("*THREAD: Trans. finished")
                self.OQ.put(1)
                while self.IQ.empty(): pass
                self.IQ.get()
                print("*THREAD: Thread end")

                self.OQ.put(1)
            except:
                print(sys.exc_info())
                sys.exit()

con = apsw.Connection('test.db')
c = con.cursor()

t = TestThr()
t.IQ.put(1)
t.start()
while t.OQ.empty(): pass
t.OQ.get()

# c.execute('begin')
def ReadLastRec():
    rec = None
    for rec in c.execute('select * from a'): pass
    print("- MAIN: Read last record", rec)

ReadLastRec()
t.IQ.put(1)
while t.OQ.empty(): pass
t.OQ.get()
ReadLastRec()
t.IQ.put(1)
while t.OQ.empty(): pass
t.OQ.get()
ReadLastRec()
t.IQ.put(1)
while t.OQ.empty(): pass
t.OQ.get()
ReadLastRec()
```

```
t.IQ.put(1)
while t.OQ.empty(): pass
# c.execute('end')

print("\n- MAIN: Finished")
```
执行后会输出：
```
*THREAD: Thread started
*THREAD: <<< Prepare database
*THREAD: >>> Prepare database
- MAIN: Read last record None
*THREAD: <<< Fillup 1000 values
*THREAD: Trans. started
*THREAD: >>> Fillup 1000 values
- MAIN: Read last record None
*THREAD: Trans. finished
- MAIN: Read last record (999,)
*THREAD: <<< Fillup 1000 values
Trans. started
*THREAD: >>> Fillup 1000 values
*THREAD: Trans. finished
- MAIN: Read last record (1999,)
*THREAD: Thread end

- MAIN: Finished
```

5.4 连接 SQL Server 数据库

在开发 Python 程序的过程中，可以使用 pymssql 模块实现和 SQL Server 数据库的连接与操作。开发者可以通过如下命令安装 pymssql。

```
pip install pymssql
```

如果不能安装，可以登录 LFD 官网下载对应的 ".whl" 文件，然后通过如下命令进行安装。

```
python -m pip install -user ".whl文件名"
```

在 Python 程序中，使用 pymssql 模块的基本流程如下。

（1）连接数据库，设置连接参数。

（2）打开 cursor，执行 SQL 操作语句。

（3）通过 cursor 获取数据，可以一次获取所有数据，也可以一次获取一行。整个结果集是元组列表，是 list 类型，且每一条记录是一个元组。

（4）如果要增加或修改数据，就要调用函数 commit() 来提交事务；否则，程序会退出，数据库中的数据不会有变化。

（5）用函数 close() 关闭数据库连接，及时释放连接资源。

下面的实例演示了使用 pymssql 模块连接并操作 SQL Server 数据库中数据的过程。

（1）编写 SQL 文件 23.sql。具体代码如下。

源码路径：daima\5\5-4\23.sql

```
create database test;
go
use test;
go
if object_id('tb') is not null
    drop table tb;
go
CREATE TABLE TB(ID INT,NAME NVARCHAR(20),SCORE NUMERIC(10,2));
INSERT INTO TB(ID,NAME,SCORE)
VALUES(1,'语文',100),
    (2,'数学',80),
    (3,'英语',900),
    (4,'政治',65),
    (5,'物理',65),
```

5.4 连接 SQL Server 数据库

```
        (6,'化学',85),
        (7,'生物',55),
        (8,'地理',100)
```

在 SQL Server 数据库中打开上述 SQL 文件，单击"执行"按钮后会生成一个名为"tset"的数据库，并在里面添加了一条数据。

（2）编写程序文件 mssql01.py，使用 pymssql 模块连接并操作 SQL Server 数据库中的数据。具体实现代码如下。

源码路径：daima\5\5-4\mssql01.py

```python
import pymssql
#数据库连接
conn=pymssql.connect(host='DESKTOP-VMVTB06',user='sa',password='guanxijing',database='test')
#打开游标
cur=conn.cursor();
if not cur:
    raise Exception('数据库连接失败！')
sSQL = 'SELECT * FROM TB'
#执行SQL，获取所有数据
cur.execute(sSQL)
result=cur.fetchall()
#1.result是列表，其中的每个元素是元组
print(type(result),type(result[0]))
#2.
print('\n\n总行数: '+ str(cur.rowcount))
#3.通过enumerate返回行号
for i,(id,name,v) in enumerate(result):
    print('第 '+str(i+1)+' 行记录->>> '+ str(id) +':'+ name+ ':' + str(v) )
#4.修改数据
cur.execute("insert into tb(id,name,score) values(9,'历史',75)")
cur.execute("update tb set score=95 where id=7")
conn.commit() #修改数据后提交事务
#再查一次
cur.execute(sSQL)
#5.一次取一条数据,cur.rowcount为-1
r=cur.fetchone()
i=1
print('\n')
while r:
    id,name,v =r #r是一个元组
    print('第 '+str(i)+' 行记录->>> '+ str(id) +':'+ name+ ':' + str(v) )
    r=cur.fetchone()
    i+= 1
conn.close()
```

程序执行后会和指定 SQL Server 数据库建立连接，实现查询、更新和统计数据库表 TB 中数据的功能，并输出整个操作过程。

```
<class 'list'> <class 'tuple'>


总行数: 8
第 1 行记录->>> 1:语文:100.00
第 2 行记录->>> 2:数学:80.00
第 3 行记录->>> 3:英语:900.00
第 4 行记录->>> 4:政治:65.00
第 5 行记录->>> 5:物理:65.00
第 6 行记录->>> 6:化学:85.00
第 7 行记录->>> 7:生物:55.00
第 8 行记录->>> 8:地理:100.00


第 1 行记录->>> 1:语文:100.00
第 2 行记录->>> 2:数学:80.00
第 3 行记录->>> 3:英语:900.00
第 4 行记录->>> 4:政治:65.00
第 5 行记录->>> 5:物理:65.00
第 6 行记录->>> 6:化学:85.00
第 7 行记录->>> 7:生物:95.00
```

第 8 行记录->>> 8:地理:100.00
第 9 行记录->>> 9:历史:75.00

在上述实例中用到了如下所示的连接参数。

- user：用户名。
- password：密码。
- trusted：布尔值，指定是否使用 Windows 身份认证登录。
- host：主机名。
- database：数据库。
- timeout：查询超时。
- login_timeout：登录超时。
- charset：数据库的字符集。
- as_dict：布尔值，指定返回值是字典还是元组。
- max_conn：最大连接数。

而下面的实例文件 mssql02.py 演示了使用 pymssq 模块创建 SQL Server 数据库表并查询里面数据的过程。在编写 Python 程序之前，需要先在 SQL Server 数据库中创建一个名为"dyt"的数据库。程序文件 mssql02.py 的具体实现代码如下。

源码路径：daima\5\5-4\mssql02.py

```
import pymssql

# server表示数据库服务器名称或IP
# user 表示用户名
# password表示密码
# database表示数据库名称
conn = pymssql.connect('DESKTOP-VMVTB06', 'sa', 'guanxijing', 'dyt')
cursor = conn.cursor()
#新建、插入操作
cursor.execute("""
IF OBJECT_ID('persons', 'U') IS NOT NULL
DROP TABLE persons
CREATE TABLE persons (
id INT NOT NULL,
name VARCHAR(100),
salesrep VARCHAR(100),
PRIMARY KEY(id)
)
""")
cursor.executemany(
    "INSERT INTO persons VALUES (%d, %s, %s)",
    [(1, 'John Smith', 'John Doe'),
     (2, 'Jane Doe', 'Joe Dog'),
     (3, 'Mike T.', 'Sarah H.')])
#如果没有指定autocommit属性为True，就需要调用commit()方法
conn.commit()

#查询操作
cursor.execute('SELECT * FROM persons WHERE salesrep=%s', 'John Doe')
row = cursor.fetchone()
while row:
    print("ID=%d, Name=%s" % (row[0], row[1]))
row = cursor.fetchone()
#也可以使用for循环来迭代查询结果
for row in cursor:print("ID=%d, Name=%s" % (row[0], row[1]))
#关闭连接
conn.close()
```

执行后会输出：

```
D=1, Name=John Smith
ID=1, Name=John Smith
ID=1, Name=John Smith
ID=1, Name=John Smith
```

```
ID=1, Name=John Smith
ID=1, Name=John Smith
ID=1, Name=John Smith
ID=1, Name=John Smith
ID=1, Name=John Smith
ID=1, Name=John Smith
```

5.5 连接 NoSQL 数据库

本章前面介绍的 MySQL、SQL Server、SQLite 和 PostgreSQL 都是关系型数据库，而本节将要介绍的 NoSQL 是非关系型数据库。随着互联网 Web 2.0 网站的兴起，传统的关系数据库在应付 Web 2.0 网站，特别是超大规模和高并发的 SNS 类型的 Web 2.0 纯动态网站已经显得力不从心，暴露了很多难以克服的问题，而非关系型的数据库则由于其本身的特点得到了非常迅速的发展。NoSQL 数据库的产生就是为了解决大规模数据集合多重数据种类带来的挑战，尤其是大数据应用难题。本节将详细讲解 Python 常用的连接 NoSQL 数据库的第三方库。

5.5.1 使用 cassandra-driver 连接 Cassandra 数据库

Cassandra 是一套开源分布式 NoSQL 数据库系统。它最初由 Facebook 开发，用于储存收件箱等简单格式的数据，集 GoogleBigTable 的数据模型与 Amazon Dynamo 的完全分布式架构于一身。Facebook 于 2008 将 Cassandra 开源，此后，由于 Cassandra 良好的可扩展性，被 Digg、Twitter 等知名 Web 2.0 网站所采用，成为一种流行的分布式结构化数据存储方案。

在安装 Cassandra 数据库后，如果想在 Python 程序中使用 Cassandra 数据库，需要先通过如下命令安装 Cluster 模块。

```
pip install Cluster
```

然后通过如下命令安装 cassandra-driver 模块。

```
pip install cassandra-driver
```

下面的实例文件 cassandra01.py 演示了使用 cassandra-driver 连接 Cassandra 数据库并创建、插入数据的过程。

源码路径：daima\5\5-5\cassandra01.py

```
from cassandra.cluster import Cluster
cluster = Cluster(['127.0.0.1'])
session = cluster.connect()
session.execute("create KEYSPACE test_cassandra WITH replication = {'class':
'SimpleStrategy', 'replication_factor': 2};")
session.execute("use test_cassandra")
session.execute("create table users(id int, name text, primary key(id));")
session.execute("insert into users(id, name) values(1, 'I loving fish!');")
session.execute("insert into users(id, name) values(2, 'Zhang zhipeng');")
session = cluster.connect("test_cassandra")
rows = session.execute("select * from users;")
type(rows)
print(rows)
row = rows[0]
print(row.id)
print(row.name)
print(row.count(1))
print(row.count('I loving fish!'))
print(row.count('I loving fish??'))
print(cluster.is_shutdown)
cluster.shutdown()
print(cluster.is_shutdown)
```

执行后会输出：

```
[Row(id=1, name=u'I loving fish!'), Row(id=2, name=u'Zhang zhipeng')]
1
[Row(id=1, name=u'I loving fish!'), Row(id=2, name=u'Zhang zhipeng')]
1
```

```
1
0
False
True
```

5.5.2 使用 PyMongo 驱动连接 MongoDB 数据库

MongoDB 是一个基于分布式文件存储的数据库，用 C++编写，旨在为 Web 应用提供可扩展的高性能数据存储解决方案。MongoDB 是一个介于关系数据库和非关系数据库之间的产品，是非关系数据库中功能最丰富并且最像关系数据库的。下面将详细讲解在 Python 程序中使用 MongoDB 数据库的知识。

1. 搭建 MongoDB 环境

MongoDB 官网提供了可用于 32 位和 64 位系统的预编译二进制包，读者可以从 MongoDB 官网下载安装包，如图 5-13 所示。

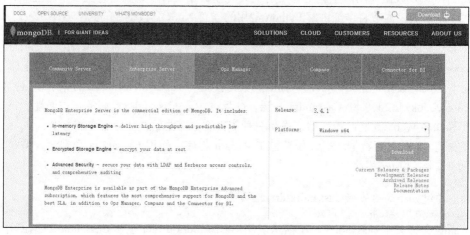

图 5-13　MongoDB 下载页面

根据当前计算机的操作系统选择下载安装包，因为作者使用的是 64 位的 Windows 系统，所以选择"Windows x64"，然后单击"Download"按钮。在弹出的界面中选择"msi"，如图 5-14 所示。

下载完成后得到一个".msi"文件，双击这个文件，然后按照操作提示进行安装即可。安装界面如图 5-15 所示。

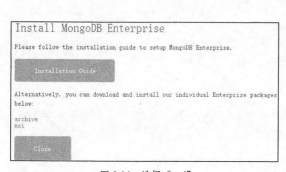

图 5-14　选择"msi"

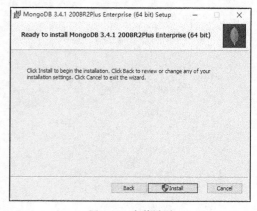

图 5-15　安装界面

5.5 连接 NoSQL 数据库

2. 在 Python 程序中使用 MongoDB 数据库

在 Python 程序中使用 MongoDB 数据库时，必须首先确保安装了 PyMongo 这个第三方库。如果下载的是 "exe" 格式的安装文件，则可以直接安装。对于压缩包形式的安装文件，可以使用以下命令进行安装。

```
pip install pymongo
```

如果没有下载安装文件，可以通过如下命令进行在线安装。

```
easy_install pymongo
```

安装完成后的界面如图 5-16 所示。

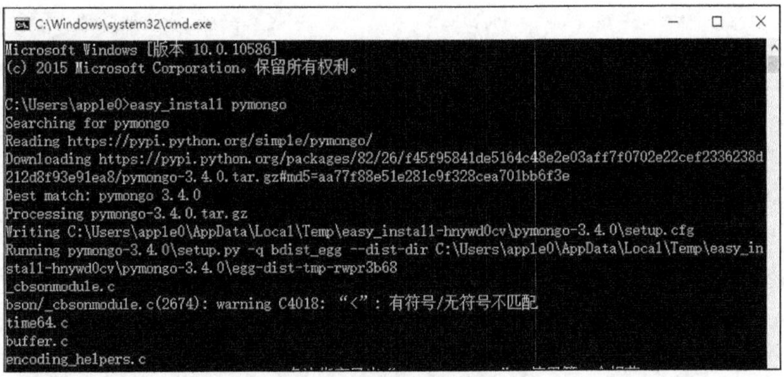

图 5-16　安装完成后的界面

下面的实例文件 mdb.py 演示了在 Python 程序中使用 MongoDB 数据库的过程。

源码路径：daima\5\5-5\mdb.py

```
from pymongo import MongoClient
import random
…省略部分代码…
if __name__ == '__main__':
    print("建立连接...")                    #输出提示信息
    stus = MongoClient().test.stu           #建立连接
    print('插入一条记录...')                 #输出提示信息
    #向表stu中插入一条数据
    stus.insert({'name':get_str(2,4),'passwd':get_str(8,12)})
    print("显示所有记录...")                 #输出提示信息
    stu = stus.find_one()                   #获取数据库的信息
    print(stu)                              #显示数据库中的数据
    print('批量插入多条记录...')             #输出提示信息
    stus.insert(get_data_list(3))           #向表stu中插入多条数据
    print('显示所有记录...')                 #输出提示信息
    for stu in stus.find():                 #遍历数据信息
        print(stu)                          #显示数据库中的数据
    print('更新一条记录...')                 #输出提示信息
    name = input('请输入记录的name:')#提示输入要修改的数据
    #修改表stu中的一条数据
    stus.update({'name':name},{'$set':{'name':'langchao'}})
    print('显示所有记录...')                 #输出提示信息
    for stu in stus.find():                 #遍历数据
        print(stu)                          #显示数据库中的数据信息
    print('删除一条记录...')                 #输出提示信息
    name = input('请输入记录的name:')#提示输入要删除的数据
    stus.remove({'name':name})              #删除表中的数据
    print('显示所有记录...')                 #输出提示信息
    for stu in stus.find():                 #遍历数据信息
        print(stu)                          #显示数据库中的数据
```

在上述实例代码中，使用两个函数生成字符串。在主程序中首先连接集合，然后使用集合对象的方法对集合中的文档进行插入、更新和删除操作。每当数据被修改后，会显示集合中所有文档，以验证操作结果的正确性。

在运行本实例时，初学者很容易遇到如下 Mongo 运行错误。

```
Failed to connect 127.0.0.1:27017,reason:errno:10061由于目标计算机拒绝，无法连接...
```

出现上述错误的原因是没有开启 MongoDB 服务。下面是开启 MongoDB 服务的命令。

```
mongod --dbpath "h:\data"
```

在上述命令中，"h:\data" 是一个保存 MongoDB 数据库中数据的目录，读者可以随意在本地计算机硬盘中创建，还可以自定义目录名字。在 CMD 控制台界面中，成功开启 MongoDB 服务时的界面如图 5-17 所示。

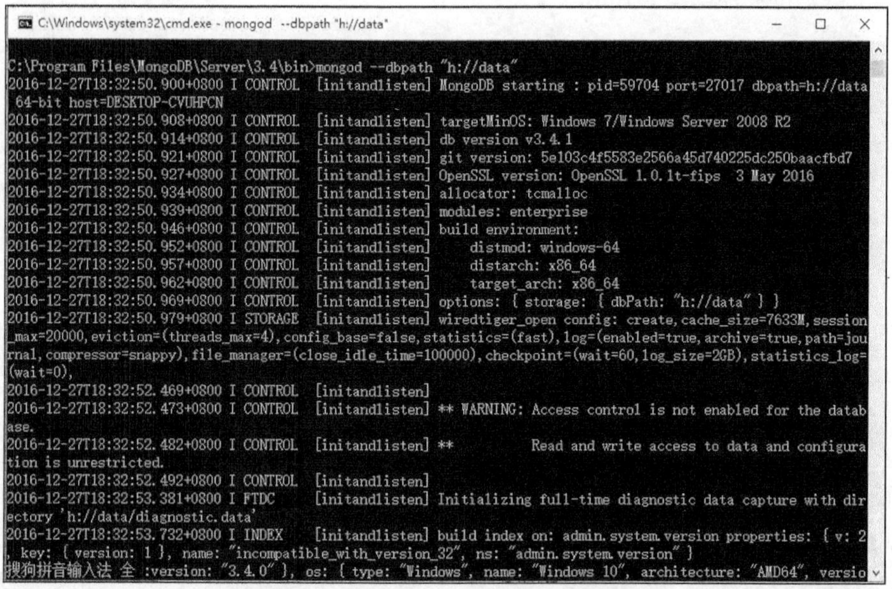

图 5-17　成功开启 MongoDB 服务时的界面

在运行本实例程序时，必须在 CMD 控制台中启动 MongoDB 服务，并且确保上述控制台界面处于打开状态。本实例执行后会输出：

```
建立连接...
插入一条记录...
显示所有记录...
{'_id': ObjectId('586243795cd071f570ed3b39'), 'name': 'vvtj', 'passwd': 'iigbddauwj'}
批量插入多条记录...
显示所有记录...
{'_id': ObjectId('586243795cd071f570ed3b39'), 'name': 'vvtj', 'passwd': 'iigbddauwj'}
{'_id': ObjectId('5862437a5cd071f570ed3b3a'), 'name': 'nh', 'passwd': 'upyufzknzgdc'}
{'_id': ObjectId('5862437a5cd071f570ed3b3b'), 'name': 'rgf', 'passwd': 'iqdlyjhztq'}
{'_id': ObjectId('5862437a5cd071f570ed3b3c'), 'name': 'dh', 'passwd': 'rgupzruqb'}
{'_id': ObjectId('586243e45cd071f570ed3b3e'), 'name': 'hcq', 'passwd': 'chiwwvxs'}
{'_id': ObjectId('586243e45cd071f570ed3b3f'), 'name': 'yrp', 'passwd': 'kiocdmeerneb'}
{'_id': ObjectId('586243e45cd071f570ed3b40'), 'name': 'hu', 'passwd': 'pknqgfnm'}
{'_id': ObjectId('5862440d5cd071f570ed3b43'), 'name': 'tlh', 'passwd': 'cikouuladgqn'}
{'_id': ObjectId('5862440d5cd071f570ed3b44'), 'name': 'qxf', 'passwd': 'jlsealrqeeel'}
{'_id': ObjectId('5862440d5cd071f570ed3b45'), 'name': 'vlzp', 'passwd': 'wolypmej'}
{'_id': ObjectId('58632e6c5cd07155543cc27a'), 'sid': 2, 'name': 'sgu', 'passwd': 'ogzvdq'}
{'_id': ObjectId('58632e6c5cd07155543cc27b'), 'sid': 3, 'name': 'jiyl', 'passwd': 'atgmhmxr'}
{'_id': ObjectId('58632e6c5cd07155543cc27c'), 'sid': 4, 'name': 'dbb', 'passwd': 'wmwoeua'}
{'_id': ObjectId('5863305b5cd07155543cc27d'), 'sid': 27, 'name': 'langchao', 'passwd': '123123'}
{'_id': ObjectId('5863305b5cd07155543cc27e'), 'sid': 28, 'name': 'oxp', 'passwd': 'acgjph'}
{'_id': ObjectId('5863305b5cd07155543cc27f'), 'sid': 29, 'name': 'sukj', 'passwd': 'hjtcjf'}
{'_id': ObjectId('5863305b5cd07155543cc280'), 'sid': 30, 'name': 'bf', 'passwd': 'cqerluvk'}
{'_id': ObjectId('5988087533fda81adc0d332f'), 'name': 'hg', 'passwd': 'gmflqxfaxxnv'}
{'_id': ObjectId('5988087533fda81adc0d3330'), 'name': 'ojb', 'passwd': 'rgxodvkprm'}
{'_id': ObjectId('5988087533fda81adc0d3331'), 'name': 'gtdj', 'passwd': 'zigavkysc'}
{'_id': ObjectId('5988087533fda81adc0d3332'), 'name': 'smgt', 'passwd': 'sizvlhdll'}
{'_id': ObjectId('5a33c1cb33fda859b82399d0'), 'name': 'dbu', 'passwd': 'ypdxtqjjafsm'}
```

```
{'_id': ObjectId('5a33c1cb33fda859b82399d1'), 'name': 'qg', 'passwd': 'frnoypez'}
{'_id': ObjectId('5a33c1cb33fda859b82399d2'), 'name': 'ky', 'passwd': 'jvzjtcfs'}
{'_id': ObjectId('5a33c1cb33fda859b82399d3'), 'name': 'glnt', 'passwd': 'ejrerztki'}
更新一条记录...
请输入记录的name:
```

5.5.3 使用 redis-py 连接 Redis

Redis 是一个 "key-value" 类型的存储系统，为开发者提供了丰富的数据结构，包括列表、集合、散列，还包括了关于这些数据结构的丰富操作。Redis 和 Memcached 类似，但是它支持存储的 value 类型相对更多，包括字符串、列表、集合、有序集合和散列类型。这些数据类型都支持 push/pop、add/remove 及取交集、并集和差集，乃至更丰富的操作，而且这些操作都是原子性的。在此基础上，Redis 支持各种不同方式的排序。与 Memcached 一样，为了保证运行效率，数据都缓存在内存中。区别是 Redis 会周期性地把更新的数据写入磁盘或者把修改操作写入追加的记录文件，并且在此基础上实现了主从(master-slave)同步。

在 Python 程序中使用 Redis 时，需要使用 redis-py 驱动连接 Redis 的 Python 客户端。具体安装命令如下。

```
pip install redis
```

也可以通过如下命令安装。

```
easy_install redis
```

Redis 提供了两个类——Redis 和 StrictRedis 用于实现 Redis 的命令。其中 StrictRedis 用于实现大部分官方命令，并使用官方的语法和命令，Redis 是 StrictRedis 的子类。下面的实例文件 r01.py 演示了使用 Redis 连接服务器的过程。

源码路径：daima\5\5-5\r01.py

```
import redis

r = redis.Redis(host='127.0.0.1', port=6379, db=0)
r.set('name', 'zhangsan')   # 添加
print(r.get('name'))        # 获取
```

运行后会输出：

```
b'zhangsan'
```

读者需要注意，在运行上述实例代码之前，必须确保已经启动了 Redis。启动命令如下。

```
redis-server redis.windows.conf
```

另外，官方还提供了针对 Windows 系统的开源版本，读者下载后可以用微软的 Visual Studio 运行。成功启动 Redis 后的界面如图 5-18 所示。

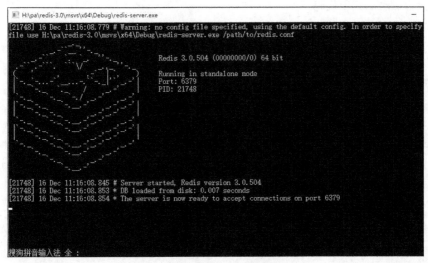

图 5-18 成功启动 Redis 后的界面

Redis 使用 ConnectionPool 来管理一个 Redis Server 的所有连接，避免每次建立、释放连接的开销。在默认情况下，每个 Redis 实例都会维护一个自己的连接池。我们可以直接建立一个连接池，然后作为参数 Redis，这样就可以使多个 Redis 实例共享一个连接池。下面的实例文件 r02.py 演示了使用 ConnectionPool 创建连接池的过程。

源码路径：daima\5\5-5\r02.py

```python
import redis

pool = redis.ConnectionPool(host='127.0.0.1', port=6379)
r = redis.Redis(connection_pool=pool)
r.set('name', 'zhangsan')
print(r.get('name'))
```

执行后会输出：

```
b'zhangsan'
```

Redis 也可以实现"发布-订阅"模式。接下来，定义一个 RedisHelper 类，连接 Redis，定义频道为 monitor，定义发布及订阅方法。具体实现代码如下。

源码路径：daima\5\5-5\RedisHelper.py

```python
import redis

class RedisHelper(object):
    def __init__(self):
        self.__conn = redis.Redis(host='127.0.0.1', port=6379)   # 连接Redis
        self.channel = 'monitor'   # 定义名称

    def publish(self, msg):   # 定义发布方法
        self.__conn.publish(self.channel, msg)
        return True

    def subscribe(self):   # 定义订阅方法
        pub = self.__conn.pubsub()
        pub.subscribe(self.channel)
        pub.parse_response()
        return pub
```

下面是发布者的实现代码。

源码路径：daima\5\5-5\r03.py

```python
from RedisHelper import RedisHelper

obj = RedisHelper()
obj.publish('hello')   # 发布
```

下面是订阅者的实现代码。

源码路径：daima\5\5-5\r04.py

```python
from RedisHelper import RedisHelper

obj = RedisHelper()
redis_sub = obj.subscribe()   # 调用订阅方法

while True:
    msg = redis_sub.parse_response()
    print(msg)
```

在 Redis 应用中，delete 命令用于删除已存在的键，不存在的键会被忽略。通过使用 exists 命令可以判断键是否存在。下面的实例文件 r05.py 演示了使用 delete 和 exists 命令的过程。

源码路径：daima\5\5-5\r05.py

```python
import redis
#这个Redis 连接不能用，请根据自己的需要修改
r =redis.Redis(host="127.0.0.1",port=6379)
print(r.set('1', '4028b2883d3f5a8b013d57228d760a93'))
print(r.get('1'))
print(r.delete('1'))
print(r.get('1'))
```

```
#设定键2的值是  4028b2883d3f5a8b013d57228d760a93
r.set('2', '4028b2883d3f5a8b013d57228d760a93')
# 存在就返回True 不存在就返回False
print(r.exists('2'))    #返回True
print(r.exists('33'))   #返回False
```
执行后会输出：
```
True
b'4028b2883d3f5a8b013d57228d760a93'
1
None
True
False
```

在 Redis 应用中，expire 命令用于设置键的过期时间，键过期后将不再可用。expireat 命令用于以 UNIX 时间戳格式设置键的过期时间，键过期后将不再可用。读者需要注意，时间精确到秒，时间戳是 10 位数字。下面的实例文件 r06.py 演示了使用 expire 和 expireat 命令的过程。

源码路径：daima\5\5-5\r06.py

```
import redis
# 这个Redis 连接不能用，请根据自己的需要修改
r = redis.Redis(host="127.0.0.1", port=6379)
r.set('2', '4028b2883d3f5a8b013d57228d760a93')
#如果成功就返回True，如果失败就返回False，下面的20表示20s
print(r.expire('2',20))
#如果没有过期我们能得到键2的值；否则，返回None
print(r.get('2'))
r.set('2', '4028b2883d3f5a8b013d57228d760a93')
#成功就返回True 失败就返回False，下面的1598033936表示在2020-08-22 02:18:56键2过期
print(r.expireat('2',1598033936))
print(r.get('2'))
```
执行后会输出：
```
True
b'4028b2883d3f5a8b013d57228d760a93'
True
b'4028b2883d3f5a8b013d57228d760a93'
```

在 Redis 应用中，keys 命令用于查找所有符合给定模式的键。persist 命令用于移除给定键的过期时间，使得键永不过期；move 命令用于将当前数据库的键移动到给定的数据库 db 中；select 可以设定当前的数据库。下面的实例文件 r07.py 演示了使用 persist、keys 和 move 命令的过程。

源码路径：daima\5\5-5\r07.py

```
import redis
# 这个Redis 连接不能用，请根据自己的需要修改
r = redis.Redis(host="127.0.0.1", port=6379)
print (r.set('111', '11'))
print (r.set('122', '12'))
print (r.set('113', '13'))
print (r.keys(pattern='11*'))
# 输出的结果是 ['111', '112']，因为键122和11*不匹配
r.move(2,1)#将键2移动到数据库 1 中去
#设定键1的值为11
print (r.set('1', '11'))
#设定键1的过期时间为100s
print (r.expire(1,100))
# 查看键1的过期时间还剩下多少
print (r.ttl('1'))
# 目的是13s后移除键1的过期时间
import time
time.sleep(3)
# 查看键1的过期时间还剩下多少
print (r.ttl('1'))
#移除键1的过期时间
r.persist(1)
# 查看键1的过期时间还剩下多少，输出的结果是 None，可以通过redis desktop manager 查看键1的过期时间
print (r.ttl('1'))
```

执行后会输出：
```
True
True
True
[b'111', b'113']
True
True
100
97
None
```

在 Redis 应用中还有很多其他重要的命令，具体说明见表 5-1。

表 5-1　　　　　　　　　　　　Redis 的常用命令

命令	描述
Redis DEL	该命令用于在键存在时删除键
Redis Dump	序列化给定键，并返回被序列化的值
Redis EXISTS	检查给定键是否存在
Redis Expire	为给定键设置过期时间
Redis Expireat	EXPIREAT 的作用和 EXPIRE 类似，都用于为键设置过期时间。不同之处在于 EXPIREAT 命令接受的时间参数是 UNIX 时间戳
Redis PEXPIREAT	设置键的过期时间，以毫秒计
Redis PEXPIREAT	设置键过期时间的时间戳，以毫秒计
Redis Keys	查找所有符合给定模式的键
Redis Move	将当前数据库的键移动到给定的数据库 db 当中
Redis PERSIST	移除键的过期时间，键将持久保存
Redis Pttl	以毫秒为单位返回键的剩余过期时间
Redis TTL	以秒为单位，返回给定键的剩余生存时间（time to live, TTL）
Redis RANDOMKEY	从当前数据库中随机返回一个键
Redis Rename	修改键的名称
Redis Renamenx	仅当新建不存在时，将键改名为新建
Redis Type	返回键所存储的值的类型

第 6 章

使用 ORM 操作数据库

ORM 是对象关系映射（Object Relational Mapping）的简称，用于实现面向对象编程语言中不同类型系统的数据之间的转换。本章将详细讲解在 Python 程序中使用 ORM 操作数据库的知识。

6.1　ORM 的背景

从实现效果来看，ORM 其实创建了一个可以在编程语言里使用的"虚拟对象数据库"。从另外的角度来看，在面向对象编程语言中使用的是对象，而对象中的数据需要保存到数据库中，或数据库中的数据用来构造对象。在从数据库中提取数据并构造对象或将对象数据存入数据库的过程中，有很多代码是可以重复使用的，如果这些重复的功能完全自己实现，那就是在"重复造轮子"。在这种情况下就诞生了 ORM，它使得从数据库中提取数据来构造对象或将对象数据保存（持久化）到数据库中更简单。

在实际应用中有很多不同的数据库工具，并且其中的大部分系统都包含 Python 接口，能够使开发者更好地利用它们的功能。这些不同数据库工具系统的唯一缺点是需要了解 SQL 语言。如果你是一个更愿意使用 Python 对象而不是 SQL 查询的程序员，并且仍然希望使用关系数据库作为程序的数据后端，那么可能会更加倾向于使用 ORM。

这些 ORM 系统的创始人将纯 SQL 语句进行了抽象化处理，将其实现为 Python 中的对象，这样开发者只要操作这些对象就能完成与生成 SQL 语句相同的任务。一些软件系统也允许一定的灵活性，可以让我们执行几行 SQL 语句。但在大多数情况下，应该避免使用普通的 SQL 语句。

在 ORM 系统中，数据库表被转化为 Python 类，其中的数据列作为属性，而数据库操作则作为方法。读者应该会发现，让应用支持 ORM 与使用标准数据库适配器有些相似。因为 ORM 需要代替你执行很多工作，所以一些事情变得更加复杂，或者比直接使用适配器需要更多的代码行。不过，值得欣慰的是，这一点额外工作可以获得更高的开发效率。

在开发过程中，最著名的 Python ORM 是 SQLAlchemy 和 SQLObject。另外一些常用的 Python ORM 还包括 Storm、PyDO/PyDO2、PDO、Dejavu、Durus、QLime 和 ForgetSQL。基于 Web 的大型系统也会包含它们自己的 ORM 组件，如 WebWare MiddleKit 和 Django 的数据库 API。读者需要注意的是，并不是所有知名的 ORM 都适用于你的应用程序，你需要根据自己的需求来选择。

6.2　使用 mysqlclient 连接数据库

在 Python 程序中，SQLAlchemy 是一种经典的 ORM。在使用 SQLAlchemy 之前需要先安装，安装命令如下。

```
easy_install SQLAlchemy
```

成功安装后的界面如图 6-1 所示。

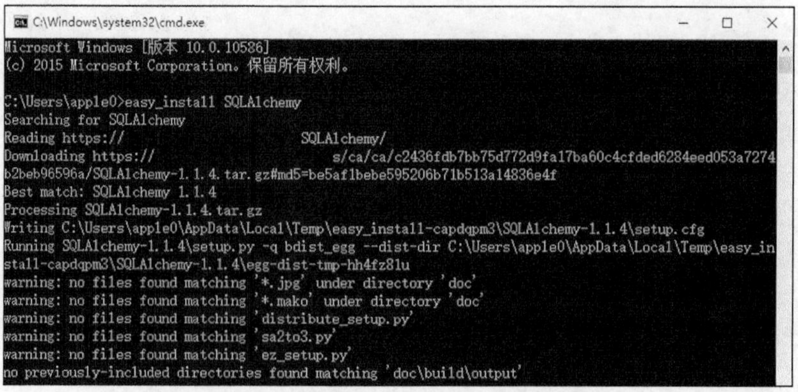

图 6-1　成功安装 SQLAlchemy 后的界面

下面的实例文件 SQLAlchemy.py 演示了在 Python 程序中使用 SQLAlchemy 操作 MySQL 和 SQLite 两种数据库的过程。

源码路径：daima\6\6-2\SQLAlchemy.py

```python
from distutils.log import warn as printf
from os.path import dirname
from random import randrange as rand
from sqlalchemy import Column, Integer, String, create_engine, exc, orm
from sqlalchemy.ext.declarative import declarative_base
from db import DBNAME, NAMELEN, randName, FIELDS, tformat, cformat, setup
DSNs = {
    'mysql': 'mysql://root@localhost/%s' % DBNAME,
    'sqlite': 'sqlite:///:memory:',
}
Base = declarative_base()
class Users(Base):
    __tablename__ = 'users'
    login  = Column(String(NAMELEN))
    userid = Column(Integer, primary_key=True)
    projid = Column(Integer)
    def __str__(self):
        return ''.join(map(tformat,
            (self.login, self.userid, self.projid)))
class SQLAlchemyTest(object):
    def __init__(self, dsn):
        try:
            eng = create_engine(dsn)
        except ImportError:
            raise RuntimeError()
        try:
            eng.connect()
        except exc.OperationalError:
            eng = create_engine(dirname(dsn))
            eng.execute('CREATE DATABASE %s' % DBNAME).close()
            eng = create_engine(dsn)
        Session = orm.sessionmaker(bind=eng)
        self.ses = Session()
        self.users = Users.__table__
        self.eng = self.users.metadata.bind = eng
    def insert(self):
        self.ses.add_all(
            Users(login=who, userid=userid, projid=rand(1,5)) \
            for who, userid in randName()
        )
        self.ses.commit()
    def update(self):
        fr = rand(1,5)
        to = rand(1,5)
        i = -1
        users = self.ses.query(
            Users).filter_by(projid=fr).all()
        for i, user in enumerate(users):
            user.projid = to
        self.ses.commit()
        return fr, to, i+1
    def delete(self):
        rm = rand(1,5)
        i = -1
        users = self.ses.query(
            Users).filter_by(projid=rm).all()
        for i, user in enumerate(users):
            self.ses.delete(user)
        self.ses.commit()
        return rm, i+1
    def dbDump(self):
        printf('\n%s' % ''.join(map(cformat, FIELDS)))
        users = self.ses.query(Users).all()
        for user in users:
```

```
                printf(user)
            self.ses.commit()
        def __getattr__(self, attr):
            return getattr(self.users, attr)
        def finish(self):
            self.ses.connection().close()
    def main():
        printf('*** Connect to %r database' % DBNAME)
        db = setup()
        if db not in DSNs:
            printf('\nERROR: %r not supported, exit' % db)
            return
        try:
            orm = SQLAlchemyTest(DSNs[db])
        except RuntimeError:
            printf('\nERROR: %r not supported, exit' % db)
            return
        printf('\n*** Create users table (drop old one if appl.)')
        orm.drop(checkfirst=True)
        orm.create()
        printf('\n*** Insert names into table')
        orm.insert()
        orm.dbDump()
        printf('\n*** Move users to a random group')
        fr, to, num = orm.update()
        printf('\t(%d users moved) from (%d) to (%d)' % (num, fr, to))
        orm.dbDump()
        printf('\n*** Randomly delete group')
        rm, num = orm.delete()
        printf('\t(group #%d; %d users removed)' % (rm, num))
        orm.dbDump()
        printf('\n*** Drop users table')
        orm.drop()
        printf('\n*** Close cxns')
        orm.finish()
    if __name__ == '__main__':
        main()
```

在上述实例代码中，首先导入了 Python 标准库中的模块（distutils、os.path、random），然后是第三方或外部模块（sqlalchemy），最后是应用的本地模块（db）。本地模块会给我们提供主要的常量和工具函数。

代码中使用了 SQLalchemy 的声明层，在使用前必须先导入 sqlalchemy.ext.declarative.declarative_base，然后使用它创建一个 Base 类，最后让数据子类继承自这个 Base 类。类定义的下一个部分包含了一个 __tablename__ 属性，它定义了映射的数据库表名。也可以显式地定义一个低级别的 sqlalchemy.Table 对象，在这种情况下需要将其写为 __table__。在大多数情况下使用对象进行数据行的访问，不过也会使用表级别的行为保存表。接下来是"列"属性，可以通过查阅文档来获取所有支持的数据类型。最后有一个 __str()__ 方法，它用来返回易于阅读的数据行的字符串格式。因为该输出是定制化的（通过 tformat()函数的协助），所以不推荐在开发过程中这样使用。

通过自定义函数分别实现行的插入、更新和删除操作。插入使用了 session.add_all()方法，这将使用迭代的方式产生一系列的插入操作。最后，还可以决定是进行提交还是进行回滚。update()和 delete()方法都存在会话查询的功能，它们使用 query.filter_by()方法进行查找。随机更新会选择一个成员，通过改变 ID 的方法，将其从一个项目组（fr）移动到另一个项目组（to）。计数器（i）会记录有多少用户受到影响。删除操作则根据 ID（rm）随机选择一个项目并假设已将其取消，因此项目中的所有员工都将被解雇。当要执行操作时，需要通过会话对象进行提交。

函数 dbDump()负责向屏幕上显示正确的输出。该方法从数据库中获取数据行，并按照 db.py 中相似的样式输出数据。

6.2 使用 mysqlclient 连接数据库

本实例执行后会输出：

```
Choose a database system:

(M)ySQL
(G)adfly
(S)SQLite

Enter choice: S

*** Create users table (drop old one if appl.)

*** Insert names into table

LOGIN       USERID      PROJID
Faye        6812        4
Serena      7003        1
Amy         7209        2
Dave        7306        3
Larry       7311        3
Mona        7404        3
Ernie       7410        3
Jim         7512        3
Angela      7603        3
Stan        7607        3
Jennifer    7608        1
Pat         7711        1
Leslie      7808        4
Davina      7902        4
Elliot      7911        1
Jess        7912        4
Aaron       8312        3
Melissa     8602        4

*** Move users to a random group
    (1 users moved) from (2) to (1)

LOGIN       USERID      PROJID
Faye        6812        4
Serena      7003        1
Amy         7209        1
Dave        7306        3
Larry       7311        3
Mona        7404        3
Ernie       7410        3
Jim         7512        3
Angela      7603        3
Stan        7607        3
Jennifer    7608        1
Pat         7711        1
Leslie      7808        4
Davina      7902        4
Elliot      7911        1
Jess        7912        4
Aaron       8312        3
Melissa     8602        4

*** Randomly delete group
    (group #1; 5 users removed)

LOGIN       USERID      PROJID
Faye        6812        4
Dave        7306        3
Larry       7311        3
Mona        7404        3
Ernie       7410        3
Jim         7512        3
Angela      7603        3
Stan        7607        3
Leslie      7808        4
```

```
Davina      7902       4
Jess        7912       4
Aaron       8312       3
Melissa     8602       4

*** Drop users table

*** Close cxns
```

6.3 使用 Peewee 连接数据库

Peewee 是一款简单、轻巧的 Python ORM，支持的数据库有 SQLite、MySQL 和 PostgreSQL。本节将详细讲解使用 Peewee 连接数据库的基本知识。

6.3.1 Peewee 的基本用法

开发者可以通过如下命令安装 Peewee。

```
pip install Peewee
```

在使用 Peewee 连接数据库时，推荐使用 playhouse 中的 db_url 模块。db_url 的 connect 方法可以通过传入的 URL 字符串生成数据库连接。方法 connect(url, **connect_params)的功能是通过传入的 url 字符串创建一个数据库实例。参数 url 的形式有以下两种。

- mysql://user:passwd@ip:port/my_db：将创建一个本地 MySQL 的 my_db 数据库的实例。
- mysql+pool://user:passwd@ip:port/my_db?charset=utf8&max_connections=20&stale_timeout=300：将创建一个本地 MySQL 的 my_db 连接池，最大连接数为 20，超时时间为 300s。

下面的实例文件 Peewee01.py 演示了在 Python 程序中使用 Peewee 操作 SQLite 数据库的过程。

源码路径：daima\6\6-3\Peewee01.py

```python
from datetime import date
from peewee import *

db = SqliteDatabase('people.db')

'''模型定义'''
class Person(Model):
    name = CharField()
    birthday = DateField()
    is_relative = BooleanField()

    class Meta:
        database = db  #这个模型使用了"people.db"数据库

class Pet(Model):
    owner = ForeignKeyField(Person, related_name='pets')
    name = CharField()
    animal_type = CharField()

    class Meta:
        database = db  #这个模型使用了"people.db"数据库

"""连接数据库"""
db.connect()
"""创建Person和Pet表"""
db.create_tables([Person, Pet])

uncle_bob = Person(name='Bob', birthday=date(1960, 1, 15), is_relative=True)
uncle_bob.save()

"""连接数据库关闭"""
db.close()
```

通过上述代码创建了一个名为"people.db"的 SQLite 数据库，并且在里面创建了表 Person 和表 Pet。下面的实例文件 Peewee02.py 演示了，在 Python 程序中使用 Peewee 更新和删除上述

"people.db"数据库中数据的过程。

源码路径：daima\6\6-3\Peewee02.py

```python
from datetime import date
from peewee import *

db = SqliteDatabase('people.db')

class Person(Model):
    name = CharField()
    birthday = DateField()
    is_relative = BooleanField()
    class Meta:
        database = db     #用了"people.db"数据库

class Pet(Model):
    owner = ForeignKeyField(Person, related_name='pets')
    name = CharField()
    animal_type = CharField()
    class Meta:
      database = db     #用了"people.db"数据库

"""----------------------------------------------------------------------"""
uncle_bob = Person(name='Bob', birthday=date(1960, 1, 15), is_relative=True)
uncle_bob.save()

grandma = Person.create(name='Grandma', birthday=date(1935, 3, 1), is_relative=True)
herb = Person.create(name='Herb', birthday=date(1950, 5, 5), is_relative=False)
grandma.name = 'Grandma L.'
grandma.save()      #更新数据库中grandma的名字

bob_kitty = Pet.create(owner=uncle_bob, name='Kitty', animal_type='cat')
herb_fido = Pet.create(owner=herb, name='Fido', animal_type='dog')
herb_mittens = Pet.create(owner=herb, name='Mittens', animal_type='cat')
herb_mittens_jr = Pet.create(owner=herb, name='Mittens Jr', animal_type='cat')
"""----------------------------------------------"""
herb_mittens.delete_instance()     #删除
""""""
herb_fido.owner = uncle_bob
herb_fido.save()
bob_fido = herb_fido
```

下面的实例文件 Peewee03.py 演示了，在 Python 程序中使用 Peewee 查询"people.db"数据库中指定范围内数据的过程。

源码路径：daima\6\6-3\Peewee03.py

```python
from datetime import date
from peewee import *

db = SqliteDatabase('people.db')

class Person(Model):
    name = CharField()
    birthday = DateField()
    is_relative = BooleanField()
    class Meta:
        database = db     #用了"people.db"数据库

class Pet(Model):
    owner = ForeignKeyField(Person, related_name='pets')
    name = CharField()
    animal_type = CharField()
    class Meta:
      database = db     #用了"people.db"数据库
"""----------------------------------------------------------------------"""

#获取单个数据记录
grandma = Person.select().where(Person.name == 'Grandma L.').get()
"""同上"""
```

```python
grandma = Person.get(Person.name == 'Grandma L.')

#获取数据列表

for person in Person.select():
    print("人名:",person.name, person.is_relative)

query = Pet.select().where(Pet.animal_type == 'cat')
for pet in query:
    print("宠物名:",pet.name, "主人名:",pet.owner.name)


#连接查询

query = (Pet.select(Pet, Person)
         .join(Person)
         .where(Pet.animal_type == 'cat'))
for pet in query:
    print(pet.name, pet.owner.name)



#获取Bob拥有的所有宠物
for pet in Pet.select().join(Person).where(Person.name == 'Bob'):
    print(pet.name)

"""同上"""
uncle_bob = Person(name='Bob', birthday=date(1960, 1, 15), is_relative=True)
for pet in Pet.select().where(Pet.owner == uncle_bob).order_by(Pet.name):
    print(pet.name)

#按日期排序
for person in Person.select().order_by(Person.birthday.desc()):
    print(person.name, person.birthday)
"""--------------------------------------------------------------------"""
for person in Person.select():
    print(person.name, person.pets.count(), 'pets')
    for pet in person.pets:
        print ('    ', pet.name, pet.animal_type)

"""--------------------------------------------------------------------"""
#查询条件
d1940 = date(1940, 1, 1)
d1960 = date(1960, 1, 1)

#查询1960年之后出生和1940年之前出生的人
query = (Person.select()
         .where((Person.birthday < d1940) | (Person.birthday > d1960)))
for person in query:
    print(person.name, person.birthday)
#查询生日在1940年与1960年之间的人
query = (Person
         .select()
         .where((Person.birthday > d1940) & (Person.birthday < d1960)))
for person in query:
    print(person.name, person.birthday)
"""-----------------------------------------------------------"""
#查询以g开头的人名
expression = (fn.Lower(fn.Substr(Person.name, 1, 1)) == 'g')
for person in Person.select().where(expression):
    print(person.name)

"""连接数据库"""
db.close()
```

执行后会输出各种条件的查询结果：

```
人名: Bob True
人名: Bob True
人名: Grandma L. True
人名: Herb False
```

```
宠物名: Kitty 主人名: Bob
宠物名: Mittens Jr 主人名: Herb
Kitty Bob
Mittens Jr Herb
Kitty
Fido
Bob 1960-01-15
Bob 1960-01-15
Herb 1950-05-05
Grandma L. 1935-03-01
Bob 0 pets
Bob 2 pets
    Kitty cat
    Fido dog
Grandma L. 0 pets
Herb 1 pets
    Mittens Jr cat
Bob 1960-01-15
Bob 1960-01-15
Grandma L. 1935-03-01
Herb 1950-05-05
Grandma L.
```

下面的实例文件 Peewee04.py 演示了，在 Python 程序中使用 Peewee 在指定 MySQL 数据库中创建表 user 和表 tweet 的过程。

源码路径：daima\6\6-3\Peewee04.py

```python
from peewee import *
import datetime

db = MySQLDatabase("test", host="127.0.0.1", port=3306, user="root", passwd="66688888")
db.connect()

class BaseModel(Model):

    class Meta:
        database = db

class User(BaseModel):
    username = CharField(unique=True)

class Tweet(BaseModel):
    user = ForeignKeyField(User, related_name='tweets')
    message = TextField()
    created_date = DateTimeField(default=datetime.datetime.now)
    is_published = BooleanField(default=True)

if __name__ == "__main__":
    #创建表
    User.create_table()    #创建User表
    Tweet.create_table()   #创建Tweet表
```

在上述代码中，首先创建 User 和 Tweet 类作为表名。在类下面定义的变量为字段名，如 username、message、created_date 等。通过 CharField、DateTimeField、BooleanField 表示字段的类型。通过 ForeignKeyField 建立外键，Peewee 会默认加上 id 并且设置主键。最后，执行 create_table()方法创建两张表。执行后会在 MySQL 数据库 "test" 中创建表 user 和表 tweet，如图 6-2 所示。另外，自动创建 id 并设置主键，如图 6-3 所示。

图 6-2 创建的两个表 user 和 tweet

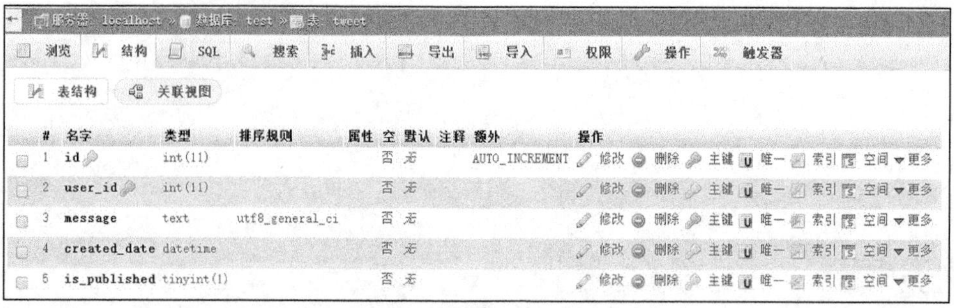

第6章 使用 ORM 操作数据库

图 6-3 自动创建 id 并设置主键

6.3.2 使用 Peewee、Flask 与 MySQL 开发一个在线留言系统

下面的步骤演示了联合使用 Peewee、Flask 和 MySQL 开发一个在线留言系统的过程。

（1）在 MySQL 中创建一个名为"flaskr"的数据库。

（2）编写程序文件 models.py，建立和数据库"flaskr"的连接，并通过类 Entries 在数据库中创建一个名为"Entries"的表。文件 models.py 的具体实现代码如下。

源码路径：daima\6\6-3\flaskr\models.py

```
#--*--编码:utf-8 --*--
from peewee import *

def get_db():
    mysql_db = MySQLDatabase("flaskr", user = "root", passwd = "66688888", charset = "utf8")
    mysql_db.connect()  #连接数据库
    return mysql_db

mysql_db = get_db()

class MySQLModel(Model):
    class Meta:
        database = mysql_db

class Entries(MySQLModel):
    title = CharField()  #字段声明
    text = TextField()

if __name__ == '__main__':
    Entries.create_table()
```

（3）编写程序文件 flaskr.py，功能是设置系统管理员的用户名和密码，根据 app.route 跳转到对应的页面，实现用户登录验证，并将用户发表的留言添加到数据库中。文件 flaskr.py 的具体实现代码如下。

源码路径：daima\6\6-3\flaskr\flaskr.py

```
# -*-编码: utf-8 -*-
from flask import Flask, request, session, g, redirect, url_for, abort, \
     render_template, flash
from socketserver import ThreadingMixIn

import os
from models import *

DEBUG = True
SECRET_KEY = '&Us\xb9\xa0\xef\xc9\xe8H\xfc\x10\xe2\xfd9\xffR\x8c\xa2\xb65\x18\xd9\xf7?'
USERNAME = 'admin'
PASSWORD = '123456'

app = Flask(__name__)
app.config.from_object(__name__)
app.config.from_envvar('FLASKR_SETTINGS', silent=True)
```

```python
@app.route('/')
def show_entries():
    entries = Entries.select()
    return render_template('show_entries.html', entries=entries)


@app.route('/add', methods=['POST'])
def add_entry():
    if not session.get('logged_in'):
        abort(401)
    entry = Entries(title=request.form['title'],text=request.form['text'])
    entry.save()
    flash('新的留言发布成功!')
    return redirect(url_for('show_entries'))


@app.route('/login', methods=['GET', 'POST'])
def login():
    error = None
    if request.method == 'POST':
        if request.form['username'] != app.config['USERNAME']:
            error = '用户名错误!'
        elif request.form['password'] != app.config['PASSWORD']:
            error = '密码错误!'
        else:
            session['logged_in'] = True
            flash('处于登录状态!')
            return redirect(url_for('show_entries'))
    return render_template('login.html', error=error)


@app.route('/logout')
def logout():
    session.pop('logged_in', None)
    flash('你已经退出系统!')
    return redirect(url_for('show_entries'))

if __name__ == '__main__':
    app.run()
```

(4) 编写页面顶部模板文件 layout.html, 功能是在页面顶部导航中显示 "登录" 或 "注销" 链接。具体实现代码如下。

源码路径: daima\6\6-3\flaskr\templates\layout.html

```html
<!doctype html>
<title>春天留言系统</title>
<link rel="shortcut icon" href="{{ url_for('static', filename='favicon.ico') }}">
<link rel=stylesheet type=text/css href="{{ url_for('static', filename='style.css') }}">
<div class=page>
  <h1>春天留言系统</h1>
  <div class=metanav>
  {% if not session.logged_in %}
    <a href="{{ url_for('login') }}">登录</a>
  {% else %}
    <a href="{{ url_for('logout') }}">注销</a>
  {% endif %}
  </div>
  {% for message in get_flashed_messages() %}
    <div class=flash>{{ message }}</div>
  {% endfor %}
  {% block body %}{% endblock %}
</div>
```

(5) 编写留言展示模板文件 show_entries.html, 功能是在页面上展示数据库中所有的留言信息。具体实现代码如下。

源码路径: daima\6\6-2\flaskr\templates\show_entries.html

```html
{% extends "layout.html" %}
{% block body %}
```

```
        {% if session.logged_in %}
          <form action="{{ url_for('add_entry') }}" method=post class=add-entry>
            <dl>
              <dt>留言标题：
              <dd><input type=text size=30 name=title>
              <dt>留言内容：
              <dd><textarea name=text rows=5 cols=40></textarea>
              <dd><input type=submit value=Share>
            </dl>
          </form>
        {% endif %}
        <ul class=entries>
        {% for entry in entries %}
          <li><h2>{{ entry.title }}</h2>{{ entry.text|safe }}
        {% else %}
          <li><em>难以置信，到目前为止还没有任何留言！</em>
        {% endfor %}
        </ul>
{% endblock %}
```

运行 "http://127.0.0.1:5000/" 后的执行效果如图 6-4 所示。

图 6-4　展示所有留言信息

（6）编写用户登录模板文件 login.html，功能是在页面展示用户登录系统表单界面。具体实现代码如下。

源码路径：daima\6\6-3\flaskr\templates\login.html

```
{% extends "layout.html" %}
{% block body %}
  <h2>登录系统</h2>
  {% if error %}<p class=error><strong>Error:</strong> {{ error }}{% endif %}
  <form action="{{ url_for('login') }}" method=post>
    <dl>
      <dt>用户名：
      <dd><input type=text name=username>
      <dt>密码：
      <dd><input type=password name=password>
      <dd><input type=submit value=Login>
    </dl>
  </form>
{% endblock %}
```

登录界面如图 6-5 所示。

图 6-5　登录界面

6.4 使用 Pony 连接数据库

Pony 是 Python 中的一种 ORM，它允许使用生成器表达式来构造查询，将生成器表达式的抽象语法树解析成 SQL 语句。Pony 的功能强大，提供了在线 ER 图编辑器工具来帮助开发者创建 Model。本节将详细讲解在 Python 程序中使用 Pony 连接数据库的核心知识。

6.4.1 Pony 的基础知识

在使用 Pony 之前需要先进行安装，具体安装命令如下。

```
pip install pony
```

请看下面的一段 Pony 程序。

```
select(p for p in Person if p.age > 20)
```

其功能等效于下面的 SQL 代码。

```
SELECT p.id, p.name, p.age, p.classtype, p.mentor, p.gpa, p.degree
FROM person p
WHERE p.classtype IN ('Student', 'Professor', 'Person')
AND p.age > 20
```

由此可见，使用 Pony 之后整个代码变得更加简洁。而下面的一段 Pony 程序。

```
select(c for c in Customer
       if sum(c.orders.price) > 1000)
```

翻译成 SQL 语句就是:

```
SELECT "c"."id"
FROM "Customer" "c"
  LEFT JOIN "Order" "order-1"
  ON "c"."id" = "order-1"."customer"
GROUP BY "c"."id"
HAVING coalesce(SUM("order-1"."total_price"), 0) > 1000
```

开发者要提高设计 Pony 模式的开发效率，可以登录 PonyORM 的官方网站，并使用官方提供的在线设计器工具，如图 6-6 所示。

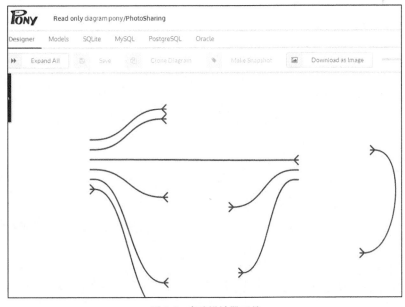

图 6-6 在线设计器工具

6.4.2 操作 SQLite 数据库

下面的实例文件 pony01.py 演示了使用 Pony 创建一个 SQLite 数据库的过程。首先创建了一个名为"music.sqlite"的数据库，然后在里面创建了表 Artist 和表 Album。文件 pony01.py 的具体实现代码如下。

源码路径：daima\6\6-4\pony01.py

```python
import datetime
import pony.orm as pny

database = pny.Database("sqlite",
                        "music.sqlite",
                        create_db=True)
########################################################################
class Artist(database.Entity):
    """
    使用Pony ORM创建表Artist
    """
    name = pny.Required(str)
    albums = pny.Set("Album")
########################################################################
class Album(database.Entity):
    """
    使用Pony ORM创建表Album
    """
    artist = pny.Required(Artist)
    title = pny.Required(str)
    release_date = pny.Required(datetime.date)
    publisher = pny.Required(str)
    media_type = pny.Required(str)

#打开调试模式
pny.sql_debug(True)
#映射模型数据库
#如果它们不存在，则创建表
database.generate_mapping(create_tables=True)
```

执行后会输出显示如下创建数据库和表的等效 SQL 语句。如果多次运行上述代码，不会重新创建数据库和表，会自动设置主键。

```
GET CONNECTION FROM THE LOCAL POOL
PRAGMA foreign_keys = false
BEGIN IMMEDIATE TRANSACTION
CREATE TABLE "Artist" (
  "id" INTEGER PRIMARY KEY AUTOINCREMENT,
  "name" TEXT NOT NULL
)

CREATE TABLE "Album" (
  "id" INTEGER PRIMARY KEY AUTOINCREMENT,
  "artist" INTEGER NOT NULL REFERENCES "Artist" ("id"),
  "title" TEXT NOT NULL,
  "release_date" DATE NOT NULL,
  "publisher" TEXT NOT NULL,
  "media_type" TEXT NOT NULL
)

CREATE INDEX "idx_album__artist" ON "Album" ("artist")

SELECT "Album"."id", "Album"."artist", "Album"."title", "Album"."release_date", "Album"."publisher", "Album"."media_type"
FROM "Album" "Album"
WHERE 0 = 1

SELECT "Artist"."id", "Artist"."name"
FROM "Artist" "Artist"
WHERE 0 = 1

COMMIT
PRAGMA foreign_keys = true
CLOSE CONNECTION
```

下面的实例文件 pony02.py 演示了，使用 Pony 向 "music.sqlite" 数据库的指定表中添加新数据的过程。

源码路径：daima\6\6-4\pony02.py

```python
import datetime
import pony.orm as pny

from pony01 import Album, Artist


# ----------------------------------------------------------------------
@pny.db_session
def add_data():
    """"""

    new_artist = Artist(name=u"Newsboys")
    bands = [u"MXPX", u"Kutless", u"Thousand Foot Krutch"]
    for band in bands:
        artist = Artist(name=band)

    album = Album(artist=new_artist,
                  title=u"Read All About It",
                  release_date=datetime.date(1988, 12, 1),
                  publisher=u"Refuge",
                  media_type=u"CD")

    albums = [{"artist": new_artist,
               "title": "Hell is for Wimps",
               "release_date": datetime.date(1990, 7, 31),
               "publisher": "Sparrow",
               "media_type": "CD"
               },
              {"artist": new_artist,
               "title": "Love Liberty Disco",
               "release_date": datetime.date(1999, 11, 16),
               "publisher": "Sparrow",
               "media_type": "CD"
               },
              {"artist": new_artist,
               "title": "Thrive",
               "release_date": datetime.date(2002, 3, 26),
               "publisher": "Sparrow",
               "media_type": "CD"}
              ]

    for album in albums:
        a = Album(**album)

if __name__ == "__main__":
    add_data()

    # use db_session as a context manager
    with pny.db_session:
        a = Artist(name="Skillet")
```

在上述代码中使用了一种名为 db_session 的装饰器来操作数据库，它负责打开连接、提交数据并关闭连接，并且还可以作为上下文管理器使用。执行后会输出和上述添加数据功能等效的 SQL 语句。

```
GET CONNECTION FROM THE LOCAL POOL
PRAGMA foreign_keys = false
BEGIN IMMEDIATE TRANSACTION
SELECT "Album"."id", "Album"."artist", "Album"."title", "Album"."release_date",
    "Album"."publisher", "Album"."media_type"
FROM "Album" "Album"
WHERE 0 = 1
```

```
SELECT "Artist"."id", "Artist"."name"
FROM "Artist" "Artist"
WHERE 0 = 1

COMMIT
PRAGMA foreign_keys = true
CLOSE CONNECTION
GET NEW CONNECTION
BEGIN IMMEDIATE TRANSACTION
INSERT INTO "Artist" ("name") VALUES (?)
['Newsboys']

INSERT INTO "Artist" ("name") VALUES (?)
['MXPX']

INSERT INTO "Artist" ("name") VALUES (?)
['Kutless']

INSERT INTO "Artist" ("name") VALUES (?)
['Thousand Foot Krutch']

INSERT INTO "Album" ("artist", "title", "release_date", "publisher", "media_type")
VALUES (?, ?, ?, ?, ?)
[1, 'Read All About It', '1988-12-01', 'Refuge', 'CD']

INSERT INTO "Album" ("artist", "title", "release_date", "publisher", "media_type")
VALUES (?, ?, ?, ?, ?)
[1, 'Hell is for Wimps', '1990-07-31', 'Sparrow', 'CD']

INSERT INTO "Album" ("artist", "title", "release_date", "publisher", "media_type")
VALUES (?, ?, ?, ?, ?)
[1, 'Love Liberty Disco', '1996-11-16', 'Sparrow', 'CD']

INSERT INTO "Album" ("artist", "title", "release_date", "publisher", "media_type")
VALUES (?, ?, ?, ?, ?)
[1, 'Thrive', '2002-03-26', 'Sparrow', 'CD']

COMMIT
RELEASE CONNECTION
GET CONNECTION FROM THE LOCAL POOL
BEGIN IMMEDIATE TRANSACTION
INSERT INTO "Artist" ("name") VALUES (?)
['Skillet']

COMMIT
RELEASE CONNECTION
```

下面的实例文件 pony03.py 演示了,使用 Pony 查询并修改"music.sqlite"数据库中指定数据的过程。

源码路径:daima\6\6-4\pony03.py

```
import pony.orm as pny

from pony01 import Artist, Album

with pny.db_session:
    band = Artist.get(name="Newsboys")
    print(band.name)
    for record in band.albums:
        print(record.title)
    # 修改数据
    band_name = Artist.get(name="Kutless")
    band_name.name = "Beach Boys"

result = pny.select(i.name for i in Artist)
```

在上述代码中使用 db_session 作为上下文管理器,在进行查询时从数据库中获取 Artist 表的一个对象并输出其名称。然后循环查找 Artist 表的 Album,这些 Album 也包含在返回的对象中。最后修改一个指定 Artist 的名字。执行后会输出与上述修改和查询数据功能等效的 SQL 语句。

```
GET CONNECTION FROM THE LOCAL POOL
PRAGMA foreign_keys = false
BEGIN IMMEDIATE TRANSACTION
SELECT "Album"."id", "Album"."artist", "Album"."title", "Album"."release_date",
    "Album"."publisher", "Album"."media_type"
FROM "Album" "Album"
WHERE 0 = 1

SELECT "Artist"."id", "Artist"."name"
FROM "Artist" "Artist"
WHERE 0 = 1

COMMIT
PRAGMA foreign_keys = true
CLOSE CONNECTION
GET NEW CONNECTION
SWITCH TO AUTOCOMMIT MODE
SELECT "id", "name"
FROM "Artist"
WHERE "name" = ?
LIMIT 2
['Newsboys']

Newsboys
SELECT "id", "artist", "title", "release_date", "publisher", "media_type"
FROM "Album"
WHERE "artist" = ?
[1]

Thrive
Read All About It
Hell is for Wimps
Love Liberty Disco
SELECT "id", "name"
FROM "Artist"
WHERE "name" = ?
LIMIT 2
['Kutless']

BEGIN IMMEDIATE TRANSACTION
UPDATE "Artist"
SET "name" = ?
WHERE "id" = ?
  AND "name" = ?
['Beach Boys', 3, 'Kutless']

COMMIT
RELEASE CONNECTION
```

下面的实例文件 pony04.py 演示了，使用 Pony 删除 "music.sqlite" 数据库的 Artist 表中 name 为 "MXPX" 的数据的过程。

源码路径：daima\6\6-4\pony04.py

```
import pony.orm as pny
from pony01 import Artist
with pny.db_session:
    band = Artist.get(name="MXPX")
    band.delete()
```

执行后会输出和上述删除数据功能等效的 SQL 语句。

```
GET CONNECTION FROM THE LOCAL POOL
PRAGMA foreign_keys = false
BEGIN IMMEDIATE TRANSACTION
SELECT "Album"."id", "Album"."artist", "Album"."title", "Album"."release_date",
    "Album"."publisher", "Album"."media_type"
FROM "Album" "Album"
```

```
WHERE 0 = 1

SELECT "Artist"."id", "Artist"."name"
FROM "Artist" "Artist"
WHERE 0 = 1

COMMIT
PRAGMA foreign_keys = true
CLOSE CONNECTION
GET NEW CONNECTION
SWITCH TO AUTOCOMMIT MODE
SELECT "id", "name"
FROM "Artist"
WHERE "name" = ?
LIMIT 2
['MXPX']

SELECT "id", "artist", "title", "release_date", "publisher", "media_type"
FROM "Album"
WHERE "artist" = ?
[2]

BEGIN IMMEDIATE TRANSACTION
DELETE FROM "Artist"
WHERE "id" = ?
  AND "name" = ?
[2, 'MXPX']

COMMIT
RELEASE CONNECTION
```

6.4.3 操作 MySQL 数据库

下面的实例文件 pony05.py 演示了使用 Pony 在指定 MySQL 数据库中创建指定表的过程。

源码路径：daima\6\6-4\pony05.py

```python
from pony.orm import *

db = Database("mysql", host="localhost",
              user="root",
              passwd="66688888",
              db="test")

class Person(db.Entity):
    name = Required(str)
    age = Required(int)

sql_debug(True)#显示调试信息
db.generate_mapping(create_tables=True)

@db_session
def create_persons():
    p1 = Person(name="Person1", age=20)
    p2 = Person(name="Person2", age=22)
    p3 = Person(name="Person3", age=12)
    print(p1.id)  #这里得不到id,没提交
    commit()
    print(p1.id)  #这里已经有id
create_persons()
```

执行后会创建指定的数据表"person"，并向表中插入 3 条数据，输出和上述创建数据功能等效的 SQL 语句。

```
GET CONNECTION FROM THE LOCAL POOL
cursor.execute("SHOW VARIABLES LIKE 'foreign_key_checks'")
SET foreign_key_checks = 0
SELECT 'person'.'id', 'person'.'name', 'person'.'age'
FROM 'person' 'person'
```

```
WHERE 0 = 1

COMMIT
SET foreign_key_checks = 1
CLOSE CONNECTION
None
GET NEW CONNECTION
INSERT INTO 'person' ('name', 'age') VALUES (%s, %s)
['Person1', 20]

INSERT INTO 'person' ('name', 'age') VALUES (%s, %s)
['Person2', 22]

INSERT INTO 'person' ('name', 'age') VALUES (%s, %s)
['Person3', 12]

COMMIT
7
RELEASE CONNECTION
```

在使用 Pony 操作 MySQL 数据库时，外键必须设置两边的字段，Django 可以只设置多对一的关系。下面是一对一的代码。

```
class User(db.Entity):
    name = Required(str)
    cart = Optional("Cart")

class Cart(db.Entity):
    user = Required("User")
```

而下面是多对一的代码。

```
class Product(db.Entity):
    tags = Set("Tag")

class Tag(db.Entity):
    products = Set(Product)
```

为了更好地说明对应关系，我们举一个通俗易懂的例子。例如，学生和教授从 Person 继承，教授有多个学生（一对多的关系），Person 包含了所有子类的属性，可以用一个 classtype 来表明子类属于哪个对象，默认是 class name，可以添加_discriminator_属性修改 classtype。下面的实例文件 pony06.py 演示了，使用 Pony 在指定 MySQL 数据库中实现一对多和继承操作的过程。

源码路径：daima\6\6-4\pony06.py

```
from pony.orm import *
from decimal import Decimal

db = Database("mysql", host="localhost",
              user="root",
              passwd="66688888",
              db="test")
db.drop_table("person", with_all_data=True)

class Person(db.Entity):
    _discriminator_ = 1
    name = Required(str)
    age = Required(int)

class Student(Person):
    _discriminator_ = 3
    gpa = Optional(Decimal)
    mentor = Optional("Professor")

class Professor(Person):
    _discriminator_ = 2
    degree = Required(str)
    students = Set("Student")
```

```
sql_debug(True)    # 显示调试信息
db.generate_mapping(create_tables=True)

# @db_session
# def create_persons():
#     p1 = Person(name="Person", age=20)
#     s = Student(name="Student", age=22, gpa=1.2)
#也可以向Professor中添加Student
#     p2 = Professor(name="Professor", age=12, degree="aaaaaa", students=[s])
#     commit()

@db_session
def create_persons():
    p1 = Person(name="Person", age=20)
    s = Student(name="Student", age=22, gpa=1.2)
    p2 = Professor(name="Professor", age=12, degree="aaaaaa", students=[s])
    commit()

create_persons()
```

执行后会输出和上述数据操作功能等效的 SQL 语句。

```
GET NEW CONNECTION
SET foreign_key_checks = 0
CREATE TABLE `person` (
  'id' INTEGER PRIMARY KEY AUTO_INCREMENT,
  'name' VARCHAR(255) NOT NULL,
  'age' INTEGER NOT NULL,
  'classtype' VARCHAR(255) NOT NULL,
  'gpa' DECIMAL(12, 2),
  'mentor' INTEGER,
  'degree' VARCHAR(255)
)

CREATE INDEX 'idx_person__mentor' ON 'person' ('mentor')

ALTER TABLE 'person' ADD CONSTRAINT 'fk_person__mentor' FOREIGN KEY ('mentor')
  REFERENCES 'person' ('id')

SELECT `person`.'id', 'person'.'name', 'person'.'age', 'person'.'classtype',
  'person'.'gpa', 'person'.'mentor', 'person'.'degree'
FROM 'person' 'person'
WHERE 0 = 1

COMMIT
SET foreign_key_checks = 1
CLOSE CONNECTION
GET NEW CONNECTION
INSERT INTO 'person' ('name', 'age', 'classtype') VALUES (%s, %s, %s)
['Person', 20, '1']

INSERT INTO 'person' ('name', 'age', 'classtype', 'degree') VALUES (%s, %s, %s, %s)
['Professor', 12, '2', 'aaaaaa']

INSERT INTO 'person' ('name', 'age', 'classtype', 'gpa', 'mentor') VALUES (%s, %s, %s, %s, %s)
['Student', 22, '3', Decimal('1.2'), 2]

COMMIT
RELEASE CONNECTION
```

6.5 使用 mongoengine 连接 MongoDB 数据库

在 Python 程序中，MongoDB 数据库的 ORM 框架是 mongoengine。在使用 mongoengine 框架之前需要先安装 mongoengine，具体安装命令如下。

```
easy_install mongoengine
```

安装成功后的界面如图 6-7 所示。

6.5 使用 mongoengine 连接 MongoDB 数据库

图 6-7　成功安装 mongoengine

在运行上述命令之前，必须先确保使用如下命令安装了 pymongo 框架。

```
easy_install pymongo
```

下面的实例文件 orm.py 演示了，在 Python 程序中使用 mongoengine 操作数据库数据的过程。

源码路径：daima\6\6-5\orm.py

```python
import random                           #导入内置模块
from mongoengine import *
connect('test')                         #连接数据库对象'test'
class Stu(Document):                    #定义ORM框架类Stu
    sid = SequenceField()
    name = StringField()
    passwd = StringField()
    def introduce(self):                #定义函数introduce()来显示自己的介绍信息
        print('序号:',self.sid,end=" ") #显示id
        print('姓名:',self.name,end=' ') #显示姓名
        print('密码:',self.passwd)#显示密码
    def set_pw(self,pw):                #定义函数set_pw()用于修改密码
        if pw:
            self.passwd = pw            #修改密码
            self.save()                 #保存修改的密码
...省略部分代码...
if __name__ == '__main__':
    print('插入一个文档:')
    stu = Stu(name='langchao',passwd='123123')#创建文档类对象实例stu,设置用户名和密码
    stu.save()                                #保存文档
    stu = Stu.objects(name='lilei').first()   #查询数据并对类进行初始化

    if stu:
        stu.introduce()                       #显示文档信息
    print('插入多个文档')                       #显示提示信息
    for i in range(3):                        #遍历操作
        Stu(name=get_str(2,4),passwd=get_str(6,8)).save() #插入3个文档
    stus = Stu.objects()                      #文档类对象实例stu
    for stu in stus:                          #遍历所有的文档信息
        stu.introduce()                       #显示所有的遍历文档
    print('修改一个文档')                       #显示提示信息
    stu = Stu.objects(name='langchao').first() #查询某个要操作的文档
    if stu:
        stu.name='daxie'                      #修改用户名属性
        stu.save()                            #保存修改
        stu.set_pw('bbbbbbbb')                #修改密码属性
        stu.introduce()                       #显示修改后的结果
    print('删除一个文档')                       #显示提示信息
    stu = Stu.objects(name='daxie').first()   #查询某个要操作的文档
    stu.delete()                              #删除这个文档
```

```
    stus = Stu.objects()
    for stu in stus:                                    #遍历所有的文档
        stu.introduce()                                 #显示删除后的结果
```

在上述代码中，在导入 mongoengine 库和连接 MongoDB 数据库后，首先定义了一个继承于类 Document 的子类 Stu。在主程序中通过创建类的实例，并调用其方法 save()将类持久化到数据库中。通过类 Stu 中的方法 objects()来查询数据库并映射为类 Stu 的实例，同时调用其自定义方法 introduce()来显示载入的信息。然后插入 3 个文档信息，并调用方法 save()存入数据库中。接下来，通过调用类中的自定义方法 set_pw()修改数据并存入数据库中。最后通过调用类中的方法 delete()从数据库中删除一个文档。

开始测试程序。在运行本实例程序时，必须在 CMD 控制台中启动 MongoDB 服务，并且确保上述控制台界面处于打开状态。下面是开启 MongoDB 服务的命令。

```
mongod --dbpath "h:\data"
```

在上述命令中，"h:\data"是一个保存 MongoDB 数据库中数据的目录。

第 7 章

特殊文本格式处理

在开发 Python 应用程序的过程中,经常需要将一些数据处理后保存成不同格式的文件,如 Office、PDF 和 CSV 等。本章将详细讲解在 Python 第三方库中将数据处理成特殊文件格式的知识。

7.1 使用 Tablib 模块

在 Python 程序中，可以使用第三方模块 Tablib 将数据导出为各种不同的格式，包括 Excel、JSON、HTML、Yaml、CSV 和 TSV 等格式。在使用之前需要先安装 Tablib，安装命令如下。

```
pip install tablib
```

接下来将详细讲解使用 Tablib 模块的知识。

7.1.1 基本用法

1. 创建数据集

在 Tablib 模块中，使用 tablib.Dataset 可以创建一个简单的数据集对象实例。

```
data = tablib.Dataset()
```

接下来，就可以填充数据集对和数据。

2. 添加行

假如我们想收集一个简单的人名列表，首先看下面的实现代码。

```
#名称的集合
names = ['Kenneth Reitz', 'Bessie Monke']

for name in names:
    #分割名称
    fname, lname = name.split()

    # 将名称添加到数据集中
    data.append([fname, lname])
```

在 Python 中可以通过下面的代码获取人名。

```
>>> data.dict
[('Kenneth', 'Reitz'), ('Bessie', 'Monke')]
```

3. 添加标题

通过下面的代码可以在数据集中添加标题。

```
data.headers = ['First Name', 'Last Name']
```

接下来，可以通过下面的代码获取数据集信息。

```
>>> data.dict
[{'Last Name': 'Reitz', 'First Name': 'Kenneth'}, {'Last Name': 'Monke', 'First Name': 'Bessie'}]
```

4. 添加列

在数据集中可以继续添加列，例如下面的代码。

```
data.append_col([22, 20], header='Age')
```

接下来，可以通过下面的代码获取数据集信息。

```
>>> data.dict
[{'Last Name': 'Reitz', 'First Name': 'Kenneth', 'Age': 22}, {'Last Name': 'Monke', 'First Name': 'Bessie', 'Age': 20}]
```

5. 导入数据

在创建 tablib.Dataset 对象实例后，可以直接导入已经存在的数据集，例如，下面的代码导入了 CSV 文件中的数据。

```
imported_data = Dataset().load(open('data.csv').read())
```

在 Tablib 模块中如果需要导入数据，只要具备适当的格式化程序导入窗口，就可以从窗口把不同类型的文件导入数据。

6. 导出数据

Tablib 模块最强大的功能是将数据集导出为不同类型的文件，例如，下面的代码将前面创建的数据集导出为 CSV 文件。

```
>>> data.export('csv')
Last Name,First Name,Age
```

```
Reitz,Kenneth,22
Monke,Bessie,20
```
通过下面的代码导出为 JSON 文件格式。
```
>>> data.export('json')
[{"Last Name": "Reitz", "First Name": "Kenneth", "Age": 22}, {"Last Name": "Monke",
"First Name": "Bessie", "Age": 20}]
```
通过下面的代码导出为 YAML 文件格式。
```
>>> data.export('yaml')
- {Age: 22, First Name: Kenneth, Last Name: Reitz}
- {Age: 20, First Name: Bessie, Last Name: Monke}
```
通过下面的代码导出为 Excel 文件格式。
```
>>> data.export('xls')
<censored binary data>
```

7.1.2 操作数据集中指定的行和列

下面的实例文件 Tablib01.py 演示了使用 Tablib 模块操作数据集中指定行和列的过程。

源码路径：daima\7\7-1\Tablib01.py

```python
import tablib
names = ['Kenneth Reitz', 'Bessie Monke']
data = tablib.Dataset()
for name in names:
    fname, lname = name.split()
    data.append([fname, lname])

data.headers = ['First Name', 'Last Name']
data.append_col([22, 20], header='Age')
#显示某条信息
print(data[0])
#显示某列的值
print(data['First Name'])
#使用索引访问列
print(data.headers)
print(data.get_col(1))
#计算平均年龄
ages = data['Age']
print(float(sum(ages)) / len(ages))
```

执行后会输出：
```
('Kenneth', 'Reitz', 22)
['Kenneth', 'Bessie']
['First Name', 'Last Name', 'Age']
['Reitz', 'Monke']
21.0
```

7.1.3 删除并导出不同格式的数据

下面的实例文件 Tablib02.py 演示了使用 Tablib 模块删除数据集中指定数据，并将数据导出为不同格式文件的过程。

源码路径：daima\7\7-1\Tablib02.py

```python
import tablib
headers = ('area', 'user', 'recharge')
data = [
    ('1', 'Rooney', 20),
    ('2', 'John', 30),
]
data = tablib.Dataset(*data, headers=headers)

#可以通过下面这些方式得到各种格式的数据
print(data.csv)
print(data.html)
print(data.xls)
print(data.ods)
print(data.json)
print(data.yaml)
print(data.tsv)
```

```
#增加行
data.append(['3', 'Keven',18])
#增加列
data.append_col([22, 20,13], header='Age')
print(data.csv)

#删除行
del data[1:3]
#删除列
del data['Age']
print(data.csv)
```

执行后会输出：

```
area,user,recharge
1,Rooney,20
2,John,30

<table>
<thead>
<tr><th>area</th>
<th>user</th>
<th>recharge</th></tr>
</thead>
<tr><td>1</td>
<td>Rooney</td>
<td>20</td></tr>
<tr><td>2</td>
<td>John</td>
<td>30</td></tr>
</table>
--------------------------------------------------

省略其他文件格式
--------------------------------------------------

[{"area": "1", "user": "Rooney", "recharge": 20}, {"area": "2", "user": "John", "recharge": 30}]
- {area: '1', recharge: 20, user: Rooney}
- {area: '2', recharge: 30, user: John}

area    user    recharge
1       Rooney       20
2       John30

area,user,recharge,Age
1,Rooney,20,22
2,John,30,20
3,Keven,18,13

area,user,recharge
1,Rooney,20
```

7.1.4 生成一个 Excel 文件

下面的实例文件 Tablib03.py 演示了将 Tablib 数据集导出到新建的 Excel 文件的过程。

源码路径：daima\7\7-1\Tablib03.py

```
import tablib
headers = ('lie1', 'lie2', 'lie3', 'lie4', 'lie5')
mylist = [('23','23','34','23','34'),('sadf','23','sdf','23','fsad')]
mylist = tablib.Dataset(*mylist, headers=headers)
with open('excel.xls', 'wb') as f:
    f.write(mylist.xls)
```

执行后会创建一个 Excel 文件 excel.xls，里面填充的是数据集中的数据，如图 7-1 所示。

	A	B	C	D	E
1	lie1	lie2	lie3	lie4	lie5
2	23	23	34	23	34
3	sadf	23	sdf	23	fsad

图 7-1 创建的 Excel 文件

7.1.5 处理多个数据集

在实际应用中，有时需要在表格中处理多个数据集，如将多个数据集导出到一个 Excel 文件中，这时候可以使用 Tablib 模块中的 Databook 实现。在下面的实例文件 Tablib04.py 中，不但演示了增加、删除数据集数据的方法，而且演示了将多个 Tablib 数据集导出到 Excel 文件的过程。

源码路径：daima\7\7-1\Tablib04.py

```
import tablib
import os

#创建数据集，方法1
dataset1 = tablib.Dataset()
header1 = ('ID', 'Name', 'Tel', 'Age')
dataset1.headers = header1
dataset1.append([1, 'zhangsan', 13711111111, 16])
dataset1.append([2, 'lisi',     13811111111, 18])
dataset1.append([3, 'wangwu',   13911111111, 20])
dataset1.append([4, 'zhaoliu',  15811111111, 25])
print('dataset1:', os.linesep, dataset1, os.linesep)

#创建数据集，方法2
header2 = ('ID', 'Name', 'Tel', 'Age')
data2 = [
    [1, 'zhangsan', 13711111111, 16],
    [2, 'lisi',     13811111111, 18],
    [3, 'wangwu',   13911111111, 20],
    [4, 'zhaoliu',  15811111111, 25]
]
dataset2 = tablib.Dataset(*data2, headers = header2)
print('dataset2: ', os.linesep, dataset2, os.linesep)

#增加行
dataset1.append([5, 'sunqi', 15911111111, 30])       #添加到最后一行的下面
dataset1.insert(0, [0, 'liuyi', 18211111111, 35])    #在指定位置添加行
print('增加行后的dataset1: ', os.linesep, dataset1, os.linesep)

#删除行
dataset1.pop()          #删除最后一行
dataset1.lpop()         #删除第一行
del dataset1[0:2]       #删除第[0,2)行数据
print('删除行后的dataset1:', os.linesep, dataset1, os.linesep)

#增加列
#现在dataset1就剩两行数据了
dataset1.append_col(('beijing', 'shenzhen'), header='city')    #增加列到最后一列
dataset1.insert_col(2, ('male', 'female'), header='sex')       #在指定位置添加列
print('增加列后的dataset1: ', os.linesep, dataset1, os.linesep)

#删除列
del dataset1['Tel']
print('删除列后的dataset1: ', os.linesep, dataset1, os.linesep)

#获取各种格式的数据
print('yaml format: ', os.linesep ,dataset1.yaml, os.linesep)
print('csv format: ' , os.linesep ,dataset1.csv , os.linesep)
print('tsv format: ' , os.linesep ,dataset1.tsv , os.linesep)

#导出到Excel表格中
dataset1.title = 'dataset1'      #设置Excel中表单的名称
dataset2.title = 'dataset2'
myfile = open('mydata.xls', 'wb')
myfile.write(dataset1.xls)
myfile.close()
```

```
#如果有多个sheet表单，使用DataBook就可以了
myDataBook = tablib.Databook((dataset1, dataset2))
myfile = open(myfile.name, 'wb')
myfile.write(myDataBook.xls)
myfile.close()
```

执行后会输出：

```
dataset1:
 ID|Name    |Tel        |Age
 --|--------|-----------|---
 1 |zhangsan|13711111111|16
 2 |lisi    |13811111111|18
 3 |wangwu  |13911111111|20
 4 |zhaoliu |15811111111|25

dataset2:
 ID|Name    |Tel        |Age
 --|--------|-----------|---
 1 |zhangsan|13711111111|16
 2 |lisi    |13811111111|18
 3 |wangwu  |13911111111|20
 4 |zhaoliu |15811111111|25
```

增加行后的dataset1：

```
 ID|Name    |Tel        |Age
 --|--------|-----------|---
 0 |liuyi   |18211111111|35
 1 |zhangsan|13711111111|16
 2 |lisi    |13811111111|18
 3 |wangwu  |13911111111|20
 4 |zhaoliu |15811111111|25
 5 |sunqi   |15911111111|30
```

删除行后的dataset1：

```
 ID|Name    |Tel        |Age
 --|--------|-----------|---
 3 |wangwu  |13911111111|20
 4 |zhaoliu |15811111111|25
```

增加列后的dataset1：

```
 ID|Name    |sex   |Tel        |Age|city
 --|--------|------|-----------|---|--------
 3 |wangwu  |male  |13911111111|20 |beijing
 4 |zhaoliu |female|15811111111|25 |shenzhen
```

删除列后的dataset1：

```
 ID|Name    |sex   |Age|city
 --|--------|------|---|--------
 3 |wangwu  |male  |20 |beijing
 4 |zhaoliu |female|25 |shenzhen

yaml format:
 - {Age: 20, ID: 3, Name: wangwu, city: beijing, sex: male}
 - {Age: 25, ID: 4, Name: zhaoliu, city: shenzhen, sex: female}

csv format:
 ID,Name,sex,Age,city
 3,wangwu,male,20,beijing
 4,zhaoliu,female,25,shenzhen

tsv format:
 ID   Name      sex     Age    city
 3    wangwu    male    20     beijing
 4    zhaoliu   female  25     shenzhen
```

执行后会创建 Excel 文件 mydata.xls，其中保存了从数据集中导出的数据，如图 7-2 所示。

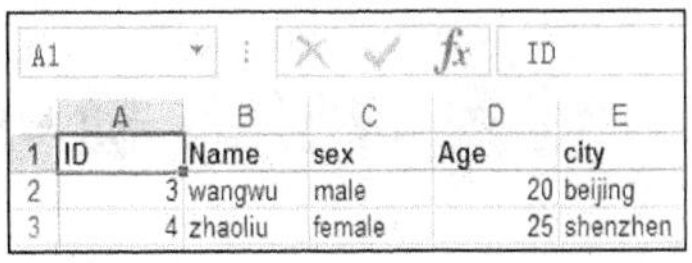

	A	B	C	D	E
1	ID	Name	sex	Age	city
2	3	wangwu	male	20	beijing
3	4	zhaoliu	female	25	shenzhen

图 7-2　Excel 文件中的数据

7.1.6　使用标签过滤数据

在使用 Tablib 数据集时，可以作为参数添加一个标签到指定的行。这样在后面的程序中，可以通过这个标签筛选数据集中基于任意条件的数据。下面的实例文件 Tablib05.py 演示了使用标签过滤 Tablib 数据集的过程。

源码路径：daima\7\7-1\Tablib05.py

```
import tablib
students = tablib.Dataset()
students.headers = ['first', 'last']
students.rpush(['Kenneth', 'Reitz'], tags=['male', 'technical'])
students.rpush(['Bessie', 'Monke'], tags=['female', 'creative'])
print(students.filter(['male']).yaml)
```

执行后会输出：

```
- {first: Kenneth, last: Reitz}
```

7.1.7　分离表格中的数据

在将 Tablib 数据导出到某个格式的文件中时，有时需要将多种数据集对象进行分类。下面的实例文件 Tablib06.py 演示了将两组数据按类别导入 Excel 文件的过程。

源码路径：daima\7\7-1\Tablib06.py

```
import tablib
daniel_tests = [
    ('11/24/09', 'Math 101 Mid-term Exam', 56.),
    ('05/24/10', 'Math 101 Final Exam', 62.)
]

suzie_tests = [
    ('11/24/09', 'Math 101 Mid-term Exam', 56.),
    ('05/24/10', 'Math 101 Final Exam', 62.)
]
tests = tablib.Dataset()
tests.headers = ['Date', 'Test Name', 'Grade']

# Daniel的数据
tests.append_separator('Daniel的得分')

for test_row in daniel_tests:
    tests.append(test_row)

# Susie的数据
tests.append_separator('Susie的得分')

for test_row in suzie_tests:
    tests.append(test_row)

# 写入Excel表格
with open('grades.xls', 'wb') as f:
    f.write(tests.export('xls'))
```

执行后会将 Tablib 按类别导入 Excel 文件中，如图 7-3 所示。

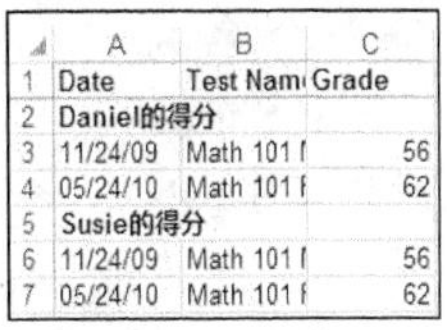

	A	B	C
1	Date	Test Nam	Grade
2	Daniel的得分		
3	11/24/09	Math 101 I	56
4	05/24/10	Math 101 F	62
5	Susie的得分		
6	11/24/09	Math 101 I	56
7	05/24/10	Math 101 F	62

图 7-3　分离的数据

7.2　使用 Office 模块/库

在 Python 程序中，可以使用第三方模块/库将数据转换成 Office 格式。本节将详细讲解这些模块/库的使用方法。

7.2.1　使用 openpyxl 模块

使用 openpyxl 模块可以读写 Excel 文件，包括 xlsx、xlsm、xltx 和 xltm 格式。在使用 openpyxl 之前需要先安装，安装命令如下。

```
pip install openpyxl
```

在 openpyxl 中主要用到如下 3 个概念。

- Workbook：代表一个 Excel 工作表。
- Worksheet：代表工作表中的一页。
- Cell：代表最简单的一个单元格。

openpyxl 是围绕着上述 3 个概念实现读写功能的，基本使用流程也是围绕上述 3 个概念进行的：首先打开 Workbook，然后定位到 Sheet，最后操作 Cell。

1. Workbook 对象

因为一个 Workbook 对象代表一个 Excel 文档，所以在操作 Excel 之前，应该先创建一个 Workbook 对象。如果想创建一个新的 Excel 文档，直接调用 Workbook 类即可。对于一个已经存在的 Excel 文档，可以使用 openpyxl 模块中的 load_workbook 函数进行读取，该函数包含多个参数，但只有 filename 参数为必需参数。参数 filename 代表一个文件名，也可以是一个打开的文件对象。

在 Workbook 对象中提供了很多属性和方法，其中大部分方法都与 sheet 有关。主要属性的具体说明如下。

- active：获取当前活跃的 Worksheet。
- worksheets：以列表的形式返回所有的 Worksheet。
- read_only：判断是否以 read_only 模式打开 Excel 文档。
- encoding：获取文档的字符集编码。
- properties：获取文档的元数据，如标题、创建者、创建日期等。
- sheetnames：获取工作簿中的表（列表）。

在其他的属性中，freeze_panes 比较特别，主要用于在表格较大时冻结顶部的行或左边的行。对于冻结的行，在用户滚动时是始终可见的。可以将其设置为一个 Cell 对象或一个单元格坐标的字符串，单元格上面的行和左边的列将会冻结（单元格所在的行和列不会被冻结）。假设存在一个名为“template.xls”的 Excel 文件，在里面有 Sheet1、Sheet2 和 Sheet3 这 3 个页，在 Sheet3 中填入图 7-4 所示的内容。

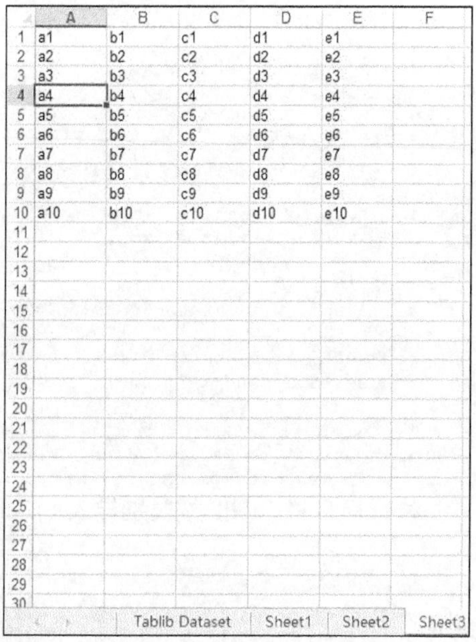

	A	B	C	D	E	F
1	a1	b1	c1	d1	e1	
2	a2	b2	c2	d2	e2	
3	a3	b3	c3	d3	e3	
4	a4	b4	c4	d4	e4	
5	a5	b5	c5	d5	e5	
6	a6	b6	c6	d6	e6	
7	a7	b7	c7	d7	e7	
8	a8	b8	c8	d8	e8	
9	a9	b9	c9	d9	e9	
10	a10	b10	c10	d10	e10	

图 7-4　名为“template.xls”的 Excel 文件

如果我们要冻结第一行，那么设置 a2 为 freeze_panes；如果要冻结第一列，那么设置 b1

为 freeze_panes；如果要同时冻结第一行和第一列，那么需要设置 b2 为 freeze_panes。当 freeze_panes 值为 none 时，表示不冻结任何列。

在 Workbook 中提供的主要方法如下。

- get_sheet_names：获取所有表格的名称(新版已经不建议使用，通过 Workbook 的 sheetnames 属性即可获取)。
- get_sheet_by_name：通过表格名称获取 Worksheet 对象(新版也不建议使用，通过 Worksheet['表名']获取)。
- get_active_sheet：获取活跃的表格(新版本建议通过 active 属性获取)。
- remove_sheet：删除一个表格。
- create_sheet：创建一个空的表格。
- copy_worksheet：在 Workbook 内复制表格。

2. Worksheet 对象

有了 Worksheet 对象以后，我们可以通过这个 Worksheet 对象获取表格的属性，得到单元格中的数据，修改表格中的内容。openpyxl 提供了非常灵活的方式来访问表格中的单元格和数据，其中常用的 Worksheet 属性如下。

- title：表示表格的标题。
- dimensions：表示表格的大小，这里的大小是指含数据的表格大小，即，左上角的坐标和右下角的坐标。
- max_row：表示表格的最大行数。
- min_row：表示表格的最小行数。
- max_column：表示表格的最大列数。
- min_column：表示表格的最小列数。
- rows：按行获取单元格。
- columns：按列获取单元格。
- freeze_panes：冻结窗格，同前面介绍的功能和含义一样。
- values：按行获取表格的内容。

常用的 Worksheet 方法如下。

- iter_rows：按行获取所有单元格，内置属性有 min_row、max_row、min_col 和 max_col。
- iter_columns：按列获取所有的单元格。
- append：在表格末尾添加数据。
- merged_cells：合并多个单元格。
- unmerged_cells：移除合并的单元格。

3. Cell 对象

Cell 对象比较简单，常用的属性如下。

- row：表示单元格所在的行。
- column：表示单元格所在的列。
- value：表示单元格的值。
- coordinate：表示单元格的坐标。

下面的实例文件 office01.py 演示了使用 openpyxl 读取指定 Excel 文件数据的过程。

源码路径：daima\7\7-2\office01.py

```
from openpyxl import load_workbook
wb = load_workbook("template.xlsx")#打开一个xlsx文件
print(wb.sheetnames)
```

```
sheet = wb.get_sheet_by_name("Sheet3")#看看打开的Excel表里面有哪些Sheet
#下面读取指定的Sheet
print(sheet["C"])
print(sheet["4"])
print(sheet["C4"].value)      # c4        <- C4格的值
print(sheet.max_row)          # 10        <-最大行数
print(sheet.max_column)       # 5         <-最大列数
for i in sheet["C"]:
    print(i.value, end=" ")   # c1 c2 c3 c4 c5 c6 c7 c8 c9 c10    <-C列中的所有值
```

执行后会输出：

```
['Sheet1', 'Sheet2', 'Sheet3']
  sheet = wb.get_sheet_by_name("Sheet3")
(<Cell 'Sheet3'.C1>, <Cell 'Sheet3'.C2>, <Cell 'Sheet3'.C3>, <Cell 'Sheet3'.C4>, <Cell 'Sheet3'.C5>, <Cell 'Sheet3'.C6>, <Cell 'Sheet3'.C7>, <Cell 'Sheet3'.C8>, <Cell 'Sheet3'.C9>, <Cell 'Sheet3'.C10>)
(<Cell 'Sheet3'.A4>, <Cell 'Sheet3'.B4>, <Cell 'Sheet3'.C4>, <Cell 'Sheet3'.D4>, <Cell 'Sheet3'.E4>)
c4
10
5
c1 c2 c3 c4 c5 c6 c7 c8 c9 c10
```

下面的实例文件 office02.py 演示了将 4 组数据导入 Excel 文件中的过程。

源码路径：daima\7\7-2\office02.py

```
import openpyxl
import time

ls = [['马坡','接入交换','192.168.1.1','G0/3','AAAA-AAAA-AAAA'],
      ['马坡','接入交换','192.168.1.2','G0/8','BBBB-BBBB-BBBB'],
      ['马坡','接入交换','192.168.1.2','G0/8','CCCC-CCCC-CCCC'],
      ['马坡','接入交换','192.168.1.2','G0/8','DDDD-DDDD-DDDD']]

#定义数据

time_format = '%Y-%m-%d__%H:%M:%S'
time_current = time.strftime(time_format)
#定义时间格式

def savetoexcel(data,sheetname,wbname):
    print("写入excel: ")
    wb=openpyxl.load_workbook(filename=wbname)
    #打开Excel文件

    sheet=wb.active #关联Excel活动的Sheet（这里关联的是Sheet1）
    max_row = sheet.max_row #获取Sheet1中最大的行数
    row = max_row + 3      #将新数据写入最大行数加3的位置
    data_len=row+len(data)   #计算当前数据的长度

    for data_row in range(row,data_len):  # 写入数据
    #轮询每一行进行写入数据
        for data_col1 in range(2,7):
        #针对每一行，下面还要执行for循环来写入列的数据
            _ =sheet.cell(row=data_row, column=1, value=str(time_current))
            #在每行的第一列写入时间
            _ =sheet.cell(row=data_row,column=data_col1,value=str(data[data_row-data_len]
            [data_col1-2]))
            #从第二列开始写入数据

    wb.save(filename=wbname)       #保存数据
    print("保存成功")

savetoexcel(ls,"Sheet1","template.xlsx")
```

执行后会在指定文件 template.xlsx 中导入 4 组数据，如图 7-5 所示。

	A	B	C	D	E	F	G
1	a1	b1	c1	d1	e1		
2	a2	b2	c2	d2	e2		
3	a3	b3	c3	d3	e3		
4	a4	b4	c4	d4	e4		
5	a5	b5	c5	d5	e5		
6	a6	b6	c6	d6	e6		
7	a7	b7	c7	d7	e7		
8	a8	b8	c8	d8	e8		
9	a9	b9	c9	d9	e9		
10	a10	b10	c10	d10	e10		
11							
12							
13	2018-04-0	马坡	接入交换	192.168.1.	G0/3	AAAA-AAAA-AAAA	
14	2018-04-0	马坡	接入交换	192.168.1.	G0/8	BBBB-BBBB-BBBB	
15	2018-04-0	马坡	接入交换	192.168.1.	G0/8	CCCC-CCCC-CCCC	
16	2018-04-0	马坡	接入交换	192.168.1.	G0/8	DDDD-DDDD-DDDD	

图 7-5　导入的 4 组数据

下面的实例文件 office03.py 演示了在指定 Excel 文件中按关键字检索数据的过程。

源码路径：daima\7\7-2\office03.py

```
import openpyxl

wb=openpyxl.load_workbook("template.xlsx")
the_list =[]

while True:
    info = input('请输入关键字：').upper().strip()
    if len(info) == 0:   # 输入的关键字不能为空，否则继续循环
        continue
    count = 0
    for line1 in wb['Sheet3'].values:   # 轮询列表
        if None not in line1:
        ##Excel中空行的数据表示None，当这里匹配None时就不会再进行for循环，所以需要匹配非
        None的数据才能进行下面的for循环
            for line2 in line1:   # 因为列表中还存在元组，所以需要将元组的内容也轮询一遍
                if info in line2:
                    count += 1   # 统计关键字被匹配了多少次
                    print(line1) #匹配关键字后输出元组信息

    else:
        print('匹配"%s"的数量统计：%s个条目被匹配' % (info, count))   # 显示查找的关键字被匹配
        #了多少次
```

执行后可以通过输入关键字的方式快速查询 Excel 文件中的数据，例如下面的检索过程。

```
请输入关键字：马坡
('2018-04-04__16:28:45', '马坡', '接入交换', '192.168.1.1', 'G0/3', 'AAAA-AAAA-AAAA')
('2018-04-04__16:28:45', '马坡', '接入交换', '192.168.1.2', 'G0/8', 'BBBB-BBBB-BBBB')
('2018-04-04__16:28:45', '马坡', '接入交换', '192.168.1.2', 'G0/8', 'CCCC-CCCC-CCCC')
('2018-04-04__16:28:45', '马坡', '接入交换', '192.168.1.2', 'G0/8', 'DDDD-DDDD-DDDD')
匹配"马坡"的数量统计：4个条目被匹配
请输入关键字：192.168.1.1
('2018-04-04__16:28:45', '马坡', '接入交换', '192.168.1.1', 'G0/3', 'AAAA-AAAA-AAAA')
匹配"192.168.1.1"的数量统计：1个条目被匹配
```

下面的实例文件 office04.py 演示了，将指定数据导入 Excel 文件中，并根据导入的数据在 Excel 文件中生成图表的过程。

源码路径：daima\7\7-2\office04.py

```
from openpyxl import Workbook
from openpyxl.chart import (
    AreaChart,
    Reference,
    Series,
)

wb = Workbook()
ws = wb.active

rows = [
```

```
        ['Number', 'Batch 1', 'Batch 2'],
        [2, 40, 30],
        [3, 40, 25],
        [4, 50, 30],
        [5, 30, 10],
        [6, 25, 5],
        [7, 50, 10],
]

for row in rows:
    ws.append(row)

chart = AreaChart()
chart.title = "Area Chart"
chart.style = 13
chart.x_axis.title = 'Test'
chart.y_axis.title = 'Percentage'

cats = Reference(ws, min_col=1, min_row=1, max_row=7)
data = Reference(ws, min_col=2, min_row=1, max_col=3, max_row=7)
chart.add_data(data, titles_from_data=True)
chart.set_categories(cats)

ws.add_chart(chart, "A10")

wb.save("area.xlsx")
```

执行后会将 rows 中的数据导入文件 area.xlsx 中，并在文件 area.xlsx 中根据数据绘制图表，如图 7-6 所示。

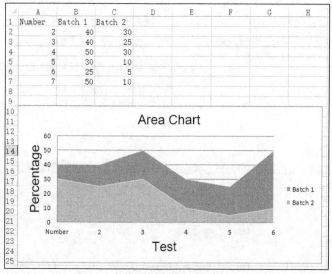

图 7-6　导入数据并绘制图表

7.2.2　使用 pyexcel 模块

使用 pyexcel 模块可以操作 Excel 和 CSV 文件。在使用 pyexcel 之前需要先安装 pyexcel_xls，安装命令如下。

```
pip install pyexcel_xls
```

然后通过如下命令安装 pyexcel。

```
pip install pyexcel
```

下面的实例文件 office05.py 演示了使用 pyexcel 读取并写入 CSV 文件的过程。

源码路径：daima\7\7-2\office05.py

```
import pyexcel as p
sheet = p.get_sheet(file_name="example.csv")
```

```
print(sheet)

with open('tab_example.csv', 'w') as f:
    unused = f.write('I\tam\ttab\tseparated\tcsv\n') # for passing doctest
    unused = f.write('You\tneed\tdelimiter\tparameter\n') # unused is added
sheet = p.get_sheet(file_name="tab_example.csv", delimiter='\t')
print(sheet)
```

在上述代码中,首先读取了文件 example.csv 中的内容,然后将指定的数据写入新建的 CSV 文件 tab_example.csv 中。输出内容和写入 Excel 中的内容如图 7-7 (a) 和 (b) 所示。

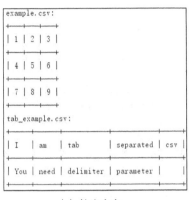

(a) 输出内容　　　　　　　　(b) 写入 Excel 中的内容

图 7-7　执行效果

下面的实例文件 read_cell_by_cell.py 演示了,使用 pyexcel 读取指定 Excel 文件中每个单元格的过程。

源码路径：daima\7\7-2\read_cell_by_cell.py

```
import os
import pyexcel as pe
def main(base_dir):
    # 读取的文件"example.xlsm"
    spreadsheet = pe.get_sheet(file_name=os.path.join(base_dir, "example.csv"))

    # 遍历每一行
    for r in spreadsheet.row_range():
        # 遍历每一列
        for c in spreadsheet.column_range():
            print(spreadsheet.cell_value(r, c))

if __name__ == '__main__':
    main(os.getcwd())
```

执行后的效果如图 7-8 所示。

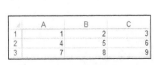 

(a) Excel 中的内容　　　　(b) 执行后的输出结果

图 7-8　执行效果

下面的实例文件 read_column_by_column.py 演示了使用 pyexcel 按列读取并显示指定 Excel

文件中每个单元格的过程。

源码路径：daima\7\7-2\read_column_by_column.py

```
def main(base_dir):
    spreadsheet = pe.get_sheet(file_name=os.path.join(base_dir, "example.xlsx"))
    for value in spreadsheet.columns():
        print(value)

if __name__ == '__main__':
    main(os.getcwd())
```

执行后的效果如图 7-9 所示。

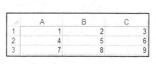
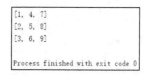

(a) Excel 中的内容　　　　(b) 读取后的输出结果

图 7-9　执行效果

如果在一个 Excel 文件中有多个 Sheet，例如，在文件 multiple-sheets-example.xls 中有 3 个 Sheet，里面的数据如图 7-10 所示。

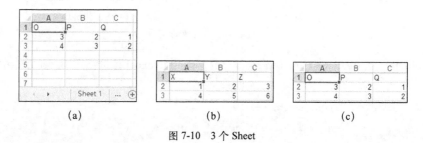

　　(a)　　　　　　　　(b)　　　　　　　　(c)

图 7-10　3 个 Sheet

通过下面的实例文件 read_excel_book.py，可以读取并显示上述 read_excel_book.py 文件中的所有数据。

源码路径：daima\7\7-2\read_excel_book.py

```
def main(base_dir):
    book = pe.get_book(file_name=os.path.join(base_dir,"multiple-sheets-example.xls"))
        # 默认的迭代器为Boo实例
    for sheet in book:
        # 每张表都有名字
        print("sheet: %s" % sheet.name)
        # 一旦拥有了一个表实例，
        #就可以将其视为一个读取器实例，可以按照期望的方式迭代它的成员
        for row in sheet:
            print(row)

if __name__ == '__main__':
    main(os.getcwd())
```

执行后会输出：

```
sheet: Sheet 1
[1, 2, 3]
[4, 5, 6]
[7, 8, 9]
sheet: Sheet 2
['X', 'Y', 'Z']
[1, 2, 3]
[4, 5, 6]
sheet: Sheet 3
```

```
['O', 'P', 'Q']
[3, 2, 1]
[4, 3, 2]
```

而通过下面的实例文件 write_excel_book.py，可以将 3 组数据导入新建的 multiple-sheets1.xls 文件中。3 组数据分别对应里面的 3 个 Sheet。

源码路径：daima\7\7-2\write_excel_book.py

```
def main(base_dir):
    data = {
        "Sheet 1": [[1, 2, 3], [4, 5, 6], [7, 8, 9]],
        "Sheet 2": [['X', 'Y', 'Z'], [1, 2, 3], [4, 5, 6]],
        "Sheet 3": [['O', 'P', 'Q'], [3, 2, 1], [4, 3, 2]]
    }
    pe.save_book_as(bookdict=data, dest_file_name="multiple-sheets1.xls")

if __name__ == '__main__':
    main(os.getcwd())
```

执行后会创建拥有 3 个 Sheet 的 Excel 文件，如图 7-11 所示。

图 7-11　创建的 Excel 文件

下面的实例文件 series.py 演示了，使用 pyexcel 以多种方式获取 Excel 数据的过程。

源码路径：daima\7\7-2\series.py

```
def main(base_dir):
    sheet = pe.get_sheet(file_name=os.path.join(base_dir,"example_series.xls"),
                         name_columns_by_row=0)
    print(json.dumps(sheet.to_dict()))
    #获取列标题
    print(sheet.colnames)
    #在一维数组中获取内容
    data = list(sheet.enumerate())
    print(data)
    #逆序获取一维数组中的内容
    data = list(sheet.reverse())
    print(data)

    #在一维数组中获取内容，但垂直地迭代
    data = list(sheet.vertical())
    print(data)
    #获取一维数组中的内容，以相反的顺序垂直迭代
    data = list(sheet.rvertical())
    print(data)

    #获取二维数组的数据
    data = list(sheet.rows())
    print(data)

    #以相反的顺序获取二维数组的行
```

```
            data = list(sheet.rrows())
            print(data)

            #获取二维数组的列
            data = list(sheet.columns())
            print(data)

            #以相反的顺序获取一个二维数组的列
            data = list(sheet.rcolumns())
            print(data)

            #可以把结果写入一个文件中
            sheet.save_as("example_series.xls")

    if __name__ == '__main__':
        main(os.getcwd())
```

通过上述代码以多种方式获取了 Excel 中的数据，包括一维数组顺序和逆序、二维数组顺序和逆序。执行后会输出：

```
{"Column 1": [1, 2, 3], "Column 2": [4, 5, 6], "Column 3": [7, 8, 9]}
['Column 1', 'Column 2', 'Column 3']
[1, 4, 7, 2, 5, 8, 3, 6, 9]
[9, 6, 3, 8, 5, 2, 7, 4, 1]
[1, 2, 3, 4, 5, 6, 7, 8, 9]
[9, 8, 7, 6, 5, 4, 3, 2, 1]
[[1, 4, 7], [2, 5, 8], [3, 6, 9]]
[[3, 6, 9], [2, 5, 8], [1, 4, 7]]
[[1, 2, 3], [4, 5, 6], [7, 8, 9]]
[[7, 8, 9], [4, 5, 6], [1, 2, 3]]
```

下面的实例文件 import_xls_into_database_via_sqlalchemy.py 演示了，使用 pyexcel 将数据分别导入 Excel 文件和 SQLite 数据库的过程。

源码路径：daima\7\7-2\import_xls_into_database_via_sqlalchemy.py

```
engine = create_engine("sqlite:///birth.db")
Base = declarative_base()
Session = sessionmaker(bind=engine)


class BirthRegister(Base):
    __tablename__ = 'birth'
    id = Column(Integer, primary_key=True)
    name = Column(String)
    weight = Column(Float)
    birth = Column(Date)

Base.metadata.create_all(engine)

#创建数据
data = [
    ["name", "weight", "birth"],
    ["Adam", 3.4, datetime.date(2017, 2, 3)],
    ["Smith", 4.2, datetime.date(2014, 11, 12)]
]
pyexcel.save_as(array=data,
                dest_file_name="birth.xls")

#导入Excel文件中
session = Session()  # obtain a sql session
pyexcel.save_as(file_name="birth.xls",
                name_columns_by_row=0,
                dest_session=session,
                dest_table=BirthRegister)

#验证结果
sheet = pyexcel.get_sheet(session=session, table=BirthRegister)
print(sheet)
session.close()
```

执行后会输出：
```
birth:
+------------+----+-------+--------+
| birth      | id | name  | weight |
+------------+----+-------+--------+
| 2017-02-03 | 1  | Adam  | 3.4    |
+------------+----+-------+--------+
| 2014-11-12 | 2  | Smith | 4.2    |
+------------+----+-------+--------+
```

下面的实例代码演示了在 Flask Web 项目中使用 pyexcel 处理数据的过程。

(1) 编写程序文件 pyexcel_server.py。首先通过函数 upload()实现文件上传功能，将上传的 Excel 文件导出为 JSON 格式并显示在页面中。然后定义 data，在里面保存了将要处理的数据。最后通过函数 download()实现文件下载功能，使用 data 中的数据生成一个 CSV 文件并下载。文件 pyexcel_server.py 的具体实现代码如下。

源码路径：daima\7\7-2\memoryfile\pyexcel_server.py

```python
app = Flask(__name__)

@app.route('/upload', methods=['GET', 'POST'])
def upload():
    if request.method == 'POST' and 'excel' in request.files:
            #处理上传的文件
            filename = request.files['excel'].filename
            extension = filename.split(".")[1]
            #获取文件扩展名和内容,传递一个元组而不是文件名
            content = request.files['excel'].read()
            if sys.version_info[0] > 2:
                    #为了支持Python,必须将字节解码为STR
                    content = content.decode('ANSI')
            sheet = pe.get_sheet(file_type=extension, file_content=content)
            #然后像往常一样使用它
            sheet.name_columns_by_row(0)
            #用JSON回应
            return jsonify({"result": sheet.dict})
    return render_template('upload.html')

data = [
    ["REVIEW_DATE", "AUTHOR", "ISBN", "DISCOUNTED_PRICE"],
    ["1985/01/21", "Douglas Adams", '0345391802', 5.95],
    ["1990/01/12", "Douglas Hofstadter", '04650265867', 9.95],
    ["1998/07/15", "Timothy \"The Parser\" Campbell", '0968411304', 18.99],
    ["1999/12/03", "Richard Friedman", '0060630353', 5.95],
    ["2004/10/04", "Randel Helms", '0879755725', 4.50]
]
@app.route('/download')
def download():
    sheet = pe.Sheet(data)
    output = make_response(sheet.csv)
    output.headers["Content-Disposition"] = "attachment; filename=export.csv"
    output.headers["Content-type"] = "text/csv"
    return output

if __name__ == "__main__":
    #启动Web服务器
    app.run()
```

(2) 编写模板文件 upload.html 实现了一个文件上传界面，具体实现代码如下。

源码路径：daima\7\7-2\memoryfile\templates\upload.html

```html
<html>
<title>Example upload : pyexcel example</title>
<body>
<form method=POST enctype=multipart/form-data action="{{ url_for('upload') }}">
    <input type=file name=excel>
     <input type=submit value="upload">
</form>
```

```
</body>
</html>
```
在运行本实例程序之前要确保已经安装了 gunicorn，然后通过如下命令运行本程序。
```
python pyexcel_server.py runserver
```
在浏览器地址栏中输入"http://127.0.0.1:5000/upload"后会显示文件上传界面，如图 7-12 所示。

图 7-12　文件上传界面

单击"上传文件"按钮后选择一个文件，单击"upload"按钮后将在页面中显示该文件的 JSON 格式，如图 7-13 所示。

在浏览器地址栏中输入"http://127.0.0.1:5000/download"后，会下载指定的 CSV 文件 export.csv。这个文件中的数据是从 data 中导入并生成的。执行效果如图 7-14 所示。

图 7-13　显示上传的文件的 JSON 格式

图 7-14　下载指定文件 export.csv

7.2.3　使用 python-docx 模块

使用 python-docx 模块可以读取、查询以及修改 Office（Word、Excel 和 PowerPoint）文件，安装命令如下。
```
pip install python-docx
```
使用模块 python-docx 操作 Office 文件的流程如下。

（1）打开文档，创建工作文档对象，方法如下。
```
from docx import Document
document = Document()
```
这将打开一个基于默认"模板"的空白文档，可以打开并使用现有的 Word 文档的 python-docx。

（2）增加一段文本。

段落是 Word 的基础，可以用于正文文本，也可以用于标题和列表项目（如项目符号）。下面的程序演示了添加一个段落的简单方法。
```
paragraph = document.add_paragraph('Lorem ipsum dolor sit amet.')
```
上述方法将返回关于段落的引用，新添加的段落在文档的结尾，新的段落引用被分配给 paragraph。还可以使用一个段落作为"光标"，并在其中直接插入一个新段落。
```
prior_paragraph = paragraph.insert_paragraph_before('Lorem ipsum')
```
以上语句允许将一个段落插入文档的中间，这在修改现有文档时通常很重要，而不是重新生成。

（3）添加标题。

在除了最短文档之外的任何内容中，正文文本被分成多个部分，每个部分以一个标题开始。

下面是添加一个标题的代码。

```
document.add_heading('The REAL meaning of the universe')
```

在默认情况下，这会添加顶级标题，在 Word 中显示为"标题 1"。当需要用到子级的标题时，指定所需的级别为 1~9 的任意整数即可，例如下面的演示代码。

```
document.add_heading('The role of dolphins', level=2)
```

如果指定级别为 0，将添加"标题"段落。这可以方便地启动一个相对较短的文档，没有单独的标题页。

（4）添加分页符。

在模块 python-docx 中通过以下代码添加分页符。

```
document.add_page_break()
```

这个分页符可能是一个标志，可以设置一个段落样式属性，也可以将标题设置为某个级别以便始终启动新页面。

（5）添加表。

下面演示了在模块 python-docx 中添加表方法。

```
table = document.add_table(rows=2, cols=2)
```

表具有几个属性和方法，我们需要填充它们。例如，通过如下代码设置可以始终按行号和列号访问单元格。

```
cell = table.cell(0, 1)
```

这样便给出了我们刚刚创建的表最上面一行右边的单元格。注意，行号和列号是基于零的，就像在列表中访问一样。

一旦拥有一个单元格，就可以往里面填写东西，例如下面的演示代码。

```
cell.text = 'parrot, possibly dead'
```

通常，一次访问一行单元格更容易。

接下来通过具体实例来演示模块 python-docx 的使用方法。

1. 创建 Word 文档

下面的实例文件python-docx01.py演示了使用python-docx创建一个简单Word文档的过程。

源码路径：daima\7\7-2\python-docx01.py

```
from docx import Document
document = Document()
document.add_paragraph('Hello,Word!')
document.save('demo.docx')
```

首先，在上述代码中，第一行引入 docx 库和 Document 类。类 Document 代表"文档"，第二行创建了类 Document 的实例 document，相当于"这篇文档"。然后，我们在文档中利用函数 add_paragraph()添加了一个段落，段落的内容是"Hello,Word!"。最后，使用函数 save()将文档保存在磁盘上。执行后会创建一个名为 demo.docx 的文件，打开后的内容如图 7-15 所示。

图 7-15　文件 demo.docx 的内容

2. 在 Word 中插入图片

下面的实例文件 python-docx02.py 演示了，使用 python-docx 向 Word 文档中插入 10 个实心圆的过程。

源码路径：dalma\7\7-2\python-docx02.py

```python
from docx import Document
from PIL import Image,ImageDraw
from io import BytesIO

document = Document()                                #新建文档
p = document.add_paragraph()                         #添加一个段落
r = p.add_run()                                      #添加一个游程
img_size = 20
for x in range(20):
    im = Image.new("RGB", (img_size,img_size), "white")
    draw_obj = ImageDraw.Draw(im)
    draw_obj.ellipse((0,0,img_size-1,img_size-1), fill=255-x)#画圆
    fake_buf_file = BytesIO()                        #用BytesIO将图片保存在内存里，减少磁盘操作
    im.save(fake_buf_file,"png")
    r.add_picture(fake_buf_file)                     #在当前游程中插入图片
    fake_buf_file.close()
document.save("demo.docx")
```

运行上述代码后，会在 Word 文档 demo.docx 中添加 20 个实心圆，颜色为红色且由浅入深。打开后的内容如图 7-16 所示。

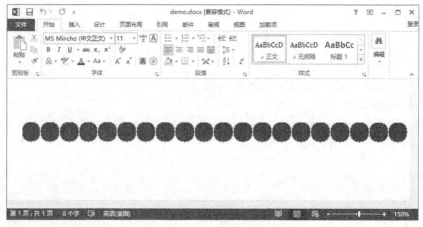

图 7-16　文件 demo.docx 的内容

3. 创建结构文档

创建结构文档，就是创建具有不同样式的段落。在类 Document 的函数 add_paragraph()中，第一个参数表示段落的文字，第二个可选参数表示段落的样式。通过这个样式参数即可设置所添加段落的样式。如果不指定这个参数，则默认样式为"正文"。函数 add_paragraph()的返回值是一个段落对象，可以通过这个对象的 style 属性得到该段落的样式，也可以重写这个属性以设置该段落的样式。

下面的实例文件 python-docx03.py 演示了，使用 python-docx 向 Word 文档中插入 10 个实心圆的过程。

源码路径：dalma\7\7-2\python-docx03.py

```python
from docx import Document

doc = Document()
doc.add_paragraph(u'Python为什么这么受欢迎？','Title')
doc.add_paragraph(u'作者','Subtitle')
```

```
doc.add_paragraph(u'摘要：本文阐明了Python的优势...','Body Text 2')
doc.add_paragraph(u'简单','Heading 1')
doc.add_paragraph(u'易学')
doc.add_paragraph(u'易用','Heading 2')
doc.add_paragraph(u'功能强')
p = doc.add_paragraph(u'贴合小年轻')
p.style = 'Heading 2'
doc.save('demo.docx')
```

通过上述代码创建了一个指定段落样式内容的文件 demo.docx，打开后的内容如图 7-17 所示。

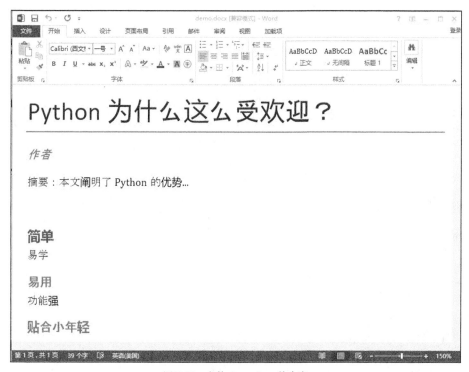

图 7-17　文件 demo.docx 的内容

在上述代码中，Title、Heading 等都是 Word 的内置样式。启动 Word 后，在"样式"选项卡中看到的样式图标就是 Word 的内置样式，如图 7-18 所示。

图 7-18　Word 的内置样式

对于英文版的 Word 来说，在样式选项卡下标注的样式名称就是在 Python 代码中可以使用的样式名称。而对于中文版的 Word 来说，内置样式仍然使用英文名称。如果我们没有安装英文版的 Word，可以用下面的实例文件 python-docx04.py 得到英文的样式名称。

源码路径：daima\7\7-2\python-docx04.py

```
from docx import Document
from docx.enum.style import WD_STYLE_TYPE
doc = Document()
styles = doc.styles
print("\n".join([s.name for s in styles if s.type == WD_STYLE_TYPE.PARAGRAPH]))
```

执行后会输出如下的英文样式名称：

```
Normal
Heading 1
```

```
Heading 2
Heading 3
Heading 4
Heading 5
Heading 6
Heading 7
Heading 8
Heading 9
No Spacing
Title
Subtitle
List Paragraph
Body Text
Body Text 2
Body Text 3
List
List 2
List 3
List Bullet
List Bullet 2
List Bullet 3
List Number
List Number 2
List Number 3
List Continue
List Continue 2
List Continue 3
macro
Quote
Caption
Intense Quote
TOC Heading
```

在上述样式列表中,文档结构常用的样式如下。

- Title:文档的标题,样式窗格里显示为"标题"。
- Subtitle:副标题。
- Heading n: n 级标题,样式窗格里显示为"标题 n"。
- Normal:正文。

对于文档标题以及 n 级标题的样式来说,可以使用类 Document 对象中的函数 add_heading() 来设置。这个函数的第一个参数表示文本内容。如果第二个参数设置为 0,则等价于 add_paragraph(text,'Title');如果第二个参数设置为大于 0 的整数 n,则等价于 add_paragraph(text, 'Heading %d' % n)。

4. 读取 Word 文档

下面的实例文件python-docx05.py演示了获取指定 Word 文档中的文本样式名称和每个样式的文字数目的过程。

源码路径:dalma\7\7-2\python-docx05.py

```
from docx import Document
import sys
path = "demo.docx"
document = Document(path)
for p in document.paragraphs:
    print(len(p.text))
    print(p.style.name)
```

对于中文文本来说,len 得到的是汉字个数,这和 Python 默认的多语言处理方法是一致的。读者可以用手里现有的 Word 文档试试(需要是 docx 格式)。如果文档是使用 Word 默认的样式创建的,则会输出 Title、Normal、Heading x 之类的样式名称。如果文档对默认样式进行了修改,那么依然会输出原有样式名称。如果文档创建了新样式,则使用新样式的段落会显示新样式的名称。在作者的计算机中执行后输出:

```
15
Title
2
Subtitle
20
Body Text 2
2
Heading 1
2
Normal
2
Heading 2
3
Normal
5
Heading 2
```

下面的实例文件 python-docx06.py 演示了获取指定 Word 文档中的文本的过程。

源码路径：daima\7\7-2\python-docx06.py

```
from docx import Document
path = "demo.docx"
document = Document(path)
for paragraph in document.paragraphs:
    print(paragraph.text)
```

执行后会输出文件 demo.docx 的内容。

```
Python为什么这么受欢迎？
作者
摘要：本文阐明了Python的优势...
简单
易学
易用
功能强
贴合小年轻
```

5. 创建表格

在一个 Word 文档中有两种表格，具体说明如下。

- 和段落同级的顶级表格：可以使用类 Document 中函数的 add_table()创建一个新的顶级表格对象，也可以使用类 Document 中的 tables 得到文档中所有的顶级表格。
- 表格里嵌套的表格：本书将主要讨论顶级表格，下面讨论的表格均指顶级表格。

我们可以把一个表格看作 M 行、N 列的矩阵。利用类 Table 中的 Cell 对象的 text 属性，可以设置、获取表格中任意单元格的文本。下面的实例文件 python-docx07.py 演示了获取在 Word 文档中创建指定表格的过程。

源码路径：daima\7\7-2\python-docx07.py

```
from docx import Document
import psutil

#获取当前计算机的配置
vmem = psutil.virtual_memory()
vmem_dict = vmem._asdict()

trow = 2
tcol = len(vmem_dict.keys())
#产生表格

document = Document()
table = document.add_table(rows=trow,cols=tcol,style = 'Table Grid')
for col,info in enumerate(vmem_dict.keys()):
    table.cell(0,col).text = info
    if info == 'percent':
        table.cell(1,col).text = str(vmem_dict[info])+'%'
    else:
        table.cell(1,col).text = str(vmem_dict[info]/(1024*1024)) + 'MB'
document.save('table.docx')
```

在上述代码中，使用库 psutil 获取当前计算机的内存信息，将获取到的这些信息作为表格中的填充数据。首先把从 psutil 得到的内存信息转化为一个 dict，它的 key 是项目（物理内存总大小、使用的内存、剩余的内存等），value 是各个项目的数值。执行后会创建文件 table.docx，在此文件中显示当前计算机的内存信息，如图 7-19 所示。

total	available	percent	used	free
16290.6875MB	8248.26953125MB	49.4%	8042.41796875MB	8248.26953125MB

图 7-19　文件 table.docx 中的内容

1）Row、Column 和 Cell 对象

在 python-docx 中，Row 和 Column 对象在遍历表格的时候非常有用。前面介绍的 Table 对象有两个属性 rows 和 columns，这两个对象的返回值实际上是对象 Rows 和 Columns。但是，Rows 和 Columns 这两个对象也等同于 Row 列表以及 Column 列表。正因为如此，列表的迭代、求长度等操作也适用于 Rows 和 Columns。例如：

```
for obj_row in table.columns:
    #obj_row是Row对象的一个实例
```

使用 Table 对象中的函数 add_row() 和 add_column() 可以动态地增加表格尺寸。函数 add_row() 的功能是在表格下方创建新列并返回新行的 Row 对象，函数 add_column() 的功能是在表格右侧创建新列并返回新行的 Column 对象。

在创建表格后，接下来可以用如下 5 种方法得到 Cell 对象。

- 使用 Table 对象中的函数 cell(row,col)，其中左上角的坐标为（0,0）。
- 使用 Table 对象中的函数 row_cells(row_index) 得到一个列表，它包含了某一行按列排序的所有 Cell。
- 在得到一个 Row 对象后，使用属性 Row.cells 得到该 Row 按列排序的所有 Cell。
- 使用 Table 对象中的函数 column_cells(column_index) 得到一个列表，其中包含了某一列按行排序的所有 Cell。
- 在得到一个 Column 对象后，使用属性 Column.cells 得到该 Column 按行排序的所有 Cell。

要遍历所有 Cell，可以先遍历所有行（table.rows），后遍历每一行所有的 Cell；也可以先遍历所有列（table.columns），后遍历每一列所有的 Cell。

Cell 对象中最常用的属性是 text。此属性的功能是设置单元格的内容，通过读取这个属性可以获取单元格的内容。另外，在 Cell 对象中还有 add_paragraph()、add_table() 等方法，利用这些方法可以形成复杂的表格，如多个段落以及嵌套的表格。

可以利用 Cell 对象中的函数 merge(other_cell) 合并单元格。合并的方式是以当前 Cell 为左上角，以 other_cell 为右下角进行合并。在下面的实例文件 python-docx08.py 中，首先创建了一个表格，然后合并其中的单元格并另存为 Word 文档。最后读取这个 Word 文档，并把每个 Cell 的坐标标注到了 Cell 里。

源码路径：daima\7\7-2\python-docx08.py

```
document = Document()
table = document.add_table(rows=9,cols=10,style = 'Table Grid')
cell_1 = table.cell(1,2)
cell_2 = table.cell(4,6)
cell_1.merge(cell_2)
document.save('table-1.docx')

document = Document('table-1.docx')
```

```
table = document.tables[0]
for row,obj_row in enumerate(table.rows):
    for col,cell in enumerate(obj_row.cells):
        cell.text = cell.text + "%d,%d " % (row,col)

document.save('table-2.docx')
```

执行后会生成两个 Word 文件 table-1.docx 和 table-2.docx，内容如图 7-20（a）和（b）所示。

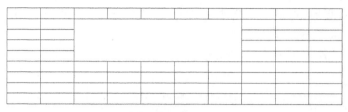

(a) 文件 table-1.docx

0,0	0,1	0,2	0,3	0,4	0,5	0,6	0,7	0,8	0,9
1,0	1,1	1,2 1,3 1,4 1,5 1,6 2,2 2,3 2,4 2,5 2,6 3,2 3,3					1,7	1,8	1,9
2,0	2,1	3,4 3,5 3,6 4,2 4,3 4,4 4,5 4,6					2,7	2,8	2,9
3,0	3,1						3,7	3,8	3,9
4,0	4,1						4,7	4,8	4,9
5,0	5,1	5,2	5,3	5,4	5,5	5,6	5,7	5,8	5,9
6,0	6,1	6,2	6,3	6,4	6,5	6,6	6,7	6,8	6,9
7,0	7,1	7,2	7,3	7,4	7,5	7,6	7,7	7,8	7,9
8,0	8,1	8,2	8,3	8,4	8,5	8,6	8,7	8,8	8,9

(b) 文件 table-2.docx

图 7-20 生成的两个 Word 文件的内容

由此可见，在合并单元格之后，可以利用合并区域的任何一个单元格的坐标指代这个合并区域。也就是说，单元格的合并并没有使 Cell 消失，只是这些 Cell 共享里面的内容而已。

2）调整宽度

行及其单元格的高度受字体限制，不能够手动调整高度，但是列及其单元格的宽度可以手动调整。下面的实例文件 python-docx09.py 演示了调整 Word 中表格宽度的过程。

源码路径：daima\7\7-2\python-docx09.py

```
document = Document()
for row in range(9):
    t = document.add_table(rows=1, cols=1, style='Table Grid')
    t.autofit = False    # 很重要，必须设置
    w = float(row) / 2.0
    t.columns[0].width = Inches(w)

document.save('table-step.docx')
```

执行后会设置文件 table-step.docx 中表格的宽度，效果如图 7-21 所示。

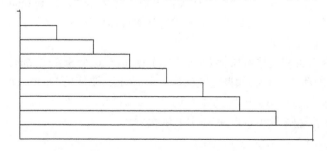

图 7-21 文件 table-step.docx 的宽度变化

3）设置表格的设计样式

在 Word 中的"表格工具"选项卡中有各种样式，通过这些样式可以将表格设置为不同的

效果。在python-docx中可以通过Document.add_table的第三个参数设定表格样式,也可以用table的属性style获取和设置样式。如果设置样式,可以直接用样式的英文名称,例如"Table Grid";如果读取样式,会得到一个 Style 对象,这个对象是可以跨文档使用的。另外,也可以使用Style.name 方法得到它的名称。

python-docx 使用独立于 Office 的样式命名体系。随着 Office 的更新,python-docx 内建的样式名称和较新版本的 Office 样式名称可能会不一致。这时可以用下面的实例文件python-docx10.py 得到 python-docx 内部的表格样式名称。

源码路径:dalma\7\7-2\python-docx10.py

```
document = Document()
styles = document.styles
table_styles = [s for s in styles if s.type == WD_STYLE_TYPE.TABLE]
for style in table_styles:
    print(style.name)
```

执行后会输出:

```
Normal Table
Table Grid
Light Shading
Light Shading Accent 1
Light Shading Accent 2
Light Shading Accent 3
Light Shading Accent 4
Light Shading Accent 5
Light Shading Accent 6
……省略后的样式列表
```

下面的实例文件 python-docx11.py 演示了使用指定样式修饰表格的过程。

源码路径:dalma\7\7-2\python-docx11.py

```
table = document.add_table(rows=trow,cols=tcol,style = 'Colorful Grid Accent 4')
```

执行上述代码后得到的表格样式如图 7-22 所示。

total	available	percent	used	free
16290.6875M	8866.4765625M	45.6%	7424.2109375M	8866.4765625M

图 7-22 新的表格样式

6. 使用样式管理器

在 Word 以及带有文本编辑功能的 Office 软件中,常用的样式类型有字符样式、段落样式、表格样式和枚举样式等。python-docx 目前主要支持上述 3 类样式,每种样式包含的内容主要有字体(包括字体名称、颜色、倾斜、加粗)和段落格式(缩进、分页模式)等。根据样式应用对象的不同,每类样式不一定包含所有的内容。例如,字符样式就没有段落格式这个内容。

Word 中的样式管理器负责管理内建样式和用户自定义样式,库 python-docx 可以操作这个管理器,把自定义的样式添加到管理器里,或者从管理器中选取一个样式加以应用。

1) 创建样式和设置字体

字符对象、段落对象和表格对象都有一个成员Style,在这个Style 中都有一个字体成员Font,在里面包含了字体的所有设置。如果凭空创建一个 Style,要使用 Document 对象的 Styles 集合的 add_style()函数实现。在下面的实例文件 python-docx12.py 中添加了 10 个段落,为每个段落创建一个新的 Style。这些 Style 的区别是字号依次增大。

源码路径:dalma\7\7-2\python-docx12.py

```
from docx import Document
from docx.shared import Pt
from docx.enum.style import WD_STYLE_TYPE

doc = Document()
```

```
for i in range(10):
    p = doc.add_paragraph(u'段落 %d' % i)
    style = doc.styles.add_style('UserStyle%d' % i, WD_STYLE_TYPE.PARAGRAPH)
    style.font.size = Pt(i + 20)
    p.style = style

doc.save('style-1.docx')
```

在上述代码中，通过 python-docx 中的 docx.shared 模块，提供了以 Pt（磅）为单位的字体。类似的单位有毫米、厘米和英寸等。执行后会创建一个包含指定样式段落文本的 Word 文件 style-1.docx，如图 7-23 所示。

图 7-23　文件 style-1.docx 的内容

下面的实例文件 python-docx13.py 演示了直接使用 Run.font 设置字体样式的过程。

源码路径：daima\7\7-2\python-docx13.py

```
doc = Document()
p = doc.add_paragraph()
text_str = u'好好学习Python，努力做开发专家，成为最牛的程序员。'
for i, ch in enumerate(text_str):
    run = p.add_run(ch)
    font = run.font
    font.name = u'微软雅黑'
    # bug of python-docx
    run._element.rPr.rFonts.set(qn('w:eastAsia'), u'微软雅黑')
    font.bold = (i % 2 == 0)
    font.italic = (i % 3 == 0)
    color = font.color
    color.rgb = RGBColor(i * 10 % 200 + 55, i * 20 % 200 + 55, i * 30 % 200 + 55)

doc.save('style-2.docx')
```

执行后会创建一个包含指定样式文本的 Word 文件 style-2.docx，如图 7-24 所示。因为本书不是彩色印刷，所以书中的执行效果不够明显，建议读者在计算机中查看文本颜色。

2）设置段落格式

通过使用 python-docx 中段落格式的成员 paragraph_format，可以设置指定的段落样式和表格样式，这相当于 Word 中的"段落"选项卡，如图 7-25 所示。

图 7-24　文件 style-2.docx 的内容　　　　图 7-25　Word 中的"段落"选项卡

在 python-docx 的段落格式对象中有各种成员，相当于 Word 的"段落"对话框中的各个要素，如图 7-26 所示。

第 7 章　特殊文本格式处理

在 python-docx 中包含了如下和段落格式相关的成员。

- ParagraphFormat.alignment：选择一种 WD_PARAGRAPH_ALIGNMENT（对齐方式）。
- ParagraphFormat.left_indent：表示左侧缩进长度。
- ParagraphFormat.right_indent：表示右侧缩进长度。
- ParagraphFormat.first_line_indent：表示特殊缩进的长度。
- ParagraphFormat.space_before：表示段前间距长度。
- ParagraphFormat.space_after：表示段后间距长度。
- ParagraphFormat.line_spacing_rule：选择一个 WD_LINE_SPACING（几倍行距）。
- ParagraphFormat.line_spacing：表示"间距"中的"设置值"。
- ParagraphFormat.widow_control：孤行控制，True 表示设置，None 表示继承 Style 设置。
- ParagraphFormat.keep_with_next：与下一段同页，True 表示设置，None 表示继承 Style 设置。
- ParagraphFormat.keep_together：段中不分页，True 表示设置，None 表示继承 Style 设置。
- ParagraphFormat.page_break_before：段前分页，True 表示设置，None 表示继承 Style 设置。

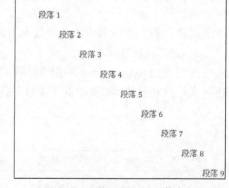

图 7-26　"段落"对话框

下面的实例文件 python-docx14.py 演示了设置段落递进的左对齐样式的过程。

源码路径：daima\7\7-2\python-docx14.py

```
doc = Document()
for i in range(10):
    p = doc.add_paragraph(u'段落 %d' % i)
    style = doc.styles.add_style('UserStyle%d' % i, WD_STYLE_TYPE.PARAGRAPH)
    style.paragraph_format.left_indent = Cm(i)
    p.style = style

doc.save('style-3.docx')
```

执行后将会创建一个包含指定段落样式文本的 Word 文件 style-3.docx，如图 7-27 所示。

3）样式管理器

在 Word 中自带了多种样式，可以使用库 python-docx 中的 Document.styles 集合来访问 builtin 属性为 True 的自带样式。当然，开发者通过 add_style()函数添加的样式，也会存放在 styles 集合中。对于开发者自己创建的样式，如果将其属性 hidden 和 quick_style 分别设置为 False 与 True，则可以将这个自建样式添加到 Word 快速样式管理器中。下面的实例文件 python-docx15.py 演示了开发者创建样式的过程。

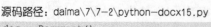

图 7-27　文件 style-3.docx 的内容

源码路径：daima\7\7-2\python-docx15.py

```
doc = Document()
for i in range(10):
```

```
        p = doc.add_paragraph(u'段落 %d' % i)
        style = doc.styles.add_style('UserStyle%d' % i, WD_STYLE_TYPE.PARAGRAPH)
        style.paragraph_format.left_indent = Cm(i)
        p.style = style
        if i == 7:
            style.hidden = False
            style.quick_style = True

for style in doc.styles:
    print(style.name, style.builtin)

doc.paragraphs[3].style = doc.styles['Subtitle']
doc.save('style-4.docx')
```

通过上述代码，在 Word 文件 style-4.docx 中创建了 9 种（UserStyle1~ UserStyle9）样式，如图 7-28（a）和（b）所示。

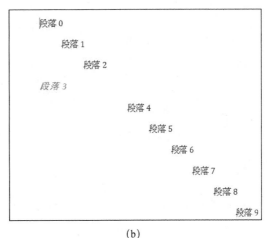

(a)　　　　　　　　　　　　　　　　(b)

图 7-28　文件 style-4.docx 的样式和内容

7.2.4　使用 xlrd 和 xlwt 库读写 Excel

1. xlrd 库

在 Python 程序中，可以使用 xlrd 库读取 Excel 文件的内容。安装 xlrd 库的命令如下。

```
pip install xlrd
```

使用 xlrd 库读取 Excel 文件的步骤如下。

（1）通过如下代码导入 xlrd。

```
import xlrd
```

（2）读取 Excel 数据，例如，下面代码用于打开 unit 表，将表中数据读进 data 中。

```
xlrd.open_workbook(excel路径)
data = xlrd.open_workbook('unit.xlsx')1
```

（3）获取一个工作表。

获取工作表的方法有两种：通过索引顺序获取和通过工作表名字获取。下面是这两种获取方式的演示代码。

```
sheet = data.sheet_by_index(0)         #通过索引顺序获取第一个工作表
sheet = data.sheet_by_name(u'Sheet1')  #通过名称获取工作表
```

在上述代码中，sheet.name 表示工作表的名字。

（4）获取每一行或每一列的信息，例如，通过如下代码获取一个工作表的总行数和总列数。

```
nrows = sheet.nrows    #行数
ncols = sheet.ncols    #列数
```

通过如下代码获取一个工作表中整行或整列的值（数组）。

```
sheet.row_values(n)    #获取第n行的值，返回一个数组
sheet.col_values(m)    #获取第m行的值，返回一个数组
```

通过如下代码获取一个单元格的值。
```
sheet.cell(i, j).value
```
假设存在一个 Excel 文件 example.xlsx，其内容如图 7-29 所示。

图 7-29　文件 example.xlsx 的内容

下面的实例文件 ex01.py 演示了使用 xlrd 库读取指定 Excel 文件内容的过程。

源码路径：daima\7\7-2\ex01.py

```
import xlrd
#打开Excel
data = xlrd.open_workbook('example.xlsx')
#查看文件中包含sheet的名称
data.sheet_names()
#得到第一个工作表
table = data.sheets()[0]
table = data.sheet_by_index(0)
table = data.sheet_by_name(u'Sheet1')
#获取行数和列数
nrows = table.nrows
ncols = table.ncols
print(nrows)
print(ncols)
#遍历行,得到索引的列表
for rownum in range(table.nrows):
    print(table.row_values(rownum))
#分别使用行与列索引
cell_A1 = table.row(0)[0].value
cell_A2 = table.col(1)[0].value
print(cell_A1)
print(cell_A2)
```

执行后会输出：
```
3
3
[1.0, 2.0, 3.0]
[4.0, 5.0, 6.0]
[7.0, 8.0, 9.0]
1.0
6.0
1.0
```

2．xlwt 库

在 Python 程序中，可以使用 xlwt 库向 Excel 文件中写入内容。安装 xlwt 库的命令如下。
```
pip install xlwt
```
使用 xlwt 库写入 Excel 文件的步骤如下。

（1）通过如下代码导入 xlwt。
```
import xlwt
```
（2）通过如下代码创建 workbook，其实就是 Excel。
```
workbook = xlwt.Workbook(encoding = 'ascii')
```
（3）通过如下代码创建表。
```
worksheet = workbook.add_sheet('My Worksheet')
```
（4）通过如下代码向单元格中写入内容。
```
worksheet.write(0, 0, label = 'Row 0, Column 0 Value')
```
（5）保存 Excel 单元格，例如下面的代码。
```
workbook.save('Excel_Workbook.xls')
```

下面的实例文件 ex02.py 演示了，使用 xlwt 库将指定内容写入 Excel 文件并创建 Excel 文件的过程。

源码路径：daima\7\7-2\ex02.py

```python
import xlwt
from datetime import datetime

style0 = xlwt.easyxf('font: name Times New Roman, color-index red, bold on',
    num_format_str='#,##0.00')
style1 = xlwt.easyxf(num_format_str='D-MMM-YY')#当前日期

wb = xlwt.Workbook()
ws = wb.add_sheet('A Test Sheet')                      #sheet的名字

ws.write(0, 0, 1234.56, style0)                        #第1个单元格的内容
ws.write(1, 0, datetime.now(), style1)                 #第2个单元格的内容
ws.write(2, 0, 1)                                      #第3个单元格的内容
ws.write(2, 1, 1)                                      #第4个单元格的内容
ws.write(2, 2, xlwt.Formula("A3+B3"))                  #第5个单元格的内容

wb.save('example02.xls')
```

执行后会将指定内容写入文件 example02.xls 中，如图 7-30 所示。

图 7-30　文件 example02.xls 的内容

7.2.5　使用 xlsxwriter 库

在 Python 程序中，可以使用 xlsxwriter 库操作 Excel 文件。安装命令如下。

```
pip install xlsxwriter
```

使用库 xlsxwriter 的基本流程如下。

（1）创建一个 Excel 文档。

```
workbook = xlsxwriter.Workbook(dir)
```

（2）在文档中创建表。

```
table_name = 'sheet1'
worksheet = workbook.add_worksheet(table_name)  #创建一个表名为sheet1的表，并返回这个表
```

（3）创建表后，就可以在表格上面进行写入操作。

```
worksheet.write_column('A1', 5)  # 在A1单元格中写入数字5
```

我们可以修改输入内容的格式，例如，设置字体颜色加粗、斜体和日期格式等，这些可以通过使用 xlsxwriter 提供的格式类实现。通过下面的代码写入了一个红色粗体的日期类。

```
import datetime
#需要先把字符串格式化成日期
date_time = datetime.datetime.striptime('2017-1-25', '%Y-%m-%d')
#定义一个格式类，粗体的红色的日期
date_format = workbook.add_format({'bold': True, 'font_color': 'red', 'num_format': 'yyyy-mm-dd'})
#写入该格式类
worksheet.write_column('A2', date_time, date_format)
```

1. 创建一个表格

下面的实例文件 xlsxwriter01.py 演示了，使用 xlsxwriter 库创建一个指定内容的 Excel 文件的过程。

源码路径：daima\7\7-2\xlsxwriter01.py

```
import xlsxwriter    #导入模板

workbook = xlsxwriter.Workbook('hello.xlsx')   #创建一个名为 hello.xlsx的workbook
worksheet = workbook.add_worksheet()   #创建一个默认工作簿
#工作簿也支持命名，
#如workbook.add_worksheet('hello')

worksheet.write('A1', 'Hello world')   #使用工作簿在A1中写入Hello world
workbook.close()    # 关闭工作簿
```

执行后会创建一个 Excel 文件 hello.xlsx，如图 7-31 所示。

下面的实例文件 xlsxwriter02.py 演示了，使用 xlsxwriter 库向指定 Excel 文件中批量写入指定内容的过程。

源码路径：daima\7\7-2\xlsxwriter02.py

```
import xlsxwriter

workbook = xlsxwriter.Workbook('Expenses01.xlsx')
worksheet = workbook.add_worksheet()

#需要写入的数据
expenses = (['Rent', 1000],
            ['Gas', 100],
            ['Food', 300],
            ['Gym', 50],
            )

#行与列的初始位置
row = 0
col = 0

for item, cost in (expenses):
    worksheet.write(row, col, item)   # 在第一列写入item
    worksheet.write(row, col + 1, cost)   # 在第二列写入cost
    row + 1    # 每次循环行数发生改变

worksheet.write(row, 0, 'Total')
worksheet.write(row, 1, '=SUM(B1:B4)')   # 写入公式
```

执行后会创建一个 Excel 文件 Expenses01.xlsx，如图 7-32 所示。

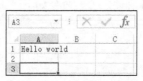

图 7-31　文件 hello.xlsx 的内容　　　　图 7-32　文件 Expenses02.xlsx 的内容

2. 设置表格样式

表格样式包含字体、颜色、模式、边框和数字格式等，设置表格样式需要使用函数 add_format()。xlsxwriter 库中包含的样式信息如表 7-1 所示。

表 7-1　　　　　　　　　　xlsxwriter 库中包含的样式信息

类别	描述	属性	方法名
字体	字体	font_name	set_font_name()
	字体大小	font_size	set_font_size()
	字体颜色	font_color	set_font_color()
	加粗	bold	set_bold()

续表

类别	描述	属性	方法名
字体	斜体	italic	set_italic()
	下划线	underline	set_underline()
	删除线	font_strikeout	set_font_strikeout()
	上标/下标	font_script	set_font_script()
数字	数字格式	num_format	set_num_format()
保护	表格锁定	locked	set_locked()
	隐藏公式	hidden	set_hidden()
对齐	水平对齐	align	set_align()
	垂直对齐	valign	set_align()
	旋转	rotation	set_rotation()
	文本包装	text_wrap	set_text_warp()
	底端对齐	text_justlast	set_text_justlast()
	中心对齐	center_across	set_center_across()
	缩进	indent	set_indent()
	缩小填充	shrink	set_shrink()
模式	表格模式	pattern	set_pattern()
	背景颜色	bg_color	set_bg_color()
	前景颜色	fg_color	set_fg_color()
边框	表格边框	border	set_border()
	底部边框	bottom	set_bottom()
	上边框	top	set_top()
	右边框	right	set_right()
	边框颜色	border_color	set_border_color()
	底部颜色	bottom_color	set_bottom_color()
	顶部颜色	top_color	set_top_color()
	左边颜色	left_color	set_left_color()
	右边颜色	right_color	set_right_color()

下面的实例文件xlsxwriter03.py演示了使用库xlsxwriter创建指定Excel格式内容的过程。

源码路径：daima\7\7-2\xlsxwriter03.py

```
# 创建文件及sheet
workbook = xlsxwriter.Workbook('Expenses03.xlsx')
worksheet = workbook.add_worksheet()
#设置粗体，默认是False
bold = workbook.add_format({'bold': True})
#定义数字格式
money = workbook.add_format({'num_format': '$#,##0'})
#自定义粗体格式
worksheet.write('A1', 'Item', bold)
worksheet.write('B1', 'Cost', bold)
#写入表中的数据
expenses = (
['Rent', 1000],
['Gas',    100],
['Food',   300],
['Gym',     50],
)

#从标题下面的第一个单元格开始
row = 1
```

```
        col = 0

        # 迭代数据并逐行地写入
        for item, cost in (expenses):
            worksheet.write(row, col,     item)      # 用默认格式写入
            worksheet.write(row, col + 1, cost, money)  # 用自定义money格式写入
            row += 1

        # 用公式计算总数
        worksheet.write(row, 0, 'Total',       bold)
        worksheet.write(row, 1, '=SUM(B2:B5)', money)

        workbook.close()
```

执行后会创建一个 Excel 文件 Expenses03.xlsx，表格中的字体样式是我们自己定义的，如图 7-33 所示。

3. 插入图像

下面的实例文件 xlsxwriter04.py 演示了使用 xlsxwriter 库向指定 Excel 文件中插入指定图像的过程。

源码路径：daima\7\7-2\xlsxwriter04.py

图 7-33 文件 Expenses03.xlsx 的内容

```
#创建一个新Excel文件并添加工作表
workbook = xlsxwriter.Workbook('demo.xlsx')
worksheet = workbook.add_worksheet()

#展开第一列，使正文更清楚
worksheet.set_column('A:A', 20)

#添加粗体格式用于突出显示单元格
bold = workbook.add_format({'bold': True})

#写一些简单的文字
worksheet.write('A1', 'Hello')

#设置文本与格式
worksheet.write('A2', 'World', bold)

#写入一些数字
worksheet.write(2, 0, 123)
worksheet.write(3, 0, 123.456)
#插入图像
worksheet.insert_image('B5', '123.png')
workbook.close()
```

执行后会创建一个包含指定图像内容的 Excel 文件 demo.xlsx，如图 7-34 所示。

图 7-34 文件 demo.xlsx 的内容

4. 插入图表

Excel 的核心功能之一是根据表格内的数据生成统计图表，使整个数据变得更加直观。通过使用 xlsxwriter 库，可以根据 Excel 表格内的数据生成图表。Excel 支持两种类型的图表。第一种类型有 9 大类。

- area：面积图。
- bar：转置直方图。
- column：柱状图。
- line：直线图。
- pie：饼状图。
- doughnut：环形图。
- scatter：散点图。
- stock：股票趋势图。
- radar：雷达图。

第二种类型则用于描述是否有连线、是否有平滑曲线等。第二种类型包含如下几种。

- area
  - stacked
  - percent_stacked
- bar
  - stacked
  - percent_stacked
- column
  - stacked
  - percent_stacked
- scatter
  - straight_with_markers
  - straight
  - smooth_with_markers
  - smooth
- radar
  - with_markers
  - filled

下面的实例文件 xlsxwriter05.py 演示了，使用 xlsxwriter 库向指定 Excel 文件中插入数据并绘制柱状图的过程。

源码路径：daima\7\7-2\xlsxwriter05.py

```
import xlsxwriter

workbook = xlsxwriter.Workbook('chart.xlsx')
worksheet = workbook.add_worksheet()

#新建图表对象
chart = workbook.add_chart({'type': 'column'})

#向Excel文件中写入数据，在建立图表时要用到
data = [
    [1, 2, 3, 4, 5],
    [2, 4, 6, 8, 10],
```

```
        [3, 6, 9, 12, 15],
]
worksheet.write_column('A1', data[0])
worksheet.write_column('B1', data[1])
worksheet.write_column('C1', data[2])

#向图表中添加数据
chart.add_series({'values': '=Sheet1!$A$1:$A$5'})
chart.add_series({'values': '=Sheet1!$B$1:$B$5'})
chart.add_series({'values': '=Sheet1!$C$1:$C$5'})

#将图标插入表单中
worksheet.insert_chart('A7', chart)

workbook.close()
```

执行后会创建一个包含指定数据内容的 Excel 文件 chart.xlsx，并根据数据内容绘制一个柱状图，如图 7-35 所示。

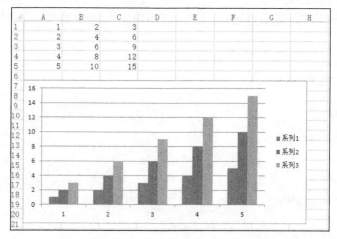

图 7-35　文件 chart.xlsx 的内容

下面的实例文件 xlsxwriter06.py 演示了，使用 xlsxwriter 库向指定 Excel 文件中插入数据并绘制散点图的过程。

源码路径：daima\7\7-2\xlsxwriter06.py

```
import xlsxwriter

workbook = xlsxwriter.Workbook('chart_scatter.xlsx')
worksheet = workbook.add_worksheet()
bold = workbook.add_format({'bold': 1})

#添加图表引用表格中的数据
headings = ['Number', 'Batch 1', 'Batch 2']
data = [
    [2, 3, 4, 5, 6, 7],
    [10, 40, 50, 20, 10, 50],
    [30, 60, 70, 50, 40, 30],
]

worksheet.write_row('A1', headings, bold)
worksheet.write_column('A2', data[0])
worksheet.write_column('B2', data[1])
worksheet.write_column('C2', data[2])

#创建一个散点图
```

```
chart1 = workbook.add_chart({'type': 'scatter'})

#配置第一个系列的散点
chart1.add_series({
    'name':       '=Sheet1!$B$1',
    'categories': '=Sheet1!$A$2:$A$7',
    'values':     '=Sheet1!$B$2:$B$7',
})

#配置第二个系列的散点,注意,使用替代语法来定义范围
chart1.add_series({
    'name':       ['Sheet1', 0, 2],
    'categories': ['Sheet1', 1, 0, 6, 0],
    'values':     ['Sheet1', 1, 2, 6, 2],
})

#添加图表标题和轴标签
chart1.set_title ({'name': 'Results of sample analysis'})
chart1.set_x_axis({'name': 'Test number'})
chart1.set_y_axis({'name': 'Sample length (mm)'})

#设置Excel图表样式
chart1.set_style(11)

#将图表插入工作表(带偏移量)中
worksheet.insert_chart('D2', chart1, {'x_offset': 25, 'y_offset': 10})
workbook.close()
```

执行后会创建一个包含指定数据内容的 Excel 文件 chart_scatter.xlsx,并根据数据内容绘制一个散点图,如图 7-36 所示。

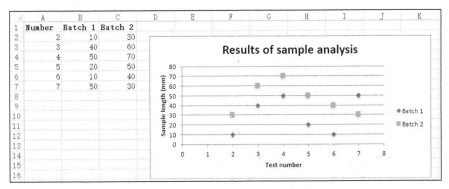

图 7-36　文件 chart_scatter.xlsx 的内容

下面的实例文件 xlsxwriter07.py 演示了,使用 xlsxwriter 库向指定 Excel 文件中插入数据并绘制柱状图和饼状图的过程。

源码路径:daima\7\7-2\xlsxwriter07.py

```
import xlsxwriter

#新建一个Excel文件,命名为expense01.xlsx
workbook = xlsxwriter.Workbook("123.xlsx")
#添加一个Sheet,不写名字,默认为Sheet1
worksheet = workbook.add_worksheet()
#准备数据
headings=["姓名","数学","语文"]
data=[["C罗张",78,60],["糖人李",98,89],["梅西徐",88,100]]
#样式
head_style = workbook.add_format({"bold":True,"bg_color":"yellow","align":"center","font":13})
#写数据
worksheet.write_row("A1",headings,head_style)
```

```python
for i in range(0,len(data)):
    worksheet.write_row("A{}".format(i+2),data[i])
#添加柱状图
chart1 = workbook.add_chart({"type":"column"})
chart1.add_series({
    "name":"=Sheet1!$B$1",#图例项
    "categories":"=Sheet1!$A$2:$A$4",
    "values":"=Sheet1!$B$2:$B$4"
})
chart1.add_series({
    "name":"=Sheet1!$C$1",
    "categories":"=Sheet1!$A$2:$A$4",
    "values":"=Sheet1!$C$2:$C$4"
})
#添加柱状图标题
chart1.set_title({"name":"柱状图"})
#y轴名称
chart1.set_y_axis({"name":"分数"})
#x轴名称
chart1.set_x_axis({"name":"人名"})
#图表样式
chart1.set_style(11)

#添加柱状图叠图子类型
chart2 = workbook.add_chart({"type":"column","subtype":"stacked"})
chart2.add_series({
    "name":"=Sheet1!$B$1",
    "categories":"=Sheet1!$A$2:$a$4",
    "values":"=Sheet1!$B$2:$B$4"
})
chart2.add_series({
    "name":"=Sheet1!$C$1",
    "categories":"=Sheet1!$A$2:$a$4",
    "values":"=Sheet1!$C$2:$C$4"
})
chart2.set_title({"name":"叠图子类型"})
chart2.set_x_axis({"name":"姓名"})
chart2.set_y_axis({"name":"成绩"})
chart2.set_style(12)

#添加饼图
chart3 = workbook.add_chart({"type":"pie"})
chart3.add_series({
    #"name":"饼形图",
    "categories":"=Sheet1!$A$2:$A$4",
    "values":"=Sheet1!$B$2:$B$4",
    #定义各饼块的颜色
    "points":[
        {"fill":{"color":"yellow"}},
        {"fill":{"color":"blue"}},
        {"fill":{"color":"red"}}
    ]
})
chart3.set_title({"name":"饼图成绩单"})
chart3.set_style(3)

#插入图表
worksheet.insert_chart("B7",chart1)
worksheet.insert_chart("B25",chart2)
worksheet.insert_chart("J2",chart3)

#关闭Excel文件
workbook.close()
```

执行后会创建一个包含指定数据内容的 Excel 文件 123.xlsx，并根据数据内容分别绘制两个柱状图和一个饼状图，如图 7-37 所示。

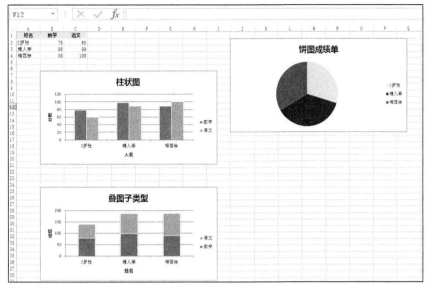

图 7-37　文件 123.xlsx 的内容

7.3　使用 PDF 模块/库

在 Python 程序中，可以使用第三方模块/库处理 PDF 文件中的数据。本节将详细讲解这些模块的使用方法。

7.3.1　使用 PDFMiner 模块

使用 PDFMiner 模块可以解析 PDF 文件，在使用 PDFMiner 之前需要先安装，安装命令如下。
```
pip install pdfminer3k
```
因为解析 PDF 是一件非常耗时和耗费内存的工作，所以 PDFMiner 使用了一种懒解析策略——只在需要的时候才去解析，以缩短时间和减少使用的内存。要使用 PDFMiner 解析 PDF 至少需要用到两个类——PDFParser 和 PDFDocument。其中 PDFParser 类用于从文件中提取数据，PDFDocument 类用于保存数据。另外，还需要 PDFPageInterprete 类去处理 PDF 页面中的内容，PDFDevice 类将其转换为我们所需要的结果。PDFResourceManager 类用于保存共享内容（如字体或图片）。

假设存在一个 PDF 文件"开发 Python 应用程序.pdf"，其内容如图 7-38 所示。

下面的实例文件 PDFMiner01.py 演示了，使用 PDFMiner 将上述 PDF 文件中的内容转换为文本文件的过程。

源码路径：daima\7\7-3\PDFMiner01.py

```python
import sys
import importlib
importlib.reload(sys)

from pdfminer.pdfparser import PDFParser,PDFDocument
from pdfminer.pdfinterp import PDFResourceManager, PDFPageInterpreter
from pdfminer.converter import PDFPageAggregator
from pdfminer.layout import LTTextBoxHorizontal,LAParams
from pdfminer.pdfinterp import PDFTextExtractionNotAllowed

'''
解析PDF文本，保存到文本文件中
'''
path = r'开发Python应用程序.pdf'
def parse():
```

```python
fp = open(path, 'rb')  # 以二进制读模式打开
#用文件对象来创建一个PDF文档分析器
praser = PDFParser(fp)
#创建一个PDF文档
doc = PDFDocument()
#连接分析器与文档对象
praser.set_document(doc)
doc.set_parser(praser)

#提供初始化密码
#如果没有密码，就创建一个空的字符串
doc.initialize()

#检测文档是否提供转换方法，不提供就忽略
if not doc.is_extractable:
    raise PDFTextExtractionNotAllowed
else:
    #创建PDF资源管理器来管理共享资源
    rsrcmgr = PDFResourceManager()
    #创建一个PDF设备对象
    laparams = LAParams()
    device = PDFPageAggregator(rsrcmgr, laparams=laparams)
    #创建一个PDF解释器对象
    interpreter = PDFPageInterpreter(rsrcmgr, device)

    #遍历列表，每次处理一页的内容
    for page in doc.get_pages():  #获取page列表
        interpreter.process_page(page)
        # 接受该页面的LTPage对象
        layout = device.get_result()
        # 这里layout是一个LTPage对象，里面存放着这个page解析出的各种对象，一般包括
        #LTTextBox、LTFigure、LTImage、LTTextBoxHorizontal等，要获取文本就需要获得对
        #象的text属性
        for x in layout:
            if (isinstance(x, LTTextBoxHorizontal)):
                with open(r'123.txt', 'a') as f:
                    results = x.get_text()
                    print(results)
                    f.write(results + '\n')

if __name__ == '__main__':
    parse()
```

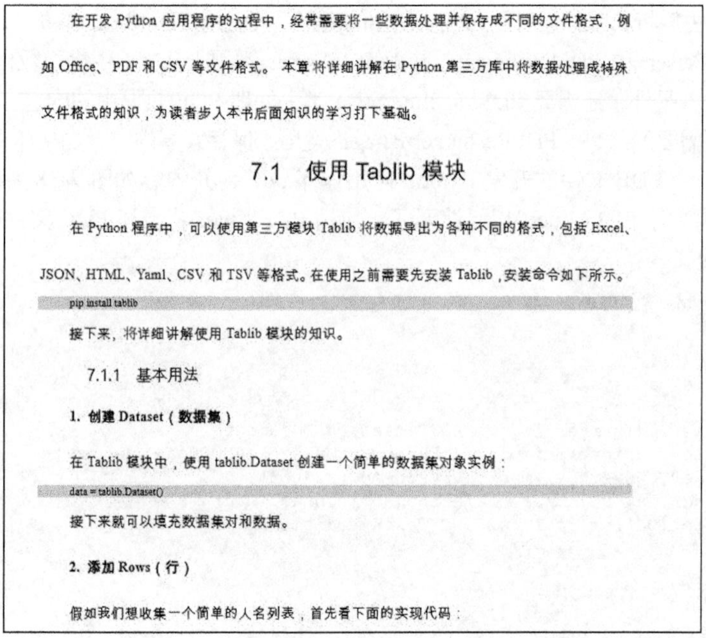

在开发 Python 应用程序的过程中，经常需要将一些数据处理并保存成不同的文件格式，例如 Office、PDF 和 CSV 等文件格式。本章将详细讲解在 Python 第三方库中将数据处理成特殊文件格式的知识，为读者步入本书后面知识的学习打下基础。

7.1 使用 Tablib 模块

在 Python 程序中，可以使用第三方模块 Tablib 将数据导出为各种不同的格式，包括 Excel、JSON、HTML、Yaml、CSV 和 TSV 等格式。在使用之前需要先安装 Tablib，安装命令如下所示。

```
pip install tablib
```

接下来，将详细讲解使用 Tablib 模块的知识。

7.1.1 基本用法

1. 创建 Dataset（数据集）

在 Tablib 模块中，使用 tablib.Dataset 创建一个简单的数据集对象实例：

```
data = tablib.Dataset()
```

接下来就可以填充数据集对和数据。

2. 添加 Rows（行）

假如我们想收集一个简单的人名列表，首先看下面的实现代码：

图 7-38 文件"开发 Python 应用程序.pdf"的内容

执行后上述代码后会将文件"开发 Python 应用程序.pdf"中的内容保存到记事本文件"123.txt"中,如图 7-39 所示,并在命令行界面中显示解析后的内容,如图 7-40 所示。

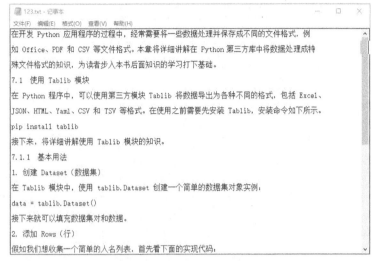

图 7-39 文件"123.txt"的内容

图 7-40 在命令行界面中显示解析后的内容

下面的实例文件 PDFMiner02.py 演示了,使用 PDFMiner 解析某个在线 PDF 文件的内容的过程。

源码路径:daima\7\7-3\PDFMiner02.py

```
import importlib
import sys
import random
from urllib.request import urlopen
from urllib.request import Request

from pdfminer.converter import PDFPageAggregator
from pdfminer.layout import LTTextBoxHorizontal, LAParams
from pdfminer.pdfinterp import PDFResourceManager, PDFPageInterpreter
from pdfminer.pdfinterp import PDFTextExtractionNotAllowed
from pdfminer.pdfparser import PDFParser, PDFDocument

'''
```

第 7 章 特殊文本格式处理

```python
解析PDF文本,保存到文本文件中
'''
importlib.reload(sys)

user_agent = ['Mozilla/5.0 (Windows NT 10.0; WOW64)', 'Mozilla/5.0 (Windows NT 6.3; WOW64)',
              'Mozilla/5.0 (Windows NT 6.1; WOW64; rv:54.0) Gecko/20100101 Firefox/54.0',
              'Mozilla/5.0 (Windows NT 6.1) AppleWebKit/537.11 (KHTML, like Gecko)
                  Chrome/23.0.1271.64 Safari/537.11',
              'Mozilla/5.0 (Windows NT 6.3; WOW64; Trident/7.0; rv:11.0) like Gecko',
              'Mozilla/5.0 (Windows NT 5.1) AppleWebKit/537.36 (KHTML, like Gecko)
                  Chrome/28.0.1500.95 Safari/537.36',
              'Mozilla/5.0 (Windows NT 6.1; WOW64; Trident/7.0; SLCC2; .NET CLR
                  2.0.50727; .NET CLR 3.5.30729; .NET CLR 3.0.30729; Media Center PC
                  6.0; .NET4.0C; rv:11.0) like Gecko)',
              'Mozilla/5.0 (Windows; U; Windows NT 5.2) Gecko/2008070208 Firefox/3.0.1',
              'Mozilla/5.0 (Windows; U; Windows NT 5.1) Gecko/20070309 Firefox/2.0.0.3',
              'Mozilla/5.0 (Windows; U; Windows NT 5.1) Gecko/20070803 Firefox/1.5.0.12',
              'Opera/9.27 (Windows NT 5.2; U; zh-cn)',
              'Mozilla/5.0 (Macintosh; PPC Mac OS X; U; en) Opera 8.0',
              'Opera/8.0 (Macintosh; PPC Mac OS X; U; en)',
              'Mozilla/5.0 (Windows; U; Windows NT 5.1; en-US; rv:1.8.1.12) Gecko/20080219
                  Firefox/2.0.0.12 Navigator/9.0.0.6',
              'Mozilla/4.0 (compatible; MSIE 8.0; Windows NT 6.1; Win64; x64; Trident/4.0)',
              'Mozilla/4.0 (compatible; MSIE 8.0; Windows NT 6.1; Trident/4.0)',
              'Mozilla/5.0 (compatible; MSIE 10.0; Windows NT 6.1; WOW64; Trident/6.0;
                  SLCC2; .NET CLR 2.0.50727; .NET CLR 3.5.30729; .NET CLR 3.0.30729;
                  Media Center PC 6.0; InfoPath.2; .NET4.0C; .NET4.0E)',
              'Mozilla/5.0 (Windows NT 6.1; WOW64) AppleWebKit/537.1 (KHTML, like Gecko)
                  Maxthon/4.0.6.2000 Chrome/26.0.1410.43 Safari/537.1 ',
              'Mozilla/5.0 (compatible; MSIE 10.0; Windows NT 6.1; WOW64; Trident/6.0;
                  SLCC2; .NET CLR 2.0.50727; .NET CLR 3.5.30729; .NET CLR 3.0.30729; Media
                  Center PC 6.0; InfoPath.2; .NET4.0C; .NET4.0E; QQBrowser/7.3.9825.400)',
              'Mozilla/5.0 (Windows NT 6.1; WOW64; rv:21.0) Gecko/20100101 Firefox/21.0 ',
              'Mozilla/5.0 (Windows NT 6.1; WOW64) AppleWebKit/537.1 (KHTML, like Gecko)
                  Chrome/21.0.1180.92 Safari/537.1 LBBROWSER',
              'Mozilla/5.0 (compatible; MSIE 10.0; Windows NT 6.1; WOW64; Trident/6.0;
                  BIDUBrowser 2.x)',
              'Mozilla/5.0 (Windows NT 6.1; WOW64) AppleWebKit/536.11 (KHTML, like Gecko)
                  Chrome/20.0.1132.11 TaoBrowser/3.0 Safari/536.11']

def parse(_path):
    # fp = open(_path, 'rb')    # rb以二进制读模式打开本地PDF文件
    # 随机从user_agent列表中抽取一个元素
    request = Request(url=_path, headers={'User-Agent': random.choice(user_agent)})
    fp = urlopen(request)   #打开在线PDF文档

    #用文件对象来创建一个PDF文档分析器
    praser_pdf = PDFParser(fp)

    #创建一个PDF文档
    doc = PDFDocument()

    #连接分析器与文档对象
    praser_pdf.set_document(doc)
    doc.set_parser(praser_pdf)

    #提供初始化密码doc.initialize("123456")
    # 如果没有密码,就创建一个空的字符串
    doc.initialize()

    #检测文档是否提供转换方法,不提供就忽略
    if not doc.is_extractable:
        raise PDFTextExtractionNotAllowed
    else:
        #创建PDF资源管理器来管理共享资源
        rsrcmgr = PDFResourceManager()

        #创建一个PDF参数分析器
```

```
                    laparams = LAParams()

                    #创建聚合器
                    device = PDFPageAggregator(rsrcmgr, laparams=laparams)

                    #创建一个PDF页面解释器对象
                    interpreter = PDFPageInterpreter(rsrcmgr, device)

                    #遍历列表，每次处理一页的内容
                    #获取page列表
                    for page in doc.get_pages():
                        #使用页面解释器来读取
                        interpreter.process_page(page)

                        #使用聚合器获取内容
                        layout = device.get_result()

                        #这里layout是一个LTPage对象，里面存放着这个page解析出的各种对象，一般包括
                        #LTTextBox、LTFigure、LTImage、LTTextBoxHorizontal等，要获取文本就需要获得
                        #对象的text属性，
                        for out in layout:
                            # 判断是否含有get_text()方法
                            # if hasattr(out,"get_text"):
                            if isinstance(out, LTTextBoxHorizontal):

                                results = out.get_text()
                                print("results: " + results)
if __name__ == '__main__':
    url = "http://***.pdf"
    parse(url)
```

执行后将会解析输出指定网址中在线 PDF 文件的内容。

7.3.2 使用 PyPDF2

使用库 PyPDF2 可以读写、分割或合并 PDF 文件。在使用 PyPDF2 库之前需要先安装，安装命令如下。

```
pip install PyPDF2
```

接下来将详细讲解 PyPDF2 库中的成员。

1. PdfFileReader 类

PdfFileReader 类的功能是初始化一个 PdfFileReader 对象并读取 PDF 文件的内容。因为 PDF 流的交叉引用表被读入内存，所以此操作可能需要一些时间。此类的构造方法如下。

```
PyPDF2.PdfFileReader(stream,strict = True,warndest = None,overwriteWarnings = True)
```

- stream：*File 对象或支持与 File 对象类似的标准读取和查找方法的对象，也可以是表示 PDF 文件路径的字符串。
- strict（bool）：确定是否应该提醒用户所出现的问题，默认是 True。
- warndest：记录警告的目标（默认是 sys.stderr）。
- overwriteWarnings(bool)：确定 warnings.py 是否用自定义实现覆盖 Python 模块（默认为 True）。

PdfFileReader 对象中的常用属性和方法如下。

- getDestinationPageNumber(destination)：检索给定目标对象的页码。
- getDocumentInfo()：检索 PDF 文件的文档信息字典。
- getFields(tree = None,retval = None,fileObj= None)：如果此 PDF 文件包含交互式表单字段，则提取字段数据。
- getFormTextFields()：从文档中检索带有文本数据的表单域。
- getNameDestinations(tree = None,retval= None) ：检索文档中的指定目标。
- getNumPages()：计算此 PDF 文件中的页数。

- getOutlines(node = None,outline = None,)：检索文档中出现的文档大纲。
- getPage(pageNumber)：从这个 PDF 文件中检索指定编号的页面。
- getPageLayout()：获取页面布局。
- getPageMode()：获取页面模式。
- getPageNumber(pageObject)：检索给定 pageObject 处于的页码。
- getXmpMetadata()：从 PDF 文档根目录中检索 XMP 数据。
- isEncrypted：显示 PDF 文件是否加密的只读布尔属性。
- namedDestinations：访问 getNamedDestinations()函数的只读属性。

下面的实例文件 PyPDF201.py 演示了使用 PyPDF2 读取指定 PDF 文件的过程。

源码路径：daima\7\7-3\PyPDF201.py

```
from PyPDF2 import PdfFileReader, PdfFileWriter

readFile = '开发Python应用程序.pdf'
# 获取 PdfFileReader 对象
pdfFileReader = PdfFileReader(readFile)
# 或者pdfFileReader = PdfFileReader(open(readFile, 'rb'))
# 获取 PDF 文件的文档信息
documentInfo = pdfFileReader.getDocumentInfo()
print('documentInfo = %s' % documentInfo)
# 获取页面布局
pageLayout = pdfFileReader.getPageLayout()
print('pageLayout = %s ' % pageLayout)

# 获取页模式
pageMode = pdfFileReader.getPageMode()
print('pageMode = %s' % pageMode)

xmpMetadata = pdfFileReader.getXmpMetadata()
print('xmpMetadata = %s ' % xmpMetadata)

# 获取PDF文件的页数
pageCount = pdfFileReader.getNumPages()

print('pageCount = %s' % pageCount)
for index in range(0, pageCount):
    # 返回指定页码的 pageObject
    pageObj = pdfFileReader.getPage(index)
    print('index = %d , pageObj = %s' % (index, type(pageObj)))
    # <class 'PyPDF2.pdf.PageObject'>
    #获取 pageObject 在PDF文档中的页码
    pageNumber = pdfFileReader.getPageNumber(pageObj)
    print('pageNumber = %s ' % pageNumber)
```

执行后将会读取并显示文件"开发 Python 应用程序.pdf"的信息。

```
documentInfo = {'/Author': 'apple', '/Creator': 'Microsoft® Word 2013', '/CreationDate':
"D:20180406164456+08'00'", '/ModDate': "D:20180406164456+08'00'", '/Producer':
'Microsoft® Word 2013'}
pageLayout = None
pageMode = None
xmpMetadata  = None
pageCount = 1
index = 0 , pageObj = <class 'PyPDF2.pdf.PageObject'>
pageNumber = 0
```

2. PdfFileWriter 类

PdfFileWriter 类提供了向 PDF 文件中写入数据的功能。该类主要包括如下所示的属性和方法。

- addAttachment(fname,fdata)：在 PDF 文件中嵌入文件。
- addBlankPage(width= None,height=None)：追加一个空白页面到这个 PDF 文件中并返回。
- addBookmark(title,pagenum,parent=None,color=None,bold=False,italic=False,fit='/fit,*args')：添

7.3 使用 PDF 模块/库

加书签。
- addJS(javascript)：在当前将要打开的 PDF 中添加启动的 JavaScript 脚本程序。
- addLink(pagenum,pagedest,rect,border=None,fit='/fit',*args)：从一个矩形区域添加一个内部链接到指定的页面中。
- addPage(page)：添加一个页面到此 PDF 文件中，该页面通常从 PdfFileReader 实例获取。
- getNumpages()：获取页数。
- getPage(pageNumber)：从这个 PDF 文件中检索一个编号的页面。
- insertBlankPage(width=None,height=None,index=0)：插入一个空白页面到这个 PDF 文件中并返回它，如果没有指定页面大小则使用最后一页的大小。
- insertPage(page,index=0)：在这个 PDF 文件中插入一个页面，该页面通常从 PdfFileReader 实例获取。
- removeLinks()：删除链接。
- removeText(ignoreByteStringObject = False)：从输出中删除图像。
- write(stream)：将添加到此对象的页面集合写入 PDF 文件。

下面的实例文件 PyPDF202.py 演示了，使用 PyPDF2 将指定 PDF 文件的内容写入另外一个 PDF 文件的过程。

源码路径：daima\7\7-3\PyPDF202.py

```
from PyPDF2 import PdfFileReader, PdfFileWriter

readFile = '开发Python应用程序.pdf'
outFile = 'copy.pdf'
pdfFileWriter = PdfFileWriter()

# 获取pdfFileReader 对象
pdfFileReader = PdfFileReader(readFile)
# 或者pdfFileReader = PdfFileReader(open(readFile, 'rb'))
numPages = pdfFileReader.getNumPages()

for index in range(0, numPages):
        pageObj = pdfFileReader.getPage(index)
        pdfFileWriter.addPage(pageObj)   # 根据每页返回的 PageObject，写入文件
        pdfFileWriter.write(open(outFile, 'wb'))

pdfFileWriter.addBlankPage()    #在文件的最后一页写入一个空白页，保存至文件中
pdfFileWriter.write(open(outFile,'wb'))
```

执行后会将文件"开发 Python 应用程序.pdf"中的内容写入文件"copy.pdf"中，新生成的文件"copy.pdf"的内容如图 7-41 所示。

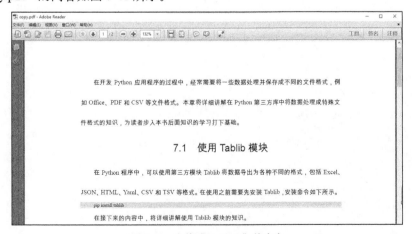

图 7-41 文件"copy.pdf"的内容

3. 类 PageObject

类 PageObject 表示 PDF 文件中的单个页面。通常这个对象是通过访问 PdfFileReader 对象的 getPage()方法得到的，也可以使用 createBlankPage()静态方法创建一个空的页面。类 PageObject 的构造原型如下。

```
PageObject(pdf=None,indirectRef=None)
```

- pdf：页面所属的 PDF 文件。
- indirectRef：将源对象的原始间接引用存储在其源 PDF 中。

PageObject 对象中的常用属性和方法如下。

- static createBlankPage(pdf=None,width=None,height=None)：返回一个新的空白页面。
- extractText()：找到所有文本绘图命令，按照它们在内容流中提供的顺序，提取文本。
- getContents()：访问页面内容，返回 Contents 对象或 None。
- rotateClockwise(angle)：顺时针旋转 90°。
- scale(sx,sy)：在内容中应用转换矩阵并更新页面的大小。

下面的实例文件 PyPDF203.py 演示了，使用 PyPDF2 将两个指定 PDF 文件内容合并为一个 PDF 文件的过程。

源码路径：daima\7\7-3\PyPDF203.py

```
from PyPDF2 import PdfFileReader, PdfFileWriter
def mergePdf(inFileList, outFile):
    '''
    合并文档
    :param inFileList: 要合并的文档的列表
    :param outFile:    合并后的输出文件
    :return:
    '''
    pdfFileWriter = PdfFileWriter()
    for inFile in inFileList:
        # 依次打开要合并的文件
        pdfReader = PdfFileReader(open(inFile, 'rb'))
        numPages = pdfReader.getNumPages()
        for index in range(0, numPages):
            pageObj = pdfReader.getPage(index)
            pdfFileWriter.addPage(pageObj)

        # 最后统一写入输出文件中
        pdfFileWriter.write(open(outFile, 'wb'))

mergePdf(['copy.pdf','123.pdf'],'456.pdf')
```

执行上述代码后，会将文件 copy.pdf 和 123.pdf 的内容合并到文件 456.pdf 中。

下面的实例文件 PyPDF204.py 演示了使用 PyPDF2 分割某个指定 PDF 文件的过程。

源码路径：daima\7\7-3\PyPDF204.py

```
from PyPDF2 import PdfFileReader, PdfFileWriter
def splitPdf():
    readFile = '123.pdf'
    outFile = '789.pdf'
    pdfFileWriter = PdfFileWriter()

    #获取pdfFileReader对象
    pdfFileReader = PdfFileReader(readFile)
    # 或者pdfFileReader = PdfFileReader(open(readFile, 'rb'))
    #文档总页数
    numPages = pdfFileReader.getNumPages()

    if numPages > 5:
        #将第5页之后的页面输出到一个新的文件中，即分割文档
        for index in range(5, numPages):
            pageObj = pdfFileReader.getPage(index)
            pdfFileWriter.addPage(pageObj)
```

```
    # 添加完之后一起保存至文件中
        pdfFileWriter.write(open(outFile, 'wb'))
splitPdf()
```

执行上述代码后，会分割文件 123.pdf 的内容，将此文件第 5 页之后的内容分离出来并另为文件 789.pdf。

下面的实例文件 PyPDF205.py 演示了使用 PyPDF2 合并 3 个 PDF 文件的过程。

源码路径：daima\7\7-3\PyPDF205.py

```
import PyPDF2

pdff1 = open("123.pdf", "rb")
pr = PyPDF2.PdfFileReader(pdff1)
print(pr.numPages)

pdff2 = open("456.pdf", "rb")
pr2 = PyPDF2.PdfFileReader(pdff2)

pdf3 = open("789.pdf", "rb")
pr3 = PyPDF2.PdfFileReader(pdf3)

pdfw = PyPDF2.PdfFileWriter()
pageobj = pr.getPage(0)
pdfw.addPage(pageobj)

for pageNum in range(pr2.numPages):
    pageobj2 = pr2.getPage(pageNum)
    pdfw.addPage(pageobj2)

pageobj3 = pr3.getPage(0)
pdfw.addPage(pageobj3)

pdfout = open("aaa.pdf", "wb")
pdfw.write(pdfout)
pdfout.close()
pdff1.close()
pdff2.close()
pdf3.close()
```

执行上述代码后会合并 3 个文件（123.pdf、456.pdf 和 789.pdf）的内容，将合并后的内容保存到新建的文件 aaa.pdf 中。

7.3.3 使用 Reportlab 库

使用 Reportlab 库可以读写、分割或合并 PDF 文件。在使用 Reportlab 库之前需要先安装，安装命令如下。

```
pip install reportlab
```

1. 写入文本

下面的实例文件 reportlab01.py 演示了，使用 Reportlab 在指定 PDF 文件中写入文本的过程。

源码路径：daima\7\7-3\reportlab01.py

```
#引入所需要的基本包
from reportlab.pdfgen import canvas
#设置绘画开始的位置
def hello(c):
    c.drawString(100, 100, "hello world!")
#定义要生成的pdf的名称
c=canvas.Canvas("1.pdf")
#调用函数进行绘画，并将canvas对象作为参数传递
hello(c)
#showPage函数：保存当前页的canvas
c.showPage()
#save函数：保存文件并关闭canvas
c.save()
```

执行后会在文件 1.pdf 中写入文本"hello world!"。

下面的实例文件 reportlab02.py 演示了，使用 Reportlab 在指定 PDF 文件中写入指定样式文本的过程。

源码路径：daima\7\7-3\reportlab02.py

```python
from reportlab.lib.styles import getSampleStyleSheet
from reportlab.platypus import Paragraph,SimpleDocTemplate
from reportlab.lib import  colors

Style=getSampleStyleSheet()

bt = Style['Normal']         #字体的样式
# bt.fontName='song'         #使用的字体
bt.fontSize=14               #字号
bt.wordWrap = 'CJK'          #该属性支持自动换行，'CJK'用于中文方式的换行，用于英文中会截断单词，造成
#阅读困难，可改为'Normal'
bt.firstLineIndent = 32      #该属性支持第一行开头的空格
bt.leading = 20              #该属性用于设置行距

ct=Style['Normal']
# ct.fontName='song'
ct.fontSize=12
ct.alignment=1               #居中

ct.textColor = colors.red

t = Paragraph('hello',bt)
pdf=SimpleDocTemplate('2.pdf')
pdf.multiBuild([t])
```

通过上述代码，在文件 2.pdf 中写入了指定格式的文本。一份 PDF 文件可以定义多种字体样式，如 bt 和 ct。字体有多种属性，例如，在上述代码中，自动换行属性 wordWrap 的参数'CJK'表示是按照中文方式换行（可以在字符之间换行），英文换行方式为'Normal'（在空格处换行）。alignment 的取值有 3 个，其中 0 表示左对齐，1 表示居中，2 表示右对齐。

2. 绘制图形

下面的实例文件 reportlab03.py 演示了，使用 Reportlab 在指定 PDF 文件中绘制矢量图形的过程。

源码路径：daima\7\7-3\reportlab03.py

```python
#导入所需要的基本包
from reportlab.pdfgen import canvas
from reportlab.lib.units import inch
#设置绘画开始的位置
def hello(c):
    #设置描边色
    c.setStrokeColorRGB(0, 0, 1.0)
    #设置填充色
    c.setFillColorRGB(1,0,1)
    # draw some lines
    c.line(0.1*inch, 0.1*inch, 0.1*inch, 1.7*inch)
    c.line(0.1*inch, 0.1*inch, 1*inch, 0.1*inch)
    # 画矩形
    c.rect(0.2*inch, 0.2*inch, 1*inch, 1.5*inch, fill=1)
#定义要生成的PDF文件的名称
c=canvas.Canvas("3.pdf")
#调用函数进行绘画，并作为参数传递canvas对象
hello(c)
#保存当前页的canvas
c.showPage()
#保存文件并关闭canvas
c.save()
```

执行上述代码后会在文件 3.pdf 中绘制一个矢量图形，如图 7-42 所示。

下面的实例文件 reportlab04.py 演示了，使用 Reportlab 在指定 PDF 文件中绘制图像的过程。

源码路径：daima\7\7-3\reportlab04.py

```
#引入所需要的基本包
from reportlab.pdfgen import canvas
from reportlab.lib.units import mm

def drawBitmap(c):
    c.drawImage("123.png", 5*mm, 5*mm, 62*mm, 88.6*mm)

#定义要生成的PDF文件的名称
c=canvas.Canvas("4.pdf")
#调用函数生成条形码和二维码,并作为参数传递canvas对象
drawBitmap(c)
#保存当前页的canvas
c.showPage()
#保存文件并关闭canvas
c.save()
```

执行上述代码后，会在文件 4.pdf 中绘制图像 123.png，如图 7-43 所示。

图 7-42　绘制的图形

图 7-43　绘制的图像

下面的实例文件 reportlab05.py 演示了，分别在指定 PDF 文件和 PNG 文件中绘制饼状图的过程。

源码路径：daima\7\7-3\reportlab05.py

```
from reportlab.graphics.charts.piecharts import Pie
from reportlab.graphics.shapes import Drawing, _DrawingEditorMixin
from reportlab.lib.colors import Color, magenta, cyan


class pietests(_DrawingEditorMixin, Drawing):
    def __init__(self, width=400, height=200, *args, **kw):
        Drawing.__init__(self, width, height, *args, **kw)
        self._add(self, Pie(), name='pie', validate=None, desc=None)
        self.pie.sideLabels = 1
        self.pie.labels = ['Label 1', 'Label 2', 'Label 3', 'Label 4', 'Label 5']
        self.pie.data = [20, 10, 5, 5, 5]
        self.pie.width = 140
        self.pie.height = 140
        self.pie.y = 35
        self.pie.x = 125


def main():
    drawing = pietests()
    # you can do all sorts of things to drawing, lets just save it as pdf and png.
    drawing.save(formats=['pdf', 'png'], outDir='.', fnRoot=None)
    return 0
```

```
if __name__ == '__main__':
    main()
```

执行后将分别在文件 pietests000.pdf 和 pietests000.png 中绘制一个饼状图，如图 7-44 所示。

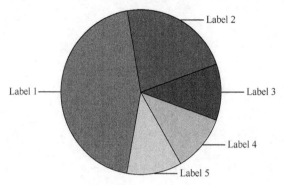

图 7-44 绘制的饼状图

3. 生成条形码和二维码

下面的实例文件 reportlab06.py 演示了，使用 Reportlab 在指定 PDF 文件中生成条形码和二维码的过程。

源码路径：daima\7\7-3\reportlab06.py

```python
#导入所需要的基本包
from reportlab.pdfgen import canvas
from reportlab.graphics.barcode import code39, code128, code93
from reportlab.graphics.barcode import eanbc, qr, usps
from reportlab.graphics.shapes import Drawing
from reportlab.lib.units import mm
from reportlab.graphics import renderPDF

#-------------------------------------------------------------------
def createBarCodes(c):
    barcode_value = "1234567890"

    barcode39 = code39.Extended39(barcode_value)
    barcode39Std = code39.Standard39(barcode_value, barHeight=20, stop=1)

    barcode93 = code93.Standard93(barcode_value)

    barcode128 = code128.Code128(barcode_value)
    #barcode128Multi = code128.MultiWidthBarcode(barcode_value)

    barcode_usps = usps.POSTNET("50158-9999")

    codes = [barcode39, barcode39Std, barcode93, barcode128, barcode_usps]

    x = 1 * mm
    y = 285 * mm

    for code in codes:
        code.drawOn(c, x, y)
        y = y - 15 * mm

    barcode_eanbc8 = eanbc.Ean8BarcodeWidget(barcode_value)
    d = Drawing(50, 10)
    d.add(barcode_eanbc8)
    renderPDF.draw(d, c, 15, 555)

    barcode_eanbc13 = eanbc.Ean13BarcodeWidget(barcode_value)
    d = Drawing(50, 10)
    d.add(barcode_eanbc13)
    renderPDF.draw(d, c, 15, 465)
```

```
    qr_code = qr.QrCodeWidget('http://***')
    bounds = qr_code.getBounds()
    width = bounds[2] - bounds[0]
    height = bounds[3] - bounds[1]
    d = Drawing(45, 45, transform=[45./width,0,0,45./height,0,0])
    d.add(qr_code)
    renderPDF.draw(d, c, 15, 405)

#定义要生成的PDF文件的名称
c=canvas.Canvas("6.pdf")
#调用函数生成条形码和二维码,并作为参数传递canvas对象
createBarCodes(c)
#保存当前页的canvas
c.showPage()
#保存文件并关闭canvas
c.save()
```

执行后将在文件 6.pdf 中生成条形码和二维码,每个条形码和二维码都有自己的含义,执行结果此处省略。

第 8 章

图像处理

在开发 Python 应用程序的过程中，经常需要使用图像文件来提高程序的美观性。本章将详细讲解使用 Python 第三方库处理图像的知识。

8.1 使用 Pillow 库

Pillow 是 Python Imaging Library 的简称,是 Python 中最常用的图像处理库。Pillow 库提供了对 Python 3 的支持,为 Python 3 解释器提供了图像处理功能。通过使用 Pillow 库,可以方便地使用 Python 程序对图片进行处理,例如,常见的尺寸、格式、色彩、旋转等处理。

8.1.1 安装 Pillow 库

安装 Pillow 库的方法与安装 Python 其他第三方库的方法相同,也可以从 PyPI 或 GitHub 网站下载 Pillow 库的压缩包。解压下载的压缩包文件后,在命令提示符下进入下载目录,然后运行如下命令即可安装。

```
python setup.py install
```

如果计算机可以联网,可以运行 "pip" 命令自动从互联网中下载并安装。

```
pip install pillow
```

在联网的计算机上也可以通过 "easy_install" 命令安装 Pillow。

```
easy_install Pillow
```

例如,在 Windows 系统中成功安装后的界面如图 8-1 所示。

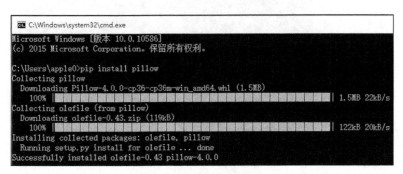

图 8-1 成功安装后的界面

8.1.2 使用 Image 模块

在 Pillow 库中,最常用的内置模块是 Image。开发者可以通过多种方法创建 Image 实例。

1. 打开和新建

在 Pillow 库中,通过使用 Image 模块,可以从文件加载图像,或者处理其他图像,或者从 scratch 中创建图像。在对图像进行处理时,首先需要打开要处理的图片。在 Image 模块中,使用函数 open() 打开一幅图片,执行后返回 Image 类的实例。当文件不存在时,会引发 IOError 错误。使用函数 open() 的语法格式如下。

```
open ( file, mode)
```

- file:要打开的图片文件的路径。
- mode:可选参数,表示打开文件的方式,通常使用默认值 r。

在 Image 模块中,可以使用函数 new() 新建图像,具体语法格式如下。

```
new (mode, size, color=0)
```

- mode:表示图片模式。
- size:表示图片尺寸,一个由 width 和 height 两个元素构成的元组。
- color:表示默认颜色(黑)。

下面的实例文件 dakai.py 演示了使用 Image 模块打开一幅图片的过程。

源码路径：daima\8\8-1\dakai.py

```
from PIL import Image                    #导入Image模块
im = Image.open("IMG_1.jpg")             #打开指定的图片
print(im.format, im.size, im.mode)       #显示图片的属性信息
im.show()                                #显示打开的这幅图片
```

在上述实例代码中，使用函数 open()打开了当前目录中的图片文件 IMG_1.jpg，然后显示了这幅图片的属性信息，最后使用函数 show()显示这幅图片。其中 format 属性标识了图像来源，如果图像不从文件读取它的值，则 format 的值是 None。属性 size 是一个二元组，包含 width 和 height（宽度和高度，单位都是 px）。属性 mode 定义了图像 bands 的数量和名称，以及像素类型和深度。如果文件打开错误，返回 IOError 错误。执行后将显示图片 IMG_1.jpg 的属性并打开这幅图片，执行效果如图 8-2 所示。

图 8-2　执行效果

2．混合

1）透明度混合处理

在 Pillow 库的 Image 模块中，可以使用函数 blend()实现透明度混合处理。具体语法格式如下。

```
blend (im1, im2, alpha)
```

- im1：参与混合的图片 1。
- im2：参与混合的图片 2。
- alpha：混合透明度，取值范围是 0～1。

通过使用函数 blend()，可以将 im1 和 im2 这两幅图片（尺寸相同）以一定的透明度进行混合处理。混合后的透明度如下。

```
(im1 ×(1-alpha) + im2×alpha)
```

当混合透明度 alpha 的值为 0 时，显示 im1。当混合透明度 alpha 的值为 1 时，显示 im2。下面的实例文件 hun.py 演示了使用 Image 模块实现图片透明度混合的过程。

源码路径：daima\8\8-1\hun.py

```
from PIL import Image                         #导入Image模块
imga = Image.open('IMG_1.jpg')                #打开指定的图片1
imgb = Image.open('IMG_2.jpg')                #打开指定的图片2
Image.blend(imga,imgb,0.3).show()             #混合两幅图片
```

原始图片 IMG_1.jpg 和 IMG_2.jpg 的效果如图 8-3 所示。

实现混合处理后的效果如图 8-4 所示。

(a) IMG_1.jpg　　　　　　　　(b) IMG_2.jpg

图 8-3　原图效果　　　　　　　　　　　　图 8-4　混合效果

2）遮罩混合处理

在 Pillow 库的 Image 模块中，可以使用函数 composite() 实现遮罩混合处理。具体语法格式如下。

```
composite (im1, im2, mask)
```

- im1：将要混合处理的图片 1。
- im2：将要混合处理的图片 2。
- mask：混合遮罩模式，可以是 "1" "L" 或 "RGBA" 等模式。

函数 composite() 的功能是使用 mask 来混合图片 im1 和图片 im2，并且要求 mask、im1 和 im2 的尺寸相同。下面的实例文件 zhe.py 演示了，使用 Image 模块实现图片遮罩混合处理的过程。

源码路径：daima\8\8-1\zhe.py

```
from PIL import Image                              #导入Image模块
imga = Image.open('IMG_1.jpg')                     #打开指定的图片1
imgb = Image.open('IMG_2.jpg')                     #打开指定的图片2
mask = Image.open('IMG_3.jpg')                     #打开指定的图片3
Image.composite(imga,imgb,mask).show()             #实现3幅图片的遮罩混合
```

原始图片 IMG_1.jpg、IMG_2.jpg 和 IMG_3.jpg 的效果如图 8-5 所示。

(a) IMG_1.jpg　　　　　　　　(b) IMG_2.jpg　　　　　　　　(c) IMG_3.jpg

图 8-5　原图效果

实现遮罩混合处理后的效果如图 8-6 所示。

3．复制和缩放

1）复制图像

在 Pillow 库的 Image 模块中，可以使用函数 Image.copy() 复制指定的图片。

2）缩放像素

在 Pillow 库的 Image 模块中，可以使用函数 eval() 实现像素缩放处理，同时使用函数 fun() 对输入图片的每个像素进行计算并返回。使用函数 eval() 的语法格式如下。

```
eval(image, fun)
```

- image：输入的图片。
- fun：给输入图片的每个像素应用此函数，fun()函数只允许接收一个整型参数。如果一幅图片含有多个通道，则每个通道都会应用这个函数。

下面的实例文件 suo.py 演示了使用 Image 模块缩放指定图片像素的过程。

源码路径：daima\8\8-1\suo.py

```
from PIL import Image                    #导入Image模块
def div2(v):                             #定义函数div2()
    return v//2                          #缩放为原来的一半
imga = Image.open('IMG_1.jpg')           #打开指定的图片
Image.eval(imga,div2).show()             #显示缩放后的图片
```

上述代码中定义了一个用于计算像素值的函数 div2()，然后打开一幅指定的图片 IMG_1.jpg，最后调用 eval()函数实现像素缩放处理并输出。执行后的效果如图 8-7 所示。

图 8-6　遮罩混合效果　　　　　　　　　　图 8-7　执行效果

3）缩放图像

在 Pillow 库的 Image 模块中，可以使用函数 thumbnail()缩放指定的图像。具体语法格式如下。

```
Image.thumbnail(size, resample=3)
```

下面的实例文件 suo1.py 演示了使用 Image 模块缩放指定图片的过程。

源码路径：daima\8\8-1\suo1.py

```
from PIL import Image                    #导入Image模块
imga = Image.open('IMG_1.jpg')           #打开指定的图片
print('图像格式：',imga.format)           #显示图像格式
print('图像模式：',imga.mode)             #显示图像模式
print('图像尺寸：',imga.size)             #显示图像尺寸
imgb = imga.copy()                       #复制打开的图片
imgb.thumbnail((224,168))                #缩放为指定的大小
imgb.show()                              #显示缩放后的图片
```

在上述代码中分别显示了指定图像"IMG_1.jpg"的格式、模式和尺寸，然后使用函数 copy() 复制图像，并将其缩放为指定的大小。执行后的效果如图 8-8 所示。

图 8-8　执行效果

4. 粘贴和裁剪

1）粘贴

在 Pillow 库的 Image 模块中，函数 paste()的功能是粘贴源图像或像素至该图像。具体语法格式如下。

```
Image.paste (im, box=None, mask=None)
```

- im：表示源图或像素值。
- box：表示粘贴区域。
- mask：表示遮罩。

参数 box 的设置可以分为以下 3 种情况。

- (x1, y1)：将源图像左上角对齐 (x1, y1) 点，舍弃其余超出被粘贴的图像的区域。
- (x1, y1, x2, y2)：源图像与此区域必须一致。
- None：源图像与被粘贴的图像大小必须一致。

2）裁剪图像

在 Pillow 库的 Image 模块中，函数 crop()的功能是剪切图片中 box 所指定的区域。具体语法格式如下。

```
Image.crop(box=None)
```

参数"box" 是一个 4 元组，分别定义了剪切区域的左、上、右、下这 4 个坐标。

下面的实例文件 jian.py 演示了使用 Image 模块对指定图片实现剪切和粘贴功能的过程。

源码路径：daima\8\8-1\jian.py

```
print('图像通道列表：',imga.getbands())    #显示图像通道列表
imgb = imga.copy()                          #复制图片
imgc = imga.copy()                          #复制图片
region = imgb.crop((5,5,120,120))           #剪切指定区域的图片
imgc.paste(region,(230,230))                #粘贴的图片
imgc.show()                                 #显示粘贴后的图像
```

执行后的效果如图 8-9 所示。图中被圈出的区域便是剪切并粘贴后的区域。

5. 格式转换

1）convert()

在 Pillow 库的 Image 模块中，函数 convert()的功能是返回模式转换后的图像实例。目前支持的模式有"L""RGB""CMYK"，参数 matrix 只支持"L"和"RGB"两种模式。具体语法格式如下。

```
Imageo.convert(mode=None, matrix=None, dither=None, palette=0, colors=256)
```

图 8-9 执行效果

- mode：转换文件为此模式。
- matrix：转换使用的矩阵（4 或 16 元素的浮点型元组）。
- dither：当取值为 None 且转换为黑白图时，非零（1～255）像素均为白色，也可以设置此参数为 FLOYDSTEINBERG。

2）transpose()

在 Pillow 库的 Image 模块中，函数 transpose()的功能是实现图像格式的转换。具体语法格式如下。

```
Image.transpose(method)
```

转换图像后，返回转换后的图像，"method"的取值如下。

- PIL.Image.FLIP_LEFT_RIGHT：左右镜像。
- PIL.Image.FLIP_TOP_BOTTOM：上下镜像。

- PIL.Image.ROTATE_90：旋转 90°。
- PIL.Image.ROTATE_180：旋转 180°。
- PIL.Image.ROTATE_270：旋转 270°。
- PIL.lmage.TRANSPOSE：颠倒顺序。

下面的实例文件 zhuan.py 演示了使用 Image 模块转换指定图片格式的过程。

源码路径：daima\8\8-1\zhuan.py

```
from PIL import Image
#设置要操作的指定图片
imga = Image.open('IMG_1.jpg')
imgb = imga.copy()
#创建新图像
img_output = Image.new('RGB',(448,168))
img_output.paste(imgb,(0,0))
img_output.show()
b = imgb.convert('CMYK')            #转换为CMYK格式的图像
img_output.paste(b,(224,0))         #粘贴转换后的图像
img_output.show()
#得到一幅左右镜像的图像
flip = b.transpose(Image.FLIP_LEFT_RIGHT)
img_output.paste(flip,(224,0))      #粘贴镜像的图像
img_output.show()
#将图像转换为灰度图
b = imgb.convert('L')
img_output.paste(b,(224,0))         #粘贴灰度图
img_output.show()                   #显示图片
```

执行后的效果此处省略。

6. 重设和旋转

1）重设

在 Pillow 库的 Image 模块中，可以使用函数 resize() 来重新设置指定图像的尺寸。函数 resize() 的语法格式如下。

```
Image.resize (size, resample=0)
```

2）旋转

在 Pillow 库的 Image 模块中，可以使用函数 rotate() 来旋转指定的图像。函数 rotate() 的语法格式如下。

```
Image.rotate (angle, resample=0, expand=0)
```

- angle：表示旋转角度，使用逆时针方式。
- expand：展开可选的扩展标志。如果为 TRUE，则扩大输出图像，使其足够大，以容纳整个旋转图像。如果是 FALSE，则省略，使输出图像和输入图像的大小相同。

下面的实例文件 xuan.py 演示了使用 Image 模块旋转指定图片的过程。

源码路径：daima\8\8-1\xuan.py

```
from PIL import Image                                  #导入Image模块
imga = Image.open('IMG_1.jpg')                         #打开指定的图片
imgb = imga.copy()                                     #复制打开的图片
img_output = Image.new('RGB',(448,168))                #创建一个新的图像区域
b = imgb.rotate(45)                                    #旋转45°
img_output.paste(b,(224,0))                            #粘贴矩形区域
img_output.show()                                      #输出显示图片
```

执行后的效果如图 8-10 所示。

7. 分离和合并

1）分离

在 Pillow 库的 Image 模块中，使用函数 split() 可以将图片分隔为多个通道列表。使用函数 split() 的语法格式如下。

```
Image.split()
```

图 8-10　执行效果

8.1 使用 Pillow 库

2) 合并

在 Pillow 库的 Image 模块中，使用函数 merge() 可以将一个通道的图像合并到更多通道的图像中。使用函数 merge() 的语法格式如下。

```
PIL.Image.merge(mode, bands)
```

- mode：输出图像的模式。
- bands：波段通道，一个序列包含一个单一的带图通道。

下面的实例文件 he.py 演示了，使用 Image 模块对指定图片分别实现合并和分离操作的过程。

源码路径：daima\8\8-1\he.py

```
chnls = imgb.split()                              #分离图像通道
b = Image.merge('RGB',chnls[::-1])                #合并R和B互换后的通道
img_output.paste(b,(224,0))                       #粘贴合并后的图像
img_output.show()                                 #显示处理结果
from PIL import ImageFilter
b = imgb.filter(ImageFilter.GaussianBlur)         #处理R通道中的每一个像素
img_output.paste(b,(224,0))                       #合并R通道中变化后的图像
img_output.show()
```

执行后的效果此处省略。

8. 滤镜

在 Pillow 库的 Image 模块中，使用函数 filter() 可以对指定的图片应用滤镜效果。在 Pillow 库中可以用的滤镜保存在 Imagefilter 模块中。使用函数 filter() 的语法格式如下。

```
Image.filter( filter)
```

通过使用函数 filter()，可以使用给定的过滤器过滤指定的图像，参数"filter"表示滤波器内核。下面的实例文件 guo.py 演示了使用 Image 模块对指定图片实现模糊处理的过程。

源码路径：daima\8\8-1\guo.py

```
from PIL import ImageFilter                       #导入Imagefilter模块
#使用函数filter()实现滤镜效果
b = imgb.filter(ImageFilter.GaussianBlur)
img_output.paste(b,(224,0))                       #粘贴指定大小的区域
img_output.show()                                 #显示图片
```

执行后的效果如图 8-11 所示。

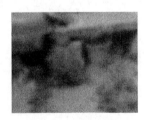

图 8-11　执行效果

8.1.3 绘制随机漫步图

下面的实例文件 photomancy.py 演示了使用 Pillow 库绘制随机漫步图的过程。

源码路径：daima\8\8-1\photomancy.py

```python
from PIL import Image, ImageDraw

import random
import math

def get_random_pixel():
    """用一个三元组的形式返回一个随机RGB值"""
    return (random.randint(0, 255), random.randint(0, 255), random.randint(0, 255))

def set_to_noise(img):
    """用随机数字替换图像的内容"""
```

```python
            img.putdata([random.randint(0, 255) for p in range(img.width * img.height)])

        def get_rgb_noise(width, height):
            """返回随机颜色像素的图像"""
            bands = [Image.new('L', (width, height)) for band in ('r', 'g', 'b')]
            for band in bands:
                set_to_noise(band)
            img = Image.merge('RGB', bands)
            return img

        def for_each_cell(func, img, cell_radius=1):
            """在源图像上对像素组调用函数并返回转换后的图像"""
            width = img.width
            height = img.height
            img2 = Image.new(img.mode, (width, height))
            draw = ImageDraw.Draw(img2)
            for y in range(cell_radius, height-cell_radius):
                for x in range(cell_radius, width-cell_radius):
                    neighbors = [img.getpixel((x_offs, y_offs)) for x_offs in range(x-cell_radius,
                        x+cell_radius+1) for y_offs in range(y-cell_radius, y+cell_radius+1)]
                    func(img, draw, x, y, neighbors)
            return img2

        def blur(img):
            """将每个非边界像素替换为其邻域的均值，并作为新的图像返回"""
            def blur_func(img, draw, x, y, neighbors):
                avg_color = get_avg_color(neighbors)
                draw.point((x,y), avg_color)

            return for_each_cell(blur_func, img)

        def get_avg_color(neighbors):
            """获取平均颜色"""
            return tuple([math.floor(sum(p[i] for p in neighbors)/len(neighbors)) for i in (0, 1, 2)])

        def get_avg_value(neighbors):
            """获取平均值"""
            return math.floor(sum(neighbors)/len(neighbors))

        def rgb_push_func(img, draw, x, y, neighbors):
            orig_color = img.getpixel((x, y))#获取对应点的像素值
            avg_color = get_avg_color(neighbors)
            max_band_val = max(avg_color)
            pushed_color = tuple(32 if v == max_band_val else -32 for v in avg_color)
            combined_color = tuple(min(255, v[0] + v[1]) for v in zip(orig_color, pushed_color))
            # combined_color = orig_color + pushed_color
            draw.point((x,y), combined_color)

        def bw_push_func(img, draw, x, y, neighbors):
            orig_value = img.getpixel((x, y))#获取对应点的像素值
            avg_value = get_avg_value(neighbors)
            # pushed_value = 32 if avg_value > 127 else -32
            # combined_value = min(max(avg_value + pushed_value, 0), 255)
            pushed_value = 255 if avg_value > 127 else 0
            draw.point((x, y), pushed_value)

        def filter_color(color):
            """颜色过滤"""
            if all([not(0 < v < 255) for v in color]):
                return tuple(0 if v == 255 else 255 for v in color)
            else:
                return color

        if __name__ == "__main__":
            width = 800
            height = 600
            img = get_rgb_noise(width, height)
            for i in range(5):
                img = for_each_cell(rgb_push_func, img)
```

```
            img = for_each_cell(lambda img, draw, x, y, neighbors: draw.point((x, y),
                filter_color(img.getpixel((x, y)))), img, cell_radius=0)
            # img = blur(img)
        img.show()

        img.save("output.png")
```

在上述代码中定义了多个图像蠢立函数,在注释中进行了详细说明,执行后可以生成随机点阵图。在现实应用中,读者可以在自己的程序中直接调用上述文件模块来定制自己的点阵图。下面的实例文件 caves.py 演示了使用 photomancy.py 模块绘制点阵图的过程。

源码路径:daima\8\8-1\caves.py

```
from PIL import Image, ImageDraw

import photomancy

import random

if __name__ == "__main__":
    width, height = (800, 600)
    img = Image.new('L', (width, height))
    photomancy.set_to_noise(img)
    for i in range(5):
        img = photomancy.for_each_cell(photomancy.bw_push_func, img, cell_radius=3)
        img.show()

    img.save("output.png")
```

执行上述两个文件都可以生成点阵图,在作者的计算机上的执行效果如图 8-12 所示。

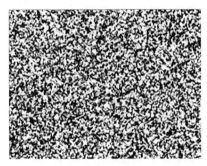

图 8-12 执行效果

8.1.4 使用 ImageChops 模块合成图片

在 Pillow 库的内置模块 ImageChops 中包含了多个用于实现图片合成的函数。这些合成功能是通过计算通道中像素值的方式来实现的,其主要用于制作特效、合成图片等操作。在模块 ImageChops 中,最常用的内置函数如下。

(1) 相加函数 add(),其功能是对两张图片进行加法运算。具体语法格式如下。

```
ImageChops.add(image1, image2, scale=1.0, offset=0)
```

在合成后的图像中,每个像素值依据下面的公式进行计算。

```
out= ((image1+image2)/scale+offset)
```

(2) 减法函数 subtract(),其功能是对两张图片进行减法运算。具体语法格式如下。

```
ImageChops.subtract(image1, image2, scale=1.0, offset=0)
```

在合成后的图像中,每个像素值依据下面的公式进行计算。

```
out= ((image1 - image2)/scale+offset)
```

(3) 变暗函数 darker(),其功能是比较两张图片的像素,并取两张图片中对应像素的较小值,所以合成后两幅图像中对应位置的暗部分得到保留,亮部分去除。具体语法格式如下。

```
ImageChops.darker(image1, image2)
```

像素的计算公式如下。

```
out - min (image1, image2)
```

(4)变亮函数 lighter(),与变暗函数 darker()相反,其功能是比较两张图片(逐像素地比较),返回一张新的图片,这张新的图片是将两张图片中较暗部分的叠加。也就使说,在某一点上,取两张图中较小的像素值。具体语法格式如下。

```
ImageChops.lighter (image1, image2)
```

函数 lighter()与函数 darker()的功能相反,计算后得到的图像是两幅图像对应位置的亮部分。像素值的计算公式如下。

```
out=max (image1, image2)
```

(5)叠加函数 multiply(),其功能是将两张图片互相叠加。如果用纯黑色与某图片进行叠加操作,会得到一张纯黑色的图片;如果用纯白色与图片叠加,则图片不受影响。具体语法格式如下。

```
ImageChops.multiply (image1, image2 )
```

叠加效果类似于两张透明的描图纸叠放在一起观看。其对应像素值的计算公式如下。

```
out=image1*image2/MAX
```

(6)屏幕函数 screen(),其功能是先反色后叠加,实现合成图像的效果,就像将两张幻灯片用两台投影机同时投射到一个屏幕上。具体语法格式如下。

```
ImageChops.screen(image1, image2)
```

其对应像素值的计算公式为:

```
out=image1*image2/MAX
```

(7)反色函数 invert(),其功能类似于集合操作中的求补集,最大值为 Max,对于每个像素做减法,取反色。在反色时用 255 减去一幅图像的每个像素值,从而得到原来图像的反相。也就是说,其得到的是"底片"性质的图像。具体语法格式如下。

```
ImageChops.invert (image)
```

像素值的计算公式如下。

```
out=MAX - image
```

(8)比较函数 difference(),计算两张图片的绝对值,逐像素地做减法操作。函数 difference()能够得到两幅图像的像素值相减后的图像,对应像素值相同的,则为黑色。函数 difference()通常用来找出图像之间的差异。具体语法格式如下。

```
ImageChops.difference (image1, image2)
```

像素值的计算公式如下。

```
out=abs (image1-image2)
```

(9)灰度填充函数 constant(),其功能是用所给的灰度等级来填充各像素,用来生成给定的灰度值图像。具体语法格式如下。

```
ImageChops.constant (image,value)
```

下面的实例文件 hecheng.py 演示了使用 ImageChops 模块实现图片合成的过程。

源码路径:daima\8\8-1\hecheng.py

```
from PIL import Image                                    #导入Image模块
from PIL import ImageChops                               #导入ImageChops模块
imga = Image.open('IMG_1.jpg')                           #打开图片1
imgb = Image.open('IMG_2.jpg')                           #打开图片2
ImageChops.add(imga,imgb,1,0).show()                     #对两张图片进行加法运算
ImageChops.subtract(imga,imgb,1,0).show()                #对两张图片进行减法运算
ImageChops.darker(imga,imgb).show()                      #使用变暗函数darker()
ImageChops.lighter(imga,imgb).show()                     #使用变亮函数lighter()
ImageChops.multiply(imga,imgb).show()                    #将两张图片互相叠加
ImageChops.screen(imga,imgb).show()                      #实现反色后叠加
ImageChops.invert(imga).show()                           #使用反色函数invert()
ImageChops.difference(imga,imga).show()                  #使用比较函数difference()
```

执行后的部分效果如图 8-13 所示。如果对一幅图片分别实现相加、相减、变暗、变亮、叠加、屏幕、反相和比较操作,其中经过最后比较操作后产生的图像为纯黑色,这是因为比较的是同一幅图片。

图 8-13　执行效果

8.1.5　使用 ImageEnhance 模块增强图像

在 Pillow 库的内置模块 ImageEnhance 中包含了多个用于增强图像效果的函数，主要用来调整图像的色彩、对比度、亮度和清晰度等，这和调整电视机的显示参数一样。在模块 ImageEnhance 中，所有的图片增强对象都实现了一个通用接口，这个接口只包含下面一个方法。

```
enhance(factor)
```

方法 enhance() 会返回一个加强过的 Image 对象，参数 factor 是一个大于 0 的浮点数，1 表示返回原始图片。

在 Python 程序中使用模块 ImageEnhance 增强图像效果时，需要先创建对应的增强调整器，然后调用调整器输出函数，根据指定的增强系数（小于 1 表示减弱，大于 1 表示增强，等于 1 表示不变）进行调整，最后输出调整后的图像。

在模块 ImageEnhance 中，最常用的内置函数如下。

- ImageEnhance.Color(image)：功能是调整图像色彩，相当于彩色电视机的色彩调整，实现了上面提到的接口的 enhance 方法。
- ImageEnhance.Contrast(image)：功能是调整图像对比度，相当于彩色电视机的对比度调整。
- ImageEnhance.Brightness(image)：功能是调整图像亮度。
- ImageEnhance.Sharpness(image)：功能是调整图像清晰度，用于锐化/钝化图片。锐化操作的 factor 是一个 0~2 的浮点数。当 factor=0.0 时，返回一个完全模糊的图片对象；当 factor=2.0 时，返回一个完全锐化的图片对象；当 factor=1.0 时，返回原始图片对象。

下面的实例文件 zeng.py 演示了使用 ImageEnhance 模块实现图像增强的过程。

源码路径：daima\8\8-1\zeng.py

```
w,h = imga.size                              #定义变量w和h的初始值
img_output = Image.new('RGB',(2*w,h))        #创建图像区域
img_output.paste(imga,(0,0))                 #将创建的部分粘贴到图片中
nhc = ImageEnhance.Color(imga)               #调整图像色彩
nhb = ImageEnhance.Brightness(imga)          #调整图像亮度
for nh in [nhc,nhb]:                         #使用内嵌循环输出调整后的图像
    for ratio in [0.6,1.8]:                  #减弱和增强两个系数
        b = nh.enhance(ratio)                #增强处理
        img_output.paste(b,(w,0))            #粘贴修改后的图像
        img_output.show()                    #显示对比的图像
```

执行后的效果如图 8-14（a）~（d）所示，分别实现色彩减弱、色彩增强、亮度减弱和亮度增强效果。

图 8-14 执行效果

下面的实例文件 123.py 演示了，使用 ImageEnhance 模块实现同时增强处理多幅图像的过程。

源码路径：daima\8\8-1\123.py

```
from PIL import Image, ImageFilter, ImageEnhance, ImageDraw

import os, sys

# 素材图片的路径，能够同时处理这个目录中的所有图片
path = "123/"
dirs = os.listdir(path)

def enhancer():
    #从文件夹中导入文件
    for image in dirs:
        if os.path.isfile(path+image):
            source = Image.open(path+image)
            f, e = os.path.splitext(path+image)

            #设置两个输入图像源，分别用DETAIL过滤器和FIND_EDGES过滤器过滤图像
            filter1 = source.filter(ImageFilter.DETAIL)
            filter2 = source.filter(ImageFilter.FIND_EDGES)

            #一幅图像用DETAIL过滤器过滤，另一幅图像用FIND_EDGES过滤器过滤，
            #两幅过滤后的图像混合在一起，把alpha设置为0.1
            compose = Image.blend(filter1, filter2, alpha=.1)

            #取第一个图像混合（合成）的结果并把它再次与SMOOTH过滤器混合，
            #作为一个新的图像源输入，把alpha设置为0.1
            filter3 = source.filter(ImageFilter.SMOOTH)
            blend = Image.blend(compose, filter3, alpha=.1)

            #最后的混合用于移动到增强阶段。
            #此阶段旨在增强图像的颜色
```

```
    # IMAGE > enhanced > COLOR
    imageColor = ImageEnhance.Color(blend)
    renderStage1 = imageColor.enhance(1.5)

    #此阶段旨在增强图像的对比度
    # IMAGE(color) > enhanced > CONTRAST
    imageContrast = ImageEnhance.Contrast(renderStage1)
    renderStage2 = imageContrast.enhance(1.1)

    #此阶段旨在增强图像的对比度
    # IMAGE(contrast) > enhanced > BRIGHTNESS

    imageBrightness = ImageEnhance.Brightness(renderStage2)
    renderFinal = imageBrightness.enhance(1.1)

    #最后,写入新图像文件
    #文件格式为jpeg,把quality设置为100
    renderFinal.save(f + '_enhanced.jpg', 'JPEG', quality=100)

enhancer()
```

在上述实例代码中,首先设置图片的保存路径——"123"目录下,然后同时对这个目录中的图片进行增强处理。例如,作者在"123"目录中准备了 3 幅图片(如图 8-15(a)、(c)和(e)所示),增加后会生成名字中有"enhanced"标识的图片,如图 8-15(b)、(d)和(f)所示。

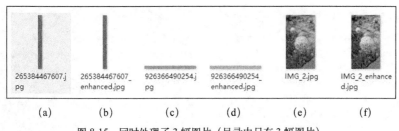

图 8-15 同时处理了 3 幅图片(目录中只有 3 幅图片)

8.1.6 使用 ImageFilter 模块实现滤镜功能

在 Pillow 库中,内置模块 ImageFilter 实现滤镜功能,可以用来创建图像特效,或以此效果作为媒介实现进一步处理。在模块 ImageFilter 中,提供了一些预定义的过滤器和自定义的过滤器函数。其中最常用的预定义过滤器如下。

- BLUR:用于模糊滤波。
- CONTOUR:用于轮廓滤波。
- DETAIL:用于细节滤波。
- EMBOSS:用于浮雕滤波。
- FIND_EDGES:用于查找边缘的滤波。
- SHARPEN:用于锐化滤波。
- SMOOTH:用于光滑滤波。
- EDGE_ENHANCE:用于边缘增强滤波。
- EDGE_ENHANCE_MORE:用于更深程度的边缘增强滤波。

在模块 ImageFilter 中,常用的自定义过滤器函数如下。

- ImageFilter.GaussianBlur(radius=2):用于高斯模糊。
- ImageFilter.UnsharpMask(radius=2, percent 150, threshold=3):用于 USM 锐化。
- ImageFilter.MedianFilter(size=3):用于中值滤波。
- ImageFilter.MinFilter(size=3):用于最小值滤波。
- ImageFilter.ModeFilter(size=3):用于模式滤波。

第 8 章 图像处理

上述过滤器函数的使用方法十分简单,把滤镜实例作为参数提供给本章前面介绍的 Image 模块中的方法 filter(),就可以返回具有滤镜特效的图像。

下面的实例文件 lv.py 演示了使用 ImageFilter 模块对指定图片实现滤镜特效的过程。

源码路径:daima\8\8-1\lv.py

```
from PIL import Image
from PIL import ImageFilter                          #导入模块ImageFilter
imga = Image.open('IMG_2.jpg')                       #打开指定的图像
w,h = imga.size                                      #图像的宽和高
img_output = Image.new('RGB',(2*w,h))                #新建指定大小的图像
img_output.paste(imga,(0,0))                         #粘贴原始图像
fltrs = []                                           #创建列表以存储滤镜
fltrs.append(ImageFilter.EDGE_ENHANCE)               #边缘强化滤镜
fltrs.append(ImageFilter.FIND_EDGES)                 #查找边缘滤镜
fltrs.append(ImageFilter.GaussianBlur(4))            #高斯模糊滤镜

for fltr in fltrs:                                   #遍历上述3种滤镜
    r = imga.filter(fltr)                            #使用滤镜
    img_output.paste(r,(w,0))                        #粘贴使用波幅后的图像
    img_output.show()                                #显示应用滤镜前后的图像
```

在上述实例代码中建立了使用滤镜的列表,然后用 for 循环遍历并输出应用滤镜前后的图像。执行后的效果如图 8-16 所示。

(a) 边缘增强

(b) 查找边缘

(c) 高斯模糊

图 8-16 执行效果

8.1.7 使用 ImageDraw 模块绘制图像

在 Pillow 库中,内置模块 ImageDraw 实现了绘图功能,可以通过创建图片的方式来绘制 2D 图像。也可以在原有图片上进行绘图,以达到修饰图片或对图片进行注释的目的。在使用

ImageDraw 模块绘图时，需要先创建一个 ImageDraw.Draw 对象，并且提供指向文件的参数，然后引用创建的 Draw 对象方法进行绘图；最后保存或直接输出绘制的图像。

```
drawObject = ImageDraw.Draw(blank)
```

在模块 Python 程序中，常用的内置函数如下。

```
drawObject.line([x1,y1,x2,y2] ,options)
```

该函数以($x1,y1$)为起始点、以($x2,y2$)为终点画一条直线。[$x1,y1,x2,y2$]也可以写为 ($x1,y1$, $x2,y2$)、[($x1,y1$),($x2,y2$)]等，options 选项包含的 fill 选项规定线条颜色。

```
drawObject.arc([x1, y1, x2, y2], startAngle, endAngle, options)
```

在左上角坐标为($x1,y1$)、右下角坐标为 ($x2,y2$)的矩形区域内，在圆 O 内，以 startAngle 为起始角度，以 endAngle 为终止角度，截取圆 O 的一部分圆弧并画出来。如果[$x1,y1,x2,y2$]区域不是正方形，则在该区域内的最大椭圆中根据角度截取片段。参数 options 设置圆弧线的颜色，具体方法与 drawObject.line 相同。

注意：[$x1,y1,x2,y2$]规定矩形框的水平中位线为 0°角，角度顺时针变大（与数学坐标系规定的方向相反）。

```
drawObject.ellipse([x1,y1,x2,y2], options)
```

用法同 arc 类似，用于画圆（或者椭圆）。options 选项中 fill 表示将圆（或者椭圆）用指定颜色填满，outlie 表示只规定圆的颜色。

```
drawObject.chord([x1, y1, x2, y2], startAngle, endAngle, options)
```

具体用法与 arc 相同，用来画圆中从 startAngle 到 endAngle 的弦。options 选项中 fill 表示将弦与圆弧之间的空间用指定颜色填满，outlie 表示只规定弦线的颜色。

```
drawObject.pieslice([x1,y1,x2,y2], startAngle, endAngle, options)
```

用法与 ellipse 相同，用于画起始角度间的扇形区域。options 选项中的 fill 选项用于将扇形区域用指定颜色填满，outline 选项表示只用指定颜色描出区域轮廓。

```
drawObject.polygon(([x1,y1,x2,y2, …],options)
```

能够根据坐标画多边形，Python 会根据第一个参数中的 x、y 坐标对，连接出整个图形。options 选项中的 fill 选项将多边形区域用指定颜色填满，outline 选项表示只用指定颜色描出区域轮廓。

```
drawObeject.rectangle([x1,y1,x2,y2],options)
```

能够在指定的区域内画一个矩形，($x1,y1$)表示矩形左上角的坐标值，($x2,y2$)表示矩形右下角的坐标值。options 选项中 fill 选项将多边形区域用指定颜色填满，outline 选项表示只用指定颜色描出区域轮廓。

```
drawObject.text(position, string, options)
```

能够在图像内添加文字。其中，参数 position 是一个二元组，用于指定字符串左上角的坐标；string 表示要写入的字符串；options 选项可以为 fill 或者 font（只能选择其中之一作为第三个参数，不能两个同时存在。要想改变字体颜色，请参阅本章后面讲解的 ImageFont 模块）。其中，fill 用于指定字的颜色，font 用于指定字体与字的尺寸，font 必须为 ImageFont 中指定的 font 类型，具体用法见 ImageFont.Truetype()。

- point (xy, fill=None)用于绘制点。
- text (xy, text, fill=Noner font=None, anchor=None)用于绘制字符串。
- setfill(fill)用于设置默认填充颜色。

下面的实例文件 huier.py 演示了使用 ImageDraw 模块绘制二维图像的过程。

源码路径：daima\8\8-1\huier.py

```
a = Image.new('RGB',(200,200),'white')      #新建白色背景的图像
drw = ImageDraw.Draw(a)                     #创建绘图对象
drw.rectangle((50,50,150,150),outline='red') #绘制矩形
drw.text((60,60),'First Draw...',fill='green') #绘制文本
a.show()                                    #显示创建二维图像
```

在上述实例代码中，在新建的图像中使用 ImageDraw 对象中的方法 rectangle()和 text()，分别绘制了一个矩形和一个字符串。执行后的效果如图 8-17 所示。

图 8-17　执行效果

8.1.8　使用 ImageFont 模块设置字体

在 Pillow 库中，内置模块 ImageFont 的功能是实现对字体和字型的处理。下面的实例文件 zi.py 演示了使用 ImageFont 模块绘制二维图像的过程。

源码路径：daima\8\8-1\zi.py

```
ft = ImageFont.truetype("C:\\WINDOWS\\Fonts\\SIMYOU.TTF", 20)        #设置本地字体目录
draw.text((30,30), u"Python图像处理库PIL",font = ft, fill = 'red')    #设置指定文本、字体和颜色
ft = ImageFont.truetype("C:\\WINDOWS\\Fonts\\SIMYOU.TTF", 40)        #设置本地字体目录
draw.text((30,100), u"Python图像处理库PIL",font = ft, fill = 'green') #设置指定文本、字体和颜色
ft = ImageFont.truetype("C:\\WINDOWS\\Fonts\\SIMYOU.TTF", 60)        #设置本地字体目录
draw.text((30,200), u"Python图像处理库PIL",font = ft, fill = 'blue')  #设置指定文本、字体和颜色
ft = ImageFont.truetype("C:\\WINDOWS\\Fonts\\SIMLI.TTF", 40)         #设置本地字体目录
draw.text((30,300), u"Python图像处理库PIL",font = ft, fill = 'red')   #设置指定文本、字体和颜色
ft = ImageFont.truetype("C:\\WINDOWS\\Fonts\\STXINGKA.TTF", 40)      #设置本地字体目录
draw.text((30,400), u"Python图像处理库PIL",font = ft, fill = 'yellow')#设置指定文本、字体和颜色
im02.show()
```

执行后的效果如图 8-18 所示。

下面的实例文件 zi01.py 演示了使用 ImageFont 模块绘制验证码的过程。

源码路径：daima\8\8-1\zi01.py

```
from PIL import Image, ImageDraw, ImageFont, ImageFilter

import random

#随机字母:
def rndChar():
    return chr(random.randint(65, 90))

#随机颜色1:
def rndColor():
    return (random.randint(64, 255), random.randint(64, 255), random.randint(64, 255))

#随机颜色2:
def rndColor2():
    return (random.randint(32, 127), random.randint(32, 127), random.randint(32, 127))

#240 * 60:
width = 60 * 4
height = 60
image = Image.new('RGB', (width, height), (255, 255, 255))
#创建Font对象:
font = ImageFont.truetype('C:\\WINDOWS\\Fonts\\Arial.ttf', 36)
#创建Draw对象:
draw = ImageDraw.Draw(image)
#填充每个像素:
for x in range(width):
    for y in range(height):
        draw.point((x, y), fill=rndColor())
#输出文字:
for t in range(4):
    draw.text((60 * t + 10, 10), rndChar(), font=font, fill=rndColor2())
```

```
#模糊：
image = image.filter(ImageFilter.BLUR)
image.save('code.jpg', 'jpeg')
```

执行后会在程序文件 zi01.py 所在目录下生成一个验证码图片文件 code.jpg，打开图片后会看到生成的验证码内容。验证码是随机的，每次执行效果都不一样，如图 8-19 所示。

图 8-18　执行效果

图 8-19　随机生成的验证码

8.1.9　绘制指定年份的日历

下面的实例文件 zi03.py 演示了同时使用模块 Image、ImageDraw、ImageFont 和 ImageOps 绘制指定年份的完整日历的过程。

源码路径：daima\8\8-1\zi03.py

```
import os
import calendar
import datetime
from PIL import Image, ImageDraw, ImageFont, ImageOps


MONTHS_LEFT = 30
MONTHS_TOP = 390
MONTH_HORIZONTAL_STEP = 220
MONTH_VERTICAL_STEP = 160
MONTHS_IN_ROW = 4
MONTH_DAYS_LINE_SPACING = 5
MONTH_WEEK_PADDING = 5
YEAR_TOP_PADDING = 10
YEAR_RIGHT_PADDING = 10

weekdays = {
    0: 'Mo',
    1: 'Tu',
    2: 'We',
    3: 'Th',
    4: 'Fr',
    5: 'Sa',
    6: 'Su',
}

MONDAY = 0
SUNDAY = 6


def get_weekdays(first_weekday=0):
    """Get weekdays as numbers [0..6], starting with first_weekday"""
    return list(list(range(0, 7)) * 2)[first_weekday: first_weekday + 7]


class TextObject(object):
    """Text object"""
```

```python
    def __init__(self, text, x, y, font, color):
        self.text = text
        self.x = x
        self.y = y
        self.font = font
        self.color = color


class MonthObject(object):
    """Month object"""

    def __init__(self, draw, year, month, **kwargs):
        """Init"""
        self.draw = draw
        self.year = year
        self.month = month
        self.month_font = kwargs['month_font']
        self.week_font = kwargs['week_font']
        self.x = kwargs['x']
        self.y = kwargs['y']
        self.month_color = kwargs['month_color']
        self.week_color = kwargs['week_color']
        self.weekend_color = kwargs['weekend_color']
        self.days_color = kwargs['days_color']
        self.first_weekday = kwargs.get('first_weekday', MONDAY)
        self.month_upper = kwargs.get('month_upper', False)
        self.week_upper = kwargs.get('week_upper', False)
        self.days_upper = kwargs.get('days_upper', False)
        self.items = []
        self._prepare(self.month_upper, self.week_upper, self.days_upper)

    def _add_month(self, x, y, color, upper=False):
        """Add month header"""
        month = '{:02x}'.format(self.month)
        if upper:
            month = month.upper()
        w, h = self.draw.textsize(month, font=self.month_font)
        text_obj = TextObject(month, x, y, self.month_font, color)
        self.items.append(text_obj)
        return w, h

    def _add_week(self, x, y, color, weekend_color, upper=False):
        """Add week"""
        space_w, space_h = self.draw.textsize(' ', font=self.week_font)
        orig_x = x
        w, h = 0, 0
        for d in get_weekdays(self.first_weekday):
            weekday_text = weekdays[d].upper() if upper else weekdays[d]
            w, h = self.draw.textsize(weekday_text, font=self.week_font)
            text_obj = TextObject(weekday_text, x, y, self.week_font, weekend_color
                if d in [5, 6] else color)
            x += w + space_w
            self.items.append(text_obj)
        return x - orig_x + w - space_w, h

    def _add_days(self, x, y, color, weekend_color, upper=False):
        """Add month days"""
        space_w, space_h = self.draw.textsize(' ', font=self.week_font)
        orig_x, orig_y = x, y
        h = 0
        for i, v in enumerate(calendar.Calendar().itermonthdays2(self.year, self.month)):
            day, wd = v
            day_str = '{:02x}'.format(day)
            if upper:
                day_str = day_str.upper()
            w, h = self.draw.textsize(day_str, font=self.week_font)
```

```python
                    if i and i % 7 == 0:
                        x = 0
                        y += h + MONTH_DAYS_LINE_SPACING
                    if day:
                        text_obj = TextObject(day_str, x, y, self.week_font, weekend_color
                            if wd in [5, 6] else color)
                        x += w + space_w
                        self.items.append(text_obj)
                    else:
                        x += w + space_w
                        continue
            return x - orig_x - space_w, y - orig_y + h

        def _prepare(self, month_upper=False, week_upper=False, days_upper=False):
            """Put all month object items together"""
            week_width, week_height = self.add_week(
                0, self.month_font.size, self.week_color, self.weekend_color, week_upper)
            month_width, _ = self.draw.textsize('{:02x}'.format(12), font=self.month_font)
            month_x, month_y = (week_width // 2) - month_width, 0 - MONTH_WEEK_PADDING
            _, month_heigth = self._add_month(month_x, month_y, self.month_color, month_upper)
            self._add_days(0, week_height * 2 + month_heigth, self.days_color, self.week
                end_color, days_upper)

        def render(self):
            """Draw items"""
            for item in self.items:
                self.draw.text((self.x + item.x, self.y + item.y), item.text, font=item.
                    font, fill=item.color)


def add_year(img, year, font, color, upper=False):
    """Add year"""
    txt_img = Image.new('RGBA', (100, 100))
    txt_img_draw = ImageDraw.Draw(txt_img)
    year_text = '{:x}'.format(year)
    if upper:
        year_text = year_text.upper()
    year_text = '0x' + year_text
    year_width, year_height = txt_img_draw.textsize(year_text, font=font)
    txt_img_draw.text((0, 0), year_text, font=font, fill=color)
    rotated_img = txt_img.rotate(90, expand=True)
    pos = (img.size[0] - txt_img.size[1] + year_width - YEAR_RIGHT_PADDING, YEAR_TOP
        _PADDING - year_height)
    img.paste(rotated_img, pos, rotated_img)


def add_image(img, image_path):
    """Add image"""
    if not os.path.isfile(image_path):
        print('File "{}" was not found!'.format(image_path))
        return
    image = Image.open(image_path, 'r')
    # resize image
    new_size = image.size
    while new_size[0] > img.size[0] or new_size[1] > MONTHS_TOP:
        new_size = (int(new_size[0] * 0.75), int(new_size[1] * 0.75))
    image = image.resize(new_size, Image.ANTIALIAS)
    # calculate coordinates
    x = (img.size[0] - image.size[0]) // 2
    y = (MONTHS_TOP - image.size[1]) // 2
    img.paste(image, (x, y))


def add_months(img, year, **kwargs):
    """Add months"""
    img_draw = ImageDraw.Draw(img)
    for month_index in range(1, 12 + 1):
```

```python
            month_object = MonthObject(
                draw=img_draw,
                year=year,
                month=month_index,
                x=MONTHS_LEFT + MONTH_HORIZONTAL_STEP * ((month_index - 1) % MONTHS_IN_ROW),
                y=MONTHS_TOP + MONTH_VERTICAL_STEP * ((month_index - 1) // MONTHS_IN_ROW),
                **kwargs
            )
            month_object.render()

def make_hex_calendar(img, year):
    """Make Hex Calendar"""
    image_path = 'spb_python_logo.png'
    add_image(img, image_path)

    month_font = ImageFont.truetype('C:\\WINDOWS\\Fonts\\Arial.ttf', 18)
    week_font = ImageFont.truetype('C:\\WINDOWS\\Fonts\\Arial.ttf', 15)
    year_font = ImageFont.truetype('C:\\WINDOWS\\Fonts\\Arial.ttf', 25)
    month_color = (70, 130, 180, 255)
    week_color = (70, 130, 180, 255)
    weekend_color = (255, 220, 75, 255)
    days_color = (255, 255, 255, 255)
    year_color = (80, 130, 180, 255)
    first_weekday = MONDAY
    month_upper = False
    week_upper = False
    days_upper = False

    add_months(
        img,
        year,
        month_font=month_font,
        week_font=week_font,
        month_color=month_color,
        week_color=week_color,
        weekend_color=weekend_color,
        days_color=days_color,
        first_weekday=first_weekday,
        month_upper=month_upper,
        week_upper=week_upper,
        days_upper=days_upper
    )

    year_upper = False
    add_year(img, year, year_font, year_color, year_upper)

def main():
    """Main"""
    width = 900
    height = 900

    img = Image.new('RGB', (width, height), 'black')

    year = 2018
    make_hex_calendar(img, year)

    img.save('hex_calendar_{}.png'.format(year))


if __name__ == '__main__':
    main()
```

在上述代码中设置的年份是 2018 年，执行后将会在素材图片 "spb_python_logo.png" 的基础上绘制 2018 年的完整日历图片 "hex_calendar_2018.png"，如图 8-20 所示。

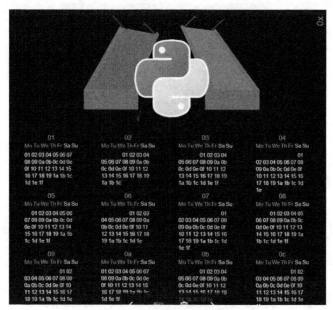

图 8-20　绘制的日历图片

8.2　使用 hmap 库

在 Python 程序中，可以使用 hmap 库实现图像直方图映射功能，能够使源图像的直方图匹配目标图像的直方图。hmap 库非常简单，只提供了两个程序文件 hmap.py 和 hmap_c.py。其中前者用于处理黑白图像，后者用于处理彩色图像。

在使用 hmap 库之前需要先确保已经安装了前面介绍的 Pillow 库，通过如下命令可以处理两幅黑白色图像。

```
python hmap.py source_image.jpg target_image.jpg
```

其中，source_image.jpg 表示一幅素材图片，target_image.jpg 表示另一幅素材图片，这两幅图片的宽度和高度（像素单位）必须相等。例如，素材图片 source_image.jpg 的效果如图 8-21 所示，素材图片 target_image.jpg 的效果如图 8-22 所示。

图 8-21　source_image.jpg 的效果

图 8-22　target_image.jpg 的效果

使用命令"python hmap_c.py source_image.jpg target_image.jpg"处理后的效果如图 8-23 所示。

图 8-23 处理后的效果

如果素材图片 source_image.jpg 和 target_image.jpg 是彩色的,可以通过如下命令进行处理。

```
python hmap_c.py source_image.jpg target_image.jpg
```

8.3 使用 pyBarcode 库创建条形码

在 Python 程序中,可以使用 pyBarcode 库创建条形码。在使用之前需要先安装。安装 pyBarcode 库的命令如下。

```
pip install pyBarcode
```

条形码的标准有多种,pyBarcode 库可以支持生成以下 9 种。

- Code 39
- PZN
- EAN-13
- EAN-8
- JAN
- ISBN-13
- ISBN-10
- ISSN
- UPC-A

在下面的实例文件 ex01.py 中,使用 pyBarcode 库创建了 EAN-13 标准的条形码,并将条形码另存为 SVG 格式的文件。

源码路径:daima\8\8-3\ex01.py

```
import barcode
ean = barcode.get('ean13', '1234567890102')         #设置条形码标准
print(ean.get_fullcode())
filename = ean.save('ean13')
print(filename)
options = dict(compress=True)
filename = ean.save('ean13', options)
print(filename)
```

执行后会生成 SVG 格式的条形码文件 ean13.svg 和 ean13.svgz,如图 8-24 所示。

在下面的实例文件 ex02.py 中,使用 pyBarcode 库创建了 EAN-13 标准的条形码,并将条形码另存为 PNG 格式的图片文件。

源码路径:daima\8\8-3\ex02.py

```
import barcode
```

```
from barcode.writer import ImageWriter
ean = barcode.get('ean13', '123456789102', writer=ImageWriter())
filename = ean.save('ean13')
print(filename)
```

执行后会生成条形码图片 ean13.png，如图 8-25 所示。

图 8-24　生成的条形码文件

图 8-25　条形码图片 ean13.png

在 pyBarcode 库中主要包含如下成员。

（1）barcode.codex.Code39(code, writer=None, add_checksum=True)：生成一个 Code39 标准的条形码。

- 参数 code：条形码的真实文本信息。
- 参数 writer：生成条形码的格式，默认值为 SVGWriter，即 SVG 格式。
- 参数 add_checksum：是否添加校验，默认值为 True。

（2）barcode.codex.PZN(pzn, writer=None)：生成一个 PZN 标准的条形码。

（3）barcode.ean.EuropeanArticleNumber13(ean, writer=None)：生成一个 EAN-13 标准的条形码。

（4）barcode.ean.EuropeanArticleNumber8(ean, writer=None)：生成一个 EAN-8 标准的条形码。

（5）barcode.ean.JapanArticleNumber(jan, writer=None)：生成一个 JAN 标准的条形码。

（6）barcode.isxn.InternationalStandardBookNumber13(isbn, writer=None)：生成一个 ISBN-13 标准的条形码。

（7）barcode.isxn.InternationalStandardBookNumber10(isbn, writer=None)：生成一个 ISBN-10 标准的条形码。

（8）barcode.isxn.InternationalStandardSerialNumber(issn, writer=None)：生成一个 ISSN 标准的条形码。

（9）barcode.upc.UniversalProductCodeA(upc, writer=None, make_ean=False)：生成一个 UPC-A 标准的条形码。

下面的实例文件 ex03.py 演示了创建两个条形码图片的过程。

源码路径：daima\8\8-3\ex03.py

```
from barcode.writer import ImageWriter
from barcode.codex import Code39
from PIL import Image, ImageDraw, ImageFont, ImageWin
from io import BytesIO

def generagteBarCode(self):
    imagewriter = ImageWriter()
    #保存到图片中
    # add_checksum : Boolean   Add the checksum to code or not (default: True)
    ean = Code39("1234567890", writer=imagewriter, add_checksum=False)
    # 不需要写后缀，ImageWriter初始化方法中默认self.format = 'PNG'
    print('保存到image2.png')
    ean.save('image2')
    img = Image.open('image2.png')
    print('展示image2.png')
    img.show()
```

```
# 写入stringio流中
i = BytesIO()
ean = Code39("0987654321", writer=imagewriter, add_checksum=False)
ean.write(i)
i = BytesIO(i.getvalue())
img1 = Image.open(i)
print('保存到stringIO中并以图片方式打开')
img1.show()

generagteBarCode('abc')
```

在上述代码中,首先使用 pyBarcode 库为信息"1234567890"创建了 Code39 标准的条形码,然后将条形码另存为 PNG 格式的图片文件,接着打开并显示这幅 PNG 条形码图片。接下来为逆序信息"0987654321"创建条形码,打开并显示这个条形码的信息。执行后会生成条形码图片 image2.png,并自动打开和显示条形码图片的信息,如图 8-26 所示。

图 8-26　条形码图片 image2.png

8.4　使用 qrcode 库创建二维码

在 Python 程序中,可以使用 qrcode 库创建二维码。在使用之前需要先确保已安装 qrcode 库,安装命令如下。

```
pip install qrcode[pil]
```

注意：在安装 qrcode 库之前需要确保已经安装了 Pillow 库。

我们只须通过如下代码即可生成一个二维码。

```
import qrcode
qrcode.run_example()
```

下面的实例文件 er01.py 演示了使用 qrcode 库将文本信息转换成二维码的过程。

源码路径：daima\8\8-4\er01.py

```
import qrcode
img=qrcode.make("some date here")          #二维码文本信息
img.save("Some.png")
```

执行后会创建一个二维码文件 Some.png,解析这个二维码图片后会得到文本信息"some date here"。

下面的实例文件 er02.py 演示了使用 qrcode 将网址信息转换成二维码的过程。

源码路径：daima\8\8-4\er02.py

```
import qrcode

data = '具体网站'
img_file = r'py_qrcode.png'

img = qrcode.make(data)
# 图片数据保存至本地文件
img.save(img_file)
# 显示二维码图片
img.show()
```

执行后会创建一个二维码文件 py_qrcode.png，解析这个二维码图片后会登录指定网址。

在使用 qrcode 库时可以设置二维码的样式属性，如设置二维码不同的纠错级别或生成不同大小的二维码图片。此功能是通过设置 QRCode 的参数实现的，包含以下 4 个参数。

(1) 参数 version：是一个 1~40 的整数，该参数用来控制二维码的尺寸（version 的最小值是 1，该 version 的尺寸是 21×21），把 version 设置为 None 且使用 fit 参数会自动生成二维码。

(2) 参数 err_correction：控制生成二维码的误差，此参数有如下 4 个可用的常量。
- ERROR_CORRECT_L：表示误差率低于 7%（包含 7%）。
- ERROR_CORRECT_M(默认值)：表示误差率低于 15%（包含 15%）。
- ERROR_CORRECT_Q：表示误差率低于 25%（包含 25%）。
- ERROR_CORRECT_H：表示误差率低于 30%（包含 30%）。

(3) 参数 box_size：用来控制二维码中每个单元格有多少像素点。

(4) 参数 border：用来控制每条边有多少个单元格，默认值是 4，这是最小值。

下面的实例文件 er03.py 演示了使用 qrcode 将网址信息转换成指定样式二维码的过程。

源码路径：daima\8\8-4\er03.py

```
import qrcode
qr = qrcode.QRCode(
    version=2,
    error_correction=qrcode.constants.ERROR_CORRECT_L,
    box_size=10,
    border=1
)
qr.add_data("具体网址")
qr.make(fit=True)
img = qr.make_image()
img.save("dhqme_qrcode.png")
```

执行后会使用设置的样式创建二维码文件 dhqme_qrcode.png，解析这个二维码图片后会登录指定网址。

在使用 qrcode 库的过程中，二维码的容错系数参数 error_correction 越高，生成的二维码允许的残缺率就越大。另外，因为二维码的数据主要保存在图片的 4 个角上，所以在二维码中间放一个小图标，对二维码的识别是不会产生多大影响的。大多数开发者倾向于将插入二维码的图标大小设置为不超过二维码长和宽的 1/4。如果太大，则意味着残缺率太高，并且会影响识别。

下面的实例文件 er04.py 演示了使用 qrcode 将网址信息转换成带有素材图片的二维码的过程。

源码路径：daima\8\8-4\er04.py

```
from PIL import Image
import qrcode

qr = qrcode.QRCode(
    version=2,
    error_correction=qrcode.constants.ERROR_CORRECT_H,
    box_size=10,
    border=1
)
qr.add_data("具体网址")
qr.make(fit=True)

img = qr.make_image()
img = img.convert("RGBA")

icon = Image.open("12345678.png")

img_w, img_h = img.size
```

```
factor = 4
size_w = int(img_w / factor)
size_h = int(img_h / factor)

icon_w, icon_h = icon.size
if icon_w > size_w:
    icon_w = size_w
if icon_h > size_h:
    icon_h = size_h
icon = icon.resize((icon_w, icon_h), Image.ANTIALIAS)

w = int((img_w - icon_w) / 2)
h = int((img_h - icon_h) / 2)
img.paste(icon, (w, h), icon)

img.save("dhqme_qrcode.png")
```

在上述代码中,结合 Python 图像库(PIL)的操作,把素材图片 12345678.png 粘贴在二维码图片的中间,最终生成了带有图标的二维码文件 dhqme_qrcode.png。

下面的实例文件 er05.py 演示了使用 qrcode 开发一个二维码生成器的过程。

源码路径:daima\8\8-4\er05.py

```
import qrcode
from PIL import Image
import os
# 生成二维码图片
def make_qr(str, save):
    qr = qrcode.QRCode(
        version=4,     # 生成的二维码尺寸
        error_correction=qrcode.constants.ERROR_CORRECT_M,
        box_size=10,
        border=2,
    )
    qr.add_data(str)
    qr.make(fit=True)
    img = qr.make_image()
    img.save(save)
#生成带logo的二维码图片
def make_logo_qr(str, logo, save):
    #参数配置
    qr = qrcode.QRCode(
        version=4,
        error_correction=qrcode.constants.ERROR_CORRECT_Q,
        box_size=8,
        border=2
    )
    #添加转换内容
    qr.add_data(str)
    qr.make(fit=True)

    #生成二维码
    img = qr.make_image()
    img = img.convert("RGBA")
    #添加logo
    if logo and os.path.exists(logo):
        icon = Image.open(logo)
        #获取二维码图片的大小
        img_w, img_h = img.size
        factor = 4
        size_w = int(img_w / factor)
        size_h = int(img_h / factor)
        #logo图片的大小不能超过二维码图片的1/4
        icon_w, icon_h = icon.size
        if icon_w > size_w:
            icon_w = size_w
        if icon_h > size_h:
            icon_h = size_h
        icon = icon.resize((icon_w, icon_h), Image.ANTIALIAS)
```

```
            #计算logo在二维码图中的位置
            w = int((img_w - icon_w) / 2)
            h = int((img_h - icon_h) / 2)
            icon = icon.convert("RGBA")
            img.paste(icon, (w, h), icon)
    #保存处理后的图片

    img.save(save)

if __name__ == '__main__':
    save_path = 'theqrcode.png'   # 生成的文件
    logo = '12345678.png'   # logo图片
    str = input('请输入要生成二维码的文本内容：')
    # make_qr(str)
    make_logo_qr(str, logo, save_path)
```

执行上述代码后将首先要求输入文内容，这里输入一个网址"http://www***"如图 8-27 所示。

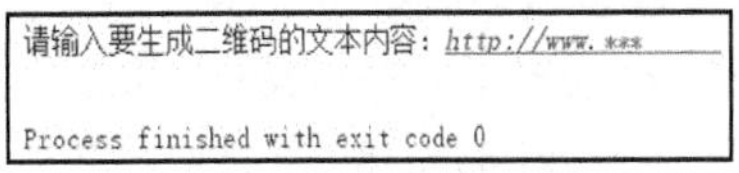

按下 Enter 键后会把素材图片 12345678.png 粘贴在二维码图片的中间，最终生成一个带有图标的二维码文件 theqrcode.png。

图 8-27　输入网址"http://www***"

8.5　使用 scikit-image 库

在 Python 程序中，可以使用 scikit-image 库实现图像的科学处理功能。因为 scikit-image 库是基于 Scipy 进行运算的，所以必须安装 Numpy 库和 scipy 库。要成功显示图片，还需要安装 Matplotlib 库。综合来说，要使用 scikit-image 库，需要安装如下库。

- Numpy，版本号大于或等于 1.6.1。
- Cython，版本号大于或等于 0.21。
- Six，版本号大于或等于 1.4。
- SciPy，版本号大于或等于 0.9。
- Matplotlib，版本号大于或等于 1.1.0。
- NetworkX，版本号大于或等于 1.8。
- Pillow，版本号大于或等于 1.7.8。
- dask[array]，版本号大于或等于 0.5.0。

安装 scikit-image 库的命令如下。

```
pip install scikit-image
```

ski-mage 的全称是 scikit-image SciKit，它实现了对 scipy.ndimage 的功能扩展，提供了更多的图片处理功能。Ski-mage 是用 Python 编写的，由 Scipy 社区开发和维护。skimage 库由许多子模块组成，各个子模块的具体功能如下。

- io：读取、保存和显示图片或视频。
- data：提供一些测试图片和样本数据。
- color：提供颜色空间变换功能。
- filters：提供图像增强、边缘检测、排序滤波器、自动阈值等功能。
- draw：提供 numpy 数组的基本图形绘制功能，包括线条、矩形、圆和文本等。
- transform：提供几何变换或其他变换功能，如旋转、拉伸和拉动变换等。
- morphology：提供形态学操作功能，如开闭运算、骨架提取等。
- exposure：提供图片强度调整功能，如亮度调整、直方图均衡等。
- feature：提供特征检测与提取等功能。

- measure：提供图像属性测量功能，如相似性或等高线等。
- segmentation：提供图像分割功能。
- restoration：提供图像恢复功能。
- util：保存通用函数。

接下来，将详细讲解 scikit-image 库中核心子模块的使用知识。

8.5.1 读取和显示

通过使用 io 和 data 子模块，可以实现图像的读取、显示和保存功能。其中 io 模块是用于图片输入/输出操作。为了便于开发者练习，scikit-image 提供了 data 模块，在里面嵌套了一些素材图片。

1. 显示图片

要读取单张彩色 RGB 图片，可以通过以下两个函数实现。
- skimage.io.imread（fname）函数，用于读取图片，参数 fname 表示需要读取的文件路径。
- skimage.io.imshow（arr）函数，用于显示图片，参数 arr 表示需要显示的 arr 数组（读取的图片以 numpy 数组形式计算）。

下面的实例文件 skimage01.py 演示了使用 skimage 从外部读入图像并显示的过程。

源码路径：daima\8\8-5\skimage01.py

```
from skimage import data,io
img = io.imread('111.jpg')
io.imshow(img)
io.show()
```

执行后会显示读取的外部图片，如图 8-28 所示。

如果要读取显示灰度图效果，可以将函数 imread() 中的 as_grey 参数设置为 True，此参数的默认值为 False。下面的实例文件 skimage02.py 演示了使用 skimage 读取并显示外部灰度图像的过程。

源码路径：daima\8\8-5\skimage02.py

```
from skimage import data,io
img = io.imread('123.jpg')
io.imshow(img)
io.show()
```

执行后会显示读取的外部灰度图片，如图 8-29 所示。

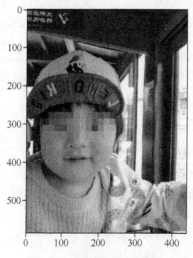

图 8-28 显示读取的外部图片

图 8-29 显示读取的外部图片

2. 显示素材图片

在 data 子模块中内置了一些素材图片,开发者可以直接使用。具体说明如下。
- astronaut:宇航员。
- coffee:咖啡。
- lena:图像处理领域常用的一幅图片。
- camera:拿相机的人。
- coins:硬币。
- moon:月亮。
- checkerboard:棋盘。
- horse:马。
- page:书页。
- chelsea:小猫。
- hubble_deep_field:星空。
- text:文字。
- clock:时钟。

下面的实例文件 skimage03.py 演示了使用 skimage 读取并显示内置的星空图片的过程。

源码路径:daima\8\8-5\skimage03.py

```
from skimage import io, data
from skimage import data_dir
image = data.hubble_deep_field()    #读取内置的星空图片
io.imshow(image)
io.show()                            #显示图片
print(data_dir)                      #显示素材图片的路径
```

在上述代码中,图片名对应的就是函数名。执行后会读取并显示星空图片,如图 8-30 所示。

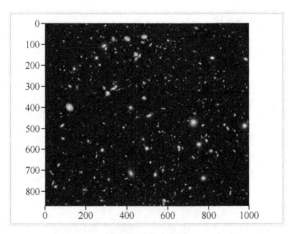

图 8-30 星空图片

在 data 子模块中,图片名对应的是函数名,例如,camera 图片对应的函数名为 camera()。这些素材图片保存在 skimage 的安装目录下,具体路径名称为 data_dir。通过上述代码中的最后一行代码,输出了 data_dir 目录的具体路径,例如,在作者的计算机中执行后会输出:

```
C:\Users\apple\AppData\Roaming\Python\Python36\site-packages\skimage\data
```

3. 保存图片

使用 io 子模块中的函数 imsave(fname,arr) 保存图片。其中参数 fname 表示保存的路径和名称，参数 arr 表示需要保存的数组变量。下面的实例文件 skimage04.py 演示了使用 skimage 读取并保存内置的星空图片的过程。

源码路径：daima\8\8-5\skimage04.py

```
from skimage import io,data
img = data.hubble_deep_field()
io.imshow(img)
io.show()
io.imsave('hubble_deep_field.jpg', img)        #保存图片
```

执行后会将读取的星空素材图片保存在本地，名称为 hubble_deep_field.jpg，如图 8-31 所示。

4. 显示图片信息

通过如下所述的成员可以获取图片的相关信息。

- type()：显示类型。
- shape()：显示尺寸、宽度、高度和通道数。
- size()：显示总像素个数。
- max()：最大像素值。
- min()：最小像素值。
- mean()：像素平均值。

下面的实例文件 skimage05.py 演示了使用 skimage 显示内置星空图片基本信息的过程。

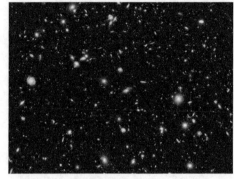

图 8-31 保存在本地的 hubble_deep_field.jpg

源码路径：daima\8\8-5\skimage05.py

```
from skimage import io,data
img=data.hubble_deep_field()
print(type(img))          #显示类型
print(img.shape)          #显示尺寸
print(img.shape[0])       #图片宽度
print(img.shape[1])       #图片高度
print(img.shape[2])       #图片通道数
print(img.size)           #显示总像素数
print(img.max())          #最大像素值
print(img.min())          #最小像素值
print(img.mean())         #平均像素值
```

执行后会输出：

```
<class 'numpy.ndarray'>
(872, 1000, 3)
872
1000
3
2616000
255
0
19.1544541284
```

8.5.2 像素操作

当将外部或内置素材图片读入程序中后，这些图片是以 numpy 数组的形式存在的。正因如此，对 numpy 数组的一切操作对这些图片也是适用的。例如，对数组元素的访问，实际上等同于对图片像素的访问。例如，下面是访问彩色图像的形式。

```
image[i, j, c]
```

其中，i 表示图片的行数，j 表示图片的列数，c 表示图片的通道数（R、G、B 三通道分别对应 0、1、2），坐标(0,0)是从左上角开始的。

下面是访问灰度图片的形式。

```
gray[i, j]
```

下面的实例文件 skimage06.py 演示了使用 skimage 显示内置的星空图片基本信息的过程。

8.5 使用 scikit-image 库

源码路径：daima\8\8-5\skimage06.py

```
from skimage import io, data
img = data.chelsea()
#输出图片G通道中第20行第30列的像素值
pixel = img[20, 30, 1]
print(pixel)
#显示猫图片红色通道的图片
R = img[:, :, 0]
io.imshow(R)
io.show()
```

执行后会输出内置猫图片 G 通道中的第 20 行第 30 列的像素值，并显示内置猫图片的红色通道，如图 8-32 所示。

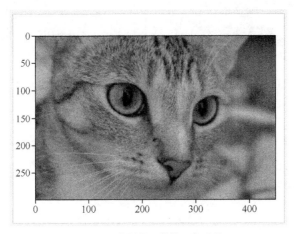

图 8-32　内置猫图片的红色通道

开发者还可以对图片进行修改。下面的实例文件 skimage07.py 演示了使用 skimage 对内置猫图片进行二值化操作的过程。

源码路径：daima\8\8-5\skimage07.py

```
from skimage import io, data, color
img=data.chelsea()
img_gray=color.rgb2gray(img)
rows,cols=img_gray.shape
for i in range(rows):
            for j in range(cols):
                if (img_gray[i,j]<=0.5):
                    img_gray[i,j]=0
                else:
                    img_gray[i,j]=1
io.imshow(img_gray)
io.show()
```

在上述代码中，使用模块 color 的函数 rgb2gray()将彩色的三通道图片转换成灰度图。转换后的结果为 float64 类型的数组，具体范围为[0,1]。执行后会输出二值化后的图片效果，如图 8-33 所示。

下面的实例文件 skimage08.py 演示了使用 skimage 对内置猫图片进行裁剪的过程。

源码路径：daima\8\8-5\skimage08.py

```
from skimage import io, data

img = data.chelsea()
roi = img[150:250, 200:300, :]
io.imshow(roi)
io.show()
```

执行后会输出裁剪之后的效果，如图 8-34 所示。

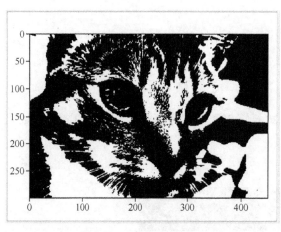

图 8-33 二值化后的效果

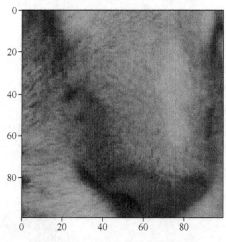

图 8-34 裁剪之后的效果

8.5.3 转换操作

通过使用 skimage 不但可以对图像类型进行转换，而且可以实现颜色转换。skimage 库中的每一张图片就是一个 numpy 数组。在 numpy 数组中可以有多种数据类型，这些类型相互之间可以进行转换。在 numpy 数组中，包含的数据类型及取值范围如下。

- unit8：0～255。
- unit16：0～65535。
- unit32：0～2^{32}。
- float：−1～1 或 0～1，一幅彩色图转换成灰度图之后会由 unit8 转换为 float。
- int8：−128～127。
- int16：−32768～32767。
- int32：−2^{32}～2^{32}−1。

因为一张图片中像素的取值范围是 0~255，所以对应的类型是 unit8。可以通过以下代码查看某图片的数据类型。

```
from skimage import io, data
img = data.chelsea()
print(img.dtype.name)
```

上述代码会输出内置猫图片的数据类型。

```
unit8
```

下面的实例文件 skimage09.py 演示了使用 skimage 将 unit8 类型转换成 float 类型的过程。

源码路径：daima\8\8-5\skimage09.py

```
from skimage import io, data, img_as_float
img = data.chelsea()
print(img.dtype.name)                    #显示原来的类型
img_grey = img_as_float(img)             #进行转换
print(img_grey.dtype.name)               #显示转换后的类型
```

执行后会分别显示转换前和转换后的类型。

```
uint8
float64
```

下面的实例文件 skimage10.py 演示了使用 skimage 将 float 转换成 unit8 类型的过程。

源码路径：daima\8\8-5\skimage10.py

```
from skimage import img_as_ubyte
import numpy as np
img = np.array([[0.2], [0.5], [0.1]], dtype=float)
```

8.5 使用 scikit-image 库

```
print(img.dtype.name)
img_unit8 = img_as_ubyte(img)
print(img_unit8.dtype.name)
```

通过上述代码将 float 类型转换为 unit8 类型，因为这个过程有可能会造成数据损失，所以执行后会显示以下告警信息。

```
float64
C:\Users\apple\AppData\Roaming\Python\Python36\site-packages\skimage\util\dtype.py:122:
    UserWarning: Possible precision loss when converting from float64 to uint8
  .format(dtypeobj_in, dtypeobj_out))
uint8
```

在 skimage 中还可以实现如下所示的数据类型转换功能。

- img_as_float：转换为 64 位 float（浮点型）。
- img_as_ubyte：转换为 8 位 uint（无符号整型）。
- img_as_uint：转换为 16 位 uint（无符号整型）。
- img_as_int：转换为 16 位 int（有符号整型）。

除了上面介绍的直接转换数据类型之外，我们还可以通过转换颜色空间的方式来实现数据类型转换功能。现实中常用的颜色空间有灰度空间、RGB 空间、HSV 空间和 CMKY 空间，在转换颜色空间以后，所有的数据类型都变成了 float 类型。下面的实例文件 skimage11.py 演示了使用 skimage 将 RGB 图转换为灰度图的过程。

源码路径：daima\8\8-5\skimage11.py

```
from skimage import io, data, color
image = data.chelsea()
image_grey = color.rgb2gray(image)
io.imshow(image_grey)
io.show()
```

执行后会显示将内置猫图片转换成灰度图的效果，如图 8-35 所示。

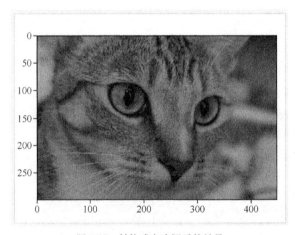

图 8-35 转换成灰度图后的效果

在 skimage 库中，如下颜色空间转换函数在 color 模块中定义。

- skimage.color.rgb2grey(rgb)：从 RGB 空间转换为灰度空间。
- skimage.color.rgb2hsv(rgb)：从 RGB 空间转换为 HSV 空间。
- skimage.color.rgb2lab(rgb)：从 RGB 空间转换为 LAB 空间。
- skimage.color.gray2rgb(image)：从灰度空间转换为 RGB 空间。
- skimage.color.hsv2rgb(hsv)：从 HSV 空间转换为 RGB 空间。
- skimage.color.label2rgb(arr)：根据标签值对图片进行着色。

其实上述所有函数的功能都可以通过如下函数来代替。

```
skimage.color.convert_colorspace(arr, fromspace, tospace)
```

上述函数用于将 arr 从 fromspace 颜色空间转换到 tospace 颜色空间。例如，下面的代码可以将内置猫图片由 RGB 颜色空间转换到 HSV 颜色空间。

```
from skimage import io, data, color
image = data.chelsea()
image_hsv = color.convert_colorspace(image, 'RGB', 'HSV')
io.imshow(image_hsv)
io.show()
```

8.5.4 绘制图像

通过使用 skimage 可以绘制图像。其实我们前面多次用到过的 io.imshow(image)函数实现的就是绘图功能。函数 io.imshow()的格式如下。

```
matplotlib.pyplot.imshow(X, cmap=None)
```

参数 X 表示要绘制的图像，参数 cmap 表示颜色图谱（通常是一个 3 列的矩阵），默认绘制为 RGB(A)颜色。参数 cmap 的颜色图谱可选值有以下几个。

- autumn：表示红-橙-黄。
- cool：表示青-洋红。
- copper：表示黑-铜。
- flag：表示红-白-蓝-黑。
- gray：表示黑-白。
- hot：表示黑-红-黄-白。
- hsv：表示 HSV 颜色空间，即红-黄-绿-青-蓝-洋红-红。
- inferno：表示黑-红-黄。
- jet：表示蓝-青-黄-红。
- magma：表示黑-红-白。
- pink：表示黑-粉-白。
- plasma：表示绿-红-黄。
- prism：表示红-黄-绿-蓝-紫-...-绿模式。
- spring：表示洋红-黄。
- summer：表示绿-黄。
- viridis：表示蓝-绿-黄。
- winter：表示蓝-绿。

1. 基本绘制

在 Python 程序中，通过如下两个函数可以分别输出 gray 和 jet 两种格式。

```
plt.imshow(image, plt.cm.gray)
plt.imshow(image, cmap = plt.cm.jet)
```

在绘制完图片后会返回一个 AxesImage 对象。要在窗口上显示这个对象，可以调用 show()函数。下面的实例文件 skimage12.py 演示了使用 skimage 绘制图片的过程。

源码路径：daima\8\8-5\skimage12.py

```
from skimage import io, data

image = data.chelsea()
axe_image = io.imshow(image)
print(type(axe_image))
io.show()
```

执行后会输出绘制图片的功能类。

```
<class 'matplotlib.image.AxesImage'>
```

Matplotlib 是一个专业的绘图库，其相关内容将在第 10 章中进行讲解。通过上述实例可知，无论利用 skimage.io.imshow()还是 matplotlib.pyplot.imshow()绘制图像，最终调用的都是 matplotlib.pyplot 模块。

2. 使用 subplot()函数绘制多个视图窗口

在使用 skimage 绘制图片的过程中，可以用 matplotlib.pyplot 模块下的 figure()函数来创建一个窗口。figure()函数的格式为：

```
matplotlib.pyplot.figure(num=None, figsize=None, dpi=None, facecolor=None, edgecolor=None)
```

figure()函数的所有参数都是可选的，因为都有默认值，所以调用该函数时可以不带任何参数。具体说明如下。

- num：整型或字符型。如果设置为整型，则该整型数字表示窗口的序号；如果设置为字符型，则该字符串表示窗口的名称。在用该参数来命名窗口时，如果两个窗口的序号或名相同，则后一个窗口会覆盖前一个窗口。
- figsize：用于设置窗口大小，是一个元组型的整数，如 figsize=（8，8）。
- dpi：一个整型数，表示窗口的分辨率。
- facecolor：用于设置窗口的背景颜色。
- edgecolor：用于设置窗口的边框颜色。

但是用 figure()函数来创建窗口时存在一个弊端，那就是只能显示一幅图片。如果要显示多幅图片，则需要将这个窗口划分为几个子图，在每个子图中显示不同的图片。此时可以使用 subplot()函数来划分子图。此函数的格式为：

```
matplotlib.pyplot.subplot(nrows, ncols, plot_number)
```

- nrows：表示子图的行数。
- ncols：表示子图的列数。
- plot_number：表示当前子图的编号。

其中以下两行不同写法的代码的执行效果是一样的。

```
plt.subplot(2, 2, 1)
plt.subplot(221)
```

当使用 subplot()函数创建多个子视图窗口后，可以用 title()函数来设置每个子图的标题，使用 axis()函数来设置是否使用坐标尺，例如：

```
plt.subplot(221)
plt.title("first window")
plt.axis('off')
```

下面的实例文件 skimage13.py 演示了使用 subplot()函数绘制多通道图像的过程。

源码路径：daima\8\8-5\skimage13.py

```python
from skimage import data,io
import matplotlib.pyplot as plt
from pylab import mpl
#下面两行代码能保证汉字正确显示
mpl.rcParams['font.sans-serif'] = ['FangSong'] #指定默认字体
mpl.rcParams['axes.unicode_minus'] = False #解决保存图像时负号显示为方块的问题
image = io.imread('111.jpg')

plt.figure(num='cat', figsize=(8, 8))   # 创建一个名为cat的窗口,并设置大小

plt.subplot(2, 2, 1)
plt.title('原始图像')
plt.imshow(image)

plt.subplot(2, 2, 2)
plt.title('R通道')
plt.imshow(image[:, :, 0])

plt.subplot(2, 2, 3)
plt.title('G通道')
```

```
plt.imshow(image[:, :, 1])

plt.subplot(2, 2, 4)
plt.title('B通道')
plt.imshow(image[:, :, 2])

plt.show()
```

执行后不但显示原始图片，而且会显示另外 3 个通道的子视图，如图 8-36 所示。

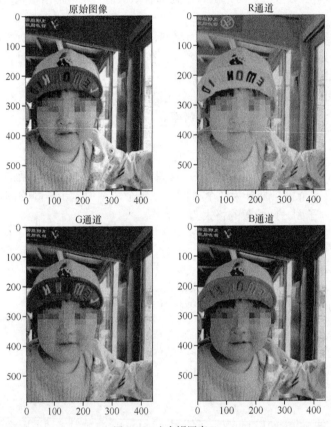

图 8-36　多个视图窗口

3．使用 subplots()函数绘制多个视图窗口

在使用 skimage 绘制图片的过程中，可以用 subplots()函数来绘制多视图窗口。函数 subplots()的格式如下。

```
matplotlib.pyplot.subplots(nrows=1, ncols=1)
```

- nrows：设置所有子图的行数，默认值为 1。
- ncols：设置所有子图的列数，默认值为 1。

函数 subplots()分别返回一个窗口 figure 和一个元组型的 ax 对象，ax 对象包含所有的子视图窗口。可以结合 ravel()函数列出所有子图。例如，通过下面的代码创建了 4 个子图，并分别命名为 ax0、ax1、ax2 和 ax3。

```
fig, axes = plt.subplots(2, 2, figsize=(8, 8))
axe0, axe1, axe2, axe3 = axes.ravel()
```

可以使用 set_title()函数设置每个子图的标题，例如下面的演示代码。

```
axe0.imshow(image)
axe0.set_title('Original Image')
```

如果有多个子视图，可以使用 tight_layout()函数来设置布局。该函数的格式为：

```
matplotlib.pyplot.tight_layout(pad=1.08, h_pad=None, w_pad=None, rect=None)
```
因为 tight_layout()函数的所有参数都是可选的,所以在调用该函数时可省略所有的参数。各个参数的具体说明如下。

- pad:设置主窗口边缘和子视图边缘间的间距,默认值为 1.08。
- h_pad:设置子视图和边缘之间的垂直间距,默认值为 pad_inches。
- w_pad:设置子视图和边缘之间的水平间距,默认值为 pad_inches。
- rect:一个矩形区域,如果设置这个值,则将所有的子视图调整到这个矩形区域内。

因为 tight_layout()函数的所有参数都可以省略,所以我们倾向于使用如下调用方法。

```
plt.tight_layout()                    #程序会自动调整subplot间的参数
```

下面的实例文件 skimage14.py 演示了使用 subplots()函数绘制多通道图像的过程。

源码路径:daima\8\8-5\skimage14.py

```python
from skimage import data,io, color
import matplotlib.pyplot as plt
from pylab import mpl
#下面两行代码能保证汉字正确显示
mpl.rcParams['font.sans-serif'] = ['FangSong'] #指定默认字体
mpl.rcParams['axes.unicode_minus'] = False #解决保存图像时负号显示为方块的问题
image = io.imread('111.jpg')
image_hsv = color.rgb2hsv(image)

fig, axes = plt.subplots(2, 2, figsize=(8, 8))
axe0, axe1, axe2, axe3 = axes.ravel()

axe0.imshow(image)
axe0.set_title('原始图像')

axe1.imshow(image_hsv[:, :, 0])
axe1.set_title('H通道')

axe2.imshow(image_hsv[:, :, 1])
axe2.set_title('S通道')

axe3.imshow(image_hsv[:, :, 2])
axe3.set_title('V通道')

for ax in axes.ravel():
    ax.axis('off')

fig.tight_layout()

plt.show()
```

执行后不但显示原始图片,而且会显示另外 3 个通道的子视图,如图 8-37 所示。

4. 使用 viewer 绘制图像

在使用 skimage 绘制图片的过程中,还可以使用其子模块 viewer 来显示图片。Viewer 使用 Qt 工具创建了一块画布,用户可以在画布上绘制图像。下面的实例文件 skimage15.py,演示了使用 viewer 绘制内置的月亮图像的过程。

源码路径:daima\8\8-5\skimage15.py

```python
from skimage import data
from skimage.viewer import ImageViewer

img = data.moon()
viewer = ImageViewer(img)
viewer.show()
```

执行后绘制输出内置的素材图片——月亮,如图 8-38 所示。

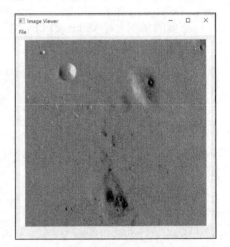

图 8-37　多个视图窗口　　　　　　图 8-38　内置的素材图片——月亮

8.5.5　图像批处理

通过使用 skimage，我们不但可以对一张图片进行处理，而且可以对一批图片进行处理。此时可以通过循环实现批处理功能，也可以调用 skimage 自带的图片集合函数 ImageCollection() 实现。函数 ImageCollection() 的格式如下。

```
skimage.io.ImageCollection(load_pattern, load_func=None)
```

函数 ImageCollection() 放在 io 模块中。各个参数的具体说明如下。

- 参数 load_pattern：图片组的路径，可以是一个 str 字符串。
- 参数 load_func：一个回调函数，当我们对图片进行批量处理时，可以使用这个回调函数。回调函数的默认值为 imread()，表示默认使用 imread() 函数批量读取图片。

下面的实例文件 skimage16.py 演示了使用 ImageCollection() 显示系统内指定素材图像的过程。

源码路径：daima\8\8-5\skimage16.py

```
from skimage import io, data_dir
data_path_str = data_dir + '/*.png'
images = io.ImageCollection(data_path_str)
print(len(images))#显示系统自带的素材图片数
io.imshow(images[1])           #显示其中的一张素材图片
io.show()
```

在上述代码中，首先使用 ImageCollection() 获取了系统内置素材图片的信息，执行后会显示 27，这说明系统内保存了 27 张素材图片。最后显示了某一张系统内置图片，如图 8-39 所示。

如果在文件夹"pic"中存放了 9 张 jpg 格式的图片，则可以通过下面的实例文件 skimage17.py 读取文件夹"pic"中 IPEG 图片的个数。

源码路径：daima\8\8-5\skimage17.py

```
import skimage.io as io
coll = io.ImageCollection('pic/*.jpg')
```

```
print(len(coll))
```
执行后会输出：
```
9
```

在使用 io.ImageCollection()函数时，如果我们不想批量读取，而想实现其他批量操作，例如，批量转换为灰度图，应该如何实现呢？下面的实例文件 skimage18.py，演示了使用 ImageCollection()将系统内指定素材图像批量转换为灰度图的过程。

源码路径：daima\8\8-5\skimage18.py
```
from skimage import io, data_dir, color
def convert_to_gray(f, **args):
    image = io.imread(f)
    image = color.rgb2gray(image)
    return image
data_path = data_dir + '/*.png'
collections = io.ImageCollection(data_path, load_func=convert_to_gray)
io.imshow(collections[1])      #显示某一张转换后的图片
io.show()
```
执行会将系统内指定素材图像批量转换为灰度图，并显示某一张转换后的图片，如图 8-40 所示。

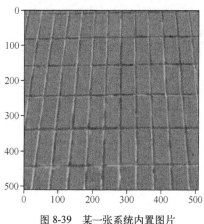

图 8-39　某一张系统内置图片

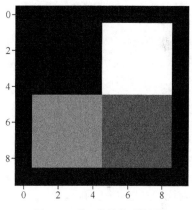

图 8-40　某一张转换后的图片

我们在得到图片集合以后，可以使用函数 concatenate_images(ic)将这些图片连接起来，构成一个维度更高的数组。此函数的格式为：
```
skimage.io.concatenate_images(ic)
```
当使用 concatenate_images(ic)函数连接图片时，需要确保连接读取的这些图片的尺寸一致，否则会出错。下面的实例文件 skimage19.py 演示了使用函数 concatenate_images(ic)连接图片的过程。

源码路径：daima\8\8-5\skimage19.py
```
from skimage import data_dir, io, color
coll = io.ImageCollection('pic/*.jpg')
print(len(coll)) #连接的图片数量
print(coll[0].shape) #连接前的图片尺寸，所有的都一样
mat=io.concatenate_images(coll)
print(mat.shape) #连接后的数组尺寸
```
执行后会输出连接前后的维度变化情况。
```
8
(4208, 2368, 3)
(8, 4208, 2368, 3)
```

8.5.6　缩放和旋转

通过使用 skimage，我们可以对指定的图片实现缩放和旋转处理。这主要是通过其内置模块 transform 实现的。在模块 transform 中主要包含如下成员。

（1）skimage.transform.resize(image,output_shape)：改变图片尺寸。

- 参数 image：需要改变尺寸的图片。
- 参数 output_shape：新的图片尺寸。

(2) skimage.transform.rescale(image, scale[, ...])：按比例缩放图片。

- 参数 image：需要缩放的图片。
- 参数 scale：可以是单个浮点数，表示缩放的倍数；也可以是一个浮点型的元组，例如，[0.2,0.5]表示将行和列分开缩放。

(3) skimage.transform.rotate(image, angle[, ...],resize=False)：旋转图像。

- 参数 image：需要旋转的图片。
- 参数 angle：一个浮点数，表示旋转的度数。
- 参数 resize：设置在旋转时是否改变大小，默认值为 False。

下面的实例文件 skimage20.py 演示了使用函数 resize()改变指定图片大小的过程。

源码路径：daima\8\8-5\skimage20.py

```python
from skimage import transform,data,io
import matplotlib.pyplot as plt
from pylab import mpl
#下面两行代码能保证汉字正确显示
mpl.rcParams['font.sans-serif'] = ['FangSong'] # 指定默认字体
mpl.rcParams['axes.unicode_minus'] = False # 解决保存图像是负号'-'显示为方块的问题
img = io.imread('111.jpg')
dst=transform.resize(img, (80, 60))
plt.figure('resize')
plt.subplot(121)
plt.title('原始图')
plt.imshow(img,plt.cm.gray)
plt.subplot(122)
plt.title('改变后')
plt.imshow(dst,plt.cm.gray)
plt.show()
```

通过上述代码，把图片 111.jpg 改变了大小。执行后会通过两个子视图显示改变图像大小前后的对比，如图 8-41 所示。

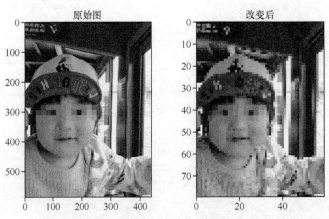

图 8-41　显示改变图像大小的前后对比

下面的实例文件 skimage21.py 演示了使用函数 rescale()缩放指定图片的过程。

源码路径：daima\8\8-5\skimage21.py

```python
from skimage import transform,data,io
img = io.imread('111.jpg')
print(img.shape)                              #图片原始大小
print(transform.rescale(img, 0.1).shape)      #缩小为原来图片的0.1
print(transform.rescale(img, [0.5,0.25]).shape)  #缩小为原来图片行数的一半，列数的1/4
print(transform.rescale(img, 2).shape)        #放大为原来图片的2倍
```

8.6 使用 face_recognition 库实现人脸识别

执行后会显示缩放后的大小。

```
(588, 441, 3)
(59, 44, 3)
(294, 110, 3)
(1176, 882, 3)
```

下面的实例文件 skimage22.py 演示了使用函数 rotate()旋转指定图片的过程。

源码路径：daima\8\8-5\skimage22.py

```
from skimage import transform,io
import matplotlib.pyplot as plt
from pylab import mpl
#下面两行代码能保证汉字正确显示
mpl.rcParams['font.sans-serif'] = ['FangSong'] #指定默认字体
mpl.rcParams['axes.unicode_minus'] = False   # 解决保存图像时负号显示为方块的问题
img=io.imread('111.jpg')
print(img.shape)                         #图片原始大小
img1=transform.rotate(img, 60)           #旋转60°，不改变大小
print(img1.shape)
img2=transform.rotate(img, 30,resize=True) #旋转30°，同时改变大小
print(img2.shape)
plt.figure('缩放')
plt.subplot(121)
plt.title('旋转60度')
plt.imshow(img1,plt.cm.gray)
plt.subplot(122)
plt.title('旋转30度')
plt.imshow(img2,plt.cm.gray)
plt.show()
```

执行后会输出显示原始图像大小、旋转 60°时的大小和旋转 30°时的大小。

```
(588, 441, 3)
(588, 441, 3)
(730, 676, 3)
```

并且还会分别显示旋转 60°后和旋转 30°后的效果，如图 8-42 所示。

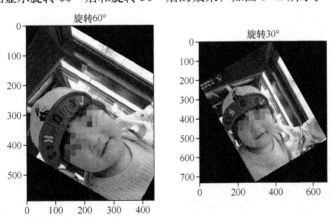

图 8-42 分别显示旋转 60°后和旋转 30°后的效果

8.6 使用 face_recognition 库实现人脸识别

通过使用第三方库 face_recognition，在 Python 程序中可以实现人脸识别。本节将详细讲解使用 face_recognition 库实现人脸识别的知识。

8.6.1 搭建开发环境

在安装 face_recognition 库之前，我们必须明白如下的依赖关系。
- 安装 face_recognition 库的必要条件是配置好 dlib 库。

- 配置好 dlib 库的必要条件是成功安装 dlib 库,并且编译。
- 安装 dlib 库的必要条件是配置好 boost 和 cmake。

在 Python 3.6 之前的版本中,开发者必须严格按照上述依赖关系进行环境搭建。而从 Python 3.6 开始,安装 face_recognition 库变得非常容易,整个过程与 boost 和 cmake 完全无关。但是我们必须注意的是,dlib 针对不同的 Python 版本提供了不同的安装文件,我们必须安装完全对应的版本,否则会出错。例如,作者安装的是 Python 3.6,所以必须安装的 dlib 版本是 19.7.0。下载 dlib-19.7.0-cp36-cp36m-win_amd64.whl 后,使用如下命令即可成功安装 dlib。

```
pip install dlib-19.7.0-cp36-cp36m-win_amd64.whl
```

接下来,通过如下命令安装 face_recognition。

```
pip install face_recognition
```

除此之外,还需要安装如下库。

```
pip install numpy
pip install scipy
pip install opencv-python
```

到此为止,在 Python 环境中安装库 face_recognition 的工作结束。

8.6.2 面部特征

face_recognition 库通过 facial_features 来处理面部特征。面部特征包含了以下 8 个特征。

- chin:表示下巴。
- left_eyebrow:表示左眉。
- right_eyebrow:表示右眉。
- nose_bridge:表示鼻梁。
- nose_tip:表示鼻尖。
- left_eye:表示左眼。
- right_eye:表示右眼。
- top_lip:表示上唇。
- bottom_lip:表示下唇。

下面的实例文件 shibie01.py 演示了显示指定人脸特征的过程。

源码路径:daima\8\8-6\shibie01.py

```python
#自动识别人脸特征

#导入pil模块
from PIL import Image, ImageDraw
#导入face_recogntion模块,可用命令安装 pip install face_recognition
import face_recognition

#将jpg文件加载到numpy数组中
image = face_recognition.load_image_file("111.jpg")

#查找图像中所有面部特征
face_landmarks_list = face_recognition.face_landmarks(image)

print("I found {} face(s) in this photograph.".format(len(face_landmarks_list)))

for face_landmarks in face_landmarks_list:

    #输出此图像中每个面部特征的位置
    facial_features = [
        'chin',
        'left_eyebrow',
        'right_eyebrow',
        'nose_bridge',
        'nose_tip',
        'left_eye',
        'right_eye',
        'top_lip',
```

8.6 使用 face_recognition 库实现人脸识别

```
            'bottom_lip'
        ]

        for facial_feature in facial_features:
            print("The {} in this face has the following points: {}".format(facial_feature,
                face_landmarks[facial_feature]))

    #在图像中描绘出每个人脸特征
    pil_image = Image.fromarray(image)
    d = ImageDraw.Draw(pil_image)

    for facial_feature in facial_features:
        d.line(face_landmarks[facial_feature], width=5)

    pil_image.show()
```

执行后会输出显示图片 111.jpg 中人像的人脸特征。

```
I found 1 face(s) in this photograph.
The chin in this face has the following points: [(35, 303), (38, 331), (42, 359), (47, 387),
(59, 411), (78, 428), (100, 441), (123, 451), (146, 454), (168, 450), (189, 439), (209,
425), (227, 407), (238, 384), (244, 358), (249, 331), (252, 304)]
The left_eyebrow in this face has the following points: [(53, 289), (66, 273), (87, 266),
(109, 269), (131, 276)]
The right_eyebrow in this face has the following points: [(162, 277), (181, 269), (203,
266), (224, 271), (236, 286)]
The nose_bridge in this face has the following points: [(144, 303), (144, 319), (144, 334),
(143, 351)]
The nose_tip in this face has the following points: [(124, 364), (134, 366), (144, 368),
(154, 366), (164, 364)]
The left_eye in this face has the following points: [(77, 304), (89, 299), (103, 300),
(115, 310), (102, 313), (87, 311)]
The right_eye in this face has the following points: [(174, 310), (185, 301), (199, 301),
(211, 305), (200, 312), (186, 313)]
The top_lip in this face has the following points: [(109, 395), (124, 391), (136, 387),
(144, 389), (153, 386), (165, 390), (180, 393), (174, 393), (153, 394), (144, 395), (136,
394), (116, 396)]
The bottom_lip in this face has the following points: [(180, 393), (165, 402), (154, 405),
(145, 406), (136, 406), (125, 404), (109, 395), (116, 396), (136, 396), (145, 396), (153,
395), (174, 393)]
```

另外，会使用 PIL 在人像中标记出人脸特征，如图 8-43 所示。

图 8-43 标记的人脸特征

再看下面的实例文件 shibie.py，该文件演示了在指定照片中标记人脸的过程。

源码路径：daima\8\8-6\shibie.py

```
#检测人脸
import face_recognition
import cv2

#读取图片并识别人脸
img = face_recognition.load_image_file("111.jpg")
face_locations = face_recognition.face_locations(img)
print(face_locations)
```

```
#调用opencv函数显示图片
img = cv2.imread("111.jpg")
cv2.namedWindow("原图")
cv2.imshow("原图", img)

#遍历每个人脸,并标注
faceNum = len(face_locations)
for i in range(0, faceNum):
    top =    face_locations[i][0]
    right =  face_locations[i][1]
    bottom = face_locations[i][2]
    left =   face_locations[i][3]

    start = (left, top)
    end = (right, bottom)

    color = (55,255,155)
    thickness = 3
    cv2.rectangle(img, start, end, color, thickness)

#显示识别结果
cv2.namedWindow("识别")
cv2.imshow("识别", img)
cv2.waitKey(0)
cv2.destroyAllWindows()
```

执行后将分别显示原始照片 111.jpg 和识别标记出人脸的效果,如图 8-44 所示。

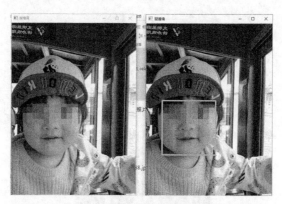

图 8-44　原始照片和标记出人脸的效果

8.6.3　识别人脸

假设我们有一张照片 888.jpg,如图 8-45 所示。

图 8-45　照片 888.jpg

8.6 使用 face_recognition 库实现人脸识别

这是一幅 3 人合影照，我们应该如何识别出这张照片中的所有人脸呢？下面的实例文件 shibie02.py 演示了识别出照片 888.jpg 中所有人脸的过程。

源码路径：daima\8\8-6\shibie02.py

```python
#识别图片中的所有人脸并显示出来
# filename : shibie02.py
#导入pil模块
from PIL import Image
#导入face_recogntion模块，可用命令安装 pip install face_recognition
import face_recognition

#将jpg文件加载到numpy数组中
image = face_recognition.load_image_file("888.jpg")

#使用默认的HOG模型查找图像中所有人脸
#这个方法已经相当准确了，但还是不如CNN模型那么准确，因为没有使用GPU加速
#另见find_faces_in_picture_cnn.py
face_locations = face_recognition.face_locations(image)

# 使用CNN模型
# face_locations = face_recognition.face_locations(image, number_of_times_to_upsample=0,
        model="cnn")

# 显示从图片中找到了多少张人脸
print("I found {} face(s) in this photograph.".format(len(face_locations)))

#循环找到的所有人脸
for face_location in face_locations:

        #显示每张脸的位置信息
        top, right, bottom, left = face_location
        print("Top: {}, Left: {}, Bottom: {}, Right: {}".format(top, left, bottom, right))
#指定人脸的位置信息，然后显示人脸图片
        face_image = image[top:bottom, left:right]
        pil_image = Image.fromarray(face_image)
        pil_image.show()
```

执行后首先会输出照片 888.jpg 中人脸的位置信息。

```
I found 3 face(s) in this photograph.
Top: 163, Left: 79, Bottom: 271, Right: 187
Top: 125, Left: 182, Bottom: 254, Right: 311
Top: 329, Left: 104, Bottom: 403, Right: 179
```

然后显示识别的 3 张人脸，如图 8-46 所示。

再看下面的例子。假设我们有一张照片 201.jpg，如图 8-47 所示。

图 8-46　识别出的 3 张人脸　　　　　　　　图 8-47　照片 201.jpg

这是一幅单人照，假设这个人的名字叫小毛毛。我们应该如何识别出照片 888.jpg 中的小毛毛呢？下面的实例文件 shibie03.py 演示了判断照片 888.jpg 中是否包含小毛毛的过程。

源码路径：daima\8\8-6\shibie03.py

```
#通过识别人脸鉴定是哪个人
import face_recognition
#将jpg文件加载到numpy数组中
chen_image = face_recognition.load_image_file("201.jpg")
#要识别的图片
unknown_image = face_recognition.load_image_file("888.jpg")
#获取每个图像文件中每个面部的编码
#因为每个图像中可能有多个人脸,所以返回一个编码列表
#但是已知每个图像只有一张脸,并且只关心每个图像中的第一个编码,所以这里取索引0
chen_face_encoding = face_recognition.face_encodings(chen_image)[0]
print("chen_face_encoding:{}".format(chen_face_encoding))
unknown_face_encoding = face_recognition.face_encodings(unknown_image)[0]
print("unknown_face_encoding :{}".format(unknown_face_encoding))

known_faces = [
    chen_face_encoding
]
#结果是布儿数组
results = face_recognition.compare_faces(known_faces, unknown_face_encoding)

print("result :{}".format(results))
print("这个未知面孔是小毛毛吗? {}".format(results[0]))
print("这个未知面孔是我们从未见过的新面孔吗? {}".format(not True in results))
```

执行后会输出如下识别结果。这说明在照片 888.jpg 中存在照片 201.jpg 所示的这个人。

```
result :[True]
这个未知面孔是小毛毛吗? True
这个未知面孔是我们从未见过的新面孔吗? False
```

假设有 3 张图片——laoguan.jpg（老管的单人照）、maomao.jpg（毛毛的单人照）和 unknown.jpg（未知的单人照，但肯定是老管或毛毛这两人之一的单人照），如图 8-48 所示。

 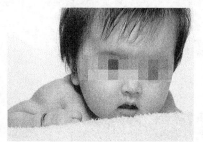 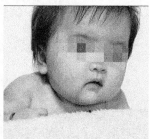

(a) laoguan.jpg　　　　　　(b) maomao.jpg　　　　　　(c) unknown.jpg

图 8-48　3 张素材图片

我们应该如何识别出照片 unknown.jpg 中的人呢？下面的实例文件 shibie04.py 演示了识别照片 unknown.jpg 中的人的过程。

源码路径：daima\8\8-6\shibie04.py

```
# 识别图片中的人脸
import face_recognition
jobs_image = face_recognition.load_image_file("laoguan.jpg");
obama_image = face_recognition.load_image_file("maomao.jpg");
unknown_image = face_recognition.load_image_file("unknown.jpg");

laoguan_encoding = face_recognition.face_encodings(jobs_image)[0]
maomao_encoding = face_recognition.face_encodings(obama_image)[0]
unknown_encoding = face_recognition.face_encodings(unknown_image)[0]

results = face_recognition.compare_faces([laoguan_encoding, maomao_encoding], unknown_encoding)
labels = ['老管', '毛毛']

print('结果:'+str(results))
```

```python
for i in range(0, len(results)):
    if results[i] == True:
        print('这个人是:'+labels[i])
```

执行后会输出识别结果。

```
结果:[False, True]
这个人是:毛毛
```

8.6.4 摄像头实时识别

假设我们保存一张"小毛毛"的照片 xiaomaomao.jpg，然后用摄像头识别不同的照片。如果是小毛毛本人的照片，则摄像区域自动识别并显示"小毛毛"；如果不是小毛毛的照片，则在摄像区域显示"unknown"。通过下面的实例文件 shibie05.py 可以实现上述识别功能。

源码路径：daima\8\8-6\shibie05.py

```python
import face_recognition
import cv2

video_capture = cv2.VideoCapture(0)#笔记本电脑的摄像头是0，外接摄像头是1

obama_img = face_recognition.load_image_file("xiaomaomao.jpg")
obama_face_encoding = face_recognition.face_encodings(obama_img)[0]

face_locations = []
face_encodings = []
face_names = []
process_this_frame = True

while True:
    ret, frame = video_capture.read()

    small_frame = cv2.resize(frame, (0, 0), fx=0.25, fy=0.25)
    if process_this_frame:
        face_locations = face_recognition.face_locations(small_frame)
        face_encodings = face_recognition.face_encodings(small_frame, face_locations)

        face_names = []
        for face_encoding in face_encodings:
            match = face_recognition.compare_faces([obama_face_encoding], face_encoding)

            if match[0]:
                name = "小毛毛"
            else:
                name = "unknown"

            face_names.append(name)

    process_this_frame = not process_this_frame

    for (top, right, bottom, left), name in zip(face_locations, face_names):
        top *= 4
        right *= 4
        bottom *= 4
        left *= 4

        cv2.rectangle(frame, (left, top), (right, bottom), (0, 0, 255), 2)

        cv2.rectangle(frame, (left, bottom - 35), (right, bottom), (0, 0, 255), 2)
        font = cv2.FONT_HERSHEY_DUPLEX
        cv2.putText(frame, name, (left+6, bottom-6), font, 1.0, (255, 255, 255), 1)

    cv2.imshow('Video', frame)

    if cv2.waitKey(1) & 0xFF == ord('q'):
        break

video_capture.release()
cv2.destroyAllWindows()
```

第 9 章

图形用户界面

图形用户界面（Graphical User Interface，GUI）是指采用图形方式显示的计算机操作用户界面。例如，Windows 系统就是一个功能强大的图形界面程序，用 C++、C#或 Visual Basic 开发的桌面程序也是图形用户界面程序。其实 Python 有一个内置的标准 GUI 库 Tkinter，通过使用 Tkinter 可以快速创建 GUI 应用程序。本章将详细讲解使用 Python 第三方库开发 GUI 程序的知识。

9.1 使用 PyQt 库

在 Python 程序中，可以使用 PyQt 库来创建美观的用户界面。在作者写作本书时，PyQt 已经发展到了 PyQt 5 版本。可以使用如下命令安装 PyQt 5 库。

```
pip install PyQt5
```

安装成功后，就可以开始我们的 PyQt 开发之旅了。

9.1.1 第一个 GUI 程序

下面的实例文件 PyQt01.py，演示了使用 PyQt 创建第一个 GUI 程序的过程。

源码路径：daima\9\9-1\PyQt01.py

```
1   import sys
2     from PyQt5.QtWidgets import QApplication, QWidget
3
4   if __name__ == '__main__':
5       app = QApplication(sys.argv)
6
7       w = QWidget()
8       w.resize(250, 150)
9       w.move(300, 300)
10      w.setWindowTitle('First')
11      w.show()
12
13      sys.exit(app.exec_())
```

在第 2 行中，导入一些必要模块，PyQt 中最基础的 widget 组件位于 PyQt5.QtWidget 模块中。

所有的 PyQt 5 程序必须创建一个应用（Application）对象。在第 5 行中，参数 sys.argv 是一个来自命令行的参数列表。Python 脚本可以在 shell 中运行。这也是开发者用来控制应用启动的一种方法。

QWidget 组件是 PyQt 5 中所有用户界面类的基础类，在第 7 行给 QWidget 提供了默认的构造方法。默认构造方法没有父类。没有父类的 widget 组件将被作为窗口来使用。

在第 8 行中，使用 resize()方法调整 widget 组件的大小，设置为 250px 宽、150px 高。

在第 9 行中，使用方法 move()将组件 widget 移动到位置（300，300）。

在第 10 行中，设置标题，这个标题显示在标题栏中。

在第 11 行中，使用方法 show()在屏幕上显示 widget，这个 widget 对象在这里第一次在内存中创建，并且之后在屏幕上显示。

在第 13 行中，整个程序进入主循环，开始执行事件处理。主循环用于接收来自窗口触发的事件，并且转发到 widget 应用上处理。如果调用 exit()方法或主 widget 组件被销毁，则主循环将退出。sys.exit()方法确保干净地退出。将会通知系统环境应用是怎样结束的。值得注意的是，方法名 exec_()有一条下划线。这是因为 exec 是 Python 中保留的关键字，所以用 exec_()来代替。

执行后将显示一个最简单的 GUI 程序，如图 9-1 所示。

下面的实例文件 PyQt02.py 演示了在 PyQt 窗体中创建一个图标的过程。

图 9-1 执行效果

源码路径：daima\9\9-1\PyQt02.py

```
import sys
from PyQt5.QtWidgets import QApplication, QWidget
from PyQt5.QtGui import QIcon      #从PyQt5.QtGui中引入类QIcon，便于图标的设定

class Ico(QWidget):
```

```
    def __init__(self):
        super().__init__()
        self.initUI()                              #用函数initUI()创建GUI

    def initUI(self):

        self.setGeometry(300, 300, 300, 220)
        self.setWindowTitle('书创文化')
        self.setWindowIcon(QIcon('123.ico'))
        self.show()
if __name__ == '__main__':

    app = QApplication(sys.argv)
    ex = Ico()
    sys.exit(app.exec_())
```

以 self 开头的前 3 行代码中的方法都是从 QWidget 类中继承的。方法 setGeometry()在屏幕上定位窗口并设置它的大小,其前两个参数是窗口的 x 和 y 坐标,第三个是宽度,第四个是窗口的高度。实际上,方法 setGeometry()在一个方法中实现了 resize()和 move()两个方法的功能。方法 setWindowIcon()用于设置应用程序图标,首先创建一个 QIcon 对象,QIcon 能够接收到我们要显示的图标的路径(和当前程序在同一个目录下)。执行后的效果如图 9-2 所示。

下面的实例文件 PyQt03.py 演示了在 PyQt 窗体中创建提示信息的过程。

图 9-2 有图标的窗体

源码路径：daima\9\9-1\PyQt03.py

```
import sys
from PyQt5.QtWidgets import (QWidget, QToolTip,
                             QPushButton, QApplication)
from PyQt5.QtGui import QFont

class Example(QWidget):

    def __init__(self):
        super().__init__()

        self.initUI()

    def initUI(self):
        QToolTip.setFont(QFont('SansSerif', 10))

        self.setToolTip('这是一个<b>QWidget</b>组件')

        btn = QPushButton('按钮', self)
        btn.setToolTip('这是一个<b>QPushButton</b>组件')
        btn.resize(btn.sizeHint())
        btn.move(50, 50)

        self.setGeometry(300, 300, 300, 200)
        self.setWindowTitle('工具栏')
        self.show()

if __name__ == '__main__':
    app = QApplication(sys.argv)
    ex = Example()
    sys.exit(app.exec_())
```

通过上述代码为两个 PyQt5 组件设置了提示信息。具体流程如下。

(1) 使用静态方法 setFont(QFont('SansSerif',10))设置了提示信息的字体,此处使用的是 10px 大小的 SansSerif 字体。

(2)调用方法 setTooltip()创建提示框。当然,我们还可以在提示框中使用更加丰富的文本格式。
(3)使用 QPushButton()创建一个按钮组件,使用 setToolTip()为按钮设置一个提示框。
(4)使用 setHint()方法给按钮设置提示信息。

执行后,当将鼠标指针放到按钮上面时会显示提示信息,如图 9-3 所示。

图 9-3　执行效果

9.1.2　菜单和工具栏

在 GUI 应用程序中,菜单是位于菜单栏的一组命令操作。工具栏是由按钮和一些常规命令操作组成的组件。在 PyQt 模块中,和菜单栏、工具栏相关的几个概念如下。

1. 主窗口

类 QMainWindow 用于实现主窗口效果,会默认创建一个拥有状态栏、工具栏和菜单栏的经典应用窗口框架。

2. 状态栏

状态栏是用来显示状态信息的组件,在 PyQt 中由 QMainWindow(依赖于 QMainWindow 组件)组件创建状态栏。下面的实例文件 PyQt04.py 演示了在 PyQt 窗体中创建状态栏信息的过程。

源码路径:daima\9\9-1\PyQt04.py

```
import sys
from PyQt5.QtWidgets import QMainWindow, QApplication
class Example(QMainWindow):

    def __init__(self):
        super().__init__()

        self.initUI()

    def initUI(self):
        self.statusBar().showMessage('这是状态信息')#调用了QtGui.QMainWindow类的statusBar()方法

        self.setGeometry(300, 300, 250, 150)
        self.setWindowTitle('状态栏演示')
        self.show()
```

在上述代码中,通过调用 QtGui.QMainWindow 类的 statusBar()方法来得到状态栏。第一次调用这个方法时会创建一个状态栏,随后方法返回状态栏对象,然后用 showMessage()方法在状态栏上显示一些信息。执行后会在状态栏中显示信息"这是状态信息",如图 9-4 所示。

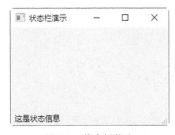

图 9-4　状态栏信息

3. 菜单栏

菜单栏是位于各种菜单中的一组命令操作，是 GUI 应用程序的重要组成部分。因为 Mac 系统对待菜单栏的方式与其他系统不同，所以为了获得全平台一致的效果，开发者通常在代码中加入如下代码：

```
menubar.setNativeMenuBar(False)
```

下面的实例文件 PyQt05.py 演示了在 PyQt 窗体中同时创建菜单栏和状态栏信息的过程。

源码路径：daima\9\9-1\PyQt05.py

```python
class Example(QMainWindow):
    def __init__(self):
        super().__init__()
        self.initUI()

    def initUI(self):
        exitAction = QAction(QIcon('123.ico'), '&退出', self)
        exitAction.setShortcut('Ctrl+Q')
        exitAction.setStatusTip('退出程序')
        exitAction.triggered.connect(qApp.quit)
        self.statusBar()
        menubar = self.menuBar()
        fileMenu = menubar.addMenu('&文件')
        fileMenu.addAction(exitAction)

        self.setGeometry(300, 300, 300, 200)
        self.setWindowTitle('菜单栏练习')
        self.show()

if __name__ == '__main__':
    app = QApplication(sys.argv)
    ex = Example()
    sys.exit(app.exec_())
```

对上述代码的具体说明如下。

- QAction 是一个用于菜单栏、工具栏或自定义快捷键的抽象动作。通过以下 3 行代码创建了一个有指定图标且文本为'Exit'的标签，还为这个动作定义了一个快捷键。下面的 3 行代码创建了一个当鼠标指针浮于菜单项之上时就会显示的状态提示。

```
exitAction = QAction(QIcon('123.ico'), '&退出', self)
exitAction.setShortcut('Ctrl+Q')
exitAction.setStatusTip('退出程序')
```

- 代码 exitAction.triggered.connect(qApp.quit)的功能是，当选中特定动作后会触发一个信号，并将信号连接到 QApplication 组件的 quit()方法，这样就会中断当前整个应用程序。
- 使用方法 menuBar()创建一个菜单栏，首先创建一个"文件"菜单，然后将退出动作添加到"文件"菜单中。

这样就通过上述代码创建了含有一个"文件"菜单的菜单栏，执行后会发现在这个菜单项中包含了一个子菜单"退出"，单击子菜单"退出"后会退出当前程序，如图 9-5 所示。

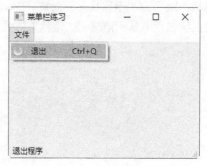

图 9-5　执行效果

4. 工具栏

前面介绍的菜单栏可以集成所有的操作命令，这样我们可以方便地在程序中使用这些集成的命令。而接下来将要讲解的工具栏则提供了一个快速访问常用命令的方式。下面的实例文件PyQt06.py演示了在 PyQt 窗体中创建工具栏的过程。

源码路径：daima\9\9-1\PyQt06.py

```
import sys
from PyQt5.QtWidgets import QMainWindow, QAction, qApp, QApplication
from PyQt5.QtGui import QIcon

class Example(QMainWindow):

    def __init__(self):
        super().__init__()
        self.initUI()

    def initUI(self):
        exitAction = QAction(QIcon('123.ico'), '退出', self)
        exitAction.setShortcut('Ctrl+Q')
        exitAction.triggered.connect(qApp.quit)
        self.toolbar = self.addToolBar('退出')
        self.toolbar.addAction(exitAction)
        self.setGeometry(300, 300, 300, 200)
        self.setWindowTitle('工具栏练习')
        self.show()

if __name__ == '__main__':
    app = QApplication(sys.argv)
    ex = Example()
    sys.exit(app.exec_())
```

在上述代码中创建了一个动作对象，整个过程和前面介绍的菜单栏中的部分代码相似。这个动作包含一个标签、一个图标和一个快捷键，将 QtGui.QMainWindow 的 quit()方法连接到了触发的信号上。执行后会创建了一个简单的工具栏图标，单击这个图标后会退出当前程序。执行效果如图 9-6 所示。

图 9-6　执行效果

9.1.3　界面布局

界面布局是 GUI 编程中的一个重要组成部分。界面布局是一种在程序窗体中排列组件的方法。在使用 PyQt 开发应用程序时，通常使用以下几种界面布局方式。

1. 绝对定位

绝对定位是指在程序中设置每一个组件的位置坐标，并且还以像素为单位设置每个组件的大小。在使用绝对定位时需要清楚以下限制。

- 如果改变了窗口大小，组件的位置和大小并不会发生改变。
- 在不同平台上，应用程序的外观可能会不同。
- 如果改变应用程序中的字体，可能会打乱整个应用程序的布局。

- 如果要修改之前的布局方式，则必须完全重写布局，这样会非常浪费时间。

下面的实例文件 PyQt07.py 演示了在 PyQt 窗体中使用绝对定位方式的过程。

源码路径：daima\9\9-1\PyQt07.py

```python
class Example(QWidget):
    def __init__(self):
        super().__init__()
        self.initUI()

    def initUI(self):
        lbl1 = QLabel('第1行文本', self)
        lbl1.move(15, 10)
        lbl2 = QLabel('第2行文本', self)
        lbl2.move(35, 40)
        lbl3 = QLabel('第3行文本', self)
        lbl3.move(55, 70)
        self.setGeometry(300, 300, 250, 150)
        self.setWindowTitle('Absolute')
        self.show()
```

在上述代码中，使用 move()方法定位了窗体中的 3 行文本。在使用 move()方法时，给 move()方法提供了 x 和 y 坐标作为参数。方法 move()使用的坐标系统是从左上角开始计算的。x 值从左到右增长，y 值从上到下增长。这种使用详细坐标定位的方式就是绝对定位。执行后的效果如图 9-7 所示。

图 9-7　执行效果

2. 箱布局

PyQt 库专门提供了内置的界面布局管理类，这是将组件定位在窗口上的首选方式。其中 QHBoxLayout 和 QVBoxLayout 是两个基础布局管理类，分别实现水平和垂直的线性布局功能。假如需要在右下角设置两个按钮，为了使用箱布局，我们将使用一个水平箱布局和垂直箱布局来实现。下面的实例文件 PyQt08.py 实现了上述布局要求。

源码路径：daima\9\9-1\PyQt08.py

```python
class Example(QWidget):
    def __init__(self):
        super().__init__()
        self.initUI()

    def initUI(self):
        okButton = QPushButton("确定")
        cancelButton = QPushButton("取消")
        hbox = QHBoxLayout()
        hbox.addStretch(1)
        hbox.addWidget(okButton)
        hbox.addWidget(cancelButton)
        vbox = QVBoxLayout()
        vbox.addStretch(1)
        vbox.addLayout(hbox)
        self.setLayout(vbox)
        self.setGeometry(300, 300, 300, 150)
        self.setWindowTitle('按钮布局')
        self.show()
```

在上述代码中使用了两个布局类 QHBoxLayout 和 QVBoxLayout，具体流程如下。

（1）使用 QPushButton()创建两个按钮。

（2）使用如下 4 行代码创建一个水平布局，并增加一个拉伸因子和两个按钮。通过拉伸因子在两个按钮的前面增加了一个可伸缩空间，这样会在窗口的右边显示按钮。

```
hbox = QHBoxLayout()
hbox.addStretch(1)
hbox.addWidget(okButton)
hbox.addWidget(cancelButton)
```

（3）通过如下 3 行代码把水平布局放置在垂直布局内，使用拉伸因子把包含两个按钮的水平布局挪到窗口的底部。

```
vbox = QVBoxLayout()
vbox.addStretch(1)
vbox.addLayout(hbox)
```

（4）使用 setLayout(vbox)设置窗体的主布局。

执行后在会在窗体右下角放置两个按钮，当改变应用窗口大小时会相对于窗口不改变按钮的位置，这也正是与绝对布局的最大区别。执行效果如图 9-8 所示。

图 9-8 执行效果

3．网格布局

网格布局是指使用单元格（行和列）来分割空间。在 PyQt 程序中可以使用类 QGridLayout 实现网格布局功能。下面的实例文件 PyQt09.py 演示了使用网格布局模拟实现一个计算器界面的过程。

源码路径：daima\9\9-1\PyQt09.py

```
class Example(QWidget):
    def __init__(self):
        super().__init__()
        self.initUI()

    def initUI(self):
        grid = QGridLayout()
        self.setLayout(grid)
        names = ['删除', '后退', '', '关闭',
                 '7', '8', '9', '除',
                 '4', '5', '6', '乘',
                 '1', '2', '3', '减',
                 '0', '.', '=', '加']

        positions = [(i, j) for i in range(5) for j in range(4)]
        for position, name in zip(positions, names):
            if name == '':
                continue
            button = QPushButton(name)
            grid.addWidget(button, *position)
        self.move(300, 150)
        self.setWindowTitle('模拟计算器')
        self.show()
```

首先，通过 QGridLayout()创建一个网格布局对象，并使用 setSpacing(10)设置各个组件之间的间距。然后，使用 grid.addWidget（reviewEdit,3,1,5,1）向网格布局中增加一个组件，设置组件的跨行和跨列参数，在此设置 reviewEdit 组件跨 5 行显示。

执行后会创建一个包含 3 个标签、两个单行编辑框和一个文本编辑框组件的窗口，并使用 QGridLayout 布局方式排列上述元素。执行效果如图 9-9 所示。

图 9-9 执行效果

4. 表单布局

在 PyQt 程序中，可以使用 QFormLayout 类来排列输入型控件和关联标签组成的那些表单。QFormLayout 是一个方便的布局类，其中的控件以两列的形式排列在表单中。左列包含标签，右列包含输入控件，如 QLineEdit、QSpinBox 和 QTextEdit 等。下面的实例文件 PyQt10.py 演示了使用表单布局实现一个留言板界面的过程。

源码路径：daima\9\9-1\PyQt10.py

```python
class Example(QWidget):
    def __init__(self):
        super().__init__()
        self.Init_UI()

    def Init_UI(self):
        self.setGeometry(300,300,300,200)
        self.setWindowTitle('书创文化留言板系统')
        formlayout = QFormLayout()
        nameLabel = QLabel("标题")
        nameLineEdit = QLineEdit("")
        introductionLabel = QLabel("内容")
        introductionLineEdit = QTextEdit("")
        formlayout.addRow(nameLabel,nameLineEdit)
        formlayout.addRow(introductionLabel,introductionLineEdit)
        self.setLayout(formlayout)
        self.show()
```

首先使用 QFormLayout()创建一个表单布局对象。然后使用 formlayout.addRow()向表单中增加一行，内容是我们定义的小部件。

执行效果如图 9-10 所示。

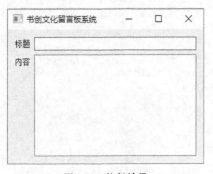

图 9-10 执行效果

9.1.4 事件处理

在计算机软件开发领域，几乎所有的 GUI 应用程序都是事件驱动的。事件主要由应用的用户操作产生的，但是事件可能由其他条件触发，如一个网络连接、一个窗口管理器、一个定时器等，这些动作都可能触发事件。当我们调用 PyQt 程序中的 exec_()方法时，程序会进入主循环。主循环用于检测事件的产生并将事件送到用于处理的对象中。

在事件模型中有如下 3 个参与者。
- 事件源：事件源是状态发生改变的对象，它产生了事件。
- 事件对象：封装了事件源中的状态变化。
- 事件目标：要通知的对象。

在 PyQt5 中通过一个独一无二的信号和槽机制来处理事件，信号和槽用于实现对象之间的通信功能。当发生指定事件时，会发射一个信号。槽可以被任何 Python 脚本调用，当发射和槽连接的信号时会调用槽。下面的实例文件 PyQt11.py 演示了使用单击按钮事件处理程序的过程。

源码路径：daima\9\9-1\PyQt11.py

```python
class Ico(QWidget):
    def __init__(self):
        super().__init__()
        self.initUI()

    def initUI(self):
        self.setGeometry(300, 300, 300, 220)
        self.setWindowTitle('书创文化')
        self.setWindowIcon(QIcon('123.ico'))
        qbtn = QPushButton('退出', self)
        qbtn.clicked.connect(QCoreApplication.instance().quit)
        qbtn.resize(70,30)
        qbtn.move(50, 50)
        self.show()
```

在上述代码中，使用 QPushButton 类创建了一个按钮，该按钮是 QPushButton 类的一个实例。构造函数的第一个参数"退出"是按钮的标签，第二个参数 self 是父窗口部件。父窗口部件是示例窗口部件，它是通过 QWidget 继承的。后面的 qbtn.clicked.connect()用于处理按钮单击事件，PyQt5 中的事件处理系统采用信号和槽机制构建。如果单击按钮，单击动作的信号被发送出去。槽可以是 Qt 槽函数或任何 Python 可调用的函数。QCoreApplication 包含主事件循环，用于处理和调度所有事件。执行后会显示拥有一个按钮的窗体，如图 9-11 所示。单击"退出"按钮后会关闭当前窗体。

图 9-11 执行效果

1. 信号和槽

下面的实例文件 PyQt12.py 演示了在 PyQt5 中使用信号和槽的过程。

源码路径：daima\9\9-1\PyQt12.py

```python
class Example(QWidget):
    def __init__(self):
        super().__init__()
        self.initUI()

    def initUI(self):
        lcd = QLCDNumber(self)
        sld = QSlider(Qt.Horizontal, self)
        vbox = QVBoxLayout()
        vbox.addWidget(lcd)
        vbox.addWidget(sld)
        self.setLayout(vbox)
        sld.valueChanged.connect(lcd.display)
        self.setGeometry(300, 300, 250, 150)
        self.setWindowTitle('信号和槽')
        self.show()
```

在上述代码中用到了一个 QtGui.QLCDNumber 和一个 QtGui.QSlider 类，当拖动滑块时上面的 lcd 数字会随之发生变化。在此将滑块的 valueChanged 信号和 lcd 数字显示的 display 槽连接在一起。发送者是一个发送信号的对象，接收者是一个接收信号的对象，槽是对信号做出反应的方法。执行后的效果如图 9-12 所示。

图 9-12　执行效果

2. 重写事件处理程序

在 PyQt 程序中，通常通过重写事件处理程序实现事件处理功能。下面的实例文件 PyQt13.py 演示了重新实现了 keyPressEvent() 事件处理程序后将要如何处理。

源码路径：daima\9\9-1\PyQt13.py

```python
class Example(QWidget):
    def __init__(self):
        super().__init__()
        self.initUI()

    def initUI(self):
        self.setGeometry(300, 300, 250, 150)
        self.setWindowTitle('事件处理')
        self.show()

    def keyPressEvent(self, e):
        if e.key() == Qt.Key_Escape:
            self.close()
```

在上述代码中，通过自定义函数重写了 keyPressEvent() 事件处理程序。执行程序后，如果按下键盘中的 Esc 键会关闭当前程序。

下面的实例文件 PyQt14.py 演示了重新实现 keyPressEvent() 事件处理程序。

源码路径：daima\9\9-1\PyQt14.py

```python
class Example(QWidget):
    def __init__(self):
        super().__init__()
```

```
        self.initUi()

    def initUi(self):
        self.setGeometry(300, 300, 350, 250)
        self.setWindowTitle('书创文化')
        self.lab = QLabel('移动方向',self)
        self.lab.setGeometry(150,100,50,50)
        self.show()

    def keyPressEvent(self, e):
        if e.key() == Qt.Key_Up:
            self.lab.setText('↑')
        elif e.key() == Qt.Key_Down:
            self.lab.setText('↓')
        elif e.key() == Qt.Key_Left:
            self.lab.setText('←')
        else:
            self.lab.setText('→')
```

在上述代码中重新实现了 **keyPressEvent()** 事件处理程序。当按住键盘中的"上""下""左""右"方向键的时候,窗口中会显示对应的方位标记。例如,按下"上"方向键时的效果如图 9-13 所示。

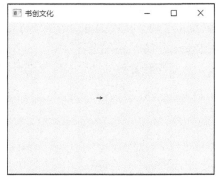

图 9-13 按下"上"方向键时的效果

3. 事件发送者

在实际的程序开发过程中,有时需要知道哪个窗口组件是信号的发送者,这一功能可以通过 PyQt 5 中的 sender() 方法实现。下面的实例文件 PyQt15.py 演示了实现一个简单的人机对战"石头剪刀布"小游戏的过程。

源码路径:daima\9\9-1\PyQt15.py

```
class Example(QWidget):

    def __init__(self):
        super().__init__()
        self.initUI()
    def initUI(self):
        self.setGeometry(200, 200, 300, 300)
        self.setWindowTitle('书创文化传播')
        bt1 = QPushButton('剪刀',self)
        bt1.setGeometry(30,180,50,50)
        bt2 = QPushButton('石头',self)
        bt2.setGeometry(100,180,50,50)
        bt3 = QPushButton('布',self)
        bt3.setGeometry(170,180,50,50)
        bt1.clicked.connect(self.buttonclicked)
        bt2.clicked.connect(self.buttonclicked)
        bt3.clicked.connect(self.buttonclicked)
        self.show()

    def buttonclicked(self):
```

```
        computer = randint(1,3)
        player = 0
        sender = self.sender()
        if sender.text() == '剪刀':
            player = 1
        elif sender.text() == '石头':
            player = 2
        else:
            player = 3
        if player == computer:
            QMessageBox.about(self, '结果', '平手')
        elif player == 1 and computer == 2:
            QMessageBox.about(self, '结果', '智者：石头，我赢了！')
        elif player == 2 and computer == 3:
            QMessageBox.about(self, '结果', '智者：布，我赢了！')
        elif player == 3 and computer == 1:
            QMessageBox.about(self,'结果','智者：剪刀，我赢了！')
        elif computer == 1 and player == 2:
            QMessageBox.about(self, '结果','智者：剪刀，人类赢了！')
        elif computer == 2 and player == 3:
            QMessageBox.about(self, '结果','智者：石头，人类赢了！')
        elif computer == 3 and player == 1:
            QMessageBox.about(self,'结果','智者：布，人类赢了！')
```

在上述代码中设置了 3 个按钮，分别代表石头、剪刀、布。在 buttonClicked()方法中，通过调用 sender()方法来确定我们单击了哪个按钮。bt1、bt2 和 bt3 按钮的 clicked 信号都连接到同一个槽 buttonclicked。通过调用 sender()方法来确定信号源，然后根据信号源确定人类究竟选择了石头、剪刀、布中的哪一个，从而与智者随机给出的数字进行比较，最终判断输赢。执行后的效果如图 9-14 所示。

图 9-14 执行效果

4. 发送自定义信号

在 PyQt 程序中，可以从 QObejct 生成的对象中发送自定义信号。下面的实例文件 PyQt16.py 演示了发送自定义信号的过程。

源码路径：daima\9\9-1\PyQt16.py

```
class Signal(QObject):
    showmouse = pyqtSignal()

class Example(QWidget):

    def __init__(self):
        super().__init__()
        self.initUI()
    def initUI(self):
        self.setGeometry(200, 200, 300, 300)
        self.setWindowTitle('书创文化')
        self.s = Signal()
        self.s.showmouse.connect(self.about)
```

```
            self.show()
        def about(self):
            QMessageBox.about(self,'警告','从实招来！你是不是按下鼠标了？')

        def mousePressEvent(self, e):
            self.s.showmouse.emit()
```

具体步骤如下。

(1) 创建一个名为 showmouse 的新信号，在 mousePress 事件发生时会发出该信号，将该信号连接到 QMainWindow 的 about() 的槽。

(2) 使用方法 pyqtSignal() 作为外部 Signal 类的属性创建一个信号。

(3) 在代码 self.s.showmouse.connect(self.about) 中，使用自定义的 showmouse 信号连接到 QMainWindow 的 about() 的槽。

(4) 定义函数 mousePressEvent()，当单击窗口时，会发出 showmouse 信号，并调用相应的槽函数。

当单击时，就会弹出对话框提醒我们按下了鼠标。执行效果如图 9-15 所示。

图 9-15　执行效果

9.1.5　对话框

对话是两个或更多人之间的会话。在计算机应用中，对话框是一个用来和应用对话的窗口。对话框是大多数主流 GUI 应用不可缺少的一部分，对话框可以用来输入数据、修改数据或改变应用设置等。

1. 输入对话框

在库 PyQt 中，类 QInputDialog 提供了实现一个简单的输入对话框功能，用于从用户处获得一个值。输入值可以是字符串、数字，或者一个列表中的列表项。下面的实例文件 PyQt17.py 演示了使用对话框获取用户名信息的过程。

源码路径：daima\9\9-1\PyQt17.py

```
class Example(QWidget):

    def __init__(self):
        super().__init__()
        self.initUI()

    def initUI(self):
        self.btn = QPushButton('Dialog', self)
        self.btn.move(20, 20)
        self.btn.clicked.connect(self.showDialog)
        self.le = QLineEdit(self)
        self.le.move(130, 22)
```

```
        self.setGeometry(300, 300, 290, 150)
        self.setWindowTitle('输入对话框')
        self.show()

    def showDialog(self):
        text, ok = QInputDialog.getText(self, '输入对话框', '用户名:')
        if ok:
            self.le.setText(str(text))
```

在上述代码中,通过 QInputDialog.getText()方法实现一个输入对话框。其第一个字符串参数 self 是对话框的标题,第二个字符串参数"输入对话框"是对话框内的消息文本。对话框返回输入的文本内容和一个布尔值。如果单击了 OK 按钮,布尔值就是 true;反之,布尔值是 false。只有在单击 OK 按钮时,返回的文本内容才会有值。

执行后会显示一个按钮和一个单行文本框组件,单击"Dialog"按钮后会显示输入对话框,用于获得用户输入的字符串值,在对话框中输入的值会在单行文本框组件中显示。执行前后的效果如图 9-16 (a) ~ (c) 所示。

(a) 初始效果　　　　　　(b) 在对话框中输入 aaa　　　　(c) 在文本框中显示 aaa

图 9-16　执行效果

2. 颜色选择对话框

在库 PyQt 中,QColorDialog 类提供了一个实现颜色选择对话框效果的组件。下面的实例文件 PyQt18.py 演示了使用颜色选择对话框设置背景颜色的过程。

源码路径:daima\9\9-1\PyQt18.py

```
class Example(QWidget):

    def __init__(self):
        super().__init__()
        self.initUI()

    def initUI(self):
        col = QColor(0, 0, 0)
        self.btn = QPushButton('选择颜色', self)
        self.btn.move(20, 20)

        self.btn.clicked.connect(self.showDialog)
        self.frm = QFrame(self)
        self.frm.setStyleSheet("QWidget { background-color: %s }"
                               % col.name())
        self.frm.setGeometry(130, 22, 100, 100)
        self.setGeometry(300, 300, 250, 180)
        self.setWindowTitle('颜色选择对话框')
        self.show()

    def showDialog(self):
        col = QColorDialog.getColor()
        if col.isValid():
            self.frm.setStyleSheet("QWidget { background-color: %s }" % col.name())
```

具体步骤如下。

(1) 使用 QColor(0,0,0)初始化 QtGuiQFrame 组件的背景颜色为黑色。

(2) 通过 QColorDialog.getColor()弹出一个颜色选择框。

(3) 通过 if col.isValid()语句进行判断，如果我们选中一个颜色并且单击了"OK"按钮，会返回一个有效的颜色值。如果单击了"Cancel"按钮，则不会返回选中的颜色值。另外，可以使用样式表来定义背景颜色。

执行后会在窗体中显示一个按钮和一个 QFrame，将 QFrame 组件的背景设置为黑色。我们可以使用颜色选择框来改变 QFrame 的背景颜色。执行前后的效果如图 9-17（a）和（b）所示。

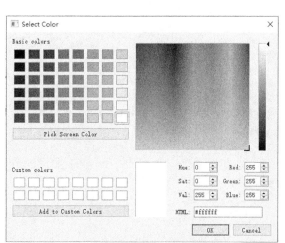

(a) 初始效果　　　　　　　　　(b) 单击"颜色选择"按钮后显示颜色对话框

图 9-17　执行效果

3. 字体选择框

在 PyQt 库中，QFontDialog 类是一个用于实现字体选择对话框效果的组件。下面的实例文件 PyQt19.py 演示了使用 Select Font 对话框设置指定字体的过程。

源码路径：daima\9\9-1\PyQt19.py

```
class Example(QWidget):

    def __init__(self):
        super().__init__()
        self.initUI()

    def initUI(self):
        vbox = QVBoxLayout()
        btn = QPushButton('选择字体', self)
        btn.setSizePolicy(QSizePolicy.Fixed,
                            QSizePolicy.Fixed)
        btn.move(20, 20)
        vbox.addWidget(btn)
        btn.clicked.connect(self.showDialog)
        self.lbl = QLabel('请看我的显示字体', self)
        self.lbl.move(130, 20)
        vbox.addWidget(self.lbl)
        self.setLayout(vbox)
        self.setGeometry(300, 300, 250, 180)
        self.setWindowTitle('字体对话框效果')
        self.show()

    def showDialog(self):
        font, ok = QFontDialog.getFont()
        if ok:
            self.lbl.setFont(font)
```

具体步骤如下。

(1) 通过 QFontDialog.getFont()弹出一个字体对话框，方法 getFont()能够返回字体名称和布尔值。如果用户单击了 OK 按钮则布尔值为 True；否则，为 False。

(2) 通过代码 if ok 进行判断，如果单击了 OK 按钮，则文本的字体会发生改变。

执行后会在窗体中显示一个按钮和一行文本。单击"选择字体"按钮后弹出 Select Font 对话框，在其中可以改变文本的字体。执行效果如图 9-18 所示。

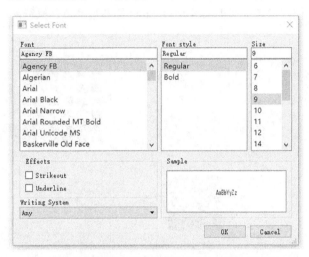

(a) 初始效果　　　　　　(b) 单击"选择字体"按钮后弹出的 Select Font 对话框

图 9-18　执行效果

4. 文件选择对话框

在 PyQt 库中，QFileDialog 类实现了文件选择对话框效果。在其中用户可以选择某个文件或目录，可以选择文件的打开和保存。下面的实例文件 PyQt20.py 演示了使用"文件对话框"打开指定文件的过程。

源码路径：daima\9\9-1\PyQt20.py

```python
class Example(QMainWindow):
    def __init__(self):
        super().__init__()
        self.initUI()

    def initUI(self):
        self.textEdit = QTextEdit()
        self.setCentralWidget(self.textEdit)
        self.statusBar()
        openFile = QAction(QIcon('123.ico'), '打开', self)
        openFile.setShortcut('Ctrl+O')
        openFile.setStatusTip('打开一个文件')
        openFile.triggered.connect(self.showDialog)
        menubar = self.menuBar()
        fileMenu = menubar.addMenu('&文件')
        fileMenu.addAction(openFile)
        self.setGeometry(300, 300, 350, 300)
        self.setWindowTitle('文件对话框')
        self.show()

    def showDialog(self):
        fname = QFileDialog.getOpenFileName(self, '打开文件', '/home')
        if fname[0]:
            f = open(fname[0], 'r')
            with f:
                data = f.read()
                self.textEdit.setText(data)
```

因为要设置一个文本编辑框组件，所以定义一个基于 QMainWindow 组件的类 Example (QMainWindow)。通过 QFileDialog.getOpenFileName(self,'打开文件','/home')实现"文件对话框"。参数"打开文件"是 getOpenFileName()方法的标题，后面的字符串参数指定了对话框的工作目录。默认将文件过滤器设置成 All files (*)，这表示所有格式的文件。选中文件后通过 data = f.read() 和 self.textEdit.setText(data)读取文件的内容，并设置成文本编辑框组件显示的文本。

执行后会在窗体中显示一个菜单栏，在中间设置了一个文本编辑框组件和一个状态栏。单击菜单项会显示 QtGui.QFileDialog（文件选择框）对话框，用于选择一个文件。会读取文件的内容并把它显示在文本编辑框组件中。执行前后的效果如图 9-19（a）和（b）所示。

 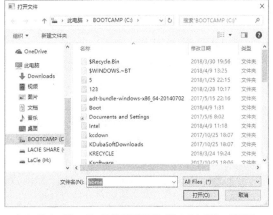

(a) 初始效果　　　　　　　　(b) 单击"打开"菜单项后弹出"打开文件"对话框

图 9-19　执行效果

9.1.6　组件

组件（widget）是构建一个 GUI 应用程序的核心模块之一，在 PyQt 库中有丰富的内置组件，如按钮、复选按钮、滑块和列表框。接下来，将详细讲解复选框（QCheckBox）、切换按钮（ToggleButton）、滑块（QSlider）、进度条（ProgressBar）和日历组件（QCalendarWidget）的使用方法。

1. 复选框

PyQt 中的复选框组件有"选中"和"未选中"两种状态，复选框是由一个选择框和一个标签组成的。QCheckBox（复选框）和 QRadioButton（单选按钮）都是选项按钮，这是因为它们都可以在开（选中）或者关（未选中）之间切换。两者的区别是对用户选择的限制：单选按钮定义了"多选一"的选择，而复选框提供的是"多选多"的选择。只要复选框被选中或者清除，就会发射一个 stateChanged()信号。如果想在复选框状态改变的时候触发一个行为，就需要连接这个信号，我们可以使用 isChecked()来查询复选框是否选中。

除了常用的"选中"和"未选中"两个状态之外，QCheckBox 还提供了第三种可选状态（半选）。如果需要用到第三种状态，可以通过 setTristate()来使它生效，并使用 checkState()来查询当前的切换状态。

下面的实例文件 PyQt21.py 演示了使用 QCheckBox 实现复选框功能的过程。

源码路径：daima\9\9-1\PyQt21.py

```
class Example(QWidget):

    def __init__(self):
        super().__init__()
```

```python
        self.initUI()

    def initUI(self):

        cb = QCheckBox('显示标题', self)
        cb.move(20, 20)
        cb.toggle()
        cb.stateChanged.connect(self.changeTitle)
        self.setGeometry(300, 300, 250, 150)
        self.setWindowTitle('QCheckBox')
        self.show()

    def changeTitle(self, state):
        if state == Qt.Checked:
            self.setWindowTitle('QCheckBox')
        else:
            self.setWindowTitle('')
```

关于上述代码注意以下几点。

- 用到了类 QCheckBox 的构造方法 QCheckBox('Show title', self)。
- 通过 cb.toggle()设置窗口标题，如果不选中复选框，默认情况下复选框不会选中，也不会设置窗口标题。
- 通过 cb.stateChanged.connect(self.changeTitle)连接自定义的槽方法 changeTitle()和信号 stateChanged，方法 changeTitle()能够切换窗口的标题。
- 定义函数 changeTitle(self, state)，复选框组件的状态会传入 changeTitle()方法的 state 参数。如果复选框被选中，就设置窗口的标题；否则，把窗口标题设置成一个空字符串。

执行后会在窗体中创建了一个复选框，通过复选框能够切换窗体的标题，切换前后的效果如图 9-20（a）和（b）所示。

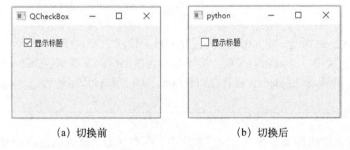

(a) 切换前　　　　　　　　　　(b) 切换后

图 9-20　窗体的标题

2. 单选按钮

QRadioButton 是一个单选按钮，可以打开（选中）或关闭（取消选中），通常为用户提供"多选一"操作。在一组单选按钮中，一次只能检查一个单选按钮。如果用户选择另一个按钮，则先前选择的按钮会关闭。

单选按钮默认为自动互斥的，如果启用了自动互斥功能，则属于同一个父窗口部件的单选按钮的行为就属于同一个互斥按钮组的一部分。

无论何时打开或关闭单选按钮，都会触发 toggled()信号。如果要在每次按钮更改状态时触发某个操作，需要连接到此信号。可以使用方法 isChecked()查看是否选择了一个特定的按钮。下面的实例文件 PyQt22.py 演示了使用 QRadioButton 实现单选按钮的过程。

源码路径：daima\9\9-1\PyQt22.py

```python
class RadioButton(QtWidgets.QWidget):
    def __init__(self, parent=None):
```

```
            QtWidgets.QWidget.__init__(self)

            self.setGeometry(300, 300, 250, 150)
            self.setWindowTitle('Check')

            self.rb = QRadioButton('Show title', self)
            self.rb.setFocusPolicy(Qt.NoFocus)

            self.rb.move(10, 10)
            self.rb.toggle()
            self.rb.toggled.connect(self.changeTitle)

        def changeTitle(self, value):
            if self.rb.isChecked():
                self.setWindowTitle('已经选择')
            else:
                self.setWindowTitle('没有选择')
```

关于上述代码，注意以下几点。
- 使用 QRadioButton('Show title', self)语句创建一个标签信息为"Show title"的单选按钮。
- 通过 self.rb.toggled.connect(self.changeTitle)将用户定义的 changeTitle()函数与单选按钮的 toggled 信号连接起来，自定义函数 changeTitle()将重置窗口的标题。
- 使用 self.rb.setFocusPolicy(Qt.NoFocus)设置无聚焦样式。
- 因为在初始化状态下设置窗口的标题，所以需要使用代码行 self.rb.toggle()将单选按钮选中。在默认情况下，单选按钮处于未选中的状态。

执行后会创建一个用来改变窗口标题的单选按钮。未选中和选中单选按钮的效果如图 9-21（a）和（b）所示。

(a) 未选中　　　　　　　　(b) 选中

图 9-21　单选按钮的效果

3．切换按钮

在 PyQt 库中，切换按钮是 QPushButton 的特殊模式。切换按钮有两种状态——按下和没有按下。可以通过单击切换按钮在两种状态之间转换。下面的实例文件 PyQt23.py 演示了使用 QPushButton 实现切换按钮功能的过程。

源码路径：daima\9\9-1\PyQt23.py

```
class Example(QWidget):

    def __init__(self):
        super().__init__()

        self.initUI()

    def initUI(self):

        self.col = QColor(0, 0, 0)

        redb = QPushButton('红', self)
        redb.setCheckable(True)
        redb.move(10, 10)
```

```
            redb.clicked[bool].connect(self.setColor)

            redb = QPushButton('绿', self)
            redb.setCheckable(True)
            redb.move(10, 60)

            redb.clicked[bool].connect(self.setColor)

            blueb = QPushButton('蓝', self)
            blueb.setCheckable(True)
            blueb.move(10, 110)

            blueb.clicked[bool].connect(self.setColor)

            self.square = QFrame(self)
            self.square.setGeometry(150, 20, 100, 100)
            self.square.setStyleSheet("QWidget { background-color: %s }" %
                                      self.col.name())

            self.setGeometry(300, 300, 280, 170)
            self.setWindowTitle('切换按钮')
            self.show()

        def setColor(self, pressed):

            source = self.sender()

            if pressed:
                val = 255
            else:
                val = 0

            if source.text() == "红":
                self.col.setRed(val)
            elif source.text() == "绿":
                self.col.setGreen(val)
            else:
                self.col.setBlue(val)

            self.square.setStyleSheet("QFrame { background-color: %s }" % self.col.name())

if __name__ == '__main__':
    app = QApplication(sys.argv)
    ex = Example()
    sys.exit(app.exec_())
```

关于上述代码，注意以下几点。

- 通过 QColor(0, 0, 0) 实现颜色初始化功能，设置 RGB 值为黑色。
- 通过 QPushButton('Red', self) 创建切换按钮，并通过 setCheckable(True) 调用 setCheckable() 方法让它可选中。
- 通过代码 redb.clicked[bool].connect(self.setColor) 把 clicked 信号连接到我们定义的方法上，使用 clicked 信号来操作布尔值。
- 使用 self.sender() 获得发生状态切换的按钮。
- 通过使用语句 if source.text() == "红"进行判断，如果发生切换的是"红"按钮，则更新 RGB 值中红色部分的颜色值。
- 通过代码 self.square.setStyleSheet("QWidget {background-color: %s}" % self.col.name()) 使用样式表来改变背景颜色。

执行后会创建 3 个切换按钮和 1 个 QWidget 组件，初始将 QWidget 组件的背景颜色设置为黑色。单击切换按钮后，QWidget 组件的背景颜色将在红色、绿色和蓝色之间进行切换。QWidget

组件的背景颜色取决于哪一个切换按钮被按下。执行效果如图 9-22（a）～（c）所示。

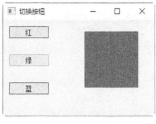
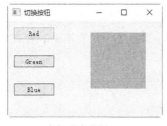
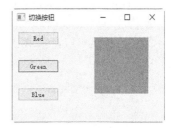

(a) 红色背景　　　　　　　(b) 蓝色背景　　　　　　　(c) 绿色背景

图 9-22　执行效果

4. 滑块

在 PyQt 库中，可以使用滑块（Qslider）组件实现滑块。在滑块中有一个可调节的手柄，我们可以前后拖动这个手柄，并且可以使用拖动的方式来选择一个具体的数值。有时使用滑块比直接输入数字或使用数值选择框更自然。下面的实例文件 PyQt24.py 演示了使用 QSlider 实现一个音量控制器的过程。

源码路径：daima\9\9-1\PyQt24.py

```python
class Example(QWidget):
    
    def __init__(self):
        super().__init__()
        
        self.initUI()
    
    def initUI(self):
        
        sld = QSlider(Qt.Horizontal, self)
        sld.setFocusPolicy(Qt.NoFocus)
        sld.setGeometry(30, 40, 100, 30)
        sld.valueChanged[int].connect(self.changeValue)
        
        self.label = QLabel(self)
        self.label.setPixmap(QPixmap('123.ico'))
        self.label.setGeometry(160, 40, 80, 30)
        
        self.setGeometry(300, 300, 280, 170)
        self.setWindowTitle('QSlider')
        self.show()
    
    def changeValue(self, value):
        
        if value == 0:
            self.label.setPixmap(QPixmap('mute.png'))
        elif value > 0 and value <= 30:
            self.label.setPixmap(QPixmap('min.png'))
        elif value > 30 and value < 80:
            self.label.setPixmap(QPixmap('med.png'))
        else:
            self.label.setPixmap(QPixmap('max.png'))
```

关于上述代码，注意以下几点。

- 使用 QSlider(Qt.Horizontal, self) 创建一个横向的滑块。
- 使用 QLabel(self) 创建一个标签组件，使用 setPixmap(QPixmap('mute.png')) 设置一幅初始无声的图片。
- 使用 valueChanged[int].connect(self.changeValue) 把 valueChanged 信号连接到自定义的 changeValue() 方法。

- 通过 if 语句根据滑块的值设置不同的标签图片，例如，代码 self.label.setPixmap(Qpixmap ('mute.png'))表示如果滑块的值等于零，则为标签显示 mute.png 图片。

执行后将显示一个滑块和一个标签，在标签中将会显示一幅图像。每幅图像代表一种音量大小，通过滑块可以显示不同的标签图像，代表不同的音量级别。执行效果如图 9-23（a）～(c) 所示。

　　(a) 静音　　　　　　　　　(b) 调大音量　　　　　　　　(c) 进一步调大音量

图 9-23　执行效果

5. 进度条

在 PyQt5 库中，使用进度条（QprogressBar）组件实行了横向和纵向的进度条选择功能。我们可以设置进度条的最大值和最小值，进度条的默认值是 0～99。下面的实例文件 PyQt25.py 演示了使用 QProgressBar 实现一个进度条效果的过程。

源码路径：dalma\9\9-1\PyQt25.py

```
class Example(QWidget):
    def __init__(self):
        super().__init__()
        self.initUI()

    def initUI(self):
        self.pbar = QProgressBar(self)
        self.pbar.setGeometry(30, 40, 200, 25)
        self.btn = QPushButton('开始', self)
        self.btn.move(40, 80)
        self.btn.clicked.connect(self.doAction)
        self.timer = QBasicTimer()
        self.step = 0
        self.setGeometry(300, 300, 280, 170)
        self.setWindowTitle('进度条练习')
        self.show()

    def timerEvent(self, e):
        if self.step >= 100:
            self.timer.stop()
            self.btn.setText('完成')
            return
        self.step = self.step + 1
        self.pbar.setValue(self.step)

    def doAction(self):
        if self.timer.isActive():
            self.timer.stop()
            self.btn.setText('开始')
        else:
            self.timer.start(100, self)
            self.btn.setText('停止')
```

关于上述代码。注意以下几点。
- 使用滑块类的构造方法 QProgressBar(self)创建一个滑块对象。
- 使用定时器对象 QtCore.QBasicTimer()激活进度条。

- 使用方法 self.timer.start(100, self) 开启定时器事件, 参数"100"表示定时时间, 参数"self"表示接收定时器事件的对象。
- 每个 QObject 类和它的子类都有用于处理定时事件的函数 timerEvent(), 为了对定时器事件做出反馈, 重新实现了这个事件处理函数 timerEvent(self, e)。
- 执行后将显示一个横向进度条和一个按钮, 通过这个按钮可以控制滑块的开始和停止。
- 编写函数 doAction(), 分别用于开始和停止定时器。

执行效果如图 9-24 所示。

图 9-24　执行效果

6. 日历组件

在 PyQt5 库中, QCalendarWidget 类提供了一个基于月的日历组件 QCalendarWidget, 允许我们通过简单直观的方式选择日期。下面的实例文件 PyQt26.py 演示了使用 QCalendarWidget 实现一个日历的过程。

源码路径: daima\9\9-1\PyQt26.py

```python
class Example(QWidget):
    def __init__(self):
        super().__init__()
        self.initUI()

    def initUI(self):
        cal = QCalendarWidget(self)
        cal.setGridVisible(True)
        cal.move(20, 20)
        cal.clicked[QDate].connect(self.showDate)

        self.lbl = QLabel(self)
        date = cal.selectedDate()
        self.lbl.setText(date.toString())
        self.lbl.move(130, 260)
        self.setGeometry(300, 300, 350, 300)
        self.setWindowTitle('日历')
        self.show()

    def showDate(self, date):
        self.lbl.setText(date.toString())
```

关于上述代码, 注意以下几点。

- 使用 QCalendarWidget(self) 创建 QCalendarWidget 对象。
- 编写代码 cal.clicked[QDate].connect(self.showDate), 如果在组件上选择了一个日期, 则会发射 clicked[QDate] 信号, 连接这个信号和自定义的 showDate() 方法。
- 定义函数 showDate(self, date), 通过方法 selectedDate() 检索选中的日期, 然后把选中的日期对象转化成字符串并显示在标签组件上。

执行后会展示一个日历组件和标签组件, 在日历中选择的日期会显示在标签组件中。执行效果如图 9-25 所示。

图 9-25 执行效果

7. 像素图

在 PyQt5 库中，像素图（QPixmap）组件是在屏幕上显示图片的最佳选择。下面的实例文件 PyQt27.py 演示了使用像素图组件在窗口中显示图片的过程。

源码路径：daima\9\9-1\PyQt27.py

```
class Example(QWidget):

    def __init__(self):
        super().__init__()
        self.initUI()

    def initUI(self):
        hbox = QHBoxLayout(self)
        pixmap = QPixmap("111.jpg")
        lbl = QLabel(self)
        lbl.setPixmap(pixmap)
        hbox.addWidget(lbl)
        self.setLayout(hbox)
        self.move(300, 200)
        self.setWindowTitle('显示图片')
        self.show()
```

关于上述代码，注意以下几点。

- 使用代码 QPixmap("111.jpg ")创建 QPixmap 对象，该对象的构造方法将以一个文件的名字作为参数。
- 通过代码 QLabel(self)把像素图对象设置给标签，使用 lbl.setPixmap(pixmap)通过标签来显示像素图。

执行后会把指定图片 111.jpg 显示在窗体中。执行效果如图 9-26 所示。

图 9-26 执行效果

8. 单行文本编辑框

在 PyQt 5 库中，允许在单行文本编辑框（QLineEdit）组件中输入单行的纯文本。此组件支持撤销、重做、剪切、粘贴、拖曳和拖动等方法。下面的实例文件 PyQt28.py 演示了使用组件 QLineEdit 创建一个单行文本编辑框的过程。

源码路径：daima\9\9-1\PyQt28.py

```python
class Example(QWidget):

    def __init__(self):
        super().__init__()
        self.initUI()

    def initUI(self):
        self.lbl = QLabel(self)
        qle = QLineEdit(self)
        qle.move(60, 100)
        self.lbl.move(60, 40)
        qle.textChanged[str].connect(self.onChanged)
        self.setGeometry(300, 300, 280, 170)
        self.setWindowTitle('单行文本框')
        self.show()

    def onChanged(self, text):
        self.lbl.setText(text)
        self.lbl.adjustSize()
```

关于上述代码，注意以下几点。
- 通过 QLineEdit(self)创建单行文本编辑框组件。
- 编写代码 qle.textChanged[str].connect(self.onChanged)，如果单行文本编辑框中的文本内容发生改变，调用 onChanged()方法进行处理。
- 编写函数 onChanged(self, text)设置标签的文本，调用 adjustSize()方法来调整标签相对于显示的文本长度。

执行后会显示一个单行编辑文本框和一个标签。在单行文本编辑框中输入文本时会在标签中同步显示文本。执行效果如图 9-27 所示。

图 9-27　执行效果

9. 分割框

在 PyQt5 库中，使用分割框（QSplitter）组件可以通过拖曳分割线的方式来控制子组件的大小。下面的实例文件 PyQt29.py 演示了由两个分割框组件控制 3 个 QFrame 组件范围的过程。

源码路径：daima\9\9-1\PyQt29.py

```python
class Example(QWidget):

    def __init__(self):
        super().__init__()
        self.initUI()
```

```python
    def initUI(self):
        hbox = QHBoxLayout(self)
        topleft = QFrame(self)
        topleft.setFrameShape(QFrame.StyledPanel)
        topright = QFrame(self)
        topright.setFrameShape(QFrame.StyledPanel)
        bottom = QFrame(self)
        bottom.setFrameShape(QFrame.StyledPanel)
        splitter1 = QSplitter(Qt.Horizontal)
        splitter1.addWidget(topleft)
        splitter1.addWidget(topright)
        splitter2 = QSplitter(Qt.Vertical)
        splitter2.addWidget(splitter1)
        splitter2.addWidget(bottom)
        hbox.addWidget(splitter2)
        self.setLayout(hbox)
        self.setGeometry(300, 300, 300, 200)
        self.setWindowTitle('分割框组件（QSplitter）')
        self.show()

    def onChanged(self, text):
        self.lbl.setText(text)
        self.lbl.adjustSize()
```

关于上述代码，注意以下几点。

- 创建一个 QFrame(self)对象，通过代码 topleft.setFrameShape(QFrame.StyledPanel)使用了一个样式框架，这样做是为了让框架组件之间的分割线显示得更加明显。
- 通过代码 QSplitter(Qt.Horizontal)创建了一个分割框组件，并分别通过 addWidget(topleft) 和 addWidget(topright)在这个分割框中添加了两个框架组件。
- 编写代码 splitter2 = QSplitter(Qt.Vertical)和 splitter2.addWidget(splitter1)，把第一个分割框添加进第二个分割框组件中。

执行后会在窗体中显示 3 个框架组件和两个分割框组件，如图 9-28 所示。读者需要注意的是，在某些样式主题下可能不会显示分割框组件。

图 9-28　执行效果

9.1.7　使用 Eric6 提高开发效率

Eric6 是 PyQt 编程的最佳 IDE（集成开发环境）。Eric6 可以到官网下载。下载后会得到一个压缩文件，解压缩后进入解压的目录文件夹，然后执行如下命令安装 Eric6。

```
python install.py
```

注意：如果读者的计算机中没有安装 QScintilla，则需要先通过如下命令安装 QScintilla，然后才能执行上面的命令。

```
pip install QScintilla
```

成功安装 Eric6 后会显示如下提示。

```
Checking dependencies
Python Version: 3.6.0
```

```
Found PyQt5
Found pyuic5
Found QScintilla2
Found QtGui
Found QtNetwork
Found QtPrintSupport
Found QtSql
Found QtSvg
Found QtWidgets
Found QtWebEngineWidgets
Qt Version: 5.10.1
sip Version: 4.19.8
PyQt Version: 5.10.1
QScintilla Version: 2.10.3
All dependencies ok.

Cleaning up old installation ...

Creating configuration file ...

Compiling user interface files ...

Compiling source files ...

Installing eric6 ...

Installation complete.
Press enter to continue...
```

Eric6 的使用方法和传统语言的 IDE 类似，拖曳对应的组件即可实现窗体界面的布局，并可以自动生成 Python 代码。Eric6 的相关用法请读者参阅其他相关资料。

9.2 使用 pyglet 库

在 Python 程序中，可以使用 pyglet 库开发跨平台窗口及多媒体程序。本节将详细讲解使用 pyglet 库的方法。

9.2.1 安装并尝试使用 pyglet

可以使用如下命令安装 pyglet 库。

```
pip install pyglet
```

也可以使用如下命令安装最新版本的 pyglet 库。

```
pip install --upgrade https://BitBucket域名.org/pyglet/pyglet/get/tip.zip
```

下面的实例文件 pyglet01.py 演示了创建第一个 pyglet 程序的过程。

源码路径：daima\9\9-2\pyglet01.py

```python
import pyglet
window = pyglet.window.Window()

label = pyglet.text.Label('Hello, world',
                          font_name='Times New Roman',
                          font_size=36,
                          x=window.width//2, y=window.height//2,
                          anchor_x='center', anchor_y='center')
@window.event
def on_draw():
    window.clear()
    label.draw()

pyglet.app.run()
```

在上述代码中，首先使用 import 语句导入库 pyglet，然后通过 pyglet.text.Label 在窗体中定义了一个标签，分别设置了标签中的文本内容及其字体、字号、位置和对齐方式。执行效果如图 9-29 所示。

图 9-29 执行效果

下面的实例文件 pyglet02.py 演示了使用 pyglet 库在窗体中显示指定图片的过程。

源码路径：daima\9\9-2\pyglet02.py

```
import pyglet

window = pyglet.window.Window()
image = pyglet.resource.image('111.jpg')

@window.event
def on_draw():
    window.clear()
    image.blit(0, 0)

pyglet.app.run()
```

在上述代码中，使用 pyglet.resource 中的 image() 函数来加载图像。此函数会自动查找图像文件与源文件（而不是工作目录）。在使用 blit() 方法绘制图像时，会告知 pyglet 在坐标 (0,0) 处绘制图像，这个坐标对应于窗体的左下角。执行后将在窗体中显示指定的图片 111.jpg，如图 9-30 所示。

图 9-30 执行效果

和其他的 GUI 库类似，pyglet 库也能够处理鼠标和键盘等事件程序。下面的实例文件 pyglet03.py 演示了使用 pyglet 库处理键盘事件的过程。

源码路径：daima\9\9-2\pyglet03.py

```
import pyglet
from pyglet.window import key
window = pyglet.window.Window()
@window.event
def on_key_press(symbol, modifiers):
```

```
        if symbol == key.A:
            print('The "A" key was pressed.')
        elif symbol == key.LEFT:
            print('The left arrow key was pressed.')
        elif symbol == key.ENTER:
            print('The enter key was pressed.')

@window.event
def on_draw():
    window.clear()

pyglet.app.run()
```

执行后将会根据用户按下的按键输出对应的提示信息。

```
The left arrow key was pressed.
The left arrow key was pressed.
The left arrow key was pressed.
The enter key was pressed.
The enter key was pressed.
The enter key was pressed.
```

9.2.2 实现 OpenGL 操作

pyglet 库提供一个接口，便于开发者开发 OpenGL 和 GLU 程序。该接口可以供所有的 pyglet 高级 API 使用，可以快速完成所有的图形渲染功能，而不是由操作系统实现。在实现 OpenGL 操作之前需要先用如下代码导入 pyglet.gl。

```
from pyglet.gl import *
```

pyglet.gl 中的所有函数名和常量都与 C 接口对应。下面的实例文件 pyglet04.py 演示了使用 pyglet 库在屏幕上画一个三角形的过程。

源码路径：daima\9\9-2\pyglet04.py

```
from pyglet.gl import *

window = pyglet.window.Window()

@window.event
def on_draw():
    glClear(GL_COLOR_BUFFER_BIT)
    glLoadIdentity()
    glBegin(GL_TRIANGLES)
    glVertex2f(0, 0)
    glVertex2f(window.width, 0)
    glVertex2f(window.width, window.height)
    glEnd()

pyglet.app.run()
```

执行后的效果如图 9-31 所示。

图 9-31　执行效果

在使用某些 OpenGL 函数实现绘图功能时需要用到一组数据，这组数据通常是由不同类型的数组构成的阵列。下面的实例文件 pyglet05.py 绘制了与前面实例相同的三角形，但是本例是使用顶点数组实现的，而不像实例文件 pyglet04.py 那样直接使用模式函数实现。

源码路径：daima\9\9-2\pyglet05.py

```python
from pyglet.gl import *

window = pyglet.window.Window()

vertices = [0, 0,
            window.width, 0,
            window.width, window.height]
vertices_gl_array = (GLfloat * len(vertices))(*vertices)

glEnableClientState(GL_VERTEX_ARRAY)
glVertexPointer(2, GL_FLOAT, 0, vertices_gl_array)

@window.event
def on_draw():
    glClear(GL_COLOR_BUFFER_BIT)
    glLoadIdentity()
    glDrawArrays(GL_TRIANGLES, 0, len(vertices) // 2)

pyglet.app.run()
```

执行后的效果与图 9-31 相同。

可以使用上述类似的阵列结构实现顶点缓冲对象，创建多边形点数据和地图。

9.2.3 开发一个 pyglet 游戏

下面的实例文件 pyglet06.py 演示了使用 pyglet 库开发一个《Minecraft》游戏的过程。《Minecraft》是一个沙盒游戏，中文非官方译名为《我的世界》。本实例使用 Python 和 pyglet 开发一个简单的《Minecraft》游戏，通过这个项目来学习 pyglet 和 Python 游戏编程。

在前期规划本项目时，设置通过如下按键控制精灵的移动。

- W 键：向前移动。
- S 键：向后移动。
- A 键：向左移动。
- D 键：向右移动。
- Space 键：跳跃
- Tab 键：切换飞行模式

另外，通过鼠标使精灵环顾四周。

在游戏中创建了 3 种方块类型——砖块、草地和沙块。按下鼠标左键会消除方块，按下鼠标右键会创造方块，按下 Esc 键会释放鼠标，然后关闭窗口。

实例文件 pyglet06.py 的具体实现流程如下。

(1) 设置在项目中需要的几个公共常量值。具体实现代码如下。

```python
TICKS_PER_SEC = 60  #每秒刷新60次

#用于减轻块负荷的扇区的大小
SECTOR_SIZE = 16

WALKING_SPEED = 5        #移动速度
FLYING_SPEED = 15        #飞行速度

GRAVITY = 20.0
MAX_JUMP_HEIGHT = 1.0    # 一个块的高度
#跳跃速度公式
JUMP_SPEED = math.sqrt(2 * GRAVITY * MAX_JUMP_HEIGHT)
TERMINAL_VELOCITY = 50   #终端的速度
```

(2) 定义函数 cube_vertices(),返回以 (x,y,z) 为中心并且边长为 2n 的正方体 6 个面的顶点坐标。具体实现代码如下所示。

```
PLAYER_HEIGHT = 2                    #玩家的高度

if sys.version_info[0] >= 3:
    xrange = range
```

(2) 定义函数 cube_vertices(),返回以 (x,y,z) 为中心并且边长为 2n 的正方体 6 个面的顶点坐标。具体实现代码如下所示。

```
def cube_vertices(x, y, z, n):
    return [
        x-n,y+n,z-n, x-n,y+n,z+n, x+n,y+n,z+n, x+n,y+n,z-n,  #上
        x-n,y-n,z-n, x+n,y-n,z-n, x+n,y-n,z+n, x-n,y-n,z+n,  #下
        x-n,y-n,z-n, x-n,y-n,z+n, x-n,y+n,z+n, x-n,y+n,z-n,  #左
        x+n,y-n,z+n, x+n,y-n,z-n, x+n,y+n,z-n, x+n,y+n,z+n,  #右
        x-n,y-n,z+n, x+n,y-n,z+n, x+n,y+n,z+n, x-n,y+n,z+n,  #前
        x+n,y-n,z-n, x-n,y-n,z-n, x-n,y+n,z-n, x+n,y+n,z-n,  #后
    ]
```

(3) 定义纹理坐标函数 tex_coord(),给出纹理图左下角的坐标,返回一个正方形纹理的 4 个顶点的坐标。因为可以将纹理图看成 4*4 的纹理 patch,所以此处的 n=4(图中实际有 6 个 patch,其他空白)。例如,欲返回左下角的那个正方形纹理 patch,如果我们输入左下角的整数坐标(0,0),则输出是 "0,0,1/4,0,1/4,1/4,0,1/4"。函数 tex_coord() 的具体实现代码如下。

```
def tex_coord(x, y, n=4):
    m = 1.0 / n
    dx = x * m
    dy = y * m
    return dx, dy, dx + m, dy, dx + m, dy + m, dx, dy + m
```

(4) 编写函数 tex_coord() 计算一个正方体 6 个面的纹理贴图坐标,将结果放入一个列表中,这 6 个面分别是 top、bottom 和 4 个侧面。具体实现代码如下。

```
def tex_coords(top, bottom, side):
    top = tex_coord(*top)#将元组(x, y)分为x和y
    bottom = tex_coord(*bottom)
    side = tex_coord(*side)
    result = []
    result.extend(top)  # extend用来连接两个列表
    result.extend(bottom)
    result.extend(side * 4)
    return result
```

(5) 设置使用的纹理图片是 texture.png,然后计算草块、沙块、砖块、石块的纹理贴图坐标(用一个列表保存)。具体实现代码如下。

```
TEXTURE_PATH = 'texture.png'
#除了草块之外,正方体6个面的贴图都一样
GRASS = tex_coords((1, 0), (0, 1), (0, 0))
SAND  = tex_coords((1, 1), (1, 1), (1, 1))
BRICK = tex_coords((2, 0), (2, 0), (2, 0))
STONE = tex_coords((2, 1), (2, 1), (2, 1))
```

(6) 设置当前位置向 6 个方向移动 1 个单位要用到的增量坐标。具体实现代码如下。

```
FACES = [
    ( 0, 1, 0),
    ( 0,-1, 0),
    (-1, 0, 0),
    ( 1, 0, 0),
    ( 0, 0, 1),
    ( 0, 0,-1),
]
```

(7) 编写函数 normalize() 对位置 x、y 和 z 取整。具体实现代码如下。

```
def normalize(position):
    x, y, z = position
    x, y, z = (int(round(x)), int(round(y)), int(round(z)))
    return (x, y, z)
```

(8) 编写函数 sectorize() 计算位置,首先对位置 x、y、z 取整,然后除以 SECTOR_SIZE,返回 (x,0,z)。这样会将许多不同的 position 映射到同一个 (x,0,z),一个 (x,0,z) 对应一个

$x*z*y=16*16*y$ 的区域内所有立方体的中心 position。具体实现代码如下。

```python
def sectorize(position):
    x, y, z = normalize(position)
    x, y, z = x // SECTOR_SIZE, y // SECTOR_SIZE, z // SECTOR_SIZE
    return (x, 0, z)  #得到的是整数坐标
```

（9）编写函数_initialize()绘制地图，大小是80*80。具体实现代码如下。

```python
def _initialize(self):
    n = 80  #地图大小
    s = 1   #步长
    y = 0   #初始高度
    for x in xrange(-n, n + 1, s):
        for z in xrange(-n, n + 1, s):
            #在地下画一层石头，上面是一层草地
            #地面从y=-2开始
            self.add_block((x, y - 2, z), GRASS, immediate=False)
            self.add_block((x, y - 3, z), STONE, immediate=False)
            #地图的四周用墙围起来
            if x in (-n, n) or z in (-n, n):
                #创建外墙
                for dy in xrange(-2, 3):
                    self.add_block((x, y + dy, z), STONE, immediate=False)

    #为了避免建到墙上，o取n-10
    o = n - 10
    #在地面上随机建造一些草块、沙块、砖块
    for _ in xrange(120):  #只想迭代120次，不需要迭代变量i，直接用 _
        a = random.randint(-o, o)   # 在[-o,o]内随机取一个整数
        b = random.randint(-o, o)   # hill的z位置
        c = -1   # hill的底部
        h = random.randint(1, 6)    # hill的高度
        s = random.randint(4, 8)    # 2 * s是hill的边长
        d = 1    #如何迅速地从hill上逐渐消失
        t = random.choice([GRASS, SAND, BRICK])
        for y in xrange(c, c + h):
            for x in xrange(a - s, a + s + 1):
                for z in xrange(b - s, b + s + 1):
                    if (x - a) ** 2 + (z - b) ** 2 > (s + 1) ** 2:
                        continue
                    if (x - 0) ** 2 + (z - 0) ** 2 < 5 ** 2:
                        continue
                    self.add_block((x, y, z), t, immediate=False)
            s -= d   #递减边的长度，hill逐渐减小
```

（10）编写函数 hit_test()检测鼠标是否能对一个立方体进行操作。此函数返回 key 和 previous。其中 key 是鼠标可操作的块（中心坐标），根据人所在位置和方向向量求出。而 previous 是与 key 处立方体相邻的空位置的中心坐标。如果返回非空值，单击删除 key 处的立方体，右击在 previous 处添加砖块。具体实现代码如下。

```python
def hit_test(self, position, vector, max_distance=8):
    m = 8
    x, y, z = position
    dx, dy, dz = vector
    previous = None
    for _ in xrange(max_distance * m):      # 迭代8*8=64次
        key = normalize((x, y, z))
        if key != previous and key in self.world:
            return key, previous
        previous = key
        x, y, z = x + dx / m, y + dy / m, z + dz / m
    return None, None
```

（11）编写函数 exposed()，只要 position 周围 6 个面有一个没有立方体，就返回真值，表示要绘制 position 处的立方体。如果 6 个面都被立方体包围，则可以不绘制 position 处的立方体，因为即使绘制了也看不到。具体实现代码如下。

```python
def exposed(self, position):
    x, y, z = position
    for dx, dy, dz in FACES:
```

```
            if (x + dx, y + dy, z + dz) not in self.world:
                return True
    return False
```

(12) 编写函数 add_block() 添加立方体。具体实现代码如下。

```
def add_block(self, position, texture, immediate=True):
    if position in self.world:#如果position已经存在于world中，要先移除它
        self.remove_block(position, immediate)
    self.world[position] = texture# 添加相应的位置和纹理
    #以区域为一组添加立方体的position到字典中
    # 16*16*y区域内的立方体都映射到一个键值，这些立方体position以tuble形式存在于一个列表中
    self.sectors.setdefault(sectorize(position), []).append(position)
    if immediate:#初始时该变量为false,不会同步绘制
        if self.exposed(position):#如果同步绘制,且该位置是显露在外的
            self.show_block(position)#绘制该立方体
        self.check_neighbors(position)
```

(13) 编写函数 remove_block() 删除立方体。具体实现代码如下。

```
def remove_block(self, position, immediate=True):
    del self.world[position]#删除world中的（position,texture）对
    self.sectors[sectorize(position)].remove(position)  #删除区域中相应的position
    if immediate: #如果同步
        if position in self.shown:# 如果position在显示列表中
            self.hide_block(position) #立即删除它
        self.check_neighbors(position)
```

(14) 编写检查函数 check_neighbors()，在删除一个立方体后检查它周围 6 个邻接的位置是否有暴露出来的立方体，如果有要把它绘制出来。具体实现代码如下。

```
def check_neighbors(self, position):

    x, y, z = position
    for dx, dy, dz in FACES:#检查周围6个位置
        key = (x + dx, y + dy, z + dz)
        if key not in self.world:#如果该处没有立方体，则跳过
            continue
        if self.exposed(key):#如果该处有立方体且暴露在外
            if key not in self.shown: #同时没有在显示列表中
                self.show_block(key) #则立即绘制出来
        else:#如果没有暴露在外，而又在显示列表中，则立即隐藏（删除）它
            if key in self.shown:
                self.hide_block(key)
```

(15) 编写函数 show_block()，功能是将 world 中显露在外的立方体绘制出来。具体实现代码如下。

```
def show_block(self, position, immediate=True):
    texture = self.world[position]#取出纹理，其实是6个面的纹理坐标信息
    self.shown[position] = texture #存入shown字典中
    if immediate:#立即绘制
        self._show_block(position, texture)
    else:#不立即绘制，进入事件队列
        self._enqueue(self._show_block, position, texture)
```

(16) 编写函数 _show_block()，将顶点列表添加到渲染对象中（on_draw()函数会负责渲染），并将 position:VertexList 对存入 _shown 中。具体实现代码如下。

```
def _show_block(self, position, texture):

    x, y, z = position
    #中心为（x,y,z）的1*1*1正方体
    #顶点坐标数据和纹理坐标数据
    vertex_data = cube_vertices(x, y, z, 0.5)
    texture_data = list(texture)
    #创建vertex 列表
    #添加顶点列表到渲染对象中
    #顶点数目count=24
    self._shown[position] = self.batch.add(24, GL_QUADS, self.group,
        ('v3f/static', vertex_data),
        ('t2f/static', texture_data))
```

(17) 编写函数 hide_block() 隐藏立方体。具体实现代码如下。

```
def hide_block(self, position, immediate=True):
```

```
            self.shown.pop(position) #从列表中移除将要隐藏的立方体中心坐标
            if immediate: #立即移除,从图上消失
                self._hide_block(position)
            else: #不立即移除,进行事件队列等待处理
                self._enqueue(self._hide_block, position)
```

(18)编写函数_hide_block()立即移除立方体,弹出 position 位置的顶点列表并删除,相应的立方体立即被移除,其实整个操作是在更新之后进行的。具体实现代码如下。

```
        def _hide_block(self, position):
            self._shown.pop(position).delete()
```

(19)编写函数 show_sector()绘制一个区域内的立方体。如果区域内的立方体位置不在显示的列表中,且位置是显露在外的,则显示立方体。具体实现代码如下。

```
        def show_sector(self, sector):
            for position in self.sectors.get(sector, []):
                if position not in self.shown and self.exposed(position):
                    self.show_block(position, False)
```

(20)编写函数 hide_sector()设置隐藏区域。如果一个立方体在显示列表中,则隐藏它。具体实现代码如下。

```
        def hide_sector(self, sector):
            for position in self.sectors.get(sector, []):
                if position in self.shown:
                    self.hide_block(position, False)  # 放入事件队列,不会立即隐藏
```

(21)编写函数 change_sectors()移动立方体,移动区域是一个连续的 x、y 子区域。具体实现代码如下。

```
        def change_sectors(self, before, after):
            before_set = set()
            after_set = set()
            pad = 4
            for dx in xrange(-pad, pad + 1):
                for dy in [0]:
                    for dz in xrange(-pad, pad + 1):
                        if dx ** 2 + dy ** 2 + dz ** 2 > (pad + 1) ** 2:
                            continue
                        if before:
                            x, y, z = before
                            before_set.add((x + dx, y + dy, z + dz))
                        if after:
                            x, y, z = after
                            after_set.add((x + dx, y + dy, z + dz))
            show = after_set - before_set
            hide = before_set - after_set
            for sector in show:
                self.show_sector(sector)
            for sector in hide:
                self.hide_sector(sector)
```

(22)编写函数_enqueue()添加事件到队列中。具体实现代码如下。

```
        def _enqueue(self, func, *args):
            self.queue.append((func, args))
```

(23)编写函数_dequeue()处理队头事件。具体实现代码如下。

```
        def _dequeue(self):
            func, args = self.queue.popleft()
            func(*args)
```

(24)编写函数 process_queue()用(1/60)s 的时间来处理队列中的事件,但是不一定要处理完。具体实现代码如下。

```
        def process_queue(self):
            start = time.clock()
            while self.queue and time.clock() - start < 1.0 / TICKS_PER_SEC:
                self._dequeue()
```

(25)编写函数 process_entire_queue()处理事件队列中的所有事件。具体实现代码如下。

```
        def process_entire_queue(self):
            while self.queue:
                self._dequeue()
```

(26) 定义 Window 类来处理窗体界面,通过函数 __init__()实现界面初始化处理。具体实现代码如下。

```python
class Window(pyglet.window.Window):

    def __init__(self, *args, **kwargs):
        super(Window, self).__init__(*args, **kwargs)

        #初始时,鼠标事件没有绑定到游戏窗口
        self.exclusive = False

        #当飞行重力没有效果时,速度增加
        self.flying = False

        #[z, x]中,z表示前后运动,x表示左右运动
        self.strafe = [0, 0]

        #开始位置在地图中间
        self.position = (0, 0, 0)
        #rotation(水平角x, 俯仰角y)
        #水平角是方向射线xoz上的投影与z轴负半轴的夹角
        #俯仰角是方向射线与xoz平面的夹角
        self.rotation = (0, 0)
        self.sector = None

        #reticle表示游戏窗口中间的那个十字
        #绘制两条直线
        self.reticle = None

        #Velocity in the y (upward) direction.
        self.dy = 0

        #可以建造的类型
        self.inventory = [BRICK, GRASS, SAND]

        #初始时右击建造的是砖块
        self.block = self.inventory[0]

        #数字键响应,用于建造类型的切换
        self.num_keys = [
            key._1, key._2, key._3, key._4, key._5,
            key._6, key._7, key._8, key._9, key._0]

        #处理地图的模型实例
        self.model = Model()

        #游戏窗口左上角的label参数设置
        self.label = pyglet.text.Label('', font_name='Arial', font_size=18,
            x=10, y=self.height - 10, anchor_x='left', anchor_y='top',
            color=(0, 0, 0, 255))

        pyglet.clock.schedule_interval(self.update, 1.0 / TICKS_PER_SEC)# 每秒刷新60次
```

(27) 编写函数 set_exclusive_mouse()设置鼠标事件是否绑定到游戏窗口。具体实现代码如下。

```python
    def set_exclusive_mouse(self, exclusive):
        super(Window, self).set_exclusive_mouse(exclusive)
        self.exclusive = exclusive
```

(28) 编写函数 get_sight_vector(),功能是根据前进方向 rotation 来决定移动 1 单位距离时各轴的移动分量是多少。具体实现代码如下。

```python
    def get_sight_vector(self):
        x, y = self.rotation
        m = math.cos(math.radians(y))
        dy = math.sin(math.radians(y))
        dx = math.cos(math.radians(x - 90)) * m
        dz = math.sin(math.radians(x - 90)) * m
        return (dx, dy, dz)
```

(29) 编写函数 get_motion_vector(),功能是在运动时计算 3 条轴的位移增量。具体实现代

码如下。

```python
def get_motion_vector(self):
    if any(self.strafe):
        x, y = self.rotation
        strafe = math.degrees(math.atan2(*self.strafe))#转换成角度
        y_angle = math.radians(y)
        x_angle = math.radians(x + strafe)
        if self.flying:#如果允许飞,那么运动时会考虑垂直方向的运动
            m = math.cos(y_angle)
            dy = math.sin(y_angle)
            if self.strafe[1]:#如果x不为0
                #左移或右移
                dy = 0.0
                m = 1
            if self.strafe[0] > 0: #如果z大于0
                #向后移动
                dy *= -1
            #当向上或向下飞行时,向左、向右的运动都比较少
            dx = math.cos(x_angle) * m
            dz = math.sin(x_angle) * m
        else:
            dy = 0.0
            dx = math.cos(x_angle)
            dz = math.sin(x_angle)
    else:
        dy = 0.0
        dx = 0.0
        dz = 0.0
    return (dx, dy, dz)
```

(30)编写函数update(),每(1/60)s调用一次以进行更新。具体实现代码如下。

```python
def update(self, dt):
    self.model.process_queue()# 用(1/60)s的时间来处理队列中的事件,不一定要处理完
    sector = sectorize(self.position)
    if sector != self.sector:#如果position的sector与当前sector不一样
        self.model.change_sectors(self.sector, sector)
        if self.sector is None:#如果sector为空
            self.model.process_entire_queue() #处理队列中的所有事件
        self.sector = sector#更新sector
    m = 8
    dt = min(dt, 0.2)
    for _ in xrange(m):
        self._update(dt / m)
```

(31)编写函数update(),更新self.dy和self.position。具体实现代码如下。

```python
def _update(self, dt):
    speed = FLYING_SPEED
    if self.flying else WALKING_SPEED#如果能飞,速度为15;否则为5
    d = dt * speed
    dx, dy, dz = self.get_motion_vector()
    #在重力面前,新的空间地位
    dx, dy, dz = dx * d, dy * d, dz * d
    #如果不能飞,则使其在y方向上符合重力规律
    if not self.flying:
        self.dy -= dt * GRAVITY
        self.dy = max(self.dy, -TERMINAL_VELOCITY)
        dy += self.dy * dt
    x, y, z = self.position
    #碰撞检测后应该移动到的位置
    x, y, z = self.collide((x + dx, y + dy, z + dz), PLAYER_HEIGHT)
    self.position = (x, y, z)# 更新位置
```

(32)编写函数collide(),实现碰撞检测,返回值p表示碰撞检测后应该移动到的位置。如果没有遇到障碍物,则p仍然是position;否则,p是新的值(会使其沿着墙走)。具体实现代码如下。

```python
def collide(self, position, height):
    pad = 0.25
```

```
        p = list(position)#将元组变为列表
        np = normalize(position)#取整
        for face in FACES:    #检查周围6个面
            for i in xrange(3):     #单独检测每一维
                if not face[i]:#如果为0,则跳过
                    continue
                #这个维度有多少重叠
                d = (p[i] - np[i]) * face[i]
                if d < pad:
                    continue
                for dy in xrange(height):   #检测每个高度
                    op = list(np)
                    op[1] -= dy
                    op[i] += face[i]
                    if tuple(op) not in self.model.world:
                        continue
                    p[i] -= (d - pad) * face[i]
                    if face == (0, -1, 0) or face == (0, 1, 0):
                        #与地面或顶部相撞,所以停止下降/上升
                        self.dy = 0
                    break
        return tuple(p)
```

（33）编写函数 on_mouse_press()，处理鼠标按下的事件。具体实现代码如下。

```
    def on_mouse_press(self, x, y, button, modifiers):
        if self.exclusive:#当鼠标事件已经绑定了此窗口时
            vector = self.get_sight_vector()
            block, previous = self.model.hit_test(self.position, vector)
            #如果按下左键且该处有block
            if (button == mouse.RIGHT) or \
                    ((button == mouse.LEFT) and (modifiers & key.MOD_CTRL)):
                if previous:
                    self.model.add_block(previous, self.block)
            #如果按下右键,且有previous位置,则在previous处增加方块
            elif button == pyglet.window.mouse.LEFT and block:
                texture = self.model.world[block]
                if texture != STONE:# 如果block不是石块,就移除它
                    self.model.remove_block(block)
        else: #否则隐藏鼠标,并绑定鼠标事件到该窗口
            self.set_exclusive_mouse(True)
```

（34）编写函数 on_mouse_press()处理鼠标移动事件，实现视角的变化，参数 dx 和 dy 分别表示鼠标从上一位置移动到当前位置时 x、y 轴上的位移。此函数能够将这个位移转换成水平角 x 和俯仰角 y 的变化，变化幅度由参数 m 控制。具体实现代码如下。

```
    def on_mouse_motion(self, x, y, dx, dy):
        if self.exclusive:#当鼠标绑定在该窗口时
            m = 0.15
            x, y = self.rotation
            x, y = x + dx * m, y + dy * m
            y = max(-90, min(90, y))#限制仰视和俯视角y只能介于-90°～90°
            self.rotation = (x, y)
```

（35）编写函数 on_key_press()处理按下按键的事件，长按 W、S、A、D 键后会不断地改变坐标。具体实现代码如下。

```
    def on_key_press(self, symbol, modifiers):
        if symbol == key.W:# OpenGL坐标系:z轴垂直平面向外,x轴向右,y轴向上
            self.strafe[0] -= 1 #向前;z坐标-1
        elif symbol == key.S:
            self.strafe[0] += 1
        elif symbol == key.A: #向左; x坐标-1
            self.strafe[1] -= 1
        elif symbol == key.D:
            self.strafe[1] += 1
        elif symbol == key.SPACE:
            if self.dy == 0:
                self.dy = JUMP_SPEED
```

```
elif symbol == key.ESCAPE:  #鼠标退出当前窗口
    self.set_exclusive_mouse(False)
elif symbol == key.TAB:  #切换是否能飞,即是否可以在垂直方向上运动
    self.flying = not self.flying
elif symbol in self.num_keys:
    index = (symbol - self.num_keys[0]) % len(self.inventory)
    self.block = self.inventory[index]  #取得相应的方块类型
```

（36）编写函数 on_key_press()，处理释放按键的事件。具体实现代码如下。

```
def on_key_release(self, symbol, modifiers):
    if symbol == key.W:  #按键释放时,各方向退回一个单位
        self.strafe[0] += 1
    elif symbol == key.S:
        self.strafe[0] -= 1
    elif symbol == key.A:
        self.strafe[1] += 1
    elif symbol == key.D:
        self.strafe[1] -= 1
```

（37）编写函数 on_resize()，处理窗口大小变化时的响应事件。具体实现代码如下。

```
def on_resize(self, width, height):
    #label的纵坐标
    self.label.y = height - 10
    #更新reticle,包含4个点,绘制成两条直线
    if self.reticle:
        self.reticle.delete()
    x, y = self.width // 2, self.height // 2
    n = 10
    self.reticle = pyglet.graphics.vertex_list(4,
        ('v2i', (x - n, y, x + n, y, x, y - n, x, y + n))
    )
```

（38）编写函数 set_2d()，在 OpenGL 中绘制三维图形。具体实现代码如下。

```
def set_2d(self):
    width, height = self.get_size()
    glDisable(GL_DEPTH_TEST)
    glViewport(0, 0, width, height)
    glMatrixMode(GL_PROJECTION)
    glLoadIdentity()
    glOrtho(0, width, 0, height, -1, 1)
    glMatrixMode(GL_MODELVIEW)
    glLoadIdentity()
```

（39）编写函数 set_3d()，在 OpenGL 中绘制三维图形，具体实现代码如下。

```
def set_3d(self):
    width, height = self.get_size()
    glEnable(GL_DEPTH_TEST)
    glViewport(0, 0, width, height)
    glMatrixMode(GL_PROJECTION)
    glLoadIdentity()
    gluPerspective(65.0, width / float(height), 0.1, 60.0)
    glMatrixMode(GL_MODELVIEW)
    glLoadIdentity()
    x, y = self.rotation
    glRotatef(x, 0, 1, 0)
    glRotatef(-y, math.cos(math.radians(x)), 0, math.sin(math.radians(x)))
    x, y, z = self.position
    glTranslatef(-x, -y, -z)
```

（40）编写函数 on_draw()，功能是重写 Window 的 on_draw 函数。当需要重绘窗口时,事件循环就会调度该事件。具体实现代码如下。

```
def on_draw(self):
    self.clear()
    self.set_3d()  #进入3D模式
    glColor3d(1, 1, 1)
    self.model.batch.draw()  #将batch中保存的顶点列表绘制出来
    self.draw_focused_block()  #绘制获得焦点的立方体的线框
    self.set_2d()  #进入2D模式
```

```
            self.draw_label()#绘制label
            self.draw_reticle()#绘制窗口中间的十字
```

(41) 编写函数 draw_focused_block(),绘制获得焦点的立方体,在它的外层画个立方体线框。具体实现代码如下。

```python
        def draw_focused_block(self):
            vector = self.get_sight_vector()
            block = self.model.hit_test(self.position, vector)[0]
            if block:
                x, y, z = block
                vertex_data = cube_vertices(x, y, z, 0.51)
                glColor3d(0, 0, 0)
                glPolygonMode(GL_FRONT_AND_BACK, GL_LINE)
                pyglet.graphics.draw(24, GL_QUADS, ('v3f/static', vertex_data))
                glPolygonMode(GL_FRONT_AND_BACK, GL_FILL)
```

(42) 编写函数 draw_label(),显示帧率、当前位置的方块数及总方块数。具体实现代码如下。

```python
        def draw_label(self):
            x, y, z = self.position
            self.label.text = '%02d (%.2f, %.2f, %.2f) %d / %d' % (
                pyglet.clock.get_fps(), x, y, z,
                len(self.model._shown), len(self.model.world))
            self.label.draw()# 绘制label的text
```

(43) 编写函数 draw_reticle(),绘制游戏窗口中间的十字,也就是一条横线加一条竖线。具体实现代码如下。

```python
        def draw_reticle(self):
            glColor3d(0, 0, 0)
            self.reticle.draw(GL_LINES)
```

(44) 编写函数 setup_fog()和 setup(),实现雾效果。具体实现代码如下。

```python
def setup_fog():
    glEnable(GL_FOG)
    glFogfv(GL_FOG_COLOR, (GLfloat * 4)(0.5, 0.69, 1.0, 1))
    glHint(GL_FOG_HINT, GL_DONT_CARE)
    glFogi(GL_FOG_MODE, GL_LINEAR)
    glFogf(GL_FOG_START, 20.0)
    glFogf(GL_FOG_END, 60.0)
#设置雾效果
def setup():
    """ 基本OpenGL配置

    """
    glClearColor(0.5, 0.69, 1.0, 1)
    glEnable(GL_CULL_FACE)
    glTexParameteri(GL_TEXTURE_2D, GL_TEXTURE_MIN_FILTER, GL_NEAREST)
    glTexParameteri(GL_TEXTURE_2D, GL_TEXTURE_MAG_FILTER, GL_NEAREST)
    setup_fog()# 设置雾效果
```

(45) 在主窗体函数 main()中调用前面的函数显示界面。具体实现代码如下。

```python
def main():
    window = Window(width=800, height=600, caption='Pyglet', resizable=True)# 创建游戏窗口
    #隐藏光标,将所有的鼠标事件都绑定到此窗口
    window.set_exclusive_mouse(True)
    setup()  #设置
    pyglet.app.run()#运行,开始监听并处理事件


if __name__ == '__main__':
    main()
```

执行效果如图 9-32 所示。

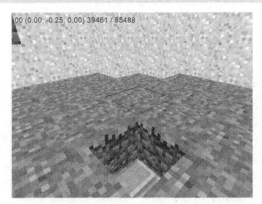

图 9-32　执行效果

9.3　使用 toga 库

toga 库是一个 Python 原生、操作系统原生的 GUI 工具包。toga 是一个跨平台工具包，可以在 Mac、Windows、Linux（GTK）、Android 系统和 iOS 中使用。本节将详细讲解使用 toga 库开发 GUI 程序的方法。

9.3.1　安装 toga 库并创建第一个 toga 示例

可以使用如下命令安装 toga 库。

```
pip install toga
```

下面的实例文件 toga01.py 演示了使用 toga 创建第一个 GUI 程序的过程。

源码路径：daima\9\9-3\toga01.py

```python
import toga

def button_handler(widget):
    print("hello")

def build(app):
    box = toga.Box()

    button = toga.Button('第一个toga程序', on_press=button_handler)
    button.style.padding = 50
    button.style.flex = 1
    box.add(button)

    return box

def main():
    return toga.App('First App', '书创文化传播', startup=build)

if __name__ == '__main__':
    main().main_loop()
```

在上述代码中，使用一个按钮构建了一个 GUI 应用程序，当按下"第一个 toga 程序"按钮时输出"hello"到控制台。其中函数 button_handler()是一个处理程序，用于处理我们按下按钮时的行为。函数 button_handler()以激活的 widget 作为第一个参数，根据正在处理的事件的类型还可以提供其他参数。然而，在只按下一个简单的按钮的情况下，没有额外的参数。

再来看函数 build(app)，它通过建立一个应用程序创建一个主窗口，在里面有主菜单。然而，toga 不知道我们想要在主窗口中包含什么内容。所以接下来定义一个用于设置在应用程序中包含的 UI 的方法，此方法是可接受的应用程序实例的可调用方法。在此我们设置了一个按钮，设置了在按钮中显示的文本，设置了按钮四周的空白都是 50 像素。

最后通过函数 main()实例化应用程序本身。应用程序有一个名称和唯一的标识符，例如，上述程序设置的是"书创文化传播"，还设置了在窗体中显示的标题是"First App"。

执行后会显示一个拥有"第一个 toga 程序"按钮的标题，单击此按钮后会在控制台中显示设置的提示信息。执行效果如图 9-33 所示。

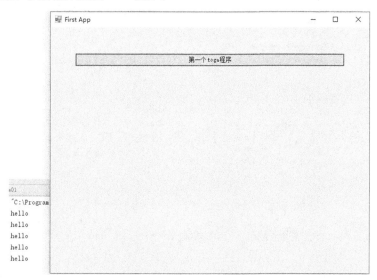

图 9-33　执行效果

9.3.2　使用基本组件

和前面介绍的库一样，toga 库也包含了按钮、文本、绘图和进度条等组件。关于各个组件的具体使用方法，请读者参阅 toga 的官方文档。下面的实例文件 toga02.py 演示了使用 toga 组件创建一个温度转换器的过程。

源码路径：daima\9\9-3\toga02.py

```
import toga
from toga.style.pack import *

def build(app):
    c_box = toga.Box()
    f_box = toga.Box()
    box = toga.Box()

    c_input = toga.TextInput(readonly=True)
    f_input = toga.TextInput()

    c_label = toga.Label('摄氏度', style=Pack(text_align=LEFT))
    f_label = toga.Label('华氏温度', style=Pack(text_align=LEFT))
    join_label = toga.Label('相当于', style=Pack(text_align=RIGHT))

    def calculate(widget):
        try:
            c_input.value = (float(f_input.value) - 32.0) * 5.0 / 9.0
        except:
            c_input.value = '???'

    button = toga.Button('转换', on_press=calculate)

    f_box.add(f_input)
    f_box.add(f_label)
```

```
        c_box.add(join_label)
        c_box.add(c_input)
        c_box.add(c_label)

        box.add(f_box)
        box.add(c_box)
        box.add(button)

        box.style.update(direction=COLUMN, padding_top=10)
        f_box.style.update(direction=ROW, padding=5)
        c_box.style.update(direction=ROW, padding=5)

        c_input.style.update(flex=1)
        f_input.style.update(flex=1, padding_left=160)
        c_label.style.update(width=100, padding_left=10)
        f_label.style.update(width=100, padding_left=10)
        join_label.style.update(width=150, padding_right=10)

        button.style.update(padding=15, flex=1)

        return box

def main():
    return toga.App('温度转换器', '书创文化传播', startup=build)


if __name__ == '__main__':
    main().main_loop()
```

通过上述代码设置了一个垂直堆叠的组件 Box。在这个 Box 中，放置了两个水平 Box 和一个按钮。因为在水平 Box 上没有宽度样式，所以会尝试将它们包含的小部件安装到可用空间中。TextInput 组件将被拉伸以适应可用的水平空间，边距和填充项确保里面的组件将垂直与水平对齐。执行后的效果如图 9-34 所示。

图 9-34 执行效果

9.3.3 使用布局组件

在 toga 库中，实现布局功能的组件有 Box、ScrollContainer、SplitContainer 和 OptionContainer。这些布局组件的用法都十分简单。例如，ScrollContainer（滚动容器）组件类似于 HTML 中的 iframe 或 div 元素，其中包含了一个滚动选择的对象。下面的实例文件 toga03.py 演示了使用组件 ScrollContainer 实现滚动功能的过程。

源码路径：daima\9\9-3\toga03.py

```python
import toga
from toga.style.pack import *

class ScrollContainerApp(toga.App):
    def startup(self):
        self.main_window = toga.MainWindow(self.name)
        box = toga.Box()
        box.style.direction = COLUMN

        for x in range(100):
            label_text = '文本 %d' % (x)
            box.add(toga.Label(label_text, style=Pack(text_align=LEFT)))

        scroller = toga.ScrollContainer()
        scroller.content = box

        self.main_window.content = scroller
        self.main_window.show()

def main():
    return ScrollContainerApp('ScrollContainer', 'toppr.net')


if __name__ == '__main__':
    app = main()
    app.main_loop()
```

通过上述代码创建了一个包含 100 行文本信息的滚动条界面，如图 9-35 所示。

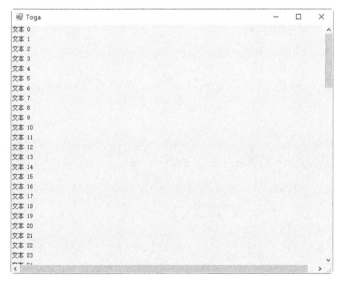

图 9-35　执行效果

9.3.4　使用绘图组件

在创建 GUI 应用程序的过程中，最常用的功能之一是绘制和操作线条、形状、文本和其他图形。在 toga 库中，可以使用组件 Canvas 实现绘图功能。下面的实例文件 toga04.py 演示了使用组件 Canvas 绘制图形的过程。需要注意，本实例只能在 GTK 环境下运行。

源码路径：daima\9\9-3\toga04.py

```python
import toga
from toga.style import Pack
```

```
import toga
from toga.style import Pack

class ExampleCanvasApp(toga.App):
    def startup(self):
        #设置主窗口
        self.main_window = toga.MainWindow(title=self.name, size=(148, 200))
        canvas = toga.Canvas(style=Pack(flex=1))
        box = toga.Box(children=[canvas])
        #在主窗口中添加内容
        self.main_window.content = box
        #显示主窗口
        self.main_window.show()
        with canvas.stroke():
            with canvas.closed_path(50, 50):
                canvas.line_to(100, 100)


def main():
    return ExampleCanvasApp('Canvas', 'toppr.net')


if __name__ == '__main__':
    main().main_loop()
```

9.4 使用 wxPython 库

wxPython 是一个流行的跨平台 GUI 工具包, 由 Robin Dunn 以及 Harri Pasanen 开发, 是一个 Python 扩展模块。

9.4.1 安装并使用 wxPython 库

可以通过如下命令安装 wxPython 库。

```
pip install -U wxPython
```

下面的实例文件 wx01.py 演示了开发第一个 wxPython 程序的过程。

源码路径：daima\9\9-4\wx01.py

```
import wx

app = wx.App()
window = wx.Frame(None, title="书创文化传播", size=(400, 300))
panel = wx.Panel(window)
label = wx.StaticText(panel, label="Hello World", pos=(100, 100))
window.Show(True)
app.MainLoop()
```

执行后的效果如图 9-36 所示。

图 9-36　执行效果

在上述代码中，使用 wx.Frame 类实现了一个窗体效果。wx.Frame 类拥有一个不带参数的默认构造函数，也有以下带参数的重载构造函数。

```
wx.Frame (parent, id, title, pos, size, style, name)
```

各个参数的具体说明如下。
- parent：窗口的父类。
- id：窗口标识，设置为−1 是为了自动生成标识符。
- title：出现在标题栏中的标题。
- pos：帧的开始位置。
- size：窗口的尺寸。
- style：窗口的外观样式。
- name：对象的内部名称。

在 Frame 类中包含如下内置函数。
- CreateStatusBar()：创建窗口底部状态栏。
- CreateToolBar()：创建位于窗口顶部或左侧的工具栏。
- GetMenuBar()：获取引用菜单栏。
- GetStatusBar()：获取引用状态栏。
- SetMenuBar()：在帧中显示菜单栏对象。
- setStatusBar()：关联状态栏对象到框架。
- SetToolBar()：关联工具栏对象到框架。
- SetStatusText()：设置在状态栏上显示的文字。
- Create()：创建提供参数的框架。
- Centre()：设置帧在中心显示。
- SetPosition()：设置在指定的屏幕坐标显示帧。
- SetSize()：根据给定的尺寸调整框架大小。
- SetTitle()：插入指定文本到标题栏中。

9.4.2 基本组件

1. StaticText 组件

在 wxPython 中，wx.StaticText 类对象实现了只读文本组件的功能。wx.StaticText 类的构造函数如下。

```
Wx.StaticText(parent, id, label, position, size, style)
```

StaticText（文本）组件中的预定义样式枚举器如表 9-1 所示。

表 9-1　　　　　　　　　　预定义样式枚举器

枚举器	控制标签的大小作用及左对齐
wx.ALIGN_RIGHT	控制标签的大小及右对齐
wx.ALIGN_CENTER	控制标签的大小及居中对齐
wx.ST_NO_AUTORESIZE	防止标签自动调整大小
wx.ST_ELLIPSIZE_START	如果文本的大小大于标签的尺寸，省略号显示在开始
wx.ST_ELLIPSIZE_MIDDLE	如果文本的大小大于标签的尺寸，省略号显示在中间
wx.ST_ELLIPSIZE_END	如果文本的大小大于标签的尺寸，省略号显示在结尾

为了设置 StaticText 中的字体，首先需要创建一个字体对象，代码如下。

```
wx.Font(pointsize, fontfamily, fontstyle, fontweight)
```

参数 fontfamily 有如下 3 个值。
- wx.FONTSTYLE_NORMAL：字体风格不使用倾斜。
- wx.FONTSTYLE_ITALIC：字体风格是斜体。
- wx.FONTSTYLE_SLANT：字体风格是倾斜，但以罗马字体风格倾斜。

参数 fontweight 有如下 3 个值。
- wx.FONTWEIGHT_NORMAL：表示普通字体。
- wx.FONTWEIGHT_LIGHT：表示高亮字体。
- wx.FONTWEIGHT_BOLD：表示粗体。

下面的实例文件 wx02.py 演示了使用 StaticText 组件在窗体中显示文本的过程。

源码路径：daima\9\9-4\wx02.py

```python
import wx
class Mywin(wx.Frame):
    def __init__(self, parent, title):
        super(Mywin, self).__init__(parent, title=title, size=(600, 200))
        panel = wx.Panel(self)
        box = wx.BoxSizer(wx.VERTICAL)
        lbl = wx.StaticText(panel, -1, style=wx.ALIGN_CENTER)

        txt1 = "Python GUI development"
        txt2 = "using wxPython"
        txt3 = " Python port of wxWidget "
        txt = txt1 + "\n" + txt2 + "\n" + txt3

        font = wx.Font(18, wx.ROMAN, wx.ITALIC, wx.NORMAL)
        lbl.SetFont(font)
        lbl.SetLabel(txt)

        box.Add(lbl, 0, wx.ALIGN_CENTER)
        lblwrap = wx.StaticText(panel, -1, style=wx.ALIGN_RIGHT)
        txt = txt1 + txt2 + txt3

        lblwrap.SetLabel(txt)
        lblwrap.Wrap(200)
        box.Add(lblwrap, 0, wx.ALIGN_LEFT)

        lbl1 = wx.StaticText(panel, -1, style=wx.ALIGN_LEFT | wx.ST_ELLIPSIZE_MIDDLE)
        lbl1.SetLabel(txt)
        lbl1.SetForegroundColour((255, 0, 0))
        lbl1.SetBackgroundColour((0, 0, 0))

        font = self.GetFont()
        lbl1.SetFont(font)

        box.Add(lbl1, 0, wx.ALIGN_LEFT)
        panel.SetSizer(box)
        self.Centre()
        self.Show()

app = wx.App()
Mywin(None, 'toppr')
app.MainLoop()
```

执行后的效果如图 9-37 所示。

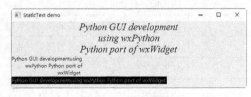

图 9-37 执行效果

9.4 使用 wxPython 库

2. TextCtrl（文本框）组件

在 wxPython 中，wx.TextCtrl 类实现文本框组件功能，用于显示文本和编辑文本。TextCtrl 中的文本可以是单行、多行或密码字段。TextCtrl 类的构造函数形式如下：

```
wx.TextCtrl(parent, id, value, pos, size, style)
```

其中，样式参数 style 接受一个或多个常量，具体说明如下。

- wx.TE_MULTILINE：允许显示多行文本，如果未指定样式，则能在该控件值中使用换行字符。
- wx.TE_PASSWORD：文本将会显示为星号。
- wx.TE_READONLY：文本将不可编辑。
- wxTE_LEFT：设置文本左对齐（默认）。
- wxTE_CENTRE：设置文本居中对齐。
- wxTE_RIGHT：设置文本右对齐。

下面的实例文件 wx03.py 演示了使用 TextCtrl 组件创建 4 种不同样式的文本框的过程。

源码路径：daima\9\9-4\wx03.py

```python
class Mywin(wx.Frame):
    def __init__(self, parent, title):
        super(Mywin, self).__init__(parent, title=title, size=(350, 250))

        panel = wx.Panel(self)
        vbox = wx.BoxSizer(wx.VERTICAL)

        hbox1 = wx.BoxSizer(wx.HORIZONTAL)
        l1 = wx.StaticText(panel, -1, "文本域")

        hbox1.Add(l1, 1, wx.EXPAND | wx.ALIGN_LEFT | wx.ALL, 5)
        self.t1 = wx.TextCtrl(panel)

        hbox1.Add(self.t1, 1, wx.EXPAND | wx.ALIGN_LEFT | wx.ALL, 5)
        self.t1.Bind(wx.EVT_TEXT, self.OnKeyTyped)
        vbox.Add(hbox1)

        hbox2 = wx.BoxSizer(wx.HORIZONTAL)
        l2 = wx.StaticText(panel, -1, "密码文本")

        hbox2.Add(l2, 1, wx.ALIGN_LEFT | wx.ALL, 5)
        self.t2 = wx.TextCtrl(panel, style=wx.TE_PASSWORD)
        self.t2.SetMaxLength(5)

        hbox2.Add(self.t2, 1, wx.EXPAND | wx.ALIGN_LEFT | wx.ALL, 5)
        vbox.Add(hbox2)
        self.t2.Bind(wx.EVT_TEXT_MAXLEN, self.OnMaxLen)

        hbox3 = wx.BoxSizer(wx.HORIZONTAL)
        l3 = wx.StaticText(panel, -1, "多行文本")

        hbox3.Add(l3, 1, wx.EXPAND | wx.ALIGN_LEFT | wx.ALL, 5)
        self.t3 = wx.TextCtrl(panel, size=(200, 100), style=wx.TE_MULTILINE)

        hbox3.Add(self.t3, 1, wx.EXPAND | wx.ALIGN_LEFT | wx.ALL, 5)
        vbox.Add(hbox3)
        self.t3.Bind(wx.EVT_TEXT_ENTER, self.OnEnterPressed)

        hbox4 = wx.BoxSizer(wx.HORIZONTAL)
        l4 = wx.StaticText(panel, -1, "只读取文本")

        hbox4.Add(l4, 1, wx.EXPAND | wx.ALIGN_LEFT | wx.ALL, 5)
        self.t4 = wx.TextCtrl(panel, value="只读文本", style=wx.TE_READONLY | wx.TE_CENTER
```

```
                hbox4.Add(self.t4, 1, wx.EXPAND | wx.ALIGN_LEFT | wx.ALL, 5)
                vbox.Add(hbox4)
                panel.SetSizer(vbox)
                self.Centre()
                self.Show()
                self.Fit()

        def OnKeyTyped(self, event):
            print(event.GetString())

        def OnEnterPressed(self, event):
            print("Enter pressed")

        def OnMaxLen(self, event):
            print("Maximum length reached")
```

在上述代码中,将 wx.TextCtrl 类的 4 个对象放置在窗体中。第一个是普通的文本框,第二个是一个密码字段,第三个是多行文本框,最后一个是不可编辑的文本框。当第一个文本框 EVT_TEXT 绑定器触发 OnKeyTyped()方法时,可以处理每个单击按键事件。设置第二个文本框的最大长度为 5,一旦用户试图输入超过 500 个字符,EVT_TEXT_MAXLEN 绑定器会运行 OnMaxLen()函数。因为设置 EVT_TEXT_ENTER 绑定器,所以多行文本框会响应按下 Enter 键的事件。执行效果如图 9-38 所示。

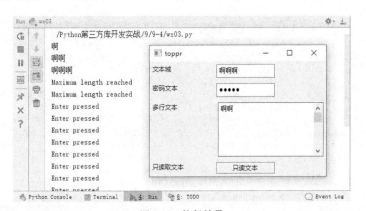

图 9-38 执行效果

3. RadioButton 组件

在 wxPython 中,wx.RadioButton 类实现单选按钮组件功能,类的一个对象会在旁边带着一个圆形按钮和一个文本标签。为了创建一组相互可选择的按钮,首先将 wx.RadioButton 对象的样式参数设置为 wx.RB_GROUP,后续按钮对象会被添加到一组中。

在 wxPython 的 API 中还包括 wx.RadioBox 类,此类的对象提供了一个边框和标签组,组中的按钮可以水平或垂直排列。wx.RadioButton 类构造如下。

```
wx.RadioButton(parent, id, label, pos, size, style)
```

参数 style 仅用于该组中的第一个按钮,它的值是 wx.RB_GROUP。对于组中随后的按钮,wx.RB_SINGLE 的 style 参数可以任选。每当单击任意组中的按钮时,wx.RadioButton 事件绑定器 wx.EVT_RADIOBUTTON 触发相关的处理程序。

在 wx.RadioButton 类中包含以下两个重要的方法。

- setValue():选择或取消选择按钮编程。
- getValue():如果选择一个按钮,则返回 true;否则,返回 false。

wx.RadioBox 类把相互排斥的按钮放在一个静态框中,该组中的每个按钮将从列表对象获取其标签,作为选择 wx.RadioBox 构造函数的参数。在 wx.RadioBox 类中包含如下内置方法。

- GetSelection()：返回所选项目的索引。
- SetSelection()：选择编程项目。
- GetString()：返回选定项的标签。
- SetString()：分配标签到所选择的项目。
- Show()：显示或隐藏指定索引的项目。

下面的实例文件 wx04.py 演示了使用 RadioBox 以及单选按钮的过程。

源码路径：daima\9\9-4\wx04.py

```python
class Example(wx.Frame):

    def __init__(self, parent, title):
        super(Example, self).__init__(parent, title=title, size=(300, 200))
        self.InitUI()

    def InitUI(self):
        pnl = wx.Panel(self)
        self.rb1 = wx.RadioButton(pnl, 11, label='Value A',
                            pos=(10, 10), style=wx.RB_GROUP)
        self.rb2 = wx.RadioButton(pnl, 22, label='Value B', pos=(10, 40))
        self.rb3 = wx.RadioButton(pnl, 33, label='Value C', pos=(10, 70))
        self.Bind(wx.EVT_RADIOBUTTON, self.OnRadiogroup)
        lblList = ['Value X', 'Value Y', 'Value Z']
        self.rbox = wx.RadioBox(pnl, label='RadioBox', pos=(80, 10), choices=lblList,
                            majorDimension=1, style=wx.RA_SPECIFY_ROWS)
        self.rbox.Bind(wx.EVT_RADIOBOX, self.onRadioBox)
        self.Centre()
        self.Show(True)

    def OnRadiogroup(self, e):
        rb = e.GetEventObject()
        print(rb.GetLabel(), ' is clicked from Radio Group')

    def onRadioBox(self, e):
        print(self.rbox.GetStringSelection(), ' is clicked from Radio Box')
```

在上述代码中，首先将 3 个单选按钮通过指定 wx.RB_GROUP 样式分组并放置在面板上。然后在 RadioBox 中从 lblList[]对象中读出标签按钮。接下来，设置两个事件绑定器，声明一个单选按钮组和其他的 RadioBox。最后，通过相应的事件处理程序确定被选择的按钮，并在控制台窗口中显示消息。

执行后的效果如图 9-39 所示。

图 9-39　执行效果

注意：在 wxPython 中还包含很多其他重要的组件，并且 wxPython 也还具有很多其他重要的知识，如界面布局、绘图、对话框和拖放处理等功能。了解详细内容，建议读者参阅官方文档。

第 10 章

数据可视化

数据可视化是指通过可视化的方式来探索数据,它与时下比较热门的数据挖掘紧密相关,而数据挖掘指的是使用代码来探索数据集的规律和关联。在软件开发领域,以图像的方式来完美呈现数据,可以让浏览者明白其中的含义,发现数据集中包含的规律和意义。本章将详细讲解使用 Python 实现数据可视化的核心知识。

10.1 使用 Matplotlib 库

Matplotlib 库是 Python 中最著名的数据可视化工具包,通过使用 Matplotlib 库,可以非常方便地实现和数据统计相关的图形,如折线图、散点图、直方图等。正因为 Matplotlib 库在绘图领域的强大功能,所以它在 Python 数据挖掘方面得到了重用。本节将详细讲解在 Python 中使用 Matplotlib 绘制统计图形的知识,为读者学习后面的知识打下基础。

10.1.1 搭建 Matplotlib 库的使用环境

在 Python 程序中使用 Matplotlib 库之前,需要先确保安装了 Matplotlib 库。在 Windows 系统中安装 Matplotlib 库之前,首先需要确保已经安装了 Visual Studio.NET。安装了 Visual Studio.NET 后,就可以安装 Matplotlib 了,其中最简单的安装方式是使用"pip"命令或"easy_install"命令。

```
easy_install matplotlib
pip install matplotlib
```

虽然上述两种安装方式比较简单,但是它们并不能保证安装的 Matplotlib 适合新版本的 Python。例如,作者在写作本书时使用的是 Python 3.6。而当时使用上述两个命令只能自动安装 Matplotlib 1.7,并不支持 Python 3.6。建议读者登录 PyPI 网站,如图 10-1 所示,在这个页面中查找与你使用的 Python 版本匹配的 wheel 文件(扩展名为".whl"的文件)。例如,如果使用的是 64 位的 Python 3.6,则需要下载 matplotlib-2.0.0rc2-cp36-cp36m-win_amd64.whl。

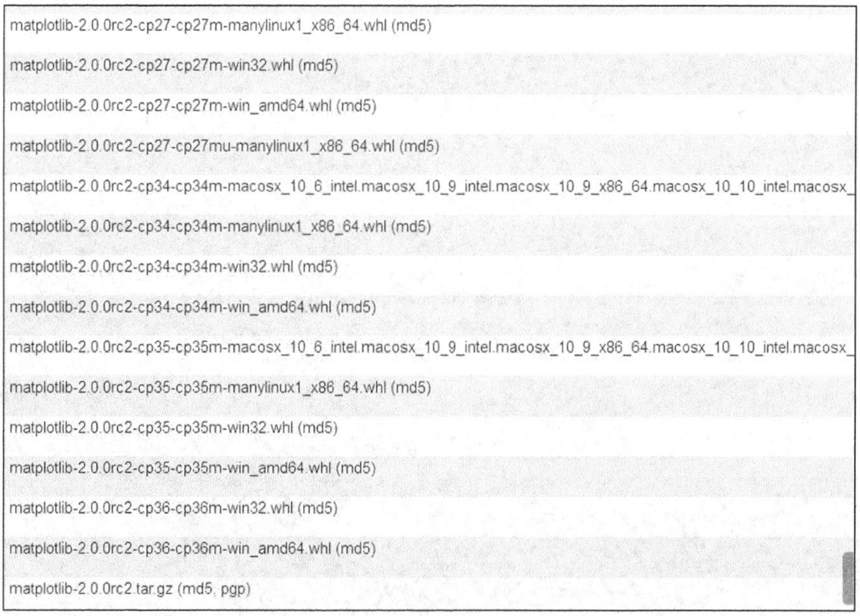

图 10-1 PyPI 网站提供 wheel 的文件

> 注意:如果登录 PyPI 网站找不到适合自己的 Matplotlib,还可以尝试登录 LFD 网站,如图 10-2 所示。这个网站发布安装程序的时间通常比 Matplotlib 官网早。

第 10 章 数据可视化

```
Matplotlib, a 2D plotting library.
Requires numpy, dateutil, pytz, pyparsing, cycler, setuptools
ffmpeg, mencoder, avconv, or imagemagick.
matplotlib‑1.5.3‑cp27‑cp27m‑win32.whl
matplotlib‑1.5.3‑cp27‑cp27m‑win_amd64.whl
matplotlib‑1.5.3‑cp34‑cp34m‑win32.whl
matplotlib‑1.5.3‑cp34‑cp34m‑win_amd64.whl
matplotlib‑1.5.3‑cp35‑cp35m‑win32.whl
matplotlib‑1.5.3‑cp35‑cp35m‑win_amd64.whl
matplotlib‑1.5.3‑cp36‑cp36m‑win32.whl
matplotlib‑1.5.3‑cp36‑cp36m‑win_amd64.whl
matplotlib‑1.5.3.chm
matplotlib‑2.0.0rc2‑cp27‑cp27m‑win32.whl
matplotlib‑2.0.0rc2‑cp27‑cp27m‑win_amd64.whl
matplotlib‑2.0.0rc2‑cp34‑cp34m‑win32.whl
matplotlib‑2.0.0rc2‑cp34‑cp34m‑win_amd64.whl
matplotlib‑2.0.0rc2‑cp35‑cp35m‑win32.whl
matplotlib‑2.0.0rc2‑cp35‑cp35m‑win_amd64.whl
matplotlib‑2.0.0rc2‑cp36‑cp36m‑win32.whl
matplotlib‑2.0.0rc2‑cp36‑cp36m‑win_amd64.whl
matplotlib‑2.0.0rc2.chm
matplotlib‑2.x‑windows‑link‑libraries.zip
matplotlib_tests‑1.5.3‑py2.py3‑none‑any.whl
matplotlib_tests‑2.0.0rc2‑py2.py3‑none‑any.whl
```

图 10-2　LFD 网站提供的文件

例如，作者当时下载到的文件是 matplotlib-2.0.0rc2-cp36-cp36m-win_amd64.whl。将这个文件保存在 "H:\matp" 目录下，然后打开一个命令窗口，并切换到该项目文件夹 "H:\matp"，使用如下所示的 "pip" 命令来安装 Matplotlib。

```
python -m pip install --user matplotlib-2.0.0rc2-cp36-cp36m-win_amd64.whl
```

具体安装过程如图 10-3 所示。

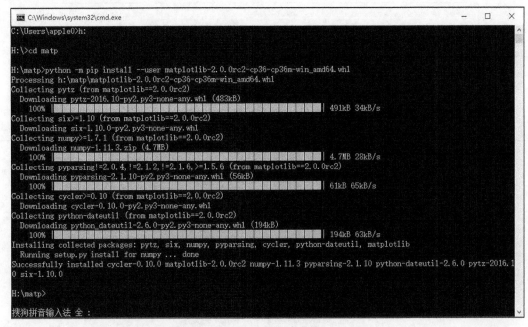

图 10-3　在 Windows 系统中安装 Matplotlib

10.1.2　初级绘图

在使用 Matplotlib 绘制图形时，有两个最常用的场景，一个是画点，另一个是画线。本节将详细讲解使用 Matplotlib 实现初级绘制的知识。

10.1 使用 Matplotlib 库

1. 绘制点

假设你有一堆的数据样本，要找出其中的异常值，最直观的方法就是将它们画成散点图。下面的实例文件 dian.py 演示了使用 Matplotlib 绘制散点图的过程。

源码路径：daima\10\10-1\dian.py

```
import matplotlib.pyplot as plt      #导入pyplot包，并缩写为plt
#定义两个点的x集合和y集合
x=[1,2]
y=[2,4]
plt.scatter(x,y)                     #绘制散点图
plt.show()                           #展示绘画框
```

在上述实例代码中绘制了拥有两个点的散点图，向函数 scatter()传递了两个分别包含 *x* 值和 *y* 值的列表。执行效果如图 10-4 所示。

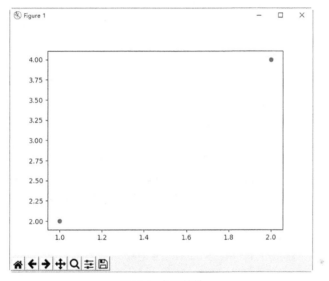

图 10-4　执行效果

在上述实例中，可以进一步调整坐标轴的样式，例如，可以加上如下代码。

```
#[]里的4个参数分别表示x轴起始点、x轴结束点、y轴起始点、y轴结束点
plt.axis([0,10,0,10])
```

2. 绘制折线

在使用 Matplotlib 绘制线形图时，最简单的是绘制折线图。在下面的实例文件 zhe.py 中，使用 Matplotlib 绘制了一个简单的折线图，并对折线样式进行了定制，这样可以实现复杂数据的可视化效果。

源码路径：daima\10\10-1\zhe.py

```
import matplotlib.pyplot as plt
squares = [1, 4, 9, 16, 25]
plt.plot(squares)
plt.show()
```

在上述实例代码中，使用平方数序列 1、4、9、16 和 25 来绘制一个折线图，在具体实现时，只要向 Matplotlib 提供这些平方数序列就能完成绘制工作。具体步骤如下。

（1）导入模块 pyplot，并给它指定别名 plt，以免反复输入 pyplot，在模块 pyplot 中包含了很多用于生成图表的函数。

（2）创建一个列表，在其中存储了前述平方数。

（3）将创建的列表传递给函数 plot()，这个函数会根据这些数字绘制出有意义的图形。

（4）通过函数 plt.show()打开 Matplotlib 查看器，并显示绘制的图形。

执行效果如图 10-5 所示。

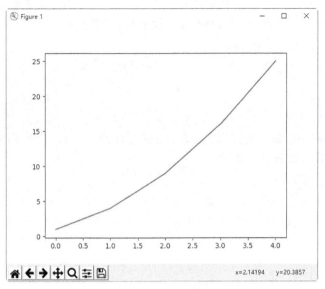

图 10-5　执行效果

3．设置标签文字和线条粗细

前面实例的界面效果不够完美，开发者可以对绘制的线条样式进行灵活设置。例如，可以设置线条的粗细、实现数据准确性校正等操作。下面的实例文件 she.py 演示了使用 Matplotlib 绘制指定样式折线图的过程。

源码路径：daima\10\10-1\she.py

```
import matplotlib.pyplot as plt          #导入模块
input_values = [1, 2, 3, 4, 5]
squares = [1, 4, 9, 16, 25]
plt.plot(input_values, squares, linewidth=5)
#设置标题，并在坐标轴上添加标签
plt.title("Numbers", fontsize=24)
plt.xlabel("Value", fontsize=14)
plt.ylabel("ARG Value", fontsize=14)
#设置单位刻度的大小
plt.tick_params(axis='both', labelsize=14)
plt.show()
```

关于上述代码，注意以下几点。

- 第 4 行代码中的"linewidth=5"。设置线条的粗细。
- 第 4 行代码中的函数 plot()。当向函数 plot()提供一系列数字时，它会假设第一个数据点对应的 x 坐标值为 0，但是实际上第一个点对应的 x 值为 1。为改变这种默认行为，可以给函数 plot()同时提供输入值和输出值，这样函数 plot()就可以正确地绘制数据。因为同时提供了输入值和输出值，所以无须对输出值的生成方式进行假设，同时最终绘制出的图形是正确的。
- 第 6 行代码中的函数 title()。设置图的标题。
- 第 6 行到第 8 行中的参数 fontsize。设置图中的文字大小。
- 第 7 行中的函数 xlabel()和第 8 行中的函数 ylabel()。分别设置 x 轴的标题和 y 轴的标题。
- 第 10 行中的函数 tick_params()。设置刻度样式，其中指定的实参将影响 x 轴和 y 轴上的刻度（axis='both'），并将刻度标记的字体大小设置为 14。

执行效果如图 10-6 所示。

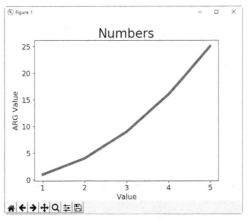

图 10-6 执行效果

10.1.3 自定义散点图样式

在现实应用中，经常需要绘制散点图并设置各个数据点的样式。例如，以一种颜色显示较小的值，而用另一种颜色显示较大的值。当绘制大型数据集时，还需要对每个点都设置同样的样式，再使用不同的样式选项重新绘制某些点，这样可以突出显示它们的效果。在 Matplotlib 库中，可以使用函数 scatter() 绘制单个点，通过传递 x 和 y 坐标的方式在指定的位置绘制一个点。

下面的实例文件 dianyang.py 演示了使用 Matplotlib 绘制指定样式散点图的过程。

源码路径：daima\10\10-1\dianyang.py

```
import matplotlib.pyplot as plt
from pylab import *
mpl.rcParams['font.sans-serif'] = ['SimHei']      #指定默认字体
mpl.rcParams['axes.unicode_minus'] = False        #解决保存图像时负号显示为方块的问题
x_values = list(range(1, 1001))
y_values = [x**2 for x in x_values]
plt.scatter(x_values, y_values, c=(0, 0, 0.8), edgecolor='none', s=40)
#设置图的标题，并设置坐标轴标签
plt.title("某地区销售统计表", fontsize=24)
plt.xlabel("节点", fontsize=14)
plt.ylabel("销售数据", fontsize=14)
#设置刻度大小
plt.tick_params(axis='both', which='major', labelsize=14)
#设置每个坐标轴的取值范围
plt.axis([0, 110, 0, 1100])
plt.show()
```

关于上述代码，注意以下几点。

- 第 2~4 行代码。导入字体库，设置中文字体，并解决负号显示为方块的问题。
- 第 5 行和第 6 行代码。使用 Python 循环实现自动计算功能。首先创建了一个包含 x 值的列表，其中包含数字 1~1000。接下来创建一个生成 y 值的列表解析，它能够遍历 x 值（for x in x_values），计算其平方值（$x**2$），并将结果存储到列表 y_values 中。
- 第 7 行代码。将输入列表和输出列表传递给函数 scatter()。另外，因为 Matplotlib 允许给散列点图中的各个点设置一个颜色，默认为蓝色点和黑色轮廓，所以当在散列点图中包含的数据点不多时效果会很好。但是当需要绘制很多个点时，这些黑色的轮廓可能会重叠在一起，因此需要删除数据点的轮廓。于是，在本行代码中，在调用函数 scatter() 时传递了实参 edgecolor='none'。为了修改数据点的颜色，在此向函数 scatter() 传递参数

c，并将其设置为要使用的颜色名称"red"。

🌲 注意：颜色映射（Colormap）是一系列颜色，它们从起始颜色渐变到结束颜色。在可视化视图模型中，颜色映射用于突出数据的规律，例如，可能需要用较浅的颜色来显示较小的值，用较深的颜色来显示较大的值。在模块 Pyplot 中内置了一组颜色映射，要使用这些颜色映射，需要告诉 Pyplot 应该如何设置数据集中每个点的颜色。

- 第 15 行代码。因为这个数据集较大，所以将点设置得较小。在本行代码中使用函数 axis() 指定了每个坐标轴的取值范围。函数 axis() 要求提供 4 个值——x 和 y 坐标轴的最小值与最大值。此处将 x 坐标轴的取值范围设置为 0～110，并将 y 坐标轴的取值范围设置为 0～1100。
- 第 16 行（最后一行）代码。使用函数 plt.show() 显示绘制的图形。当然，也可以让程序自动将图表保存到一个文件中，此时只须将对 plt.show() 函数的调用替换为对 plt.savefig() 函数的调用即可。

```
plt.savefig (' plot.png' , bbox_inches='tight' )
```

在上述代码中，第一个实参用于指定要以什么样的文件名保存图，这个文件将存储到当前实例文件 dianyang.py 所在的目录中。第二个实参用于指定将图多余的空白区域裁剪掉。如果要保留图表周围多余的空白区域，可省略这个实参。

执行效果如图 10-7 所示。

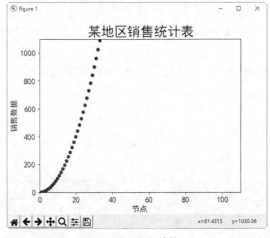

图 10-7　执行效果

10.1.4　绘制柱状图

在现实应用中，柱状图经常用于数据统计领域。在 Python 程序中，使用 Matplotlib 很容易绘制柱状图。只须使用下面的 3 行代码就可以绘制一个柱状图。

```
import matplotlib.pyplot as plt
plt.bar(left = 0,height = 1)
plt.show()
```

在上述代码中，首先使用 import 导入了 matplotlib.pyplot，然后直接调用其 bar() 函数绘制柱状图，最后用 show() 函数显示图像。其中在函数 bar() 中存在如下两个参数。

- left：柱状图左边缘的位置，如果指定为 1，那么当前柱状图左边缘的 x 值就是 1.0。
- height：柱状图的高度，也就是 y 轴的值。

执行上述代码后会绘制一个柱状图，如图 10-8 所示。

10.1 使用 Matplotlib 库

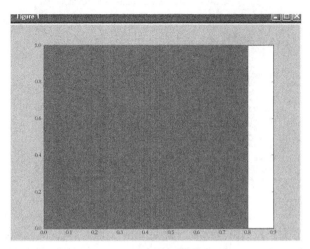

图 10-8 执行效果

虽然通过上述代码绘制了一个柱状图，但是效果不够直观。在绘制函数 bar()中，参数 left 和 height 除了可以使用单独的值（此时是一个柱形）外，还可以使用元组来替换（此时代表多个柱状条）。下面的实例文件 zhu.py 演示了使用 Matplotlib 绘制多个柱状图效果的过程。

源码路径：daima\10\10-1\zhu.py

```
import matplotlib.pyplot as plt        #导入模块
plt.bar(left = (0,1),height = (1,0.5)) #绘制两个柱形图
plt.show()                              #显示绘制的图
```

执行效果如图 10-9 所示。

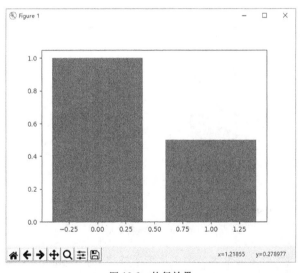

图 10-9 执行效果

在上述实例代码中，left = (0,1)的意思是共有两个柱状条。其中第一个的左边缘为 0，第二个的左边缘为 1。参数 height 的含义也类似。当然，此时有的读者可能觉得这两个柱状条"太宽"了，不够美观。可以通过指定函数 bar()中的 width 参数来设置它们的宽度。例如，通过下面的代码设置柱状条的宽度，执行效果如图 10-10 所示。

```
import matplotlib.pyplot as plt
plt.bar(left = (0,1),height = (1,0.5),width = 0.35)
plt.show()
```

347

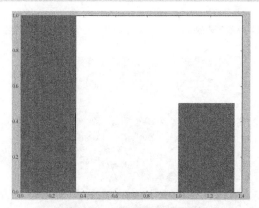

图 10-10　设置柱状图的宽度

这时可能有的读者会问："如何标明 *x* 轴和 *y* 轴的信息,如使用 *x* 轴表示性别,使用 *y* 轴表示人数？"下面的实例文件 shuo.py 演示了使用 Matplotlib 绘制有说明信息的柱状图的过程。

源码路径：daima\10\10-1\shuo.py

```
import matplotlib.pyplot as plt
from pylab import *
mpl.rcParams['font.sans-serif'] = ['SimHei']        #指定默认字体
mpl.rcParams['axes.unicode_minus'] = False          #解决保存图像时负号显示为方块的问题
plt.xlabel(u'性别')      #x轴的说明信息
plt.ylabel(u'人数')      #y轴的说明信息
plt.bar(left = (0,1),height = (1,0.5),width = 0.35)
plt.show()
```

上述代码执行后的效果如图 10-11 所示。

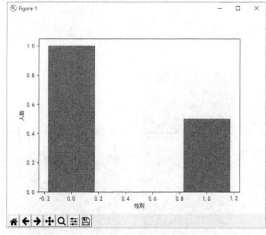

图 10-11　执行效果

> 注意：在 Python 2.7 中使用中文时一定要用字符"u",Python 3.0 以上不用。

接下来可以对 *x* 轴上的每个柱状条进行说明,例如,设置第一个柱状条是"男",第二个柱状条是"女"。此时可以通过以下代码实现。

```
plt.xlabel(u'性别')
plt.ylabel(u'人数')
plt.xticks((0,1),(u'男',u'女'))
plt.bar(left = (0,1),height = (1,0.5),width = 0.35)
plt.show()
```

在上述代码中,函数 plt.xticks()的用法和前面使用的 left 与 height 的用法差不多。你有几个柱状条,函数 plt.xticks()就对应几维的元组,其中第一个参数表示文字的位置,第二个参数表示具体的文字说明。不过这里有一个问题：有时我们指定的位置有些"偏移"，最理想的状态应该在每个柱状条的中间。可以通过直接指定函数 bar()里面的 align="center"让文字居中。

```
plt.xlabel(u'性别')
plt.ylabel(u'人数')
plt.xticks((0,1),(u'男',u'女'))
plt.bar(left = (0,1),height = (1,0.5),width = 0.35,align="center")
plt.show()
```

执行效果如图 10-12 所示。

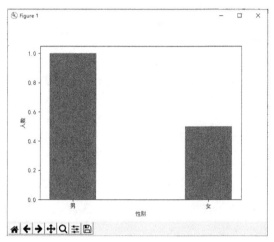

图 10-12　执行效果

接下来，可以通过如下代码给柱状图添加一个标题。

```
plt.title(u"性别比例分析")
```

为了使整个程序显得更加科学合理，我们通过如下代码设置一个图例。

```
plt.xlabel(u'性别')
plt.ylabel(u'人数')
plt.title(u"性别比例分析")
plt.xticks((0,1),(u'男',u'女'))
rect = plt.bar(left = (0,1),height = (1,0.5),width = 0.35,align="center")
plt.legend((rect,),(u"图例",))
plt.show()
```

在上述代码中用到了函数 legend()，里面的参数必须是元组。即使只有一个图例，它也必须是元组；否则，显示不正确。执行效果如图 10-13 所示。

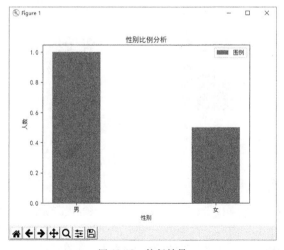

图 10-13　执行效果

接下来，还可以在每个柱状条的上面标注对应的 y 值，这需要使用以下通用的方法实现。

```
def autolabel(rects):
    for rect in rects:
        height = rect.get_height()
        plt.text(rect.get_x()+rect.get_width()/2., 1.03*height, '%s' % float(height))
```

在上述实例代码中，plt.text 有 3 个参数，分别是 x 坐标、y 坐标和要显示的文字。调用函数 autolabel() 的具体实现代码如下所示。

```
autolabel(rect)
```

为了避免绘制的柱状图紧靠着顶部，最好能够空出一段距离，这可以通过函数 bar() 的属性参数 yerr 来设置。一旦设置了这个参数，对应的柱状条上面就会有一条竖着的线。当把 yerr 这个值设置得很小时，上面的空白就自动空出来了。

```
rect = plt.bar(left = (0,1),height = (1,0.5),width = 0.35,align="center",yerr=0.0001)
```

到此，一个比较美观的柱状图就绘制完毕了，将代码整理并保存在以下实例中。实例文件 xinxi.py 的具体实现代码如下。

源码路径：daima\10\10-1\xinxi.py

```
import matplotlib.pyplot as plt
from pylab import *
mpl.rcParams['font.sans-serif'] = ['SimHei']    #指定默认字体
mpl.rcParams['axes.unicode_minus'] = False      #解决保存图像时负号'-'显示为方块的问题
def autolabel(rects):
    for rect in rects:
        height = rect.get_height()
        plt.text(rect.get_x()+rect.get_width()/2., 1.03*height, '%s' % float(height))
plt.xlabel(u'性别')
plt.ylabel(u'人数')
plt.title(u"性别比例分析")
plt.xticks((0,1),(u'男',u'女'))
#绘制柱形图
rect = plt.bar(left = (0,1),height = (1,0.5),width = 0.35,align="center",yerr=0.0001)
plt.legend((rect,),(u"图例",))
autolabel(rect)
plt.show()
```

上述代码执行后的效果如图 10-14 所示。

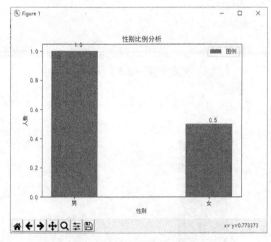

图 10-14　执行效果

10.1.5　绘制多幅子图

在 Matplotlib 绘图系统中，可以显式地控制图像、子图和坐标轴。Matplotlib 中的"图像"指的是用户界面看到的整个窗口内容。在图像里面有"子图"。子图的位置是由坐标网格确定的，而"坐标轴"却不受此限制，可以放在图像的任意位置。当调用 plot() 函数的时候，Matplotlib 调用 gca()函数以及 gcf()函数来获取当前的坐标轴和图像。如果无法获取图像，则会调用 figure()函数来创建一个。从严格意义上来说，这相当于使用 subplot(1,1,1) 创建一个只有一个子图的图像。

在 Matplotlib 绘图系统中，"图像"就是 GUI 里面以"Figure #"为标题的窗口。图像编号从"1"开始，与 MATLAB 中的风格一致，而与 Python 中从"0"开始编号的风格不同。表 10-1 中的参数是图像的属性。

表 10-1　　　　　　　　　　　　　　图像的属性

参数	默认值	描述
num	1	图像的数量
figsize	figure.figsize	图像的长和宽（英寸）
dpi	figure.dpi	分辨率（点/英寸）
facecolor	figure.facecolor	绘图区域的背景颜色
edgecolor	figure.edgecolor	绘图区域边缘的颜色
frameon	True	表示是否绘制图像边缘

在图形界面中可以单击右上角的"×"来关闭窗口（OS X 系统中是左上角）。Matplotlib 也提供了名为 close()的函数来关闭这个窗口。函数 close()的具体行为取决于以下参数。

- 不传递参数：关闭当前窗口。
- 作为参数传递窗口编号或窗口实例（instance）：关闭指定的窗口。
- all：关闭所有窗口。

和其他对象一样，开发者可以使用 setp 或者 set_something 这样的方法来设置图像的属性。

下面的实例文件 lia.py 演示了让一幅折线图和一幅散点图同时出现在一个绘画框中的过程。

源码路径：daima\10\10-1\lia.py

```
import matplotlib.pyplot as plt  #将绘画框进行对象化
fig=plt.figure()  #将p1定义为绘画框的子图，211表示将绘画框划分为2行1列，最后的1表示第一幅图
p1=fig.add_subplot(211)
x=[1,2,3,4,5,6,7,8]
y=[2,1,3,5,2,6,12,7]
p1.plot(x,y)  #将p2定义为绘画框的子图，212表示将绘画框划分为2行1列，最后的2表示第二幅图
p2=fig.add_subplot(212)
a=[1,2]
b=[2,4]
p2.scatter(a,b)
plt.show()
```

上述代码执行后的效果如图 10-15 所示。

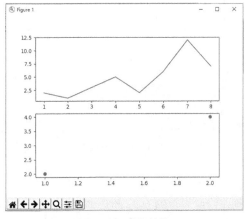

图 10-15　执行效果

在 Python 程序中，如果需要同时绘制多幅图，可以给 figure 传递一个整数参数来指定图的序号。如果指定序号的绘图对象已经存在，将不会创建新的对象，而只是让它成为当前绘图对象。例如下面的演示代码。

```
fig1 = pl.figure(1)
pl.subplot(211)
```

在上述代码中，代码"subplot(211)"把绘图区域等分为 2 行 1 列（这共有两个区域），然后在区域 1（上区域）中创建一个轴对象。代码"pl.subplot(212)"在区域 2（下区域）中创建一个轴对象。

当绘图对象中有多条轴的时候，可以通过工具栏中的 Configure Subplots 按钮，交互式地调节各个轴之间的间距和轴与边框之间的距离。如果希望在程序中调节，可以调用 subplots_adjust() 函数。此函数有 left、right、bottom、top、wspace 和 hspace 等关键字参数，这些参数的值都是 0~1 的小数，它们是以绘图区域的宽高比为 1 进行归一化之后的坐标或长度。例如下面的演示代码。

```
pl.subplots_adjust(left=0.08, right=0.95, wspace=0.25, hspace=0.45)
```

下面的实例文件 liazhe.py 演示了在一个坐标系中绘制两幅折线图的过程。

源码路径：daima\10\10-1\liazhe.py

```
import numpy as np
import pylab as pl
x1 = [1, 2, 3, 4, 5]
y1 = [1, 4, 9, 16, 25]
x2 = [1, 2, 4, 6, 8]
y2 = [2, 4, 8, 12, 16]
pl.plot(x1, y1, 'r')
pl.plot(x2, y2, 'g')
pl.title('Plot of y vs. x')
pl.xlabel('x axis')
pl.ylabel('y axis')
pl.xlim(0.0, 9.0)
pl.ylim(0.0, 30.)
pl.show()
```

上述代码执行后的效果如图 10-16 所示。

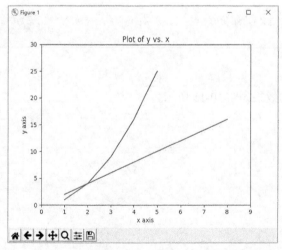

图 10-16　执行效果

10.1.6　绘制曲线

在 Python 程序中，绘制曲线最简单的方式是使用数学中的正弦函数或余弦函数。下面的实例文件 qu.py 演示了使用正弦函数和余弦函数绘制曲线的过程。

源码路径：daima\10\10-1\qu.py

```
from pylab import *
X = np.linspace(-np.pi, np.pi, 256,endpoint=True)
C,S = np.cos(X), np.sin(X)
plot(X,C)
plot(X,S)
show()
```

执行后的效果如图 10-17 所示。

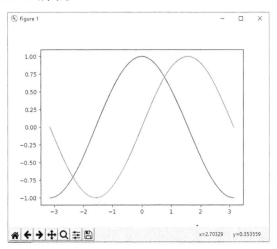

图 10-17　执行效果

上述实例展示的是使用的 Matplotlib 默认配置的效果。其实开发者可以调整大多数的默认配置，如图片大小和分辨率（dpi）、线宽、颜色、风格、坐标轴、坐标轴以及网格的属性、文字与字体属性等。但是，Matplotlib 的默认配置在大多数情况下已经做得足够好，开发人员可能只在很少的情况下才会想更改这些默认配置，下面的实例文件 zi.py 展示了使用 Matplotlib 的默认配置和自定义绘图的过程。

源码路径：daima\10\10-1\zi.py

```
# 导入Matplotlib 的所有内容（nympy 可以用 np 这个名字）
from pylab import *
#创建一个 8 * 6点的图，并设置分辨率为80dpi
figure(figsize=(8,6), dpi=80)
#创建一个新的 1 * 1 的子图，接下来的图绘制在其中的第一块（也是唯一的一块）中
subplot(1,1,1)
X = np.linspace(-np.pi, np.pi, 256,endpoint=True)
C,S = np.cos(X), np.sin(X)
#绘制余弦曲线，使用蓝色、连续、宽度为 1（像素）的线条
plot(X, C, color="blue", linewidth=1.0, linestyle="-")
#绘制正弦曲线，使用绿色、连续、宽度为 1（像素）的线条
plot(X, S, color="green", linewidth=1.0, linestyle="-")
#设置横轴的上下限
xlim(-4.0,4.0)
#设置横轴记号
xticks(np.linspace(-4,4,9,endpoint=True))
#设置纵轴的上下限
ylim(-1.0,1.0)
#设置纵轴记号
yticks(np.linspace(-1,1,5,endpoint=True))
#以72dpi的分辨率来保存图片
# savefig("exercice_2.png",dpi=72)
#在屏幕上显示
show()
```

上述实例代码中的配置与默认配置完全相同，我们可以在交互模式中修改其中的值来观察效果。执行后的效果如图 10-18 所示。

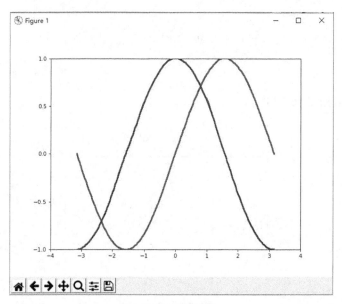

图 10-18　执行效果

在绘制曲线时可以改变线条的颜色和粗细，例如，以蓝色和红色分别表示余弦与正弦函数，然后将线条变粗一点，接着在水平方向拉伸一下整个图。具体代码如下。

```
...
figure(figsize=(10,6), dpi=80)
plot(X, C, color="blue", linewidth=2.5, linestyle="-")
plot(X, S, color="red",  linewidth=2.5, linestyle="-")
...
```

执行效果如图 10-19 所示。

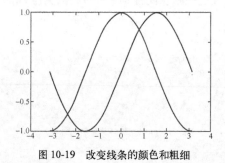

图 10-19　改变线条的颜色和粗细

在绘制曲线时也可以设置图片的边界。例如以下代码设置的当前图片的边界不合理，所以有些地方会看得不是很清楚。

```
...
xlim(X.min()*1.1, X.max()*1.1)
ylim(C.min()*1.1, C.max()*1.1)
...
```

更好的设置方式如下。

```
xmin ,xmax = X.min(), X.max()
ymin, ymax = Y.min(), Y.max()
dx = (xmax - xmin) * 0.2
dy = (ymax - ymin) * 0.2
xlim(xmin - dx, xmax + dx)
ylim(ymin - dy, ymax + dy)
```

此时的执行效果如图 10-20 所示。

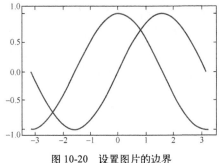

图 10-20 设置图片的边界

在绘制曲线时需要设置正确的刻度记号，例如，在使用正弦和余弦函数的时候，通常希望知道函数在 $\pm\pi$ 和 $\pm\pi/2$ 时的值。这样看来，当前的设置就不那么理想了，所以可以对代码进行如下更改。

```
...
xticks( [-np.pi, -np.pi/2, 0, np.pi/2, np.pi])
yticks([-1, 0, +1])
...
```

此时的执行效果如图 10-21 所示。

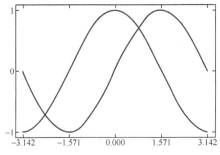

图 10-21 设置刻度记号

在绘制曲线时需要设置记号的标签，例如，我们可以把 3.142 当作 π，但这毕竟不够精确。当我们设置记号的时候，可以同时设置记号的标签。在下面的代码中使用了 LaTeX（一种基于 TEX 的排版系统）。

```
...
xticks([-np.pi, -np.pi/2, 0, np.pi/2, np.pi],
       [r'$-\pi$', r'$-\pi/2$', r'$0$', r'$+\pi/2$', r'$+\pi$'])

yticks([-1, 0, +1],
       [r'$-1$', r'$0$', r'$+1$'])
...
```

此时的执行效果如图 10-22 所示。

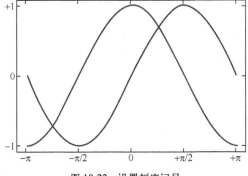

图 10-22 设置刻度记号

在绘制曲线时可以移动脊柱。坐标轴线和上面的记号连在一起就形成了脊柱（Spines，一条线段上有一系列的凸起，效果很像脊柱骨）。脊柱记录了数据区域的范围，可以放在任意位置。不过至今为止，都把它放在图的四边。实际上每幅图都有4条脊柱（上、下、左、右），为了将脊柱放在图的中间，我们必须将其中的两条（上和右）设置为无色，然后调整剩下的两条到合适的位置（数据空间的零点）。

```
...
ax = gca()
ax.spines['right'].set_color('none')
ax.spines['top'].set_color('none')
ax.xaxis.set_ticks_position('bottom')
ax.spines['bottom'].set_position(('data',0))
ax.yaxis.set_ticks_position('left')
ax.spines['left'].set_position(('data',0))
...
```

此时的执行效果如图10-23所示。

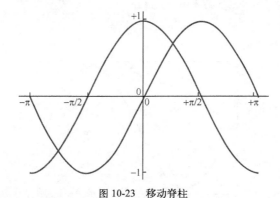

图10-23　移动脊柱

在绘制曲线时可以添加图例，例如，可以在图的左上角添加一个图例。此时只需要在plot()函数中以"键/值"对的形式增加一个参数即可。具体代码如下。

```
...
plot(X, C, color="blue", linewidth=2.5, linestyle="-", label="cosine")
plot(X, S, color="red",  linewidth=2.5, linestyle="-", label="sine")

legend(loc='upper left')
...
```

此时的执行效果如图10-24所示。

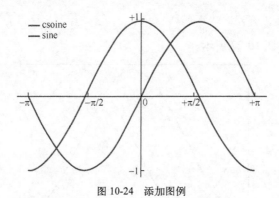

图10-24　添加图例

在绘制曲线时可以在曲线中添加一些特殊点作为注释，例如，要在$2\pi/3$的位置给两条函数曲线加上一个注释。首先，需要在对应的函数图像位置上画一个点。然后，向横轴引一条垂线，以虚线做标记。最后，写上标签。

```
...
t = 2*np.pi/3
plot([t,t],[0,np.cos(t)], color ='blue', linewidth=2.5, linestyle="--")
scatter([t,],[np.cos(t),], 50, color ='blue')

annotate(r'$\sin(\frac{2\pi}{3})=\frac{\sqrt{3}}{2}$',
         xy=(t, np.sin(t)), xycoords='data',
         xytext=(+10, +30), textcoords='offset points', fontsize=16,
         arrowprops=dict(arrowstyle="->", connectionstyle="arc3,rad=.2"))
plot([t,t],[0,np.sin(t)], color ='red', linewidth=2.5, linestyle="--")
scatter([t,],[np.sin(t),], 50, color ='red')
annotate(r'$\cos(\frac{2\pi}{3})=-\frac{1}{2}$',
         xy=(t, np.cos(t)), xycoords='data',
         xytext=(-90, -50), textcoords='offset points', fontsize=16,
         arrowprops=dict(arrowstyle="->", connectionstyle="arc3,rad=.2"))
...
```

执行效果如图 10-25 所示。

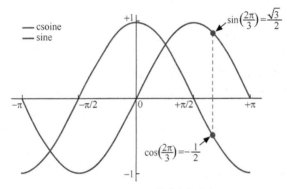

图 10-25 给一些特殊点加注释

10.1.7 绘制随机漫步图

随机漫步（random walk）是一种数学统计模型。它由一连串轨迹组成。其中每一次的轨迹都是随机的。随机漫步模型用来表示不规则的变动形式。气体或液体中分子活动的轨迹等可作为随机漫步的模型。在 1903 年由卡尔·皮尔逊首次提出随机漫步这一概念，目前已经广泛应用于生态学、经济学、心理学、计算科学、物理学、化学和生物学等领域，用来说明这些领域内观察到的行为和过程，因而随机漫步模型是记录随机活动的基本模型。

1. 在 Python 程序中生成随机漫步数据

在 Python 程序中生成随机漫步数据后，可以使用 Matplotlib 以灵活的形式将这些数据展现出来。随机漫步的行走路径很有自己的特色，每次行走都完全是随机的，没有任何明确的方向，漫步结果是由一系列随机决策决定的。例如，漂浮在水滴上的花粉因不断受到水分子的挤压而在水面上移动。水滴中的分子运动是随机的，因此花粉在水面上的运动路径犹如随机漫步。

为了在 Python 程序中模拟随机漫步的过程，在下面的实例文件 random_walk.py 中创建了一个名为 RandomA 的类，此类可以随机地选择前进方向。RandomA 类需要用到 3 个属性。其中一个是存储随机漫步次数的变量，其他两个是列表，分别用于存储随机漫步经过的每个点的 x 坐标和 y 坐标。

源码路径：daima\10\10-1\random_walk.py

```
from random import choice
class RandomA():
    """能够随机生成漫步数据的类"""
    def __init__(self, num_points=5100):
        """初始化随机漫步属性"""
```

```
            self.num_points = num_points
            #所有的随机漫步开始于 (0, 0).
            self.x_values = [0]
            self.y_values = [0]
        def shibai(self):
            """计算在随机漫步中包含的所有点"""
            #继续漫步,直到达到所需长度为止
            while len(self.x_values) < self.num_points:
                #决定前进的方向,沿着这个方向前进的距离
                x_direction = choice([1, -1])
                x_distance = choice([0, 1, 2, 3, 4])
                x_step = x_direction * x_distance
                y_direction = choice([1, -1])
                y_distance = choice([0, 1, 2, 3, 4])
                y_step = y_direction * y_distance
                #不能原地踏步
                if x_step == 0 and y_step == 0:
                    continue
                #计算下一个点的坐标,即x值和y值
                next_x = self.x_values[-1] + x_step
                next_y = self.y_values[-1] + y_step
                self.x_values.append(next_x)
                self.y_values.append(next_y)
```

在上述代码中,RandomA 类包含两个函数——__init__()和 shibai()。

函数__init__()用于实现初始化处理。为了能够做出随机决策,首先将所有可能的选择都存储在一个列表中。在每次做出具体决策时,通过"from random import choice"代码使用函数 choice()来决定使用哪种选择。然后,将随机漫步包含的默认点数设置为 5100,这个数值能够确保足以生成有趣的模式,同时也能够确保快速地模拟随机漫步。接下来,在第 8 行和第 9 行代码中创建了两个用于存储 x 值与 y 值的列表,并设置每次漫步都从点(0,0)开始出发。

函数 shibai()的功能是生成漫步包含的点,并决定每次漫步的方向。关于该函数,注意以下几点。

- 第 13 行。使用 while 语句建立了一个循环,这个循环可以不断运行,直到漫步包含所需数量的点为止。这个函数的主要功能是告知 Python 应该如何模拟 4 种漫步决定:向右走还是向左走,沿指定的方向走多远,向上走还是向下走,沿选定的方向走多远。
- 第 15 行。使用 choice([1, −1])给 x_direction 设置一个值,在漫步时要么表示向右走的"1",要么表示向左走的"−1"。
- 第 16 行。使用 choice([0, 1, 2, 3, 4])随机地选择 0~4 的一个整数,告诉 Python 沿指定的方向走的距离(x_distance)。通过包含 0,不但可以沿两个轴移动,还可以沿着 y 轴移动。
- 第 17~20 行。将移动方向乘以移动距离,以确定沿 x 轴移动的距离。如果 x_step 为正,则向右移动;如果为负,则向左移动;如果为 0,则垂直移动。如果 y_step 为正,则向上移动;如果为负,则向下移动;如果为 0,则水平移动。
- 第 22 行和第 23 行。开始执行下一次循环。如果 x_step 和 y_step 都为 0,则原地踏步。在程序中必须杜绝这种原地踏步的情况发生。
- 第 25~28 行。为了获取漫步中下一个点的 x 值,将 x_step 与 x_values 中的最后一个值相加,对 y 值进行相同的处理。获得下一个点的 x 值和 y 值之后,将它们分别附加到列表 x_values 和 y_values 的末尾。

2. 在 Python 程序中绘制随机漫步图

在前面的实例文件 random_walk.py 中已经创建了一个名为 RandomA 的类,在下面的实例文件 yun.py 中,将借助于 Matplotlib 将 RandomA 类中生成的漫步数据绘制出来,最终生成一个随机漫步图。

源码路径：daima\10\10-1\yun.py

```
import matplotlib.pyplot as plt
from random_walk import RandomA
#只要当前程序是活动的，就要不断模拟随机漫步过程
while True:
    #创建一个随机漫步实例，将包含的点都绘制出来
    rw = RandomA(51000)
    rw.shibai()
    #设置绘图窗口的尺寸大小
    plt.figure(dpi=128, figsize=(10, 6))
    point_numbers = list(range(rw.num_points))
    plt.scatter(rw.x_values, rw.y_values, c=point_numbers, cmap=plt.cm.Blues,
        edgecolors='none', s=1)
    #用特别的样式（红色、绿色与粗点）突出起点和终点
    plt.scatter(0, 0, c='green', edgecolors='none', s=100)
    plt.scatter(rw.x_values[-1], rw.y_values[-1], c='red', edgecolors='none',
        s=100)
    #隐藏坐标轴
    plt.axes().get_xaxis().set_visible(False)
    plt.axes().get_yaxis().set_visible(False)
    plt.show()
    keep_running = input("哥，还继续漫步吗？ (y/n): ")
    if keep_running == 'n':
        break
```

关于上述代码，注意以下几点。

- 第 1 行和第 2 行。分别导入模块 pyplot 与前面编写的类 RandomA。
- 第 6 行。创建一个 RandomA 实例，并将它存储到 rw 中。
- 第 7 行。调用函数 shibai()。
- 第 9 行。使用函数 figure()设置图表的宽度、高度、分辨率。
- 第 10 行。使用颜色映射来指出漫步中各点的先后顺序，并删除每个点的黑色轮廓，这样可以让它们的颜色显示更加明显。为了根据漫步中各点的先后顺序进行着色，需要传递参数 c，并将它设置为一个列表，其中包含各点的先后顺序。由于这些点是按顺序绘制的，因此给参数 c 指定的列表只要包含数字 1～51000 即可。使用函数 range()生成一个数字列表，其中包含的数字个数与漫步包含的点数相同。接下来，我们将这个列表存储在 point_numbers 中，以便后面使用它来设置每个漫步点的颜色。
- 第 11 行和第 12 行。将随机漫步包含的 x 值和 y 值传递给函数 scatter()，并选择合适的点尺寸。将参数 c 设置为在第 10 行中创建的 point_numbers，用于设置使用颜色映射 Blues，并传递实参 edgecolors='none'删除每个点周围的轮廓。
- 第 14～16 行。在绘制随机漫步图后重新绘制起点和终点，目的是突出显示随机漫步过程中的起点和终点，在程序中让起点和终点变得更大，并用不同的颜色显示。为了实现突出显示的功能，使用绿色绘制点（0,0），设置这个点比其他的点都粗大（设置 s=100）。在突出显示终点时，在漫步包含的最后一个坐标的 x 值和 y 值的位置绘制一个点，并设置它的颜色是红色，并将其粗大值 s 设置为 100。
- 第 18 行和第 19 行。隐藏图表中的坐标轴，使用函数 plt.axes()将每条坐标轴的可见性都设置为 False。
- 第 21～23 行。实现模拟多次随机漫步功能，因为每次随机漫步都不同，要在不多次运行程序的情况下使用前面的代码实现模拟多次随机漫步的功能，最简单的办法是将这些代码放在一个 while 循环中。这样通过本实例模拟一次随机漫步后，在 Matplotlib 查看器中可以浏览漫步结果。接下来可以在不关闭查看器的情况下暂停程序的执行，并询问你是否要再模拟一次随机漫步。如果输入 y，则可以模拟多次随机漫步。这些随机漫步都在起点附近进行，大多数沿着特定方向偏离起点，漫步点分布不均匀。要结束程序，只要输入 n 即可。

本实例最终执行后的效果如图 10-26 所示。

图 10-26　执行效果

10.1.8　大数据分析某年的最高温度和最低温度

在文件 death_valley_2014.csv 中保存了 2014 年每一天各个时段的温度。根据该文件，编写文件 high_lows.py，使用 Matplotlib 绘制出温度曲线图，统计出 2014 年的最高温度和最低温度。文件 high_lows.py 的具体实现代码如下。

源码路径：daima\10\10-1\high_lows.py

```
import csv
from matplotlib import pyplot as plt
from datetime import datetime

file = './csv/death_valley_2014.csv'
with open(file) as f:
    reader = csv.reader(f)
    header_row = next(reader)
    #从文件中获取最高气温
    highs,dates,lows = [], [], []
    for row in reader:
        try:
            date = datetime.strptime(row[0],"%Y-%m-%d")
            high = int(row[1])
            low = int(row[3])
        except ValueError:
            print(date,'missing data')
        else:
            highs.append(high)
            dates.append(date)
            lows.append(low)

#根据数据绘制图形
fig = plt.figure(figsize=(10,6))
plt.plot(dates,highs,c='r',alpha=0.5)
plt.plot(dates,lows,c='b',alpha=0.5)
plt.fill_between(dates,highs,lows,facecolor='b',alpha=0.2)
# 设置图形的格式
plt.title('Daily high and low temperatures-2014',fontsize=16)
plt.xlabel('',fontsize=12)
fig.autofmt_xdate()
plt.ylabel('Temperature(F)',fontsize=12)
plt.tick_params(axis='both',which='major',labelsize=20)
plt.show()
```

执行后的效果如图 10-27 所示。

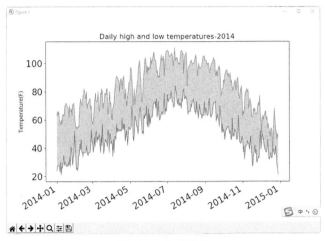

图 10-27　执行效果

10.1.9　在 Tkinter 中使用 Matplotlib 库绘制图表

下面的实例文件 123.py 演示了在标准 GUI 程序 Tkinter 中使用 Matplotlib 绘制图表的过程。

源码路径： daima\10\10-1\123.py

```python
class App(tk.Tk):
    def __init__(self, parent=None):
        tk.Tk.__init__(self, parent)
        self.parent = parent
        self.initialize()

    def initialize(self):
        self.title("在Tkinter中使用Matplotlib! ")
        button = tk.Button(self, text="退出", command=self.on_click)
        button.grid(row=1, column=0)
        self.mu = tk.DoubleVar()
        self.mu.set(5.0)    #参数的默认值是"mu"
        slider_mu = tk.Scale(self,
                             from_=7, to=0, resolution=0.1,
                             label='mu', variable=self.mu,
                             command=self.on_change
                             )
        slider_mu.grid(row=0, column=0)
        self.n = tk.IntVar()
        self.n.set(512)    #参数的默认值是"n"
        slider_n = tk.Scale(self,
                            from_=512, to=2,
                            label='n', variable=self.n, command=self.on_change
                            )
        slider_n.grid(row=0, column=1)

        fig = Figure(figsize=(6, 4), dpi=96)
        ax = fig.add_subplot(111)
        x, y = self.data(self.n.get(), self.mu.get())
        self.line1, = ax.plot(x, y)
        self.graph = FigureCanvasTkAgg(fig, master=self)
        canvas = self.graph.get_tk_widget()
        canvas.grid(row=0, column=2)

    def on_click(self):
        self.quit()

    def on_change(self, value):
        x, y = self.data(self.n.get(), self.mu.get())
```

```
                    self.line1.set_data(x, y)    #更新数据
                    #更新graph
                    self.graph.draw()

            def data(self, n, mu):
                lst_y = []
                for i in range(n):
                    lst_y.append(mu * random.random())
                return range(n), lst_y

        if __name__ == "__main__":
            app = App()
            app.mainloop()
```

执行后可以拖动左侧的滑块来控制绘制的图表，执行效果如图 10-28 所示。

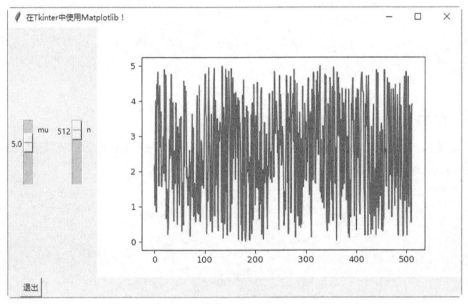

图 10-28　执行效果

10.2　使用 pygal 库

在 Python 程序中，可以使用 pygal 库实现数据的可视化操作功能，也可以生成 SVG 文件。SVG 是一种矢量图格式，全称是 Scalable Vector Graphics，被翻译为可缩放矢量图形。我们可以使用浏览器打开 SVG 文件，从而方便地与之交互。对于需要在尺寸不同的屏幕上显示图表，SVG 会变得很有用，它可以自动缩放，适应观看者的屏幕。本节将详细讲解使用 pygal 库实现数据可视化的知识。

10.2.1　安装 pygal 库

使用 pygal 库，可以在用户与图表交互时突出元素并调整元素的大小，还可以轻松地调整整个图表的尺寸，使它适合在微型智能手表或巨型显示器上显示。安装 pygal 库的命令格式如下。

```
pip install pygal
```

也可以从 GitHub 网站下载，具体命令格式如下。

```
git clone git://GitHub域名.com/Kozea/pygal.git
pip install pygal
```

具体安装过程如图 10-29 所示。

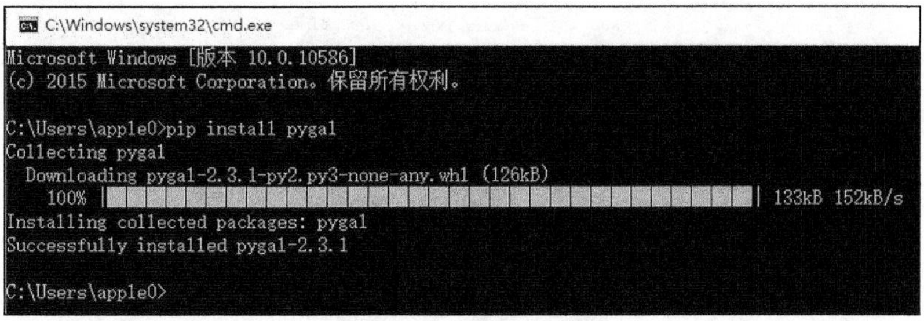

10.2 使用 pygal 库

图 10-29 安装 pygal 库

10.2.2 使用 pygal 库模拟掷骰子

下面的实例文件 01.py 演示了使用 pygal 库模拟掷骰子的过程。首先定义了骰子类 Die，然后使用函数 range()模拟掷骰子 1000 次，接下来统计每个点数出现的次数，最后在柱状图中显示统计结果。文件 01.py 的具体实现代码如下。

源码路径：daima\10\10-2\01.py

```python
import random

class Die:
    """
    一个骰子类
    """
    def __init__(self, num_sides=6):
        self.num_sides = num_sides

    def roll(self):
        return random.randint(1, self.num_sides)

import pygal

die = Die()
result_list = []
#掷1000次
for roll_num in range(1000):
    result = die.roll()
    result_list.append(result)

frequencies = []
#统计每个点数出现的次数
for value in range(1, die.num_sides + 1):
    frequency = result_list.count(value)
    frequencies.append(frequency)

#柱状图
hist = pygal.Bar()
hist.title = 'Results of rolling one D6 1000 times'
# x轴坐标
hist.x_labels = [1, 2, 3, 4, 5, 6]
# x、y轴的描述
hist.x_title = 'Result'
hist.y_title = 'Frequency of Result'
#添加数据，第一个参数是数据的标题
hist.add('D6', frequencies)
#保存到本地，格式必须是svg
hist.render_to_file('die_visual.svg')
```

执行后会生成一个名为"die_visual.svg"的文件，我们可以用浏览器打开这个 SVG 文件，打开后会显示统计柱状图。执行效果如图 10-30 所示。如果将鼠标指针指向数据，可以看到标题"D6"、x 轴的坐标以及 y 轴坐标。6 个数字出现的频次是差不多的，其实理论上概率是 1/6，随着实验次数的增加，趋势将越来越明显。

第 10 章 数据可视化

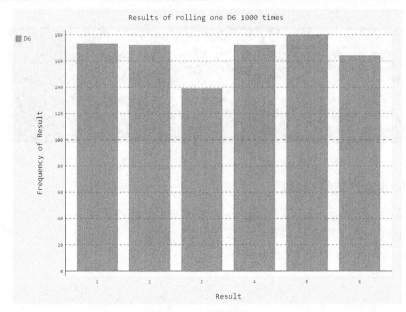

图 10-30 执行效果

我们可以对上面的实例进行升级，例如，同时掷两个骰子，这可以通过下面的实例文件 02.py 实现。在具体实现时，首先定义骰子类 Die，然后使用函数 range()模拟掷两个骰子 5000 次，接下来统计每次掷两个骰子出现的最大点数的次数，最后在柱状图中显示统计结果。具体实现代码如下。

源码路径：daima\10\10-2\02.py

```
class Die:
    """
    一个骰子类
    """
    def __init__(self, num_sides=6):
        self.num_sides = num_sides

    def roll(self):
        return random.randint(1, self.num_sides)
die_1 = Die()
die_2 = Die()

result_list = []
for roll_num in range(5000):
    #两个骰子的点数和
    result = die_1.roll() + die_2.roll()
    result_list.append(result)

frequencies = []
#能掷出的最大点数
max_result = die_1.num_sides + die_2.num_sides

for value in range(2, max_result + 1):
    frequency = result_list.count(value)
    frequencies.append(frequency)

#可视化
hist = pygal.Bar()
hist.title = 'Results of rolling two D6 dice 5000 times'
hist.x_labels = [x for x in range(2, max_result + 1)]
hist.x_title = 'Result'
hist.y_title = 'Frequency of Result'
#添加数据
```

```
hist.add('two D6', frequencies)
#格式必须是svg
hist.render_to_file('2_die_visual.svg')
```

执行后会生成一个名为"2_die_visual.svg"的文件，我们可以用浏览器打开这个 SVG 文件，打开后会显示柱状图。执行效果如图 10-31 所示。由此可以看出，两个骰子之和为 7 的次数最多，和为 2 的次数最少。因为能掷出 2 的只有（1,1）这一种情况，而掷出 7 有（1,6）、（2,5）、（3,4）、（4,3）、（5,2）、（6,1）这 6 种情况，掷出其余数字的情况都没有 7 多，故掷得 7 的概率最大。

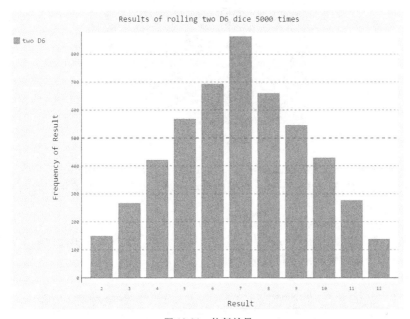

图 10-31　执行效果

10.3　使用 csvkit 库处理 CSV 文件

CSV 是逗号分隔值（Comma-Separated Value）的缩写。CSV 文件以纯文本形式存储数据。在 Python 程序中，通常使用其内置模块 csv 来处理 CSV 文件。在实际应用中，我们还可以使用第三方库 csvkit 来处理 CSV 文件。可以使用如下命令来安装 csvkit。

```
pip install csvkit
```

csvkit 库是以命令行的方式运行的，建议读者在 Linux 系统下使用。当然，在 Windows 系统中也可以使用。

在 csvkit 库的官网中提供了测试数据文件，并可以下载和使用它。

可使用如下命令将一个 Excel 文件转换为 CSV 文件。

```
in2csv ne_1033_data.xlsx > data.csv
```

通过上述命令将官方测试文件 ne_1033_data.xlsx 中的数据转换为名为 data.csv 的 CSV 文件。在 Linux 系统中，可以通过如下命令查看文件 data.csv 中的内容。

```
cat data.csv
```

1. 查看数据

在 csvkit 库中，可以使用如下命令查看文件 data.csv 的内容。

```
csvlook data.csv
```

如果想分页查看，可以使用 UNIX 系统中的 less 命令实现。

```
csvlook data.csv | less -S
```

2. 分割数据

在 csvkit 库中，可以使用 csvcut 命令来分割数据。此命令不会修改输入（原始文件），只会影响输出。例如，通过如下命令可以查看在文件 data.csv 中有哪些列，如图 10-32 所示。

```
csvcut -n data.csv
```

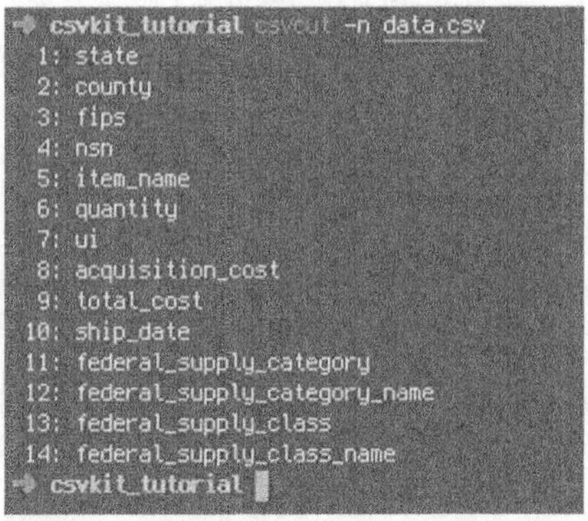

图 10-32　查看文件 data.csv 中有哪些列

由图 10-32 可见，在文件 data.csv 中一共有 14 列。在图 10-35 所示的结果中，前面的数字是数字索引位置，默认从 1 开始计数，后面的是列名称。如果只需要看感兴趣的几列数据，可以通过下面的命令实现。

```
csvcut -c 2,5,6 data.csv
```

其中，数字 2、5、6 表示列的索引。另外，我们还可以设置只显示几行数据。例如，可以使用如下 head 命令来设置输出 4 行数据，如图 10-33 所示。

```
csvcut -c 2,5,6 data.csv | head -n 5
```

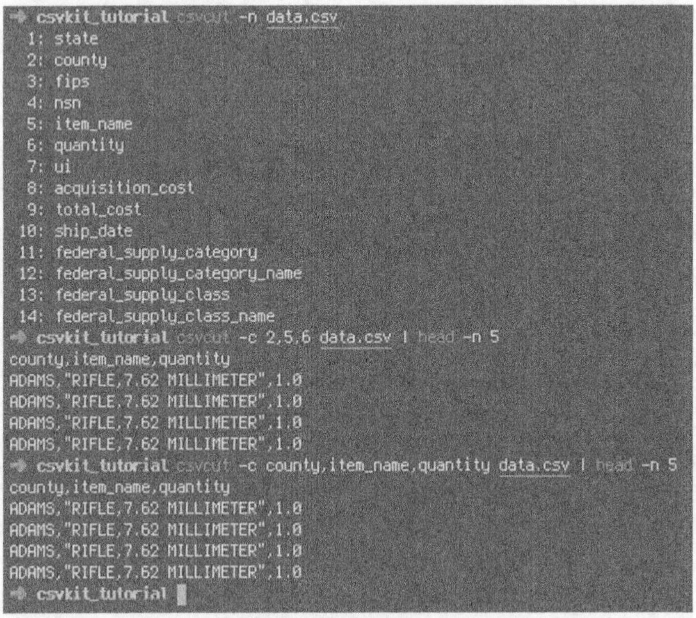

图 10-33　执行效果

在使用 csvcut 命令时，既可以使用数字作为索引，也可以使用列名作为索引，所以上面的指令和下面这条指令是等价的。

```
csvcut -c county,item_name,quantity data.csv | head -n 5
```

10.3 使用 csvkit 库处理 CSV 文件

上面的输出结果已经很美观了,能否再进一步优化呢?答案是肯定的,只要将命令用管道组合起来即可。例如,下面的命令会输出图 10-34 所示的结果。

```
csvcut -c county,item_name,quantity data.csv | csvlook | head
```

图 10-34 执行效果

如果不关心中间结果,我们可以从头到尾都用管道进行连接。例如,将上面的命令改成如下形式,也会获得一样的结果。这样就省去了生成 data.csv 文件的中间过程。

```
in2csv ne_1033_data.xlsx | csvcut -c county,item_name,quantity | csvlook | head
```

3. 数据分析

使用 csvlook 和 csvcut 查看数据的切片只是探索数据的开始。在 csvkit 库中,可以使用 csvstat 实现数据统计功能。例如,通过如下命令可以统计出文件中的唯一值、最小值、最大值、总计值、均值、中位值和标准差等信息,执行效果如图 10-35 所示。

```
csvcut -c county,acquisition_cost,ship_date data.csv | csvstat
```

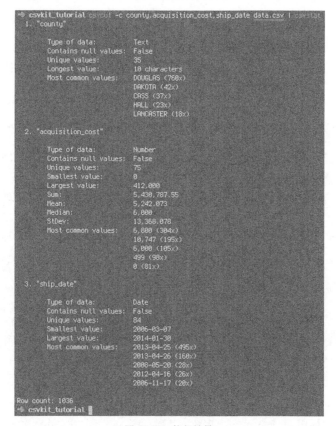

图 10-35 执行效果

4. 内容过滤

除了按照列的方式过滤之外，有时还需要按照内容进行过滤。在 csvkit 库中，可以使用 csvgrep 过滤出需要的数据。例如，通过下面的指令可以只看兰卡斯特（LANCASTER，英国的一个城市）的数据，执行效果如图 10-36 所示。

```
csvcut -c county,item_name,total_cost data.csv | csvgrep -c county -m LANCASTER | csvlook
```

图 10-36　执行效果

5. 排序处理

前面显示出来的内容都很少，但是实际项目中的数据往往有成千上万行，所以排序功能十分重要。在 csvkit 库中，使用 csvsort 对 CSV 数据进行排序处理。例如，使用下面的 csvsort 命令可以对 total_cost 列进行降序排列，如图 10-37 所示。

```
csvcut -c county,item_name,total_cost data.csv | csvgrep -c county -m LANCASTER | csvsort -c total
```

图 10-37　执行效果

6. 关联数据

在 csvkit 库中，使用 csvjoin 命令实现数据关联。接下来以官方文件 acs2012_5yr_population.csv 为例进行介绍，这是一个包含人口信息的数据文件。通过如下命令进行数据统计，结果如图 10-38 所示。

```
csvstat acs2012_5yr_population.csv
```

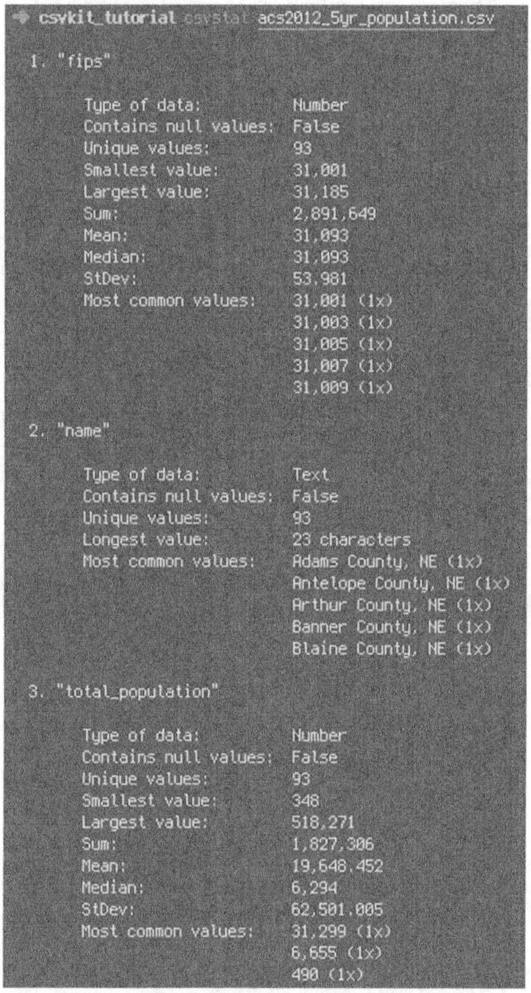

图 10-38　查看 acs2012_5yr_population.csv 的信息

由此可见，在前面的文件 data.csv 和现在的文件 acs2012_5yr_population.csv 中都包含 fips 字段，我们可以用这个字段来连接这两个 CSV 文件数据。例如，使用下面的命令进行连接后的效果如图 10-39 所示。

```
csvjoin -c fips data.csv acs2012_5yr_population.csv > joined.csv csvcut -c
  county,item_name,total_population joined.csv | csvsort -c total_population | csvlook
```

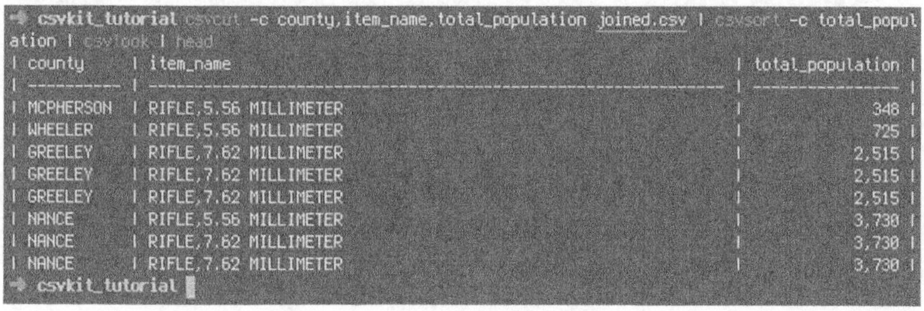

图 10-39　连接两个 CSV 文件的数据

7. 数据转换

仅仅用命令行操作是不够的，有时需要用 SQL 来实现数据的操作。在 csvkit 库中，csvsql 和 sql2csv 提供了数据库与 CSV 之间的桥梁。例如，通过如下命令可以将文件 joined.csv 转换为数据库表，执行效果如图 10-40 所示。

```
csvsql -i sqlite joined.csv
```

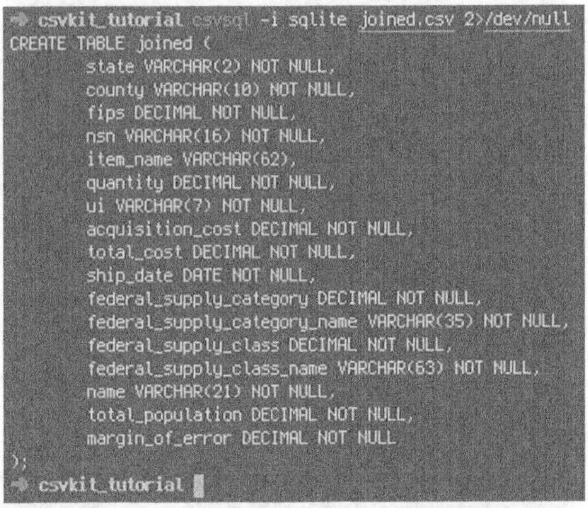

图 10-40　将文件 joined.csv 转换为数据库表

我们甚至可以一步到位，使用如下命令直接创建一个名为"leso.db"的本地数据库，如图 10-41 所示。

```
csvsql --db sqlite:///leso.db --insert joined.csv
```

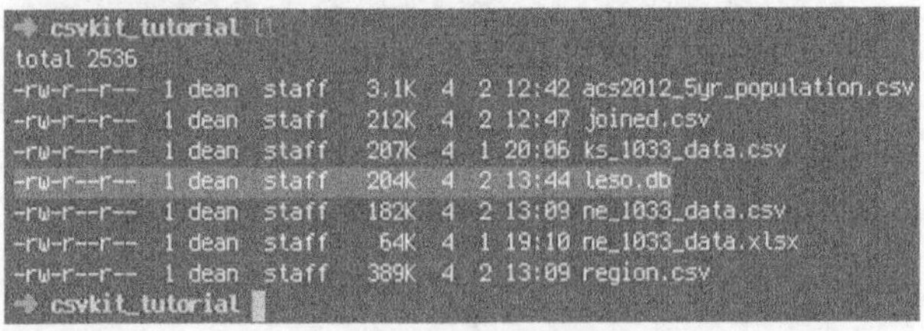

图 10-41　创建数据库

接下来，就可以利用数据库软件对这个数据库执行 SQL 查询操作，也可以用 csvkit 提供的小工具 sql2csv 来进行查询。例如，通过下面的指令查询表 joined 中的数据，查询结果如图 10-42 所示。

```
sql2csv --db sqlite:///leso.db --query "select * from joined where total_population<1000;"
```

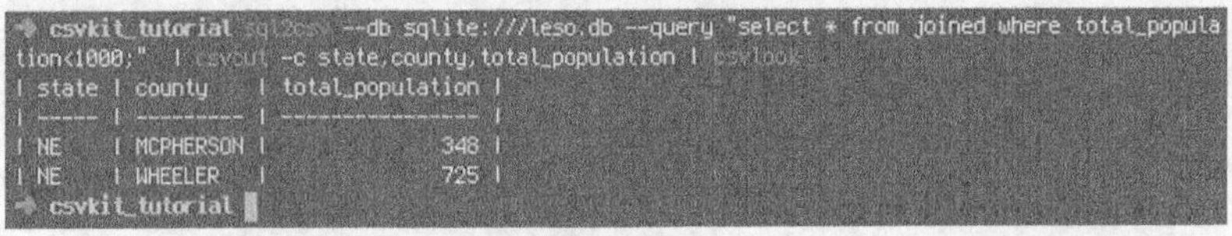

图 10-42　查询结果

注意：为了节省篇幅，csvkit 库的知识介绍完毕，详细信息请读者参阅其官方文档。

10.4 使用 Pandas 库

Pandas 是基于 NumPy 的一种工具,是为了解决数据分析任务而创建的。Pandas 纳入了大量库和一些标准的数据模型,提供了高效地操作大型数据集所需的工具。本节将详细讲解在 Python 程序中使用 Pandas 库的过程。

10.4.1 安装 Pandas 库

在计算机系统中,我们无须安装即可使用 Pandas。这时需要使用 Wakari 免费服务,以在云中提供托管的 IPython Notebook 服务。开发者只要创建一个账户,就可在几分钟内通过 IPython Notebook 在浏览器中访问并使用 Pandas。

对于大多数开发者来说,建议使用如下命令来安装 Pandas。

```
pip install pandas
```

然后通过如下实例文件 001.py 来测试是否安装成功并成功运行。

源码路径:daima\10\10-4\001.py

```
import pandas as pd
print(pd.test())
```

根据机器的配置差异,执行效果会有所区别,在作者的计算机中执行后会输出:

```
running: pytest --skip-slow --skip-network C:\Users\apple\AppData\Local\Programs\Python\Python36\lib\site-packages\pandas
=============================== test session starts ===============================
platform win32 -- Python 3.6.2, pytest-3.3.1, py-1.5.2, pluggy-0.6.0
rootdir: H:\daima\10\10-4, inifile:
collected 10360 items / 3 skipped

pandas\tests\test_algos.py ........................................... [  0%]
..............s.......................................                  [  0%]
pandas\tests\test_base.py ..............................                 [  1%]
pandas\tests\test_categorical.py .................s....                  [  1%]
........................................                                 [  2%]
......                                                                   [  2%]
pandas\tests\test_common.py .........                                    [  2%]
###为节省本书篇幅,后面省略好多信息
```

> 注意:为了节省本书的篇幅,这里不再详细讲解 Pandas API 的语法知识,关于这方面的知识,请读者阅读 Pandas 的官方文档。

10.4.2 从 CSV 文件读取数据

在 Pandas 库中,可以使用方法 read_csv()读取 CSV 文件中的数据。在默认情况下,read_csv()方法会假设 CSV 文件中的字段是用逗号进行分隔的。假设存在一个名为"bikes.csv"的 CSV 文件,在里面保存了每天在蒙特利尔 7 条不同的道路上有多少人骑自行车。下面的实例文件 002.py 读取并显示了文件 bikes.csv 中的前 3 条数据。

源码路径:daima\10\10-4\002.py

```
import pandas as pd
broken_df = pd.read_csv('bikes.csv')
print(broken_df[:3])
```

执行后会输出:

```
Date;Berri 1;Brébeuf (données non disponibles);Côte-Sainte-Catherine;Maisonneuve 1;Maisonneuve
    2;du Parc;Pierre-Dupuy;Rachel1;St-Urbain (données non disponibles)
0                01/01/2012;35;;0;38;51;26;10;16;
1                02/01/2012;83;;1;68;153;53;6;43;
2                03/01/2012;135;;2;104;248;89;3;58;
```

读者会发现上述执行效果显得比较凌乱。此时可以用方法 read_csv()中的参数进行设置。方法 read_csv()的语法格式如下。

第10章 数据可视化

```
pandas.read_csv(filepath_or_buffer, sep=', ', delimiter=None, header='infer', names=None,
    index_col=None, usecols=None, squeeze=False, prefix=None, mangle_dupe_cols=True,
    dtype=None, engine=None, converters=None, true_values=None, false_values=None,
    skipinitialspace=False, skiprows=None, nrows=None, na_values=None, keep_default_na=True,
    na_filter=True, verbose=False, skip_blank_lines=True, parse_dates=False,
    infer_datetime_format=False, keep_date_col=False, date_parser=None, dayfirst=False,
    iterator=False, chunksize=None, compression='infer', thousands=None, decimal='.',
    lineterminator=None, quotechar='"', quoting=0, escapechar=None, comment=None,
    encoding=None, dialect=None, tupleize_cols=False, error_bad_lines=True,
    warn_bad_lines=True, skipfooter=0, skip_footer=0, doublequote=True,
    delim_whitespace=False, as_recarray=False, compact_ints=False, use_unsigned=False,
    low_memory=True, buffer_lines=None, memory_map=False, float_precision=None)[source]
```

各个参数的具体说明如下。

- filepath_or_buffer。其值可以是 str、pathlib.Path、py._path.local.LocalPath 或任何具有 read() 方法的对象（如文件句柄或 StringIO）。字符串可以是 URL，有效的 URL 方案包括 http、ftp、s3 和 file。文件 URL 需要主机，例如，本地文件可以是 file：//localhost/path/to/table.csv。
- sep。表示分隔符，可以是 str，默认为逗号。长度大于 1 个字符且与'\s+'不同的分隔符将被解释为正则表达式，将强制使用 Python 解析引擎，并忽略数据中的引号。例如，正则表达式'\r\t'。
- delimiter。str，默认为 None，表示 sep 的备用参数名称。
- delim_whitespace。可以是布尔值，默认为 False，用于指定是否以空格作为分隔符，相当于设置 sep='\s+'。如果将此选项设置为 True，则不应该向 delimiter 参数传入任何内容。
- header。可以是 int 或 ints 列表，默认为'infer'，表示用作列名称的行号，以及数据的开始。如果未传递 names，默认行为就好像设置为 0；否则为 None。如果显式传递 header=0，则能够替换原来的列表。header 可以是整数列表，这个列表表示将文件中的这些仍作为列标题（即每一列有多个标题），介于中间的标题将被忽略。
- names。可以是 array-like 或 default，表示要使用的列名称列表。如果文件不包含标题行，则应明确传递 header = None。除非 mangle_dupe_cols = True（这是默认值），否则不允许在此列表中重复。
- index_col。可以是 int 或序列或 False，默认值是 None，表示用作 DataFrame 的行标签的列。如果给出序列，则使用 MultiIndex。如果在每行结尾处都有带分隔符的格式不正确的文件，则可以考虑通过 index_col = False 强制 pandas _not_ 使用第一列作为索引（行名称）。
- usecols。可以是 array-like，默认值是 None，用于返回列的子集。此数组中的所有元素必须是位置（即文档列中的整数索引）或对应于用户在名称中提供或从文档标题行推断的列名称的字符串。例如，有效的 usecols 参数将是[0，1，2]或['foo'，'bar'，'baz']。使用此参数会加快解析并降低内存使用率。
- as_recarray。可以是布尔型，默认值为 False。
- DEPRECATED。此参数将在以后的版本中删除，需要用 pd.read_csv (...) 和 to_records() 代替。在解析数据后，返回 NumPy recarray 而不是 DataFrame。如果设置为 True，此选项优先于 squeeze 参数。此外，由于行索引在此类格式中不可用，因此将忽略 index_col 参数。
- squeeze。可以是布尔型，默认为 False。如果解析的数据只包含一列，则返回一个序列。
- prefix。可以是 str，默认值是 None，表示在没有标题时添加到列号的前缀，如'X'代表 X0，X1，…
- mangle_dupe_cols。可以是布尔型，默认值为 True，用于将重复的列指定为"*X*.0"，…，"*X.N*"，而不是"*X*"…"*X*"。如果在列中存在重复的名称，则传入 False 将覆盖数据。

- dtype。表示输入列的名称或字典的类型，默认值是 None。
- engine。可以是{'c', 'python'}，可选参数，供解析器引擎使用。C 引擎速度更快，而 Python 引擎目前更加完善。
- converters。可以是 dict，默认值是 None，表示转换某些列中的值的函数，键可以是整数或列标签。
- skipinitialspace。可以是布尔值，默认为 False，用于跳过分隔符后的空格。
- skiprows。可以是 list-like 或 integer，默认值是 None，表示要跳过的行号（0 索引）或要跳过的行数（int）在文件的开头。
- skipfooter。可以是 int，默认值是 0，表示跳过文件底部的行数（不支持 engine ='c'）。
- nrows。可以是 int，默认值无，表示要读取的文件的行数，适用于读取大文件中的片段。
- na_values。可以是 scalar、str、list-like 或 dict，默认值是 None，表示可识别为 NA / NaN 的其他字符串。对于 dict，需要设置特定列的 NA 值。
- keep_default_na。可以是 bool，默认值是 True。如果指定了 na_values 并且 keep_default_na 为 False，则将覆盖默认 NaN 值；否则，将追加它们。
- na_filter。可以是布尔值，默认值是 True。用于检测缺失值标记（空字符串和 na_values 的值）。在没有任何 NA 的数据中，传递 na_filter = False 可以提高读取大文件的性能。
- verbose。可以是布尔值，默认值是 False，用于指示放置在非数字列中的 NA 值的数量。
- skip_blank_lines。可以是布尔值，默认值是 True。如果为 True，请跳过空白行，而不是解释为 NaN 值。
- parse_dates。可以是布尔值、列表、名称、dict 列表，默认值为 False。如果为 True，则尝试解析索引。
- infer_datetime_format。可以是布尔值，默认值是 False。如果启用了 True 和 parse_dates，pandas 将尝试推断列中 datetime 字符串的格式，如果可以推断，则可以切换到更快的解析方式。在某些情况下，这可以将解析速度提高 5~10 倍。
- keep_date_col。可以是布尔值，默认值是 False。如果 True 和 parse_dates 指定合并多个列，则保留原始列。
- date_parser。一个函数，用于将字符串列序列转换为 datetime 实例数组的函数。默认使用 dateutil.parser.parser 进行转换。Pandas 将尝试以 3 种不同的方式调用 date_parser：将一个或多个数组（由 parse_dates 定义）作为参数传递；将由 parse_dates 定义的列中的字符串值连接（逐行）到单个数组中并传递；对于每一行，使用一个或多个字符串（对应于由 parse_dates 定义的列）作为参数调用 date_parser 一次。
- dayfirst。布尔类型，默认值是 False，用于返回 TextFileReader 对象以进行迭代或使用 get_chunk()获取块。
- chunksize。int 类型，用于返回 TextFileReader 对象以进行迭代。在 iterator 和 chunksize 中查看 IO 工具文档了解更多信息。
- compression。{'infer', 'gzip', 'bz2', 'zip', 'xz', None}类型，用于将磁盘上的数据即时进行解压缩。如果设置为 infer，则使用 gzip、bz2、zip 或 xz；如果 filepath_or_buffer 是分别以.gz、.bz2、.zip 或 xz 结尾的字符串，则不进行解压缩；如果使用'zip'，ZIP 文件必须只包含一个要读入的数据文件。如果设置为 None，表示不解压缩。
- float_precision。string 类型，用于指定 C 引擎应该为浮点值使用哪个转换器。对于普通转换器的设置为 None；对于高精度转换器设置为 high，对于往返转换器设置为 round_trip。

- lineterminator。str（length 1）类型，用于将文件拆分成行的字符。只有 C 解析器有效。
- quotechar。str（length 1）类型，用于表示带引号项目开始和结束的字符。引号项可以包含分隔符，它将被忽略。
- doublequote。布尔类型，默认值是 True。当指定 quotechar 且引用不是 QUOTE_NONE 时，指示是否将一个字段中两个连续的元素解释为单个 quotechar 元素。
- escapechar。str（length 1）类型，用于转义分隔符的单字符字符串为 QUOTE_NONE。
- encoding。str 类型，在读/写时用于 UTF 的编码（例如，'utf-8'）。
- tupleize_cols。布尔类型，默认值是 False，用于将列上的元组列表保留为原样（默认将列转换为多索引）。
- error_bad_lines。布尔类型，默认值是 True。在默认情况下，具有太多字段的行（如具有太多逗号的 csv 行）将引发异常，并且不会返回 DataFrame。如果为 False，那么这些"坏行"将从返回的 DataFrame 中删除。注意，只有 C 解析器有效。
- warn_bad_lines。布尔类型，默认值为 True。如果 error_bad_lines 为 False，并且 warn_bad_lines 为 True，则将输出每个"坏行"的警告。注意，只有 C 解析器有效。
- low_memory。布尔类型，默认值是 True。在内部以块的方式处理文件，以便在解析时内存使用量较少，但可能出现混合类型推断。要确保没有混合类型，请设置 False，或使用 dtype 参数指定类型。请注意，无论如何，整个文件都读入单个 DataFrame，请使用 chunksize 或迭代器参数以块形式返回数据。注意，只有 C 解析器有效。
- compact_ints。布尔类型，默认值是 False。如果 compact_ints 为 True，则对于任何整数为 dtype 的列，解析器将尝试将其转换为可能的最小整数 dtype，根据 use_unsigned 参数的规范，可以是有符号数或无符号数。
- use_unsigned。布尔类型，默认值是 False。如果整数列被压缩（即 compact_ints = True），请指定该列是否应压缩为最小有符号或无符号整数。
- memory_map。布尔类型，默认值是 False。如果为 filepath_or_buffer 提供了文件路径，则将文件对象直接映射到内存上，并从中直接访问数据。使用此选项可以提高性能，因为不再有任何 I/O 开销。

下面的实例文件 003.py 使用更加规整的格式，读取并显示了文件 bikes.csv 中的前 3 条数据。

源码路径：daima\10\10-4\003.py

```
import pandas as pd
fixed_df = pd.read_csv('bikes.csv', sep=';', encoding='latin1', parse_dates=['Date'],
    dayfirst=True, index_col='Date')
print(fixed_df[:3])
```

执行后会输出：

```
            Berri 1  Brébeuf (données non disponibles)  \
Date
2010-01-01      35                                NaN
2010-01-02      83                                NaN
2010-01-03     135                                NaN

            Côte-Sainte-Catherine  Maisonneuve 1  Maisonneuve 2  du Parc  \
Date
2010-01-01                      0             38             51       26
2010-01-02                      1             68            153       53
2010-01-03                      2            104            248       89

            Pierre-Dupuy  Rachel1  St-Urbain (données non disponibles)
Date
2010-01-01            10       16                                  NaN
2010-01-02             6       43                                  NaN
2010-01-03             3       58                                  NaN
```

在读取 CSV 文件时，得到的是一种由行和列组成的数据帧，我们可以列出帧中的元素。下面的实例文件 004.py 读取并显示了文件 bikes.csv 中 "Berri 1" 列的数据。

源码路径：daima\10\10-4\004.py

```
import pandas as pd
fixed_df = pd.read_csv('bikes.csv', sep=';', encoding='latin1', parse_dates=['Date'],
    dayfirst=True, index_col='Date')
print(fixed_df['Berri 1'])
```

执行后会输出：

```
Date
2010-01-01      35
2010-01-02      83
2010-01-03     135
……省略部分行数
2010-10-23    4177
2010-10-24    3744
2010-10-25    3735
2010-10-26    4290
2010-10-27    1857
2010-10-28    1310
2010-10-29    2919
2010-10-30    2887
2010-10-31    2634
2010-11-01    2405
2010-11-02    1582
2010-11-03     844
2010-11-04     966
2010-11-05    2247
Name: Berri 1, Length: 310, dtype: int64
```

为了使应用程序更加美观，在下面的实例文件 005.py 中加入了 Matplotlib 功能，以统计图的方式展示了文件 bikes.csv 中 "Berri 1" 列的数据。

源码路径：daima\10\10-4\005.py

```
import pandas as pd
import matplotlib.pyplot as plt
plt.rcParams['figure.figsize'] = (15, 5)
fixed_df = pd.read_csv('bikes.csv', sep=';', encoding='latin1', parse_dates=['Date'],
    dayfirst=True, index_col='Date')
fixed_df['Berri 1'].plot()
plt.show()
```

执行后会显示每个月的骑行数据统计结果，执行效果如图 10-43 所示。

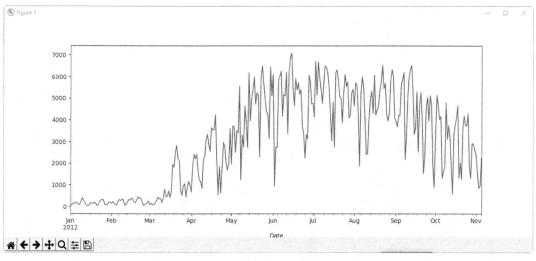

图 10-43　执行效果

10.4.3 选择指定数据

下面的实例文件 006.py 用于处理一个更大的数据集文件 311-service-requests.csv，并输出这个文件中的数据信息。文件 311-service-requests.csv 是 311 的服务请求从纽约开放数据的一个子集，完整文件有 52MB，此处只截取了一小部分，完整文件可以从网络中获取。

源码路径：daima\10\10-4\006.py

```
import pandas as pd
complaints = pd.read_csv('311-service-requests.csv')
print(complaints)
```

执行后会显示读取文件 311-service-requests.csv 的结果，并在最后统计数据数目。

而在下面的实例文件 007.py 中，首先输出显示了文件 311-service-requests.csv 中"Complaint Type"列的信息，接着输出了文件 311-service-requests.csv 中的前 5 行信息，然后输出了文件 311-service-requests.csv 中前 5 行的"Complaint Type"列的信息，接下来输出了文件 311-service-requests.csv 中"Complaint Type"和"Borough"这两列的信息，最后输出了文件 311-service-requests.csv 中"Complaint Type"和"Borough"这两列的前 10 行信息。

源码路径：daima\10\10-4\007.py

```
import pandas as pd
complaints = pd.read_csv('311-service-requests.csv')
print(complaints['Complaint Type'])
print(complaints[:5])
print(complaints[:5]['Complaint Type'])
print(complaints[['Complaint Type', 'Borough']])
print(complaints[['Complaint Type', 'Borough']][:10])
```

执行后会输出：

```
//下面首先输出"Complaint Type"列的信息
0           Noise - Street/Sidewalk
1                   Illegal Parking
2                 Noise - Commercial
3                    Noise - Vehicle
4                             Rodent
5                 Noise - Commercial
6                  Blocked Driveway
7                 Noise - Commercial
8                 Noise - Commercial
9                 Noise - Commercial
10          Noise - House of Worship
11                Noise - Commercial
12                   Illegal Parking
13                   Noise - Vehicle
14                            Rodent
15          Noise - House of Worship
16          Noise - Street/Sidewalk
17                   Illegal Parking
18             Street Light Condition
19                Noise - Commercial
20          Noise - House of Worship
21                Noise - Commercial
22                   Noise - Vehicle
23                Noise - Commercial
24                  Blocked Driveway
25          Noise - Street/Sidewalk
26             Street Light Condition
27             Harboring Bees/Wasps
28          Noise - Street/Sidewalk
29             Street Light Condition
                   ...
233                   Noise-Commercial
234                     Taxi Complaint
235                 Sanitation Condition
```

```
236             Noise - Street/Sidewalk
237                 Consumer Complaint
238            Traffic Signal Condition
239              DOF Literature Request
240             Litter Basket / Request
241                    Blocked Driveway
242             Violation of Park Rules
243              Collection Truck Noise
244                      Taxi Complaint
245                      Taxi Complaint
246              DOF Literature Request
247             Noise - Street/Sidewalk
248                     Illegal Parking
249                     Illegal Parking
250                    Blocked Driveway
251             Maintenance or Facility
252                   Noise - Commercial
253                     Illegal Parking
254                               Noise
255                              Rodent
256                     Illegal Parking
257                               Noise
258             Street Light Condition
259                        Noise - Park
260                    Blocked Driveway
261                     Illegal Parking
262                  Noise - Commercial
Name: Complaint Type, Length: 263, dtype: object
//下面输出前5行信息
   Unique Key         Created Date          Closed Date Agency  \
0    26589651  10/31/2013 02:08:41 AM                  NaN   NYPD
1    26593698  10/31/2013 02:01:04 AM                  NaN   NYPD
2    26594139  10/31/2013 02:00:24 AM  10/31/2013 02:40:32 AM   NYPD
3    26595721  10/31/2013 01:56:23 AM  10/31/2013 02:21:48 AM   NYPD
4    26590930  10/31/2013 01:53:44 AM                  NaN  DOHMH

                              Agency Name         Complaint Type  \
0             New York City Police Department  Noise - Street/Sidewalk
1             New York City Police Department          Illegal Parking
2             New York City Police Department       Noise - Commercial
3             New York City Police Department          Noise - Vehicle
4  Department of Health and Mental Hygiene                      Rodent

                     Descriptor       Location Type  Incident Zip  \
0                  Loud Talking     Street/Sidewalk       11432.0
1   Commercial Overnight Parking    Street/Sidewalk       11378.0
2             Loud Music/Party  Club/Bar/Restaurant       10032.0
3                Car/Truck Horn    Street/Sidewalk       10023.0
4   Condition Attracting Rodents         Vacant Lot       10027.0

    Incident Address                ...                          \
0   90-03 169 STREET                ...
1         58 AVENUE                 ...
2       4060 BROADWAY               ...
3      WEST 72 STREET               ...
4     WEST 124 STREET               ...

  Bridge Highway Name Bridge Highway Direction Road Ramp  \
0                 NaN                      NaN       NaN
1                 NaN                      NaN       NaN
2                 NaN                      NaN       NaN
3                 NaN                      NaN       NaN
4                 NaN                      NaN       NaN

  Bridge Highway Segment Garage Lot Name Ferry Direction Ferry Terminal Name  \
0                    NaN            NaN             NaN                 NaN
```

```
1                      NaN                   NaN                   NaN               NaN
2                      NaN                   NaN                   NaN               NaN
3                      NaN                   NaN                   NaN               NaN
4                      NaN                   NaN                   NaN               NaN

   Latitude  Longitude                                       Location
0  40.708275 -73.791604  (40.70827532593202, -73.79160395779721)
1  40.721041 -73.909453  (40.721040535628305, -73.90945306791765)
2  40.843330 -73.939144  (40.84332975466513, -73.93914371913482)
3  40.778009 -73.980213  (40.7780087446372, -73.98021349023975)
4  40.807691 -73.947387  (40.80769092704951, -73.94738703491433)

[5 rows x 52 columns]
//下面输出前5行中"Complaint Type"列的信息
[5 rows x 52 columns]
0        Noise - Street/Sidewalk
1              Illegal Parking
2              Noise - Commercial
3              Noise - Vehicle
4                       Rodent
…..省略部分
259              Noise - Park       BROOKLYN
260          Blocked Driveway         QUEENS
261           Illegal Parking       BROOKLYN
262         Noise - Commercial      MANHATTAN
[263 rows x 2 columns]
//下面输出了"Complaint Type"和"Borough"这两列的信息
          Complaint Type      Borough
0    Noise - Street/Sidewalk    QUEENS
1           Illegal Parking     QUEENS
2         Noise - Commercial  MANHATTAN
3            Noise - Vehicle  MANHATTAN
4                     Rodent  MANHATTAN
5         Noise - Commercial     QUEENS
//下面输出了"Complaint Type"和"Borough"这两列的前10行信息
          Complaint Type      Borough
0    Noise - Street/Sidewalk    QUEENS
1           Illegal Parking     QUEENS
2         Noise - Commercial  MANHATTAN
3            Noise - Vehicle  MANHATTAN
4                     Rodent  MANHATTAN
5         Noise - Commercial     QUEENS
6          Blocked Driveway      QUEENS
7         Noise - Commercial     QUEENS
8         Noise - Commercial  MANHATTAN
9         Noise - Commercial   BROOKLYN
```

在下面的实例文件 008.py 中，首先输出了文件 311-service-requests.csv 中"Complaint Type"列中前 10 条信息，然后在图表中显示前 10 条信息。

源码路径：daima\10\10-4\008.py

```python
import pandas as pd
import matplotlib.pyplot as plt

pd.set_option('display.width', 5000)
pd.set_option('display.max_columns', 60)

plt.rcParams['figure.figsize'] = (10, 6)

complaints = pd.read_csv('311-service-requests.csv')
complaint_counts = complaints['Complaint Type'].value_counts()
print(complaint_counts[:10])#输出"Complaint Type"列中前10条信息
complaint_counts[:10].plot(kind='bar')#绘制"Complaint Type"列中前10条信息
plt.show()
```

执行后会在控制台中显示"Complaint Type"列中前 10 条信息。

```
Noise - Commercial        51
Noise                     27
Noise - Street/Sidewalk   22
Blocked Driveway          21
Illegal Parking           18
Taxi Complaint            13
Traffic Signal Condition  10
Rodent                    10
Water System               9
Noise - Vehicle            7
Name: Complaint Type, dtype: int64
```

并且会在 Matplotlib 图表中统计前 10 条信息，如图 10-44 所示。

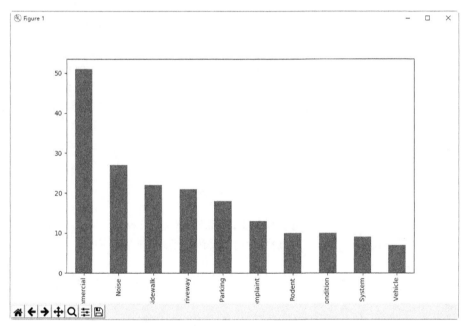

图 10-44　执行效果

10.4.4　与日期相关的操作

在进行数据统计分析时，时间通常是一个重要的因素。在下面的实例文件 009.py 中，可以使用 Matplotlib 统计出文件 bikes.csv 中每个月的骑行数据信息。

源码路径：daima\10\10-4\009.py

```
import pandas as pd
import matplotlib.pyplot as plt

plt.rcParams['figure.figsize'] = (10, 8)
plt.rcParams['font.family'] = 'sans-serif'

pd.set_option('display.width', 5000)
pd.set_option('display.max_columns', 60)

bikes = pd.read_csv('bikes.csv', sep=';', encoding='latin1', parse_dates=['Date'],
    dayfirst=True, index_col='Date')
bikes['Berri 1'].plot()
plt.show()
```

执行后的效果如图 10-45 所示。

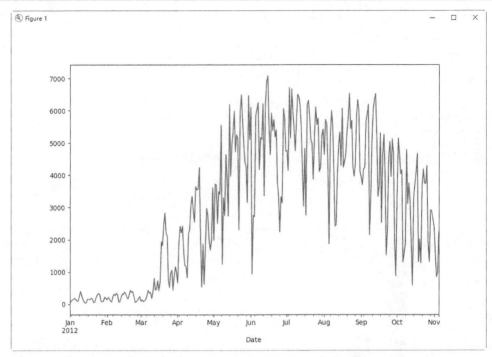

图 10-45　执行效果

在下面的实例文件 010.py 中，首先显示文件 bikes.csv 中"Berri 1"街道前 5 天的骑行数据信息，然后使用"print(berri_bikes.index)"输出了具体日期。

源码路径：daima\10\10-4\010.py

```
import pandas as pd

bikes = pd.read_csv('bikes.csv', sep=';', encoding='latin1', parse_dates=['Date'],
    dayfirst=True, index_col='Date')
berri_bikes = bikes[['Berri 1']].copy()
print(berri_bikes[:5])
print(berri_bikes.index)
```

执行后会输出：

```
            Berri 1
Date
2010-01-01       35
2010-01-02       83
2010-01-03      135
2010-01-04      144
2010-01-05      197
DatetimeIndex(['2010-01-01', '2010-01-02', '2010-01-03', '2010-01-04',
               '2010-01-05', '2010-01-06', '2010-01-07', '2010-01-08',
               '2010-01-09', '2010-01-10',
               ...
               '2010-10-27', '2010-10-28', '2010-10-29', '2010-10-30',
               '2010-10-31', '2010-11-01', '2010-11-02', '2010-11-03',
               '2010-11-04', '2010-11-05'],
              dtype='datetime64[ns]', name='Date', length=310, freq=None)
```

由上述执行效果可知，只输出显示了 310 天的统计数据。其实 Pandas 有一系列非常好的时间序列功能，所以如果我们想得到每一行的月份，可以通过文件 011.py 实现。

源码路径：daima\10\10-4\011.py

```
import pandas as pd

bikes = pd.read_csv('bikes.csv', sep=';', encoding='latin1', parse_dates=['Date'],
    dayfirst=True, index_col='Date')
```

```
berri_bikes = bikes[['Berri 1']].copy()
print(berri_bikes.index.day)
print(berri_bikes.index.weekday)
```

执行后会输出：

```
Int64Index([ 1,  2,  3,  4,  5,  6,  7,  8,  9, 10,
            ...
            27, 28, 29, 30, 31,  1,  2,  3,  4,  5],
           dtype='int64', name='Date', length=310)
Int64Index([6, 0, 1, 2, 3, 4, 5, 6, 0, 1,
            ...
            5, 6, 0, 1, 2, 3, 4, 5, 6, 0],
           dtype='int64', name='Date', length=310)
```

在上述输出结果中，0 表示周一。我们可以使用 Pandas 灵活地判断某一天是周几。请查看实例文件 012.py。

源码路径：daima\10\10-4\012.py

```
import pandas as pd

bikes = pd.read_csv('bikes.csv', sep=';', encoding='latin1', parse_dates=['Date'],
    dayfirst=True, index_col='Date')
berri_bikes = bikes[['Berri 1']].copy()
berri_bikes.loc[:,'weekday'] = berri_bikes.index.weekday
print(berri_bikes[:5])
```

执行后会输出：

```
            Berri 1  weekday
Date
2010-01-01       35        6
2010-01-02       83        0
2010-01-03      135        1
2010-01-04      144        2
2010-01-05      197        3
```

在实际应用中，我们当然也可以统计周一到周日每天的数据。在下面的实例文件 013.py 中，统计了周一到周日每天的骑行数据。

源码路径：daima\10\10-4\013.py

```
import pandas as pd

bikes = pd.read_csv('bikes.csv', sep=';', encoding='latin1', parse_dates=['Date'],
    dayfirst=True, index_col='Date')
berri_bikes = bikes[['Berri 1']].copy()

berri_bikes.loc[:,'weekday'] = berri_bikes.index.weekday

weekday_counts = berri_bikes.groupby('weekday').aggregate(sum)
print(weekday_counts)

weekday_counts.index = ['Monday','Tuesday','Wednesday','Thursday','Friday','Saturday','Sunday']
print(weekday_counts)
```

执行后会输出：

```
         Berri 1
weekday
0         134298
1         135305
2         152972
3         160131
4         141771
5         101578
6          99310
           Berri 1
Monday     134298
Tuesday    135305
Wednesday  152972
Thursday   160131
```

```
Friday      141771
Saturday    101578
Sunday       99310
```

当然，为了使统计数据更加直观，我们可以在程序中使用 Matplotlib。在下面的实例文件 014.py 中，使用 Matplotlib 图表统计了周一到周日每天的骑行数据。

源码路径：daima\10\10-4\014.py

```
import pandas as pd
import matplotlib.pyplot as plt
plt.rcParams['figure.figsize'] = (15, 5)
bikes = pd.read_csv('bikes.csv',
                    sep=';', encoding='latin1',
                    parse_dates=['Date'], dayfirst=True,
                    index_col='Date')
# 添加标识
berri_bikes = bikes[['Berri 1']].copy()
berri_bikes.loc[:,'weekday'] = berri_bikes.index.weekday

# 开始统计
weekday_counts = berri_bikes.groupby('weekday').aggregate(sum)
weekday_counts.index = ['Monday','Tuesday','Wednesday','Thursday','Friday','Saturday','Sunday']
weekday_counts.plot(kind='bar')

plt.show()
```

执行效果如图 10-46 所示。

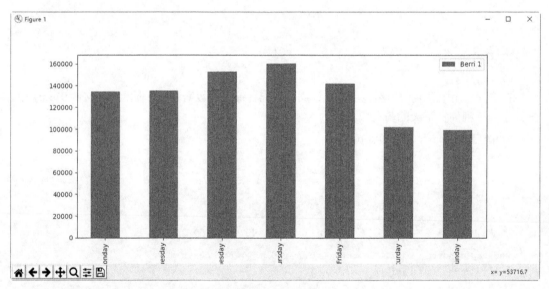

图 10-46 执行效果

再看下面的实例文件 015.py，借助于素材文件 weather_2012.csv，使用 Matplotlib 统计了加拿大 2012 年全年的天气数据。

源码路径：daima\10\10-4\015.py

```
import pandas as pd
import matplotlib.pyplot as plt
import numpy as np

plt.rcParams['figure.figsize'] = (15, 3)
plt.rcParams['font.family'] = 'sans-serif'
weather_2012_final = pd.read_csv('weather_2012.csv', index_col='Date/Time')
weather_2012_final['Temp (C)'].plot(figsize=(15, 6))
plt.show()
```

执行效果如图 10-47 所示。

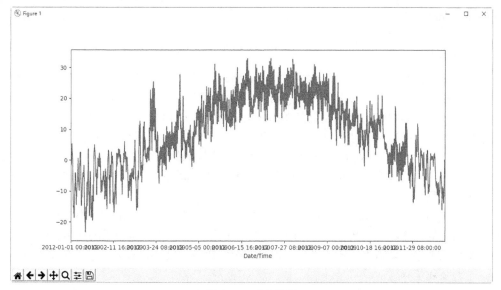

图 10-47　执行效果

通过简单的实例文件 016.py，即可输出文件 weather_2012.csv 中的全部天气信息。

源码路径：daima\10\10-4\016.py

```
import pandas as pd

weather_2012_final = pd.read_csv('weather_2012.csv', index_col='Date/Time')
print(weather_2012_final)
```

执行后会输出：

```
                     Temp (C)  Dew Point Temp (C)  Rel Hum (%)  \
Date/Time
2010-01-01 00:00:00      -1.8                -3.9           86
2010-01-01 01:00:00      -1.8                -3.7           87
2010-01-01 02:00:00      -1.8                -3.4           89
2010-01-01 03:00:00      -1.5                -3.2           88
2010-01-01 04:00:00      -1.5                -3.3           88
2010-01-01 05:00:00      -1.4                -3.3           87
2010-01-01 06:00:00      -1.5                -3.1           89
2010-01-01 07:00:00      -1.4                -3.6           85
2010-01-01 08:00:00      -1.4                -3.6           85
#在此省略好多输出结果
2010-10-31 19:00:00                 Snow
2010-10-31 20:00:00                 Snow
2010-10-31 21:00:00                 Snow
2010-10-31 22:00:00                 Snow
2010-10-31 23:00:00                 Snow

[8784 rows x 7 columns]
```

10.5　使用 NumPy 库

NumPy 库是用 Python 实现科学计算的一个库。它提供了一个多维数组对象、各种派生对象（如屏蔽的数组和矩阵）以及一系列用于数组快速操作的例程，包括数学、逻辑、形状操作、排序、选择、I/O、离散傅里叶变换、基本线性代数、基本统计操作和随机模拟等。本节将详细讲解在 Python 程序中使用 NumPy 库的核心知识。

10.5.1　安装 NumPy 库

标准的 Python 发行版不会与 NumPy 库捆绑在一起，一个轻量级的替代方法是使用如下命令来安装 NumPy 库。

```
pip install numpy
```

10.5.2 数组对象

在 NumPy 库中提供了一个 N 维数组类型 ndarray，用于描述相同类型的"元素"的集合，我们可以使用 N 个整数来对元素进行索引。在 NumPy 库中，所有 ndarrays 都是同质的：每个元素占用相同大小的内存块，并且所有块都以完全相同的方式解释。如何解释数组中的每个元素由单独的数据类型对象指定，每个数组与其中一个对象相关联。除了基本类型（整数、浮点等）之外，数据类型对象也可以表示数据结构。从数组中提取的元素（如通过索引）由一个 Python 对象表示，该对象的类型为 NumPy 中内置的数组标量类型之一，数组标量允许简单地处理更复杂的数据布局。

在 NumPy 库中，ndarray 是（通常大小固定的）一个多维容器，由相同类型和大小的元素组成。数组中的维度和元素数量由其 shape 定义，它是由 N 个正整数组成的元组，每个整数指定一个维度的大小。数组中元素的类型由单独的数据类型对象（dtype）指定，每个 ndarray 与其中一个对象相关联。

与 Python 中的其他容器对象一样，ndarray 的内容可以通过索引或切片（例如，使用 N 个整数）以及 ndarray 的方法和属性访问与修改数组。不同的 ndarrays 可以共享相同的数据，使得在一个 ndarray 中进行的改变在另一个中也可见。也就是说，ndarray 可以是到另一个 ndarray 的"视图"，并且其引用的数据由基础 ndarray 处理。ndarray 还可以是由 Python 字符串或 buffer 或 array 接口的对象拥有的内存的视图。

在下面的实例文件 001.py 中，创建了一个 2×3 的二维数组，该数组由 4 字节的整数元素组成。

源码路径：daima\10\10-5\001.py

```python
import numpy as np
x = np.array([[1, 2, 3], [4, 5, 6]], np.int32)
print(type(x))
print(x.shape)
print(x.dtype)
```

执行后会输出：

```
<class 'numpy.ndarray'>
(2, 3)
int32
```

在 NumPy 库中，数组可以使用类似 Python 容器的语法进行索引，并且切片可以生成数组的视图。下面的实例文件 002.py 演示了上述两种用法。

源码路径：daima\10\10-5\002.py

```python
import numpy as np
x = np.array([[1, 2, 3], [4, 5, 6]], np.int32)
print(x[1, 2])

y = x[:,1]
print(y)
y[0] = 9 # 这也改变了x中的对应元素
print(y)
print(x)
```

执行后会输出：

```
6
[2 5]
[9 5]
[[1 9 3]
 [4 5 6]]
```

1. 构造数组

在 NumPy 库中，可以使用如下数组创建函数来构建并操作新数组。

- empty（shape [, dtype, order]）。返回给定形状和类型的新数组，而不初始化条目。
- empty_like（a [, dtype, order, subok]）。返回具有与给定数组形状和类型相同的新数组。
- eye（N [, M, k, dtype]）。返回一个 2D 数组，其中 1 在对角线上，0 在其他地方。
- identity（n [, dtype]）。返回身份数组。
- ones（shape [, dtype, order]）。返回给定形状和类型的新数组，用数字填充。
- ones_like（a [, dtype, order, subok]）。返回与给定数组具有相同形状和类型的数组。
- zeros（shape [, dtype, order]）。返回给定形状和类型的新数组，用零填充。
- zeros_like（a [, dtype, order, subok]）。返回与给定数组形状和类型相同的零数组。
- full（shape, fill_value [, dtype, order]）。返回给定形状和类型的新数组，用 fill_value 填充。
- full_like（a, fill_value [, dtype, order, subok]）。返回与给定数组形状和类型相同的完整数组。
- array(object[, dtype, copy, order, subok, ndmin])。创建数组。
- asarray(a[, dtype, order]) 。将输入转换为数组。
- asanyarray(a[, dtype, order]) 。将输入转换为 ndarray，但传递 ndarray 子类。
- ascontiguousarray（a [, dtype]）。返回内存中的连续数组。
- asmatrix（data [, dtype]）。将输入解释为矩阵。
- copy（a [, order]）。返回给定对象的数组副本。
- frombuffer（buffer [, dtype, count, offset]）。将缓冲区解释为一维数组。
- fromfile（file [, dtype, count, sep]）。根据文本或二进制文件中的数据构造数组。
- fromfunction（function, shape, **kwargs）。通过在每个坐标上执行函数来构造数组。
- fromiter（iterable, dtype [, count]）。从可迭代对象创建新的一维数组。
- fromstring（string [, dtype, count, sep]）。根据字符串中的原始二进制或文本数据初始化的新 1D 数组。
- loadtxt（fname [, dtype, comments, delimiter, ...]）。从文本文件加载数据。
- core.records.array（obj [, dtype, shape, ...]）。从各种各样的对象构造一个记录数组。
- core.records.fromarrays（arrayList [, dtype, ...]）。从数组的（平面）列表创建一个记录数组。
- core.records.fromrecords(recList [,dtype,...])。从文本形式的记录列表创建一个 recarray。
- core.records.fromstring（datastring [, dtype, ...]）。从包含在其中的二进制数据创建（只读）记录数组。
- core.records.fromfile（fd [, dtype, shape, ...]）。从二进制文件数据创建数组。
- core.defchararray.array（obj [, itemsize, ...]）。创建 chararray。
- core.defchararray.asarray（obj [, itemsize, ...]）。将输入转换为 chararray，只在必要时复制数据。
- arange（[start,] stop [, step,] [, dtype]）。在给定间隔内返回均匀间隔的值。
- linspace（start, stop [, num, endpoint, ...]）。在指定的间隔内返回均匀间隔的数字。
- logspace（start, stop [, num, endpoint, base, ...]）。返回以对数刻度均匀分布的数字。
- meshgrid（* xi, ** kwargs）。从坐标向量返回坐标矩阵。
- mgrid。nd_grid 实例，返回密集的多维"网格"。
- ogrid。nd_grid 实例，返回一个打开的多维"meshgrid"。

- diag（*v* [，*k*]）。提取对角线上的元素或构造对角数组。
- diagflat（*v* [，*k*]）。创建一个二维数组，以扁平输入作为对角线。
- tri（*N* [，*M*，*k*，dtype]）。数组，在给定的对角线和其他地方元素为零。
- tril（*m* [，*k*]）。数组的下三角形。
- triu（*m* [，*k*]）。数组的上三角形。
- vander（*x* [，*N*，increasing]）。生成范德蒙德矩阵。
- mat（data [，dtype]）。将输入解释为矩阵。
- bmat（obj [，ldict，gdict]）。从字符串、嵌套序列或数组构建一个矩阵对象。

2. 索引数组

在 NumPy 库中，数组可以使用扩展的 Python 切片语法 array[selection] 来索引。其中类似的语法也用于访问结构化数组中的字段。

3. 数组操作函数

在 NumPy 库中，和数组操作相关的内置函数如下。

- ndarray.item（\ * args）。提取数组中的元素。
- ndarray.tolist()。将数组作为（可能是嵌套的）列表返回。
- ndarray.itemset（\ * args）。在指定位置插入标量值（修改原始数组）。
- ndarray.tostring（[order]）。在数组中构造包含原始数据字符的 Python 字节。
- ndarray.tobytes（[order]）。在数组中构造包含原始数据字节的 Python 字节。
- ndarray.tofile（fid [，sep，format]）。将数组作为文本或二进制（默认）写入文件。
- ndarray.dump（file）。将数组的 pickle 转储到指定的文件中。
- ndarray.dumps()。以字符串形式返回数组的 pickle。
- ndarray.astype（dtype [，order，casting，...]）。数组的副本，强制转换为指定的类型。
- ndarray.byteswap（inplace）。交换数组元素的字节。
- ndarray.copy（[order]）。返回数组的副本。
- ndarray.view（[dtype，type]）。数组的新视图与相同的数据。
- ndarray.getfield（dtype [，offset]）。对于给定数组的字段返回特定类型。
- ndarray.setflags（[write，align，uic]）。分别设置数组标志 WRITEABLE、ALIGNED 和 UPDATEIFCOPY。
- ndarray.fill（value）。使用标量值填充数组。
- ndarray.reshape（shape [，order]）。返回包含新形状的相同数据的数组。
- ndarray.resize（new_shape [，refcheck]）。就地更改数组的形状和大小。
- ndarray.transpose（\ * axes）。返回具有轴转置的数组的视图。
- ndarray.swapaxes（axis1，axis2）。返回数组的视图，其中 axis1 和 axis2 互换。
- ndarray.flatten（[order]）。对于折叠的数组的副本返回一个维度。
- ndarray.ravel（[order]）。返回展平的数组。
- ndarray.squeeze（[axis]）。从 *a* 形状删除单维条目。
- ndarray.take（indices [，axis，out，mode]）。返回由给定索引处的 *a* 元素组成的数组。
- ndarray.put（indices，values [，mode]）。对于所有 *n*，设置 *a*.flat [*n*] = values [*n*]。
- ndarray.repeat（repeat[，axis]）。重复数组的元素。
- ndarray.choose（choices [，out，mode]）。使用索引数组从一组选项中构造新的数组。
- ndarray.sort（[axis，kind，order]）。就地对数组进行排序。
- ndarray.argsort（[axis，kind，order]）。返回按此数组排序的索引。

- ndarray.partition (kth [, axis, kind, order])。重新排列数组中的元素，使得第 k 个位置的元素的值位于排序数组中指定的位置。
- ndarray.argpartition (kth [, axis, kind, order])。返回将此数组进行分区的索引。
- ndarray.searchsorted (v [, side, sorter])。查找应在其中插入 v 的元素以维持顺序。
- ndarray.nonzero()。返回非零元素的索引。
- ndarray.compress (condition [, axis, out])。沿给定轴返回此数组的选定切片。
- ndarray.diagonal ([offset, axis1, axis2])。返回指定的对角线。

下面的实例文件 003.py 演示了使用内置函数 arange() 和 reshape() 操作数组的过程。

源码路径：daima\10\10-5\003.py

```
import numpy as np
#一维数组
a = np.arange(24)
a.ndim
#现在调整其大小
b = a.reshape(2,4,3)
print(b)
```

执行后会输出：

```
[[[ 0  1  2]
  [ 3  4  5]
  [ 6  7  8]
  [ 9 10 11]]

 [[12 13 14]
  [15 16 17]
  [18 19 20]
  [21 22 23]]]
```

下面的实例文件 004.py 演示了，使用 arange() 函数创建一个 3×4 数组，并使用 nditer 对它进行迭代的过程。

源码路径：daima\10\10-5\004.py

```
import numpy as np
a = np.arange(0,60,5)
a = a.reshape(3,4)
print('原始数组是：')
print(a)
print('\n')
print('修改后的数组是：')
for x in np.nditer(a):
    print(x)
```

执行后会输出：

```
原始数组是：
[[ 0  5 10 15]
 [20 25 30 35]
 [40 45 50 55]]

修改后的数组是：
0 5 10 15 20 25 30 35 40 45 50 55
```

迭代的顺序匹配数组的内容，而不考虑特定的排序。这可以通过迭代上述数组的转置来看到，下面的实例文件 005.py 演示了这一用法。

源码路径：daima\10\10-5\005.py

```
import numpy as np
a = np.arange(0,60,5)
a = a.reshape(3,4)
print ('原始数组是：')
print(a)
print ('\n')
print ('原始数组的转置是：')
b = a.T
print(b)
```

```
print ('\n')
print( '修改后的数组是：')
for x in np.nditer(b):
    print(x,)
```

执行后会输出：

```
原始数组是：
[[ 0  5 10 15]
 [20 25 30 35]
 [40 45 50 55]]

原始数组的转置是：
[[ 0 20 40]
 [ 5 25 45]
 [10 30 50]
 [15 35 55]]

修改后的数组是：
0 5 10 15 20 25 30 35 40 45 50 55
```

下面的实例文件 006.py 演示了，使用函数 flatten()返回折叠为一维的数组副本的过程。

源码路径：daima\10\10-5\006.py

```
import numpy as np
a = np.arange(8).reshape(2,4)

print('原数组：')
print(a)
print('\n' )

print('展开的数组：')
print(a.flatten())
print('\n' )

print('以F风格顺序展开的数组：')
print(a.flatten(order = 'F'))
```

执行后会输出：

```
原数组：
[[0 1 2 3]
 [4 5 6 7]]

展开的数组：
[0 1 2 3 4 5 6 7]

以F风格顺序展开的数组：
[0 4 1 5 2 6 3 7]
```

10.5.3 使用通用函数

1. 字符串函数

在 NumPy 库中，通过使用如下函数，可以对 dtype 为 numpy.string_ 或 numpy.unicode_ 的数组执行向量化字符串操作。这些操作基于 Python 内置库中的标准字符串函数来完成。

- add()。返回两个 str 或 Unicode 数组中按元素字符串连接形式。
- multiply()。返回按元素多重连接后的字符串。
- center()。返回给定字符串的副本，其中元素位于特定字符串的中央。
- capitalize()。返回给定字符串的副本，其中只有第一个字母大写。
- title()。返回字符串按元素标题转换的版本。
- lower()。返回一个数组，其元素转换为小写。
- upper()。返回一个数组，其元素转换为大写。
- split()。返回字符串中的单词列表，并使用分隔符来分割。
- splitlines()。返回元素中的行列表，以换行符分割。
- strip()。返回数组副本，其中元素移除了开头或者结尾处的特定字符。

- join()。返回一个字符串,它是序列中字符串的连接形式。
- replace()。返回字符串的副本,其中所有子字符串的出现位置都被新字符串取代。
- decode()。按元素调用 str.decode。
- encode()。按元素调用 str.encode。

下面的实例文件 007.py 演示了使用上述字符串函数的过程。

源码路径:daima\10\10-5\007.py

```
import numpy as np
print('连接两个字符串: ')
print(np.char.add(['hello'],[' xyz']) )
print('连接示例: ')
print(np.char.add(['hello', 'hi'],[' abc', ' xyz']))

print(np.char.multiply('Hello ',3))

print(np.char.center('hello', 20,fillchar = '*'))

print(np.char.capitalize('hello world'))

print(np.char.title('hello how are you?'))

print(np.char.splitlines('hello\nhow are you?') )
print(np.char.splitlines('hello\rhow are you?'))

print(np.char.replace ('He is a good boy', 'is', 'was'))
```

上述代码的执行流程如下。

1)使用函数 add()实现了元素的字符串连接。

2)使用函数 multiply()实现了多重连接。

3)使用函数 center()返回所需宽度的数组,以便输入字符串位于中心,并使用 fillchar 在左侧和右侧进行填充。

4)使用函数 capitalize()使字符串中的第一个字母大写。

5)使用函数 title()使字符串中每个单词的首字母都大写。

6)使用函数 splitlines()返回数组中元素的单词列表,并以换行符进行分割。

7)使用函数 replace()使所有字符序列的出现位置都被另一个给定的字符序列取代。

执行后会输出:

```
连接两个字符串:
['hello xyz']
连接示例:
['hello abc' 'hi xyz']
Hello Hello Hello
*******hello********
Hello world
Hello How Are You?
['hello', 'how are you?']
['hello', 'how are you?']
```

2. 算术运算函数

在 NumPy 库中包含了大量实现各种数学运算功能的函数,如三角函数、算术运算函数和复数处理函数等。下面的实例文件 008.py 演示了使用正弦、余弦和正切函数的过程。

源码路径:daima\10\10-5\008.py

```
import numpy as np
a = np.array([0,30,45,60,90])
print ('不同角度的正弦值: ')
#通过乘以pi/180 转化为弧度
print(np.sin(a*np.pi/180)  )
print ('数组中角度的余弦值: ')
print(np.cos(a*np.pi/180))
print ('数组中角度的正切值: ')
print(np.tan(a*np.pi/180))
```

执行后会输出：

```
不同角度的正弦值：
[ 0.          0.5         0.70710678  0.8660254   1.        ]
数组中角度的余弦值：
[  1.00000000e+00   8.66025404e-01   7.07106781e-01   5.00000000e-01
   6.12323400e-17]
数组中角度的正切值：
[  0.00000000e+00   5.77350269e-01   1.00000000e+00   1.73205081e+00
   1.63312394e+16]
```

下面的实例文件 009.py 演示了使用算术函数 add()、subtract()、multiply() 和 divide() 实现四则运算的过程。在使用这 4 个函数时，要求输入的数组必须具有相同的形状或符合数组广播规则。

源码路径：daima\10\10-5\009.py

```python
import numpy as np
a = np.arange(9, dtype = np.float_).reshape(3,3)
print ('第一个数组：')
print(a )
print ('\n')
print ('第二个数组：' )
b = np.array([10,10,10])
print(b )
print ('\n' )
print ('两个数组相加：')
print(np.add(a,b))
print ('\n')
print ('两个数组相减：')
print(np.subtract(a,b)  )
print ('\n' )
print ('两个数组相乘：' )
print(np.multiply(a,b)   )
print ('\n'  )
print ('两个数组相除：' )
print(np.divide(a,b))
```

执行后会输出：

```
第一个数组：
[[ 0.  1.  2.]
 [ 3.  4.  5.]
 [ 6.  7.  8.]]


第二个数组：
[10 10 10]


两个数组相加：
[[ 10.  11.  12.]
 [ 13.  14.  15.]
 [ 16.  17.  18.]]


两个数组相减：
[[-10.  -9.  -8.]
 [ -7.  -6.  -5.]
 [ -4.  -3.  -2.]]


两个数组相乘：
[[  0.  10.  20.]
 [ 30.  40.  50.]
 [ 60.  70.  80.]]


两个数组相除：
[[ 0.   0.1  0.2]
 [ 0.3  0.4  0.5]
 [ 0.6  0.7  0.8]]
```

3. 统计函数

在 NumPy 库中有很多有用的统计函数，用于从给定的数组元素中求最小值、最大值、标准差和方差等。下面的实例文件 010.py 演示了使用算术函数从给定数组元素沿指定轴返回最小值和最大值的过程。

源码路径：daima\10\10-5\010.py

```
import numpy as np
a = np.array([[3,7,5],[8,4,3],[2,4,9]])
print('我们的数组是：')
print(a )
print('\n' )
print('调用 amin() 函数：')
print(np.amin(a,1) )
print('\n')
print('再次调用 amin() 函数：' )
print(np.amin(a,0) )
print('\n' )
print('调用 amax() 函数：')
print(np.amax(a) )
print('\n')
print('再次调用 amax() 函数：' )
print(np.amax(a, axis =  0))
```

执行后会输出：

```
我们的数组是：
[[3 7 5]
 [8 4 3]
 [2 4 9]]

调用 amin() 函数：
[3 3 2]

再次调用 amin() 函数：
[2 4 3]

调用 amax() 函数：
9

再次调用 amax() 函数：
[8 7 9]
```

4. 排序、搜索和计数函数

NumPy 库提供了各种排序功能。这些排序函数实现了不同的排序算法。每个排序算法的执行速度、最坏情况下的性能、所需的工作空间和算法的稳定性各不相同。表 10-1 对比了 3 种排序算法。

表 10-1 3 种排序算法的比较

种类	执行速度	最坏情况下的性能	工作空间	稳定性
'quicksort'(快速排序)	最快	$O(n2)$	0	否
'mergesort'(归并排序)	不快不慢	$O(n \lg(n))$	$n/2$	是
'heapsort'(堆排序)	最慢	$O(n*\lg(n))$	0	否

下面的实例文件 011.py 演示了使用算术函数 sort() 实现快速排序的过程。

源码路径：daima\10\10-5\011.py

```
import numpy as np
a = np.array([[3,7],[9,1]])
print('我们的数组是：')
print(a )
print('\n' )
print('调用 sort() 函数：' )
print(np.sort(a)  )
print('\n' )
```

```
print('沿轴 0 排序: ')
print(np.sort(a, axis = 0) )
print('\n')
#在sort函数中对字段排序
dt = np.dtype([('name', 'S10'),('age', int)])
a = np.array([("raju",21),("anil",25),("ravi", 17), ("amar",27)], dtype = dt)
print('我们的数组是: ')
print(a)
print('\n')
print('按 name 排序:')
print(np.sort(a, order = 'name'))
```

执行后会输出：

```
我们的数组是:
[[3 7]
 [9 1]]

调用 sort() 函数:
[[3 7]
 [1 9]]

沿轴 0 排序:
[[3 1]
 [9 7]]

我们的数组是:
[(b'raju', 21) (b'anil', 25) (b'ravi', 17) (b'amar', 27)]

按 name 排序:
[(b'amar', 27) (b'anil', 25) (b'raju', 21) (b'ravi', 17)]
```

5. 字节交换

存储在计算机内存中的数据取决于 CPU 使用的架构，存储形式可以是小端（最低有效位存储在最小的地址中）或大端（最低有效字节存储在最大的地址中）。在 NumPy 库中，通过函数 byteswap() 实现字节在大端和小端之间进行切换。下面的实例文件 012.py 演示了使用函数 byteswap() 实现字节交换的过程。

源码路径：daima\10\10-5\012.py

```
import numpy as np
a = np.array([1, 256, 8755], dtype = np.int16)
print('我们的数组是: ')
print(a)
print('以十六进制表示内存中的数据: ' )
print(map(hex,a) )
# byteswap() 函数通过传入 true 来原地交换
print('调用 byteswap() 函数: ' )
print(a.byteswap(True))
print('十六进制形式: ' )
print(map(hex,a) )
#我们可以看到字节已经交换了
```

执行后会输出：

```
我们的数组是:
[   1  256 8755]
以十六进制表示内存中的数据:
<map object at 0x000001C778D27668>
调用 byteswap() 函数:
[  256     1 13090]
十六进制形式:
<map object at 0x000001C778C64CC0>
```

6. 矩阵库

在 NumPy 库中包含了一个 Matrix 模块 numpy.matlib。此模块中的函数能够返回矩阵，而不是返回 ndarray 对象。例如，empty()函数能够返回一个新的矩阵，而不初始化元素。下面的实例文件 013.py 演示了使用函数 empty()返回一个矩阵的过程。

源码路径：daima\10\10-5\013.py

```
import numpy.matlib
import numpy as np
print(np.matlib.empty((2,2))  )
#填充为随机数据
```

执行后会输出：

```
[[  9.90263869e+067   8.01304531e+262]
 [  2.60799828e-310   9.48818959e+077]]
```

10.5.4 使用 Matplotlib 库

在 NumPy 库中可以使用 Matplotlib 库。下面的实例文件 014.py 演示了在 NumPy 库中使用 Matplotlib 库的过程。

源码路径：daima\10\10-5\014.py

```
import numpy as np
from matplotlib import pyplot as plt

x = np.arange(1,11)
y =  2  * x +  5
plt.title("Matplotlib demo")
plt.xlabel("x axis caption")
plt.ylabel("y axis caption")
plt.plot(x,y)
plt.show()
```

在上述代码中，ndarray 对象 x 由 np.arange()函数创建，它表示 x 轴上的值。y 轴上的对应值存储在另一个数组对象 y 中。这些值使用 Matplotlib 库的 pyplot 子模块的 plot()函数绘制。最后绘制的图形由 show()函数展示。执行效果如图 10-48 所示。

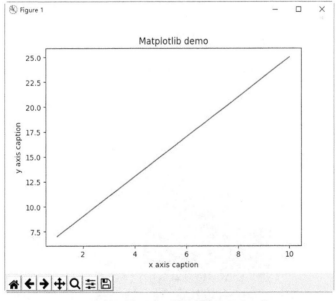

图 10-48　执行效果

下面的实例文件 015.py 演示了在 NumPy 库中使用 Matplotlib 绘制正弦波的过程。

源码路径：daima\10\10-5\015.py

```
import numpy as np
import matplotlib.pyplot as plt
```

```
#计算正弦曲线上点的 x 和 y 坐标
x = np.arange(0, 3 * np.pi, 0.1)
y = np.sin(x)
plt.title("sine wave form")
#使用Matplotlib 来绘制
plt.plot(x, y)
plt.show()
```
执行效果如图 10-49 所示。

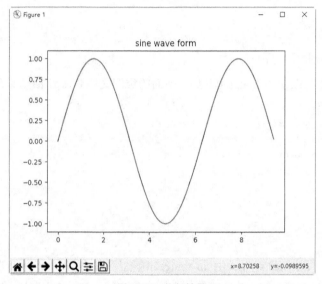

图 10-49　执行效果

下面的实例文件 016.py 演示了在 NumPy 库中使用 Matplotlib 库绘制直方图的过程。

源码路径：daima\10\10-5\016.py

```
a = np.array([22,87,5,43,56,73,55,54,11,20,51,5,79,31,27])
plt.hist(a, bins = [0,20,40,60,80,100])
plt.title("histogram")
plt.show()
```
执行效果如图 10-50 所示。

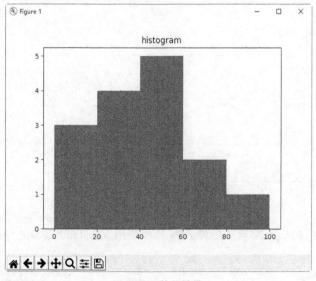

图 10-50　执行效果

第 11 章

第三方多媒体库

在软件程序的开发过程中，为了满足某些项目程序的需要，经常需要开发音频和视频等多媒体程序。本章将详细讲解使用 Python 第三方库开发多媒体应用程序的知识。

11.1 使用 audiolazy 库处理数字信号

在 Python 程序中，可以使用 audiolazy 库处理数字信号。本节将详细讲解 audiolazy 库的使用知识。

11.1.1 安装并尝试使用 audiolazy 库

安装库 audiolazy 的命令如下。

```
pip install audiolazy
```

为了使用 audiolazy 库，读者可能需要安装如下几个库。当然，这些库并不是必需的依赖库。

- pyaudio：在播放和录制音频时需要用到。
- NumPy：在做一些和数学相关的音频处理工作时用到，例如，从一个多项式求根 LSF 滤波器等。
- Matplotlib：处理和所有绘图操作相关的功能。
- SciPy（只用到测试和例子）：使用 LTI 过滤器测试 Oracle 和 Butterworth 滤波器。
- wxPython 和 Tkinter：在实现交互式 GUI FM 合成时用到。
- music21：音乐资料库合成和播放。
- Sphinx：可以用几种不同的文件格式创建软件文档。

下面的实例文件 audiolazy01.py 演示了使用 audiolazy 库处理流对象的过程。

源码路径：daima\11\11-1\audiolazy01.py

```
from audiolazy import *
from audiolazy.lazy_stream import Stream
a = Stream(2, -2, -1) #周期
b = Stream(3, 7, 5, 4) #周期
c = a + b #对应元素的总和
print(c.take(15)) #流对象的前15个元素
```

执行后会输出流对象 c 中的前 15 个元素。

```
[5, 5, 4, 6, 1, 6, 7, 2, 2, 9, 3, 3, 5, 5, 4]
```

11.1.2 实现巴特沃斯滤波器

下面的实例文件 audiolazy02.py 演示了使用 audiolazy 库实现巴特沃斯滤波器的过程。本实例用到了 audiolazy 库中的 ZFilter 接口，并用到了库 SciPy。

源码路径：daima\11\11-1\audiolazy02.py

```
rate = 44100
s, Hz = sHz(rate)
wp = pylab.array([100 * Hz, 240 * Hz]) # Bandpass range in rad/sample
ws = pylab.array([80 * Hz, 260 * Hz]) # Bandstop range in rad/sample

#使用wp/pi，因为scipy默认频率为0～1 Hz（奈奎斯特频率）
order, new_wp_divpi = buttord(wp/pi, ws/pi, gpass=dB10(.6), gstop=dB10(.4))
ssfilt = butter(order, new_wp_divpi, btype="bandpass")
filt_butter = ZFilter(ssfilt[0].tolist(), ssfilt[1].tolist())

#一些调试信息
new_wp = new_wp_divpi * pi
print("Butterworth filter order:", order) # Should be 3
print("Bandpass ~3dB range (in Hz):", new_wp / Hz)

#使用巴特沃斯滤波器中频率和带宽的谐振器
freq = new_wp.mean()
bw = new_wp[1] - new_wp[0]
filt_reson = resonator.z_exp(freq, bw)

#使用Matplotlib绘图
```

11.2 使用 audioread 库实现音频解码

```
kwargs = {
  "min_freq": 10 * Hz,
  "max_freq": 800 * Hz,
  "rate": rate, #频率的单位是赫兹
}
filt_butter.plot(pylab.figure("From scipy.signal.butter"), **kwargs)
filt_reson.plot(pylab.figure("From audiolazy.resonator.z_exp"), **kwargs)
filt_butter.zplot(pylab.figure("Zeros/Poles from scipy.signal.butter"))
filt_reson.zplot(pylab.figure("Zeros/Poles from audiolazy.resonator.z_exp"))
pylab.ioff()
pylab.show()
```

执行后会输出绘制的滤波器，如图 11-1 所示。

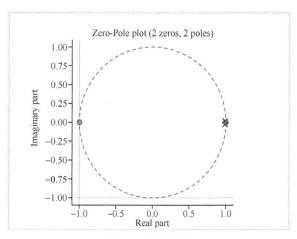

图 11-1　执行效果

11.2　使用 audioread 库实现音频解码

在 Python 程序中，可以使用 audioread 库实现音频解码功能。目前支持的库有 Gstreamer、Core Audio、MAD 和 ffmpeg。安装 audioread 库的命令如下。

```
pip install audioread
```

下面的实例文件 decode.py 演示了使用 audioread 库将音频文件解码为 WAV 文件的过程。

源码路径：daima\11\11-2\decode.py

```python
from __future__ import print_function
import audioread
import sys
import os
import wave
import contextlib


def decode(filename):
    filename = os.path.abspath(os.path.expanduser(filename))
    if not os.path.exists(filename):
        print("File not found.", file=sys.stderr)
        sys.exit(1)

    try:
        with audioread.audio_open(filename) as f:
            print('Input file: %i channels at %i Hz; %.1f seconds.' %
                  (f.channels, f.samplerate, f.duration),
                  file=sys.stderr)
            print('Backend:', str(type(f).__module__).split('.')[1],
                  file=sys.stderr)
```

```
                    with contextlib.closing(wave.open(filename + '.wav', 'w')) as of:
                        of.setnchannels(f.channels)
                        of.setframerate(f.samplerate)
                        of.setsampwidth(2)

                        for buf in f:
                            of.writeframes(buf)

    except audioread.DecodeError:
        print("File could not be decoded.", file=sys.stderr)
        sys.exit(1)


if __name__ == '__main__':
    decode(sys.argv[1])
```

进入命令行终端，输入如下命令进行解码处理。

```
python decode.py xxx
```

其中，"xxx" 表示将要处理的音频文件名。

11.3 使用 eyeD3 库处理音频

在 Python 程序中，可以使用 eyeD3 库来操作音频文件。具体来讲，就是操作包含 ID3 元信息的 MP3 音频。

11.3.1 安装并尝试使用 eyeD3 库

文件安装 eyeD3 库的命令如下。

```
pip install eyeD3
```

可能在安装 eyeD3 库后提示如下错误。

```
importerror: failed to find libmagic, check you installation
```

这时需要单独下载 python_magi 的轮子文件，例如，64 位 Python 3.6 需要下载的轮子文件是：

```
python_magic_bin-0.4.14-py2.py3-none-win_amd64.whl
```

然后通过如下命令来安装 python_magi。

```
python -m pip install --user python_magic_bin-0.4.14-py2.py3-none-win_amd64.whl
```

在安装 eyeD3 库后，既可以在终端使用命令行来操作音频文件，也可以使用编程的方式来操作音频文件。

通过如下命令可以查看音频文件 111.mp3 的基本信息。

```
eyeD3 song.mp3
```

11.3.2 使用 eyeD3 库编程

接下来，将详细讲解使用编程的方式来操作音频文件。下面的实例文件 eyeD301.py 演示了使用 eyeD3 库处理指定音频文件的过程。

源码路径：daima\11\11-3\eyeD301.py

```
import eyed3

audiofile = eyed3.load("111.mp3")
audiofile.tag.artist = u"Integrity"
audiofile.tag.album = u"Humanity"
audiofile.tag.album_artist = u"Integrity"
audiofile.tag.title = u"Hollow"
audiofile.tag.track_num = 2

audiofile.tag.save()
```

通过上述代码，为指定音频文件 111.mp3 设置了指定的信息，我们可以通过查看文件 111.mp3 的属性信息的方式来验证上述操作是否成功。音频文件 111.mp3 的属性信息如图 11-2 所示。

11.3 使用 eyeD3 库处理音频

图 11-2 文件 111.mp3 的属性信息

下面的实例文件 eyeD302.py 演示了使用 eyeD3 库批量处理指定目录下音频文件信息的过程。

源码路径：daima\11\11-3\eyeD302.py

```
import eyed3

mp3list = open(r'mlist','r').read()
for i in  mp3list.split('\n'):
    if i != '':
            artist =i.split('\t')[0]
            album=i.split('\t')[1]
            filename=i.split('\t')[2]
            title=filename.split('.')[0].split('/')[-1]
            print(filename+ " " + title + " " + artist + " " +album)

            audiofile = eyed3.load(filename)
            audiofile.initTag()
            audiofile.tag.artist = artist.decode("utf-8")
            audiofile.tag.album = album.decode("utf-8")
            audiofile.tag.album_artist = artist.decode("utf-8")
            audiofile.tag.title = title.decode("utf-8")
            #audiofile.tag.track_num = int(track_num)
            audiofile.tag.save()
```

上述代码要求有一个文件是 mlist，里面的内容格式如下（列之间用制表符进行分割）。

歌手名字　歌曲专辑　mp3文件位置

假如文件 mlist 中的内容是：

茶靡　　周杰伦　　/Users/apple/Music/Popular/周杰伦/yes周杰伦.mp3

则执行后会输出：

mlist1/111.mp3 111 茶靡　周杰伦

11.3.3 MP3 文件编辑器

下面的实例文件 eyeD303.py 演示了使用"eyeD3+tkinter+PIL"开发一个 MP3 文件编辑器的过程。文件 eyeD303.py 的具体实现流程如下。

源码路径：dalma\11\11-3\eyeD303.py

(1) 设置在 tkinter 窗体和菜单栏中显示的文本信息。具体实现代码如下。

```python
app_title = "EyeD3Tk"

# eyeD3库不使用这些枚举类
ID3_IMG_TYPES = ("OTHER", "ICON", "OTHER_ICON", "FRONT_COVER", "BACK_COVER", "LEAFLET",
                 "MEDIA", "LEAD_ARTIST",
                 "ARTIST", "CONDUCTOR", "BAND", "COMPOSER", "LYRICIST", "RECORDING_LOCATION",
                 "DURING_RECORDING",
                 "DURING_PERFORMANCE", "VIDEO", "BRIGHT_COLORED_FISH", "ILLUSTRATION",
                 "BAND_LOGO", "PUBLISHER_LOGO")


class MainWindow:
    no_img_txt = "封面：没有图像"

    mp3_file_types = [('MP3音频文件', '*.mp3 *.MP3'),
                      ('所有文件', "*")]

    img_file_types = [('ID3-兼容图像', '*.jpg *.JPG *.jpeg *.JPEG *.png *.PNG'),
                      ('所有文件', "*")]

    jpg_file_types = [('JPEG图像', '*.jpg *.JPG *.jpeg *.JPEG')]

    id3_gui_fields = (('title', "Title:"),
                      ('artist', "Artist:"),
                      ('composer', "Composer:"),
                      ('album', "Album:"),
                      ('album_artist', "Album Artist:"),
                      ('original_release_date', "Original Release Date:"),
                      ('release_date', "Release Date:"),
                      ('recording_date', "Recording Date:"),
                      ('track_num', "Track:"),
                      ('num_tracks', "Tracks in Album:"),
                      ('genre', "Genre:"),
                      ('comments', "Comments:"))
```

(2) 编写初始化函数 __init__()，设置在窗体中显示的各类控件的初始值，包括文本框控件、按钮控件、图像控件和文件选择控件。具体实现代码如下。

```python
    def __init__(self, root, cmd_line_mp3_file):
        self.cmd_line_mp3_file = cmd_line_mp3_file
        self.master = root
        self.master.title(app_title)

        self.file_frame,self.file_button,self.mp3_file_sv,self.file_entry = None,None,None,None
        self.build_mp3_file_frame()

        self.id3_section_frame, self.id3_section_label = None, None
        self.build_id3_section_frame()

        self.id3_frame = dict()
        self.id3_entry = dict()
        self.id3_label = dict()
        for field, text in self.id3_gui_fields:
            self.create_id3_field_gui_element(field, text)

        self.front_cover_frame, self.image_description_sv = None, None
        self.image_dimension_label, self.extract_image_button = None, None
        self.build_front_cover_frame()

        self.new_front_cover_frame,self.new_front_cover_button,self.remove_button=None,None,None
        self.new_front_cover_sv, self.new_front_cover_entry = None, None
        self.build_new_front_cover_frame()

        self.save_button = None
        self.build_save_button()
```

```
        self.tk_label_for_img = None
        self.audio_file = None
        self.tk_img = None
        self.image_file = None

        self.fld_val = dict()

        self.open_cmd_line_file()
```

(3) 编写函数 build_mp3_file_frame()，当单击"选择 MP3 文件"按钮时弹出文件选择框，我们可以选择一个将要处理的 MP3 文件。具体实现代码如下。

```
    def build_mp3_file_frame(self):
        self.file_frame = Frame(self.master)
        self.file_frame.pack(fill=X)

        self.file_button = Button(self.file_frame, text="选择MP3文件",
            command=self.file_select_button_action)
        self.file_button.pack(side=LEFT)

        self.mp3_file_sv = StringVar()
        self.file_entry = Entry(self.file_frame,width=75,textvariable=self.mp3_file_sv)
        self.file_entry.bind('<Return>', lambda _: self.file_entry_return_key_action())
        self.file_entry.pack(fill=X)
        self.mp3_file_sv.set("...选择一个文件...")
```

(4) 编写函数 build_id3_section_frame()，构建 eyeD3 处理框架，在窗体顶部显示提示文本"ID3 标签"。具体实现代码如下。

```
    def build_id3_section_frame(self):
        self.id3_section_frame = Frame(self.master)
        self.id3_section_frame.pack(fill=X)

        self.id3_section_label=Label(self.id3_section_frame,text="---ID3标签---",justify=CENTER)
        self.id3_section_label.pack(fill=X)
```

(5) 编写函数 build_front_cover_frame()，在窗体中设置"将所有图像提取到文件"按钮。具体实现代码如下。

```
    def build_front_cover_frame(self):
        self.front_cover_frame = Frame(self.master)
        self.front_cover_frame.pack(fill=X)

        self.image_description_sv = StringVar()
        self.image_description_sv.set(self.no_img_txt)
        self.image_dimension_label = Label(self.front_cover_frame, textvariable=self.
            image_description_sv)
        self.image_dimension_label.pack()
        self.extract_image_button = Button(self.front_cover_frame, text="将所有图像提取到文件",
                            state=DISABLED, command=self.extract_images_
                            button_action)
        self.extract_image_button.pack()
```

(6) 编写函数 build_new_front_cover_frame()，在窗体中设置"新的封面"和"删除所有图像"按钮。具体实现代码如下。

```
    def build_new_front_cover_frame(self):
        self.new_front_cover_frame = Frame(self.master)
        self.new_front_cover_frame.pack(fill=X)

        self.new_front_cover_button = Button(self.new_front_cover_frame,text="新的封面 ...",
                            command=self.new_front_cover_button_action)
        self.new_front_cover_button.pack(side=LEFT)
        self.remove_button = Button(self.new_front_cover_frame, text="删除所有图像",
                            command=self.remove_button_action)
        self.remove_button.pack(side=RIGHT)
        self.new_front_cover_sv = StringVar()
        self.new_front_cover_entry = Entry(self.new_front_cover_frame, width=50,
            textvariable=self.new_front_cover_sv)
        self.new_front_cover_entry.bind('<Return>',lambda _:self.img_entry_return_key_action())
        self.new_front_cover_entry.pack(fill=X)
```

（7）编写函数 build_save_button()，在窗体中设置"保存为 MP3"按钮。具体实现代码如下。

```
def build_save_button(self):
    self.save_button=Button(self.master,text="保存为MP3",command=self.save_button_action)
    self.save_button.pack(fill=X)
```

（8）编写函数 open_cmd_line_file()，在窗体中设置菜单"选择一个文件"或"无效的文件路径"提示信息。具体实现代码如下。

```
def open_cmd_line_file(self):
    if len(self.cmd_line_mp3_file) == 0:
        self.new_front_cover_sv.set("... 选择一个文件 ...")
    else:
        if isfile(self.cmd_line_mp3_file):
            self.mp3_file_sv.set(self.cmd_line_mp3_file)
            self.open_mp3_file()
        else:
            self.mp3_file_sv.set("无效的文件路径 : '" + self.cmd_line_mp3_file + "'")
```

（9）编写函数 create_id3_field_gui_element()创建 eyeD3 界面元素。具体实现代码如下。

```
def create_id3_field_gui_element(self, name, text):
    self.id3_frame[name] = Frame(self.master)
    self.id3_frame[name].pack(fill=X)
    self.id3_label[name] = Label(self.id3_frame[name], text=text, anchor='e', width=17)
    self.id3_label[name].pack(side=LEFT)
    self.id3_entry[name] = Entry(self.id3_frame[name])
    self.id3_entry[name].pack(fill=X)
```

（10）编写函数 new_front_cover_button_action()实现单击"选择 MP3 文件"按钮后的事件处理程序。具体实现代码如下。

```
def file_select_button_action(self):
    new_path = filedialog.askopenfilename(parent=self.file_frame, filetypes=
        self.mp3_file_types)
    self.mp3_file_sv.set(new_path)
    self.open_mp3_file()
```

（11）编写函数 new_front_cover_button_action()，实现单击"新的封面"按钮后的事件处理程序。具体实现代码如下。

```
def new_front_cover_button_action(self):
    new_path = filedialog.askopenfilename(parent=self.new_front_cover_frame,
        filetypes=self.img_file_types)
    self.new_front_cover_sv.set(new_path)
    with open(new_path, 'rb') as self.image_file:
        self.display_image_file()
        self.put_new_image_into_tag()
```

（12）编写如下所示的 6 个函数，实现将所有图像提取到文件的功能。具体实现代码如下。

```
def extract_images_button_action(self):
    self.try_to_extract_id3_images_to_files()

def try_to_extract_id3_images_to_files(self):
    try:
        self.extract_id3_images_to_files()
    except AttributeError:
        pass

def extract_id3_images_to_files(self):
    for info in self.audio_file.tag.images:
        self.extract_id3_image_to_file(info)

def extract_id3_image_to_file(self, info):
    def_ext, ftypes = self.get_image_file_extension(info)
    path = filedialog.asksaveasfilename(parent=self.front_cover_frame,
                                        defaultextension=def_ext,
                                        initialfile=self.get_initial_image_
                                        file_name(info),
                                        filetypes=ftypes)
    
    if path is not None and path != "":
        with open(path, 'wb') as img_file:
            img_file.write(info.image_data)
```

```python
    def get_image_file_extension(self, info):
        img = Image.open(BytesIO(info.image_data))
        default_extension = "." + str(img.format).lower()
        file_types = []
        if default_extension == ".jpeg":
            file_types = self.jpg_file_types
        elif default_extension == ".png":
            file_types = self.png_file_types
        return default_extension, file_types

    def get_initial_image_file_name(self, info):
        name = ID3_IMG_TYPES[info.picture_type]
        if info.description != "":
            name = info.description + '.' + name
        return name
```

（13）编写函数 open_mp3_file()和 try_to_open_mp3_file()，打开一个指定的 MP3 文件。具体实现代码如下。

```python
    def open_mp3_file(self):
        self.clear_gui_tag_entry_elements()
        self.try_to_open_mp3_file()
        if self.audio_file:
            self.load_tag_into_gui()

    def try_to_open_mp3_file(self):
        try:
            self.audio_file = load(self.file_entry.get())
        except IOError:
            self.audio_file = None
            self.mp3_file_sv.set("没有选择文件!")
```

（14）编写函数 load_tag_into_gui()，将音频标签加载到窗体中。具体实现代码如下。

```python
    def load_tag_into_gui(self):
        self.init_id3_tag()
        self.put_tag_fields_in_gui_entries()
        self.try_to_open_id3_tag_image_as_file_io()
        if self.image_file is None:
            self.clear_image_from_gui()
        else:
            self.display_image_file()
```

（15）编写函数 save_button_action()，实现单击"保存 MP3"按钮后的事件处理程序，将在表单中设置的各个信息编辑为指定的音频文件。具体实现代码如下。

```python
    def save_button_action(self):
        self.gui_fields_to_fld_val()

        for key, val in self.fld_val.items():
            print(key, ":", val)

        tag = self.audio_file.tag
        tag.version = ID3_DEFAULT_VERSION

        tag.title = self.fld_val['title']
        tag.artist = self.fld_val['artist']
        tag.composer = self.fld_val['composer']
        tag.album = self.fld_val['album']
        tag.album_artist = self.fld_val['album_artist']
        tag.original_release_date = self.fld_val['original_release_date']
        tag.release_date = self.fld_val['release_date']
        tag.recording_date = self.fld_val['recording_date']

        tag.genre = Genre(self.fld_val['genre'])
        tag.track_num = (self.fld_val['track_num'], self.fld_val['num_tracks'])
        if self.fld_val['comments'] is not None and self.fld_val['comments'] != "":
            tag.comments.set(self.fld_val['comments'])

        tag.save(encoding="utf_8")
```

（16）编写函数 init_id3_tag()，初始化标签信息。具体实现代码如下。

```python
    def init_id3_tag(self):
        if self.audio_file.tag is None:
            self.audio_file.initTag()
```

（17）编写函数 clear_gui_tag_entry_element()，清除窗体中的标签信息。具体实现代码如下。

```python
    def clear_gui_tag_entry_elements(self):
        for key, entry in self.id3_entry.items():
            entry.delete(0, END)
        self.clear_image_from_gui()
```

（18）编写函数 clear_image_from_gui() 清除音频封面图像信息。具体实现代码如下。

```python
    def clear_image_from_gui(self):
        self.image_description_sv.set(self.no_img_txt)
        self.extract_image_button['state'] = DISABLED
        if self.tk_label_for_img is not None:
            self.tk_label_for_img.pack_forget()
```

（19）编写函数 put_tag_fields_in_gui_entries()，将音频标签放入窗体文本框中。具体实现代码如下。

```python
    def put_tag_fields_in_gui_entries(self):
        self.id3_tag_to_fld_val()
        self.fld_val_to_gui_fields()
```

（20）编写函数 id3_tag_to_fld_val()，根据在文本框中设置的标签信息进行修改。具体实现代码如下。

```python
    def id3_tag_to_fld_val(self):
        tag = self.audio_file.tag
        self.fld_val['title'] = self.tag_to_str(tag.title)
        self.fld_val['artist'] = self.tag_to_str(tag.artist)
        self.fld_val['composer'] = self.tag_to_str(tag.composer)
        self.fld_val['album'] = self.tag_to_str(tag.album)
        self.fld_val['album_artist'] = self.tag_to_str(tag.album_artist)
        self.fld_val['original_release_date'] = self.tag_to_str(tag.original_release_date)
        self.fld_val['release_date'] = self.tag_to_str(tag.release_date)
        self.fld_val['recording_date'] = self.tag_to_str(tag.recording_date)

        self.fld_val['genre'] = "" if tag.genre is None else self.tag_to_str(tag.genre.name)
        self.fld_val['track_num'] = "" if tag.track_num is None or len(tag.track_num) < 1 \
            else self.tag_to_str(tag.track_num[0])
        self.fld_val['num_tracks'] = "" if tag.track_num is None or len(tag.track_num) < 2 \
            else self.tag_to_str(tag.track_num[1])
        self.id3_comments_to_fld_val()

    def tag_to_str(self, tag_element):
        return "" if tag_element is None else str(tag_element)
```

（21）编写函数 id3_comments_to_fld_val() 处理标签中的 comments 信息。具体实现代码如下。

```python
    def id3_comments_to_fld_val(self):
        self.fld_val['comments'] = ""
        for comment_accessor in self.audio_file.tag.comments:
            if comment_accessor.description != "":
                self.fld_val['comments'] += comment_accessor.description + ": "
            self.fld_val['comments'] += self.tag_to_str(comment_accessor.text)
```

（22）编写函数 display_image_file()，在窗体中预览显示用户选择的封面图像。具体实现代码如下。

```python
    def display_image_file(self):
        img = Image.open(self.image_file)
        original_dimensions = img.size
        img = img.resize((200, 200), Image.ANTIALIAS)
        self.tk_img = ImageTk.PhotoImage(img)
        if self.tk_label_for_img is None:
            self.tk_label_for_img = Label(self.front_cover_frame, image=self.tk_img)
        else:
            self.tk_label_for_img.configure(image=self.tk_img)
        self.tk_label_for_img.pack()
        self.image_description_sv.set("FRONT_COVER: {} x {}".format(original_dimensions[0],
            original_dimensions[1]))
        self.extract_image_button['state'] = NORMAL
```

（23）编写函数 remove_button_action()和 remove_all_images_from_id3_tag()，实现"删除所有图像"按钮的事件处理程序。具体实现代码如下。

```python
def remove_button_action(self):
    self.image_file = None
    self.clear_image_from_gui()
    self.remove_all_images_from_id3_tag()

def remove_all_images_from_id3_tag(self):
    for description in [info.description for info in self.audio_file.tag.images]:
        self.audio_file.tag.images.remove(description)
```

（24）编写函数 put_new_image_into_tag()将新的封面图像添加到音频标签中。具体实现代码如下。

```python
def put_new_image_into_tag(self):
    if isfile(self.new_front_cover_sv.get()):
        image_data = open(self.new_front_cover_sv.get(), 'rb').read()
        self.audio_file.tag.images.set(ImageFrame.FRONT_COVER, image_data,
            self.get_new_front_cover_mime_type())
```

（25）编写函数 should_display_image()，显示当前封面图像。具体实现代码如下。

```python
def should_display_image(self, image_idx, img_info):
    is_front_cover = img_info.picture_type == ImageFrame.FRONT_COVER
    is_last_picture = image_idx + 1 == len(self.audio_file.tag.images)
    return is_front_cover or is_last_picture
```

（26）编写函数 get_new_front_cover_mime_type()，获取新的当前封面的 MIME 类型。具体实现代码如下。

```python
def get_new_front_cover_mime_type(self):
    mime = Magic(mime=True)
    return mime.from_file(self.new_front_cover_sv.get())
```

执行后会显示一个 MP3 标签编辑器窗体程序，我们可以在表单中设置指定音频文件的标签信息。执行效果如图 11-3 所示。

图 11-3　执行效果

11.4　使用 m3u8 库

在 Python 程序中，可以使用 m3u8 库来解析 m3u8 文件。本节将详细讲解 m3u8 库的安装和使用方法。

11.4.1 m3u8 库的介绍和安装

m3u8 是苹果公司推出的一种视频播放标准,是 m3u 的一种。但是 m3u8 的编码方式是 utf-8,是一种文件检索格式,它将视频切割成一小段一小段的 ts 格式的视频文件,然后存放在服务器中(现在为了减少 I/O 访问次数,一般存储在服务器的内存中)。通过 m3u8 解析路径,然后去请求,这样每次请求很小一段视频,可以实现近似于实时播放的效果。

安装 m3u8 库的命令如下。

```
pip install m3u8
```

11.4.2 下载 m3u8 视频并转换为 MP4 文件

下面的实例文件 m3u801.py 演示了,下载指定网址的 m3u8 视频文件并将下载的 m3u8 文件转换为 MP4 文件的过程。

源码路径:daima\11\11-4\m3u801.py

```python
import m3u8
import os
import requests
import subprocess
"""
获取m3u8ts文件
"""
def getM3u8(url):
    m3u8_obj = m3u8.load(url)

    ts_url_list = []

    base_uri = m3u8_obj.base_uri

    ts_list = m3u8_obj.files

    for _ts in ts_list:

        ts_url = base_uri + _ts

        ts_url_list.append(ts_url)

    # print ts_url

    # response = requests.head(ts_url)

    # if response.status_code == 200:
    #     print "URL 没问题"

    return ts_url_list

"""
下载ts文件
"""
def download_movie(movie_url, _path):
    os.chdir(_path)
    print('>>>[+] downloading...')
    print('-' * 60)
    error_get = []

    for _url in movie_url:
        # ts视频的名称
        movie_name = _url.split('/')[-1][-6:]
        #movie_name = str(_url.split("/")[7]).split("?")[0]
        try:
            # 'Connection':'close'  防止请求端口的占用
            # timeout=30        防止请求超时
            movie = requests.get(_url, headers = {'Connection':'close'}, timeout=60)
            with open(movie_name, 'wb') as movie_content:
                movie_content.writelines(movie)
```

```
                print('>>>[+] File ' + movie_name + ' done')
            #捕获异常，记录失败的请求
            except:
                error_get.append(_url)
                continue
    #如果没有不成功的请求，就结束
    if error_get:
        # print u'共有%d个请求失败' % len(file_list)
        print('-' * 60)
        download_movie(error_get, _path)
    else:
        print('>>>[+] Download successfully!!!')
"""
合并ts文件,输出执行语句
ls * | perl -nale 'chomp;push @a,$_;END{printf "ffmpeg -i \"concat:%s\" -acodec copy -vcodec
copy -absf aac_adtstoasc out.mp4\n", join("|",@a)}'
"""
def hebing(path,outfile):
    filelist = []
    for file in os.listdir(path):
        if len(file.split(".")) == 2:
            if file.split(".")[1] == 'ts':
                filelist.append(path + file)
    str = '|'.join(filelist)

    cmd_str = 'ffmpeg -i \"concat:' + str + '\" ' + '-acodec copy -vcodec copy -absf
    aac_adtstoasc ' + path + outfile
    print(cmd_str)
    return cmd_str

"""
运行ffmpeg
"""
def runConvertMp4(cmd_str):
    str_env = "C:\ffmpeg"
    str_cmd = str_env + cmd_str
    print(str_cmd)
    subprocess.call(str_cmd, shell=True)

if __name__ == "__main__":
    url = "http://指定网址/1.m3u8"
    path = "ts"
    ts_url_list = getM3u8(url)
    download_movie(ts_url_list,path)
    cmd_str = hebing("123","out.mp4")
    runConvertMp4(cmd_str)
```

在上述代码中用到了开源软件 ffmpeg，读者需要安装 ffmpeg 并为其设置环境变量，只有这样才能调用 ffmpeg 将下载的 m3u8 文件转换为 MP4 文件。

11.5 使用 mutagen 库

在 Python 程序中，可以使用 mutagen 库来处理音频元数据。本节将详细讲解 mutagen 库的安装和使用方法。

11.5.1 安装并尝试使用 mutagen 库

mutagen 库支持 ASF、FLAC、MP4、MP3、OGG、OGG FLAC、Ogg Speex、Ogg Theora 和 Ogg Vorbis 格式的数据。安装 mutagen 库的命令如下。

```
pip install mutagen
```

下面的实例文件 mutagen_ID3.py 演示了使用 mutagen 库获取指定音频文件元数据信息的过程。

源码路径：daima\11\11-5\mutagen_ID3.py

```
from mutagen.id3 import ID3

#输出元数据
metadata = ID3("example.mp3")
print(metadata.pprint())
```

在上述代码中，通过使用 mutagen 库中的 id3 模块，获取了音频文件 example.mp3 的元数据信息。执行后会输出：

```
APIC=artist,  (image/jpeg, 140022 bytes)
POPM=MusicBee=0 186/255
SYLT=[unrepresentable data]
TALB=Aerosmith's Greatest Hits
TBPM=82
TCON=Rock
TDRC=1993
TIPL=[unrepresentable data]
TIT2=Last Child
TLAN=English
TPE1=Aerosmith
TPE2=Aerosmith
TPUB=Sony
TRCK=05
TSO2=Aerosmith
TSSE=Lame3.99
TXXX=Acoustid Id=406e3b2f-b30f-499a-9df5-cd3940c5ed5c
TXXX=Dynamic Range (DR)=8
TXXX=Dynamic Range (R128)=8.675729751586914
TXXX=MOOD=Funky / Sexy
TXXX=MusicBrainz Album Artist Id=3d2b98e5-556f-4451-a3ff-c50ea18d57cb
TXXX=MusicBrainz Album Status=bootleg
TXXX=MusicBrainz Album Type=album/compilation
TXXX=MusicBrainz Artist Id=3d2b98e5-556f-4451-a3ff-c50ea18d57cb
TXXX=MusicBrainz Release Group Id=e2255b98-9ccb-3991-821e-2a3ac6cb44d9
TXXX=Peak Level (R128)=+0.4 dBTP; +0.3 Left; +0.4 Right
TXXX=Peak Level (Sample)=+0.0 dB; +0.0 Left; +0.0 Right
TXXX=STATUS=1
TXXX=VALENCE=3.4
TXXX=Volume Level (R128)=-12.7938604354858398
TXXX=Volume Level (ReplayGain)=-7.7938599586486816
TXXX=added_timestamp=130376042744720543
TXXX=replaygain_album_gain=-6.00 dB
TXXX=replaygain_album_peak=1.058177
TXXX=replaygain_track_gain=-7.79 dB
TXXX=replaygain_track_peak=1.000000
```

下面的实例文件 mutagen_MP3.py 演示了，使用 mutagen 库获取指定音频文件属性信息的过程。

源码路径：daima\11\11-5\mutagen_MP3.py

```
from mutagen.mp3 import MP3

#输出MP3的属性信息
audio = MP3("example.mp3")
print(audio.info.length, audio.info.bitrate)
```

在上述代码中，通过使用 mutagen 库中的 mp3 模块，获取了音频文件 example.mp3 的长度和比特率信息。执行后会输出：

```
199.32009070294785 256000
```

11.5.2 获取指定音频文件的标签信息

下面的实例文件 mutagen_EasyMP3.py 演示了，使用 mutagen 库获取指定音频文件标签信息的过程。

源码路径：daima\11\11-5\mutagen_EasyMP3.py

```
from mutagen.easyid3 import EasyID3

# EasyID3使用文本字符串显示音频的标签
```

```
#注意，并不支持所有的ID3标准标签

audio = EasyID3("example.mp3")
print("---------------------------------")
print("音频文件example.mp3的标签信息:")
print("---------------------------------")
print(audio.pprint())    #以可读形式输出所有可用标记

#使用TXXX 键格式创建自定义标签
print("---------------------------------")
print("有效的标签:")
audio.RegisterTXXXKey('valence', 'VALENCE')
print(EasyID3.valid_keys.keys())
print("---------------------------------")

audio["title"] = "Last Child"
audio.save()
print(audio["title"])

audio["valence"] = "3.4"
audio.save()
print(audio["valence"])
```

在上述代码中，通过使用 mutagen 库中的 easyid3 模块，获取了音频文件 example.mp3 的标签信息。执行后会输出：

```
---------------------------------
音频文件example.mp3的标签信息:
---------------------------------
acoustid_id=406e3b2f-b30f-499a-9df5-cd3940c5ed5c
album=Aerosmith's Greatest Hits
albumartist=Aerosmith
artist=Aerosmith
bpm=82
date=1993
genre=Rock
language=English
musicbrainz_albumartistid=3d2b98e5-556f-4451-a3ff-c50ea18d57cb
musicbrainz_albumstatus=bootleg
musicbrainz_albumtype=album/compilation
musicbrainz_artistid=3d2b98e5-556f-4451-a3ff-c50ea18d57cb
musicbrainz_releasegroupid=e2255b98-9ccb-3991-821e-2a3ac6cb44d9
organization=Sony
title=Last Child
tracknumber=05
---------------------------------
有效的标签:
dict_keys(['album', 'bpm', 'compilation', 'composer', 'copyright', 'encodedby', 'lyricist',
    'length', 'media', 'mood', 'title', 'version', 'artist', 'albumartist', 'conductor',
    'arranger', 'discnumber', 'organization', 'tracknumber', 'author', 'albumartistsort',
    'albumsort', 'composersort', 'artistsort', 'titlesort', 'isrc', 'discsubtitle',
    'language', 'genre', 'date', 'originaldate', 'performer:*', 'musicbrainz_trackid',
    'website', 'replaygain_*_gain', 'replaygain_*_peak', 'musicbrainz_artistid', 'musicbrainz_
    albumid', 'musicbrainz_albumartistid', 'musicbrainz_trmid', 'musicip_puid', 'musicip_
    fingerprint', 'musicbrainz_albumstatus', 'musicbrainz_albumtype', 'releasecountry',
    'musicbrainz_discid', 'asin', 'performer', 'barcode', 'catalognumber', 'musicbrainz_
    releasetrackid', 'musicbrainz_releasegroupid', 'musicbrainz_workid', 'acoustid_
    fingerprint', 'acoustid_id', 'valence'])
---------------------------------
['Last Child']
['3.4']
```

11.5.3 批量设置视频文件的封面图片

下面的实例文件 mutagen01.py 演示了，使用 mutagen+imdbpie+tmdbsimple 库批量设置视频文件的封面图片的过程。文件 mutagen01.py 的具体实现流程如下。

源码路径：daima\11\11-5\mutagen01.py

（1）编写函数 collect_stream_metadata() 收集数据流信息。具体实现代码如下。

```python
def collect_stream_metadata(filename):
    """
    返回指定参数（文件名）传递的媒体文件中存在的流元数据列表
    """
    command = 'ffprobe -i "{}" -show_streams -of json'.format(filename)
    args = shlex.split(command)
    p = subprocess.Popen(args, stdout=subprocess.PIPE, stderr=subprocess.PIPE,
                         universal_newlines=True)
    out, err = p.communicate()

    json_data = JSONDecoder().decode(out)

    return json_data
```

（2）编写函数 PrintException() 输出异常信息。具体实现代码如下。

```python
def PrintException():
    exc_type, exc_obj, tb = sys.exc_info()
    f = tb.tb_frame
    lineno = tb.tb_lineno
    fname = f.f_code.co_filename
    linecache.checkcache(fname)
    line = linecache.getline(fname, lineno, f.f_globals)
    print('\nEXCEPTION IN ({}, LINE {} "{}"): {}'.format(fname,
                                                        lineno,
                                                        line.strip(),
                                                        exc_obj))
```

（3）编写函数 collect_files() 返回当前目录中的文件列表，例如，collect_files('txt') 能够返回一个列表中的所有文件。具体实现代码如下。

```python
def collect_files(file_type):

    filenames = []
    for filename in os.listdir(os.getcwd()):
        if filename.endswith(file_type):
            filenames.append(filename)
    return filenames
```

（4）编写函数 get_common_files() 获取一个文件名列表。参数 mediafile_list 和 strfile_list 是两个列表类型，前者表示视频文件，后者表示字幕文件。具体实现代码如下。

```python
def get_common_files(mediafile_list, srtfile_list):
    media_filenames = [i[:-4] for i in mediafile_list]
    subtitle_filenames = [i[:-4] for i in srtfile_list]
    media_type = mediafile_list[0][-4:]
    media_set = set(media_filenames)
    srt_set = set(subtitle_filenames)
    common_files = list(media_set & srt_set)
    common_files = [i + media_type for i in common_files]
    common_files.sort()
    return common_files
```

（5）编写函数 remove_common_files() 删除列表中的同名文件。具体实现代码如下。

```python
def remove_common_files(list1, list2):
    new_list1 = list(set(list1) - set(list2))
    new_list1.sort()
    return new_list1
```

（6）编写函数 start_process() 启动操作线程，其中，1 表示普通的 MP4 文件，2 表示包含字幕的 MP4，3 表示 MKV 文件，4 表示带有 MP4 字幕的 MKV。具体实现代码如下。

```python
def start_process(filenames, mode):
    for filename in filenames:
        try:
            title = filename[:-4]

            stream_md = collect_stream_metadata(filename)
            streams_to_process = []
            dvdsub_exists = False
            for stream in stream_md['streams']:
                if not stream['codec_name'] in ("dvdsub", "pgssub"):
                    streams_to_process.append(stream['index'])
```

```
            else:
                    dvdsub_exists = True

        print('\nSearching IMDb for "{}"'.format(title))

        imdb = Imdb()
        movie_results = []
        results = imdb.search_for_title(title)
        for result in results:
            if result['type'] == "feature":
                movie_results.append(result)

        if not movie_results:
            while not movie_results:
                title = input('\nNo results for "' + title +
                              '" Enter alternate/correct movie title >>')

                results = imdb.search_for_title(title)
                for result in results:
                    if result['type'] == "feature":
                        movie_results.append(result)

#最突出的结果是第一个MPR
mpr = movie_results[0]
print('\nFetching data for {} ({})'.format(mpr['title'],
                                            mpr['year']))

#电影信息
imdb_movie = imdb.get_title(mpr['imdb_id'])

imdb_movie_title = imdb_movie['base']['title']
imdb_movie_year = imdb_movie['base']['year']
imdb_movie_id = mpr['imdb_id']

imdb_movie_rating = imdb_movie['ratings']['rating']

if not 'outline' in imdb_movie['plot']:
    imdb_movie_plot_outline = (imdb_movie['plot']['summaries'][0]
    ['text'])
    print("\nPlot outline does not exist. Fetching plot summary "
          "instead.\n\n")
else:
    imdb_movie_plot_outline = imdb_movie['plot']['outline']['text']

#组成一个字符串，以获得电影的评级和情节，这将进入MP4文件的"评论"元数据
imdb_rating_and_plot = str('IMDb rating ['
                           + str(float(imdb_movie_rating))
                           + '/10] - '
                           + imdb_movie_plot_outline)

imdb_movie_genres = imdb.get_title_genres(imdb_movie_id)['genres']

#制作电影的"类型"字符串，用分号来分离多个类型的值
genre = ';'.join(imdb_movie_genres)

newfilename = (imdb_movie_title
               + ' ('
               + str(imdb_movie_year)
               + ').mp4')

#禁止在文件名中出现的字符
newfilename = (newfilename
               .replace(':', ' -')
               .replace('/', ' ')
               .replace('?', ''))

command = ""
stream_map = []
for f in streams_to_process:
```

```python
            stream_map.append("-map 0:{}".format(f))
    stream_map_str = ' '.join(stream_map)

    if mode == 1:
        #重命名一个已经存在的MP4文件，不做解码处理
        os.rename(filename, newfilename)
    if mode == 2 or mode == 4:
        command = ('ffmpeg -i "'
                   + filename
                   + '" -sub_charenc UTF-8 -i "'
                   + filename[:-4]
                   + '.srt" '
                   + stream_map_str
                   + ' -map 1 -c copy -c:s mov_text '
                   '"' + newfilename + '"')
        subprocess.run(shlex.split(command))
    if mode == 3:
        command = ('ffmpeg -i '
                   + '"' + filename + '" '
                   + stream_map_str
                   + ' -c copy -c:s mov_text '
                   '"' + newfilename + '"')
        subprocess.run(shlex.split(command))

    if dvdsub_exists:
        print("\nRemoved DVD Subtitles due to uncompatibility with "
              "mp4 file format")

    #海报取自TMDB，如果没有文件名，则在工作目录中命名为"文件名+"jpg"
    #这样用户可以提供自己的海报图像
    poster_filename = filename[:-4] + '.jpg'
    if not os.path.isfile(poster_filename):
        print('\nFetching the movie poster...')
        tmdb_find = tmdb.Find(imdb_movie_id)
        tmdb_find.info(external_source='imdb_id')

        path = tmdb_find.movie_results[0]['poster_path']
        complete_path = r'https://电影网址' + path

        uo = urllib.request.urlopen(complete_path)
        with open(poster_filename, "wb") as poster_file:
            poster_file.write(uo.read())
            poster_file.close()

    video = MP4(newfilename)
    with open(poster_filename, "rb") as f:
        video["covr"] = [MP4Cover(
            f.read(),
            imageformat=MP4Cover.FORMAT_JPEG)]
    video['\xa9day'] = str(imdb_movie_year)
    video['\xa9nam'] = imdb_movie_title
    video['\xa9cmt'] = imdb_rating_and_plot
    video['\xa9gen'] = genre
    print('\nAdding poster and tagging file...')

    try:
        video.save()
        #删除元数据即可解决这个问题
    except OverflowError:
        remove_meta_command = ('ffmpeg -i "' + newfilename
                               + '" -codec copy -map_metadata -1 "'
                               + newfilename[:-4] + 'new.mp4"')
        subprocess.run(shlex.split(remove_meta_command))
        video_new = MP4(newfilename[:-4] + 'new.mp4')
        with open(poster_filename, "rb") as f:
            video_new["covr"] = [MP4Cover(
                f.read(),
                imageformat=MP4Cover.FORMAT_JPEG)]
            video_new['\xa9day'] = str(imdb_movie_year)
```

```python
                        video_new['\xa9nam'] = imdb_movie_title
                        video_new['\xa9cmt'] = imdb_rating_and_plot
                        video_new['\xa9gen'] = genre
                        print('\nAdding poster and tagging file...')

                    try:
                        video_new.save()
                        if not os.path.exists('auto fixed files'):
                            os.makedirs('auto fixed files')
                        os.rename(newfilename[:-4]
                                  + 'new.mp4', 'auto fixed files\\'
                                  + newfilename[:-4] + '.mp4')
                        os.remove(newfilename)

                    except OverflowError:
                        errored_files.append(filename
                                             + (' - Could not save even after'
                                                'striping metadata'))
                        continue

                os.remove(poster_filename)
                print('\n' + filename
                      + (' was proccesed successfuly!\n\n===================='
                         '===================='))
        except Exception as e:
            print('\nSome error occured while processing '
                  + filename
                  + '\n\n======================================================')
            errored_files.append(filename + ' - ' + str(e))
            PrintException()
mp4_filenames = []
mkv_filenames = []
srt_filenames = []
mp4_with_srt_filenames = []
mkv_with_srt_filenames = []
errored_files = []

mp4_filenames = collect_files('mp4')
mkv_filenames = collect_files('mkv')
srt_filenames = collect_files('srt')
```

（7）通过如下代码检查是否有 SRT 字幕文件。如果有，则得到对应的 MP4 文件。如果有 MP4 字幕文件，则将 SRT 格式的字幕文件与对应的视频关联。

```python
#单独处理字幕
if not len(mp4_filenames) == 0:
    if not len(srt_filenames) == 0:
        mp4_with_srt_filenames = get_common_files(mp4_filenames,
                                                  srt_filenames)
        if not len(mp4_with_srt_filenames) == 0:
            mp4_filenames = remove_common_files(mp4_filenames,
                                                mp4_with_srt_filenames)

if not len(mkv_filenames) == 0:
    if not len(srt_filenames) == 0:
        mkv_with_srt_filenames = get_common_files(mkv_filenames, srt_filenames)
        if not len(mkv_with_srt_filenames) == 0:
            mkv_filenames = remove_common_files(mkv_filenames,
                                                mkv_with_srt_filenames)

#开始转换操作，若文件列表不为空，则开始执行
#具体转换过程取决于目标对象是什么类型
if not len(mp4_filenames) == 0:
    start_process(mp4_filenames, 1)

if not len(mp4_with_srt_filenames) == 0:
    start_process(mp4_with_srt_filenames, 2)

if not len(mkv_filenames) == 0:
    start_process(mkv_filenames, 3)
```

```
        if not len(mkv_with_srt_filenames) == 0:
            start_process(mkv_with_srt_filenames, 4)

    if (len(mp4_filenames) == 0 and len(mkv_filenames) == 0
            and len(mp4_with_srt_filenames) == 0
            and len(mkv_with_srt_filenames) == 0):
        print('There were no MP4 or MKV files found in the directory')
    else:
        #检查是否有导致溢出错误的文件,如果有,则输出
        if len(errored_files) == 0:
            print('\n\n\nAll files proccessed successfuly!')
        else:
            print('\n\n\nThe files that were not proccessed: \n')
            for er in errored_files:
                print(er)
```

执行后会检查当前目录中的视频文件,假如当前目录中有两个 MP4 文件——111.mp4 和 222.mp4,则在 IMDb 电影库检索是否有名称为"111"和"222"的视频文件。如果有,则将 IMDb 电影库中的影片封面加入到本地这两个 MP4 视频文件中。执行后会输出如下操作过程,并且成功为本地视频文件添加封面图像。

```
Searching IMDb for "111"

Fetching data for 11:14 (2003)

Fetching the movie poster...

Adding poster and tagging file...

111.mp4 was proccesed successfuly!

==========================================================

Searching IMDb for "222"

Fetching data for 2:22 (2017)

Fetching the movie poster...

Adding poster and tagging file...

222.mp4 was proccesed successfuly!
```

11.6 使用 pydub 库

在 Python 程序中,可以使用 pydub 库来处理音频文件。Pydub 库直接读取 wav 格式的音频文件,如果需要处理其他格式音频(如 MP3、OGG 等),则需要安装 ffmpeg。本节将详细讲解 pydub 库的安装和使用方法。

11.6.1 安装并尝试使用 pydub 库

安装 pydub 库的命令如下。

```
pip install pydub
```

可以通过如下代码打开 wav 文件。

```
from pydub import AudioSegment
song = AudioSegment.from_wav("never_gonna_give_you_up.wav")
```

可以通过如下代码打开 MP3 文件。

```
song = AudioSegment.from_mp3("never_gonna_give_you_up.mp3")
```

可以通过如下代码打开 ogg、flv 或 ffmpeg 支持的文件。

```
ogg_version = AudioSegment.from_ogg("never_gonna_give_you_up.ogg")
flv_version = AudioSegment.from_flv("never_gonna_give_you_up.flv")

mp4_version = AudioSegment.from_file("never_gonna_give_you_up.mp4", "mp4")
wma_version = AudioSegment.from_file("never_gonna_give_you_up.wma", "wma")
aac_version = AudioSegment.from_file("never_gonna_give_you_up.aiff", "aac")
```

使用 pydub 库还可以实现音频切片功能，例如下面的代码。

```
#设置pydub的切片单位是10ms
ten_seconds = 10 * 1000
first_10_seconds = song[:ten_seconds]
last_5_seconds = song[-5000:]
```

可以通过如下代码放大开头音量、减小结尾音量。

```
#放大音量
beginning = first_10_seconds + 6

#减小音量
end = last_5_seconds - 31
```

可以通过如下代码连接音频，将一个文件添加到另一个文件的末尾。

```
without_the_middle = beginning + end1
```

可以通过如下代码设置音频时长。

```
without_the_middle.duration_seconds == 15.0
```

可以通过如下代码实现交叉淡化（开头和结束不修改）。

```
with_style = beginning.append(end, crossfade = 1500)
```

可以通过如下代码实现重复剪辑两次。

```
do_it_over = with_style * 2
```

可以通过如下代码实现淡出。

```
#2s的淡入，3s的淡出
awesome = do_it_over.fade_in(2000).fade_out(3000)
```

可以通过如下代码保存结果（可能需要 ffmpeg）。

```
awesome.export(" mashup.mp3 ", format = " mp3 ")
```

可以通过如下代码保存结果（使用标签/元数据）。

```
awesome.export("mashup.mp3", format="mp3", tags={'artist': 'Various artists',
    'album': 'Best of 2011', 'comments': 'This album is awesome!'})
```

可以使用 ffmpeg 支持的任何语法将可选的 bitrate 参数传递给 export。

```
awesome.export ("mashup.mp3 ", format = "mp3 ", bitrate = ""192k ")
```

11.6.2 使用 AudioSegment

AudioSegment 以任何方式组合多个对象的任何操作，并且将首先确保它们具有相同的通道、帧速率、采样率、比特深度等。当这些量不匹配时，较低质量的声音被修改以匹配更高质量的声音（使得质量不会降低），把单声道转换为立体声，位深度和帧速率/采样率根据需要增加。如果不想要这种行为，你可以使用适当的 AudioSegment 方法显式减少通道数、位数等。

下面的实例文件 pydub01.py 演示了使用 pydub 库处理 wav 音频文件的过程。

源码路径：daima\11\11-6\pydub01.py

```
from pydub import AudioSegment
sound1 = AudioSegment.from_file("example001.wav", format="wav")
sound2 = AudioSegment.from_file("example002.wav", format="wav")

# sound1增大6dB,之后衰减3.5dB
louder = sound1 + 6
quieter = sound1 - 3.5

#连接sound1与sound2
combined = sound1 + sound2

# sound1重复三次
```

```
repeated = sound1 * 3

#持续时间
duration_in_milliseconds = len(sound1)

# sound1的前5s
beginning = sound1[:5000]

# sound1的后5s
end = sound1[-5000:]

#高级用法：原始数据处理
sound = AudioSegment(
    #音频原始数据
    data='…',
    # 2字节的采样
    sample_width=2,
    #采样率
    frame_rate=44100,
    #声道
    channels=16
)
```

执行后会处理两个音频文件 example001.wav 和 example002.wav，分别实现增音、减音、合并和重复等功能。

下面讨论 AudioSegment 中的一些函数。

1. 函数 from_file()

函数 from_file() 能够将音频文件作为 AudioSegment 实例打开并返回。下面的实例文件 pydub02.py 演示了使用 from_file() 处理音频文件的过程。

源码路径：daima\11\11-6\pydub02.py

```
from pydub import AudioSegment

#对于wave和raw格式，不使用ffmpeg
wav_audio = AudioSegment.from_file("example001.wav", format="wav")
raw_audio = AudioSegment.from_file("example002.wav", format="wav",
                                    frame_rate=44100, channels=2, sample_width=2)

#对于其他格式，均使用ffmpeg
mp3_audio = AudioSegment.from_file("111.mp3", format="mp3")

#使用一个已经打开的文件句柄
with open("example001.wav", "rb") as wav_file:
    audio_segment = AudioSegment.from_file(wav_file, format="wav")

from pathlib import Path
wav_path = Path("example001.wav")
wav_audio = AudioSegment.from_file(wav_path)
```

在上述代码中，函数 from_file() 中的第一个参数是要读取的文件或要读取的文件句柄的路径（作为字符串）。

2. 函数 export()

函数 export() 能够将 AudioSegment 对象写入文件中，返回输出文件的句柄。下面的实例文件 pydub03.py 演示了使用 export() 函数处理音频文件的过程。

源码路径：daima\11\11-6\pydub03.py

```
from pydub import AudioSegment
sound = AudioSegment.from_file("/path/to/sound.wav", format="wav")

#简单的输出
file_handle = sound.export("/path/to/output.mp3", format="mp3")

#复杂一些的输出
```

```
file_handle = sound.export("/path/to/output.mp3",
                            format="mp3",
                            bitrate="192k",
                            tags={"album": "The Bends", "artist": "Radiohead"},
                            cover="/path/to/albumcovers/radioheadthebends.jpg")

for i, chunk in enumerate(sound[::5000]):
    with open("sound-%s.mp3" % i, "wb") as f:
        chunk.export(f, format="mp3")
```

函数 export()的第一个参数是写入输出的位置（作为字符串），或要写入的文件句柄。如果不传递输出文件或路径，则会生成一个临时文件。

3. 函数 empty()

函数 empty()的功能是创建空的 AudioSegment 对象，这在使用循环合成长音频时很有用。下面的实例文件 pydub04.py 演示了使用 empty()函数处理音频文件的过程。

源码路径：daima\11\11-6\pydub04.py

```
from pydub import AudioSegment

sounds = [
   AudioSegment.from_wav("example001.wav"),
   AudioSegment.from_wav("example002.wav"),
]

playlist = AudioSegment.empty()
for sound in sounds:
    playlist += sound
```

4. 函数 silent()

使用函数 silent()可以创建无声的 Audiosegment 对象，这可以作为一个占位符间隔或作为一个画布覆盖其他声音的顶部。例如下面的演示代码。

```
from pydub import AudioSegment
ten_second_silence = AudioSegment.silent(duration=10000)
```

5. 函数 from_mono_audiosegments()

函数 from_mono_audiosegments()的功能是从多个单声道音频创建多声道音频，传入的每个单声道音频片段应该具有完全相同的长度。例如下面的演示代码。

```
from pydub import AudioSegment

left_channel = AudioSegment.from_wav("sound1.wav")
right_channel = AudioSegment.from_wav("sound1.wav")

stereo_sound = AudioSegment.from_mono_audiosegments(left_channel, right_channel)
```

6. 函数 frame_count()

函数 frame_count()的功能是返回 AudioSegment 的帧数，或者可以传递 ms 键控参数以检索在该 AudioSegment 切片的毫秒数中的帧数。例如下面的演示代码。

```
from pydub import AudioSegment
sound = AudioSegment.from_file("sound1.wav")

number_of_frames_in_sound = sound.frame_count()

number_of_frames_in_200ms_of_sound = sound.frame_count(ms=200)
```

7. 函数 append()

函数 append()的功能是返回一个新的 AudioSegment。函数 append()在 AudioSegment 与 "+" 操作符一起添加对象时在内部使用。默认情况下，使用 100ms（0.1s）交叉淡入淡出来消除爆裂声。例如下面的演示代码。

```
from pydub import AudioSegment
sound1 = AudioSegment.from_file("sound1.wav")
sound2 = AudioSegment.from_file("sound2.wav")
#默认100ms淡入淡出
combined = sound1.append(sound2)
```

```
# 5000ms交叉淡入淡出
combined_with_5_sec_crossfade = sound1.append(sound2, crossfade=5000)

#无交叉淡入淡出
no_crossfade1 = sound1.append(sound2, crossfade=0)

#无交叉淡入淡出
no_crossfade2 = sound1 + sound2
no_crossfade1 = sound1.append(sound2, crossfade=0)
```

8. 函数 overlay()

函数 overlay()的功能是叠加其他 AudioSegment 到这一个 AudioSegment 中，这么做的结果是同时播放 AudioSegment。如果叠加的 AudioSegment 较长，则结果将被截断（因此叠加声音的结尾将被截断）。AudioSegment 即使使用 loop 和 times 关键字参数，结果也总是与此相同的长度。由于 AudioSegment 对象是不可变的，因此可以通过将较短的声音覆盖在较长的声音上，或创建一个具有适当持续时间的静音的 AudioSegment，并将两个声音覆盖到这个声音上来解决这个问题。例如下面的演示代码。

```
from pydub import AudioSegment
sound1 = AudioSegment.from_file("sound1.wav")
sound2 = AudioSegment.from_file("sound2.wav")

played_togther = sound1.overlay(sound2)

sound2_starts_after_delay = sound1.overlay(sound2, position=5000)

volume_of_sound1_reduced_during_overlay = sound1.overlay(sound2, gain_during_overlay=-8)

sound2_repeats_until_sound1_ends = sound1.overlay(sound2, loop=true)

sound2_plays_twice = sound1.overlay(sound2, times=2)
#假设sound1为30s长，sound2为5s长：
sound2_plays_a_lot = sound1.overlay(sound2, times=10000)
len(sound1) == len(sound2_plays_a_lot)
```

9. 函数 apply_gain（gain）

函数 apply_gain（gain）的功能是设置 AudioSegment 改变的幅度。例如下面的演示代码。

```
from pydub import AudioSegment
sound1 = AudioSegment.from_file("sound1.wav")

#增大3.5 dB
louder_via_method = sound1.apply_gain(+3.5)
louder_via_operator = sound1 + 3.5

#减小5.7 dB
quieter_via_method = sound1.apply_gain(-5.7)
quieter_via_operator = sound1 - 5.7
```

10. 函数 fade()

函数 fade()的功能是实现更加通用（更灵活）的淡入淡出方法。可以指定参数 start 和 end，或两者之一与持续时间（如 start 和 duration）。例如下面的演示代码。

```
from pydub import AudioSegment
sound1 = AudioSegment.from_file("sound1.wav")

fade_louder_for_3_seconds_in_middle = sound1.fade(to_gain=+6.0, start=7500, duration=3000)

fade_quieter_beteen_2_and_3_seconds = sound1.fade(to_gain=-3.5, start=2000, end=3000)

#简单的方法是使用.fade_in()。注意，-120dB基本上就是静音
fade_in_the_hard_way = sound1.fade(from_gain=-120.0, start=0, duration=5000)
fade_out_the_hard_way = sound1.fade(to_gain=-120.0, end=0, duration=5000)
```

11. 函数 fade_out()

函数 fade_out()的功能是淡出（静默）AudioSegment 的结尾。

12. 函数 fade_in()

功能是淡入（从沉默）AudioSegment 的开头。

13. 函数 reverse()

功能是制作一个 AudioSegment 向后播放的副本。它适用于 Pink Floyd、Screwing Around 和一些音频处理算法。

14. 函数 set_sample_width()

AudioSegment 使用指定的样本宽度（以字节为单位）创建此版本的等效版本。增加该值通常不会导致质量下降。减小该值肯定会导致质量的损失。较高的样本宽度意味着更大的动态范围。

15. 函数 set_frame_rate()

AudioSegment 使用指定的帧速率（以赫兹为单位）创建此版本的等效版本。增加该值通常不会导致质量下降。减小该值肯定会导致质量的损失。较高的帧速率意味着较大的频率响应（可以表示较高的频率）。

16. 函数 set_channels()

创建具有指定通道数量的等效版本（1 是单声道，2 是立体声）。从单声道转换为立体声不会导致任何可听到的变化。从立体声转换为单声道可能会导致质量损失（但只有左右通道不同）。

17. 函数 split_to_mono()

将立体声 AudioSegment 分为两个，每个通道一个（左/右）。返回一个包含新 AudioSegment 对象的列表，左声道在索引 0，右声道在索引 1。

18. 函数 apply_gain_stereo()

将增益应用于立体声的左右声道 AudioSegment。如果 AudioSegment 是单声道，则将在应用增益之前转换为立体声。

19. 函数 get_array_of_samples()

以（数值）样本数组的形式返回原始音频数据。如果音频具有多个通道，则每个通道的采样将被序列化。

20. 函数 get_dc_offset()

返回−1.0～1.0 的值，表示通道的直流偏移量。这是使用 audioop.avg()并通过样本最大值对结果进行归一化计算出来的。

21. 函数 remove_dc_offset()

从通道中删除直流偏移量，这是通过调用 audioop.bias()实现的，所以需要注意溢出。

22. 函数 invert_phase()

制作一个 AudioSegment 副本并反转信号的相位，可以产生反相波，用于抑制或消除噪声。

11.6.3 截取指定的 MP3 文件

下面的实例文件 pydub005.py 演示了使用 pydub 库截取指定 MP3 文件的过程。

源码路径：daima\11\11-6\pydub005.py

```
from pydub import AudioSegment
file_name = "example.mp3"
sound = AudioSegment.from_mp3(file_name)
start_time = "0:00"
stop_time = "0:42"
print("time:",start_time,"~",stop_time)

start_time = (int(start_time.split(':')[0])*60+int(start_time.split(':')[1]))*1000
stop_time = (int(stop_time.split(':')[0])*60+int(stop_time.split(':')[1]))*1000

print("ms:",start_time,"~",stop_time)
```

```
word = sound[start_time:stop_time]
save_name = "word"+file_name[6:]
print(save_name)

word.export(save_name, format="mp3",tags={'artist': 'AppLeU0', 'album': save_name[:-4]})
```
运行上述代码后，将截取音频文件 example.mp3 的前 42s，并另存为 worde.mp3。执行结果如下。

```
python pydub005.py
time: 0:00 ~ 0:42
ms: 0 ~ 42000
worde.mp3
```

11.7 使用 tinytag 库

在 Python 程序中，可以使用 tinytag 库来读取 MP3、OGG、FLAC 格式文件的元数据。本节将详细讲解 tinytag 库的安装和使用方法。

11.7.1 安装并尝试使用 tinytag 库

安装 tinytag 库的命令如下。

```
pip install tinytag
```

通过如下属性可以获取音频文件的标签信息。

- tag.album：表示专辑、文本类型。
- tag.albumartist：表示专辑艺术家、文本类型。
- tag.artist：表示艺术家、文本类型。
- tag.audio_offset：表示音频偏移。
- tag.bitrate：表示比特率。
- tag.disc：表示光盘号。
- tag.disc_total：表示光盘总数。
- tag.duration：表示音频长度、单位是秒。
- tag.filesize：表示光盘大小、单位是字节。
- tag.genre：表示音频类型。
- tag.samplerate：表示每秒采样次数。
- tag.title：表示标题。
- tag.track：表示轨道号。
- tag.track_total：表示总轨道号。
- tag.year：表示年份。

下面的实例文件 tinytag01.py 演示了使用 tinytag 库获取指定音频信息的过程。

源码路径：daima\11\11-7\tinytag01.py

```
from tinytag import TinyTag
tag = TinyTag.get('111.mp3')
print('参与创作的艺术家是：%s.' % tag.artist)
print('长度是：%f 秒.' % tag.duration)
```

通过上述代码，获取了音频文件 111.mp3 的艺术家和长度信息，执行后会输出：

```
参与创作的艺术家是：Integrity.
长度是：377.037848 秒.
```

11.7.2 开发一个 MP3 播放器

下面的实例文件演示了使用 pygame、PyQt5、tinytag 和 mutagen 开发一个 MP3 播放器的过程。

(1) 看文件 gui.py，其功能是使用 PyQt5 实现整个播放器的界面布局功能。具体实现流程如下。

源码路径：daima\11\11-7\gui.py

① 写核心函数 setupUi()，规划布局在播放器中需要的各个组件，设置组件的排列方式、样式和位置定位。主要相关代码如下。

```python
def setupUi(self, Hell_Player):
    Hell_Player.setObjectName("音乐播放器")
    #Hell_Player.resize(549, 440)
    Hell_Player.setGeometry(600, 220, 635, 340)
    Hell_Player.setMinimumSize(QtCore.QSize(500, 340))
    Hell_Player.setContextMenuPolicy(QtCore.Qt.ActionsContextMenu)
    Hell_Player.setLayoutDirection(QtCore.Qt.LeftToRight)
    self.gridLayout = QtWidgets.QGridLayout(Hell_Player)
    self.gridLayout.setContentsMargins(0, 0, 0, 0)
    self.gridLayout.setObjectName("gridLayout")
    self.playlist_frame = QtWidgets.QFrame(Hell_Player)
    sizePolicy = QtWidgets.QSizePolicy(QtWidgets.QSizePolicy.Preferred,
        QtWidgets.QSizePolicy.Ignored)
    sizePolicy.setHorizontalStretch(0)
    sizePolicy.setVerticalStretch(0)
    sizePolicy.setHeightForWidth(self.playlist_frame.sizePolicy().hasHeightForWidth())
    self.playlist_frame.setSizePolicy(sizePolicy)
    self.playlist_frame.setFrameShape(QtWidgets.QFrame.Panel)
    self.playlist_frame.setFrameShadow(QtWidgets.QFrame.Sunken)
    self.playlist_frame.setProperty("setVisible", False)
    self.playlist_frame.setObjectName("playlist_frame")
    self.gridLayout_11 = QtWidgets.QGridLayout(self.playlist_frame)
    self.gridLayout_11.setObjectName("gridLayout_11")
    self.tableWidget = QtWidgets.QTableWidget(self.playlist_frame)
    self.tableWidget.setMouseTracking(True)
    self.tableWidget.setEditTriggers(QtWidgets.QAbstractItemView.NoEditTriggers)
    self.tableWidget.setDragDropMode(QtWidgets.QAbstractItemView.DropOnly)
    self.tableWidget.setSelectionMode(QtWidgets.QAbstractItemView.SingleSelection)
    self.tableWidget.setSelectionBehavior(QtWidgets.QAbstractItemView.SelectRows)
    self.tableWidget.setShowGrid(False)
    self.tableWidget.setRowCount(0)
    self.tableWidget.setColumnCount(5)
    self.tableWidget.setObjectName("tableWidget")
    item = QtWidgets.QTableWidgetItem()
    self.tableWidget.setHorizontalHeaderItem(0, item)
    item = QtWidgets.QTableWidgetItem()
    self.tableWidget.setHorizontalHeaderItem(1, item)
    item = QtWidgets.QTableWidgetItem()
    self.tableWidget.setHorizontalHeaderItem(2, item)
    item = QtWidgets.QTableWidgetItem()
    self.tableWidget.setHorizontalHeaderItem(3, item)
    item = QtWidgets.QTableWidgetItem()
    self.tableWidget.setHorizontalHeaderItem(4, item)
    self.tableWidget.horizontalHeader().setMinimumSectionSize(20)
    self.tableWidget.horizontalHeader().setStretchLastSection(True)
    self.tableWidget.verticalHeader().setCascadingSectionResizes(True)
    self.gridLayout_11.addWidget(self.tableWidget, 1, 0, 1, 1)
    self.gridLayout.addWidget(self.playlist_frame, 4, 1, 2, 2)
    self.frame_5 = QtWidgets.QFrame(Hell_Player)
    self.frame_5.setMinimumSize(QtCore.QSize(0, 339))
    self.frame_5.setMaximumSize(QtCore.QSize(16777215, 16777215))
    self.frame_5.setFrameShape(QtWidgets.QFrame.Panel)
    self.frame_5.setFrameShadow(QtWidgets.QFrame.Raised)
    self.frame_5.setObjectName("frame_5")
    self.gridLayout_4 = QtWidgets.QGridLayout(self.frame_5)
    self.gridLayout_4.setContentsMargins(0, 0, 0, 0)
    self.gridLayout_4.setSpacing(0)
    self.gridLayout_4.setObjectName("gridLayout_4")
    self.frame_2 = QtWidgets.QFrame(self.frame_5)
    sizePolicy = QtWidgets.QSizePolicy(QtWidgets.QSizePolicy.Preferred,
        QtWidgets.QSizePolicy.Fixed)
    sizePolicy.setHorizontalStretch(0)
    sizePolicy.setVerticalStretch(0)
    sizePolicy.setHeightForWidth(self.frame_2.sizePolicy().hasHeightForWidth())
```

```python
        self.frame_2.setSizePolicy(sizePolicy)
        self.frame_2.setMinimumSize(QtCore.QSize(0, 85))
        self.frame_2.setMaximumSize(QtCore.QSize(16777215, 110))
        self.frame_2.setFrameShape(QtWidgets.QFrame.NoFrame)
        self.frame_2.setFrameShadow(QtWidgets.QFrame.Raised)
        self.frame_2.setObjectName("frame_2")
        self.gridLayout_12 = QtWidgets.QGridLayout(self.frame_2)
        self.gridLayout_12.setContentsMargins(0, 0, 0, 0)
        self.gridLayout_12.setSpacing(0)
        self.gridLayout_12.setObjectName("gridLayout_12")
        self.ShowPL = QtWidgets.QPushButton(self.frame_2)
        self.ShowPL.setMaximumSize(QtCore.QSize(16777215, 15))
        self.ShowPL.setObjectName("ShowPL")
        self.gridLayout_12.addWidget(self.ShowPL, 2, 0, 1, 1)
        self.HidePL = QtWidgets.QPushButton(self.frame_2)
        self.HidePL.setMaximumSize(QtCore.QSize(16777215, 15))
        self.HidePL.setProperty("setVisible", False)
        self.HidePL.setObjectName("HidePL")
        self.gridLayout_12.addWidget(self.HidePL, 3, 0, 1, 1)
        self.frame_21 = QtWidgets.QFrame(self.frame_2)
        sizePolicy = QtWidgets.QSizePolicy(QtWidgets.QSizePolicy.Preferred,
            QtWidgets.QSizePolicy.Fixed)
        sizePolicy.setHorizontalStretch(0)
        sizePolicy.setVerticalStretch(0)
        sizePolicy.setHeightForWidth(self.frame_21.sizePolicy().hasHeightForWidth())
        self.frame_21.setSizePolicy(sizePolicy)
        self.frame_21.setMinimumSize(QtCore.QSize(0, 49))
        self.frame_21.setMaximumSize(QtCore.QSize(16777215, 49))
        self.frame_21.setFrameShape(QtWidgets.QFrame.NoFrame)
        self.frame_21.setObjectName("frame_21")
        self.horizontalLayout = QtWidgets.QHBoxLayout(self.frame_21)
        self.horizontalLayout.setObjectName("horizontalLayout")
        self.frame_4 = QtWidgets.QFrame(self.frame_21)
        sizePolicy = QtWidgets.QSizePolicy(QtWidgets.QSizePolicy.Fixed, QtWidgets.
            QSizePolicy.Preferred)
        sizePolicy.setHorizontalStretch(0)
        sizePolicy.setVerticalStretch(0)
        sizePolicy.setHeightForWidth(self.frame_4.sizePolicy().hasHeightForWidth())
        self.frame_4.setSizePolicy(sizePolicy)
        self.frame_4.setMinimumSize(QtCore.QSize(90, 0))
        self.frame_4.setFrameShape(QtWidgets.QFrame.Panel)
        self.frame_4.setFrameShadow(QtWidgets.QFrame.Sunken)
        self.frame_4.setObjectName("frame_4")
        self.song_sec = QtWidgets.QLCDNumber(self.frame_4)
        self.song_sec.setGeometry(QtCore.QRect(40, 0, 61, 31))
        sizePolicy = QtWidgets.QSizePolicy(QtWidgets.QSizePolicy.Fixed, QtWidgets.
            QSizePolicy.Minimum)
        sizePolicy.setHorizontalStretch(0)
#为节省本书篇幅，省略后面的代码
```

② 编写核心函数 retranslateUi()，设置在界面中各个组件的文本。具体实现代码如下。

```python
    def retranslateUi(self, Hell_Player):
        _translate = QtCore.QCoreApplication.translate
        Hell_Player.setWindowTitle(_translate("Hell_Player", "Hell Player"))
        self.tableWidget.setSortingEnabled(True)
        item = self.tableWidget.horizontalHeaderItem(0)
        item.setText(_translate("Hell_Player", "#"))
        item = self.tableWidget.horizontalHeaderItem(1)
        item.setText(_translate("Hell_Player", "Title"))
        item = self.tableWidget.horizontalHeaderItem(2)
        item.setText(_translate("Hell_Player", "Artist"))
        item = self.tableWidget.horizontalHeaderItem(3)
        item.setText(_translate("Hell_Player", "Album"))
        item = self.tableWidget.horizontalHeaderItem(4)
        item.setText(_translate("Hell_Player", "Year"))
        self.ShowPL.setText(_translate("Hell_Player", "显示"))
        self.HidePL.setText(_translate("Hell_Player", "隐藏"))
        self.nextButton.setText(_translate("Hell_Player", ">>"))
        self.playButton.setText(_translate("Hell_Player", ">"))
        self.pauseButton.setText(_translate("Hell_Player", "||"))
```

```
        self.prevButton.setText(_translate("Hell_Player", "<<"))
        self.label.setText(_translate("Hell_Player", "音\n""量\n""大\n""小"))
        self.muteCheckBox.setText(_translate("Hell_Player", "静音"))
        self.open_folder.setText(_translate("Hell_Player", "文件"))
        self.playlistButton.setText(_translate("Hell_Player", "播放列表"))
        self.addfilesButton.setText(_translate("Hell_Player", "添加文件"))
        self.shuffle_box.setText(_translate("Hell_Player", "刷新"))
```

(2) 看核心程序文件 main.py，功能是监听对播放器界面中组件的操作，通过调用对应的函数，实现播放器的各个功能。具体实现流程如下。

源码路径：daima\11\11-7\main.py

① 编写类 Timer，功能是监听播放器的播放进度，根据播放进度循环显示当前进度的时间，单位是秒。具体实现代码如下。

```
class Timer(QtCore.QTimer):
    def __init__(self, parent = None):
        super().__init__(parent)
        self.current_timer_test = None
        self.ind = 1
        self.current_position = 0
        self.timer()
        self.current_position = Hell_Player.position * 5

    def progress(self):
        if Hell_Player.pause_state == False and Hell_Player.play_state:


            Hell_Player.progressBar.setValue(self.current_position)
            self.current_position += 1

    def timer(self):
        if self.current_timer_test:
            self.current_timer_test.stop()
            self.current_timer_test.deleteLater()
        self.current_timer_test = QtCore.QTimer()
        self.current_timer_test.timeout.connect(self.timer)
        self.current_timer_test.setSingleShot(True)
        self.current_timer_test.start(200)
        self.progress()
```

② 编写本文件的核心功能类 MyFirstPlayer。首先通过初始化函数 __init__()显示播放界面，包括播放进度、状态、声音、列表显示等功能。具体实现代码如下。

```
    def __init__(self, parent = None):
        super().__init__(parent)
        self.position = 0
        self.setupUi(self)
        self.new_playlist = True
        self.pause_state = False
        self.play_state = False
        self.width = 635
        self.height = 370
        self.playlist_height = 590
        self.playlist = []
        self.test_var = 1
        self.current_timer = None
        self.current_timer_progress_bar = None
#       self.current_position = 0
        self.counter = 0
        self.current_sec = 0
        self.current_min = 0

        # 将按钮与自定义功能连接
        self.playButton.clicked.connect(self.play_button)
        self.pauseButton.clicked.connect(self.pause)
        self.nextButton.clicked.connect(self.next)
        self.prevButton.clicked.connect(self.prev)

        self.open_folder.clicked.connect(self.dir_choosing)
```

```
        self.playlistButton.clicked.connect(self.open_playlist)
        self.ShowPL.clicked.connect(self.increase_height)
        self.HidePL.clicked.connect(self.reduce_height)
        self.tableWidget.cellDoubleClicked["int", "int"].connect(self.get_item_clicked)
        self.volumeSlider.valueChanged["int"].connect(self.set_volume)
        self.muteCheckBox.stateChanged["int"].connect(self.mute)
        self.song_min.display("00")
        self.song_sec.display("00")
        self.progressBar.valueChanged["int"].connect(self.repaint)

        self.progressBar.sliderMoved['int'].connect(self.set_position)

        self.HidePL.setVisible(False)
        self.pauseButton.setVisible(False)
        self.tableWidget.setColumnWidth(0, 1)
        self.tableWidget.setColumnWidth(1, 237)
        self.tableWidget.setColumnWidth(2, 129)
        self.tableWidget.setColumnWidth(3, 129)
        self.tableWidget.setColumnWidth(4, 51)
```

③ 写函数 set_position()，监听进度条事件。具体实现代码如下。

```
def set_position(self, posi):
    if self.check_playlist():
        self.current_sec = int(self.timer_object.current_position / 5) - 1
        while self.current_sec > 60:
            self.current_sec = self.current_sec - 60
        self.current_min = int(self.timer_object.current_position / 300)
        if self.play_state:
            self.position = int(posi / 5)
            mus.pause()
            mus.play(0, self.position)
            self.timer_object.current_position = self.position * 5
```

④ 编写函数 set_volume()，监听单击"音量调节"按钮后的事件。具体实现代码如下。

```
    def set_volume(self, vol):
if self.muteCheckBox.isChecked:
    mus.set_volume(vol / 50)
        self.muteCheckBox.setChecked(False)
else:
       mus.set_volume(vol / 50)
```

⑤ 编写函数 mute()，监听勾选"静音"复选框后的事件。具体实现代码如下。

```
def mute(self, state):
    if state == 2:     #state 2 = unmute
        self.volume = mus.get_volume()
        mus.set_volume(0)
        self.volumeSlider.setProperty("value", 0)
        self.muteCheckBox.setChecked(True)
    elif state == 0:    #state 0 = mute
        mus.set_volume(self.volume)
        self.volumeSlider.setProperty("value", self.volume * 50)
```

⑥ 函数 get_item_clicked()，监听选择播放列表中某首音乐的事件。具体实现代码如下。

```
def get_item_clicked(self, row, column):
    self.pause_state = False
    self.index = row
    self.play_music()
    self.pauseButton.show()
    self.playButton.hide()
    self.timer_object.current_position = 0
```

⑦ 编写函数 show_hidden_playlist()，监听单击"显示"按钮后显示播放列表的事件。具体实现代码如下。

```
def show_hidden_playlist(self):
    self.playlist_frame.setVisible(True)
    self.increase_height()
```

⑧ 编写函数 dir_choosing()，监听单击"文件"按钮后的事件，在弹出的对话框中可以选择某个目录下的所有音频文件。具体实现代码如下。

```
    def dir_choosing(self):
```

```python
        self.index = 0
        self.new_playlist = True
        directory = QFileDialog.getExistingDirectory(None, "Open Directory", "/",
            QFileDialog.ShowDirsOnly)
        if directory:
            os.chdir(directory)
            self.tableWidget.setRowCount(0)
            self.playlist = []
            filters = QDir.Files
            nameFilters = ["*.mp3", "*.MP3", "*.Mp3", "*.mP3"]
            qDirIterator = QDirIterator(directory, nameFilters, filters, QDirIterator.
            Subdirectories)
            while qDirIterator.hasNext():
                qDirIterator.next()
                fileInfo = qDirIterator.fileInfo()
                fdir = fileInfo.absoluteDir().absolutePath()
                song = qDirIterator.filePath()
                self.playlist.append(song)
                try:
                    song_tags = TinyTag.get(song)
                except:
                    print("Corrupted file:\n", len(self.playlist),
                        song, "\nCan't get the tags")
                    self.add_corrupted_files(song)
                else:
                    self.add_items_to_list(song_tags, song)
            self.playlist_action()
```

⑨ 编写函数 open_playlist()，监听单击"播放列表"按钮后的事件，在弹出的对话框中可以选择多个音频文件。具体实现代码如下。

```python
    def open_playlist(self):
        self.index = 0
        self.new_playlist = True
        fileName = QFileDialog.getOpenFileName(
                    None, "Open Playlist (m3u, pls)",
                    "/")[0]
        if fileName:
            self.tableWidget.setRowCount(0)
            self.playlist = []
            playlist_file = open(fileName, "r")
            for items in playlist_file:
                #Removing \n symbols from the end of the file names
                #Making .pls and .m3u playlist files playable
                items = "/" + (items.rpartition("///")[2])[:-1]
                items = re.sub("%20", " ", items)
                if os.path.isfile(items):
                    self.playlist.append(items)
                    try:
                        song_tags = TinyTag.get(items)
                    except:
                        print("Corrupted file:", items, "\nCan't get the tags")
                        self.add_corrupted_files(items)
                    else:
                        self.add_items_to_list(song_tags, items)
            playlist_file.close()
            self.playlist_action()
```

⑩ 编写函数 add_items_to_list()向组件中添加歌曲。具体实现代码如下。

```python
    def add_items_to_list(self, song_info, file_name):
        self.rowPosition = self.tableWidget.rowCount()
        self.tableWidget.insertRow(self.rowPosition)
        self.tableWidget.setRowHeight(self.rowPosition, 18)

        song_title = song_info.title
        try:
            song_title = song_title.encode("latin-1").decode("cp1251")
        except:
            song_title = song_info.title
        if song_title == "":
```

```
            song_title = file_name.split("/")[-1]

        song_artist = song_info.artist
        try:
            song_artist = song_artist.encode("latin-1").decode("cp1251")
        except:
            song_artist = song_info.artist

        song_album = song_info.album
        try:
            song_album = song_album.encode("latin-1").decode("cp1251")
        except:
            song_album = song_info.album

        song_year = song_info.year

        #tooltip = song_title + "::" + song_artist + "::" + song_album + "::" + song_year
        #self.tableWidget.setToolTip(tooltip)
        self.tableWidget.setItem(self.rowPosition,1,QtWidgets.QTableWidgetItem(song_title))
        self.tableWidget.setItem(self.rowPosition,2,QtWidgets.QTableWidgetItem(song_artist))
        self.tableWidget.setItem(self.rowPosition,3,QtWidgets.QTableWidgetItem(song_album))
        self.tableWidget.setItem(self.rowPosition,4,QtWidgets.QTableWidgetItem(song_year))
```

⑪ 编写函数 song_info_displaying() 显示当前播放歌曲的标题。具体实现代码如下。

```
    def song_info_displaying(self):
        self.song_title_field.clear()
        current_song = self.playlist[self.index]
        current_display_info = TinyTag.get(current_song)
        title = str(current_display_info.title)
        try:
            title = title.encode("latin-1").decode("cp1251")
        except:
            title = str(current_display_info.title)

        album = str(current_display_info.album)
        try:
            album = album.encode("latin-1").decode("cp1251")
        except:
            album = str(current_display_info.album)

        artist = str(current_display_info.artist)
        try:
            artist = artist.encode("latin-1").decode("cp1251")
        except:
            artist = str(current_display_info.artist)

        year = str(current_display_info.year)
        if year == "None":
            album_to_display = album
        else:
            album_to_display = album + " (" + year + ")"
        self.song_title_field.setText(title)
        self.album_name_field.setText(album_to_display)
        self.artist_name_field.setText(artist)
        self.tableWidget.selectRow(self.index)
```

⑫ 编写函数 check_playlist() 检查播放列表是否为空。具体实现代码如下。

```
    def check_playlist(self):
        if len(self.playlist) == 0:
            self.song_title_field.clear()
            self.song_title_field.setText("Playlist is empty!")
            self.pauseButton.setVisible(False)
            self.playButton.setVisible(True)
            check = False
        else:
            check = True
        return check
```

⑬ 编写函数 play_button()，用于在单击"播放"按钮后播放列表中的音乐。具体实现代码如下。

```
        def play_button(self):
            if self.check_playlist():
                    if (self.play_state == False
                                and self.shuffle_box.isChecked()
                                and self.new_playlist):
                            self.index_generate()
                            self.play_music()
                    else:
                            self.play_music()
```

⑭ 编写函数 play_music()实现播放音乐的功能。具体实现代码如下。

```
        def play_music(self):
            if self.pause_state == False:
                    self.current_sec = 0
                    self.current_min = 0
                    self.song_sec.display("00")
                    self.song_min.display("00")

            current = TinyTag.get(self.playlist[self.index])
            self.duration = int(current.duration * 5)
            #print(self.duration)
                #current = MP3(self.playlist[self.index])
                #self.duration = int(current.info.length) * 5
            self.progressBar.setMaximum(self.duration)

            self.play_state = True
            self.new_playlist = False

            if self.check_playlist():
                    song = self.playlist[self.index]
                    if self.pause_state:
                            self.pause_state = False
                            mus.unpause()
                    else:
                            try:
                                    self.song_info_displaying()
                                    mus.load(song)
                                    mus.play()
                                    self.wait_for_end()
                            except:
                                    print("无法打开文件！")
                                    self.index += 1
                                    self.play_music()
                                    self.wait_for_end()
                            else:
                                    self.song_info_displaying()
                                    mus.load(song)
                                    mus.play()
                                    self.wait_for_end()
```

⑮ 编写函数 time_calculating_crazy_method()，计算并显示在播放音乐的过程中的时间。具体实现代码如下。

```
        def time_calculating_crazy_method(self):
            if self.current_sec < 10 and self.current_min == 0:
                    self.song_sec.display("0" + str(self.current_sec))
            elif self.current_sec < 10 and (0 < self.current_min < 10):
                    self.song_min.display("0" + str(self.current_min))
                    self.song_sec.display("0" + str(self.current_sec))
            elif self.current_sec < 10 and self.current_min >= 10:
                    self.song_min.display(str(self.current_min))
                    self.song_sec.display("0" + str(self.current_sec))
            elif (10 <= self.current_sec < 60) and self.current_min == 0:
                    self.song_sec.display(str(self.current_sec))
            elif (10 <= self.current_sec < 60) and (0 < self.current_min < 10):
                    self.song_min.display("0" + str(self.current_min))
                    self.song_sec.display(str(self.current_sec))
            elif (10 <= self.current_sec < 60) and self.current_min >= 10:
                    self.song_min.display(str(self.current_min))
                    self.song_sec.display(str(self.current_sec))
```

⑯ 编写函数 time_display(),显示播放进度中的时间。具体实现代码如下。

```
def time_display(self):
    if self.current_sec < 60:
        self.time_calculating_crazy_method()
        self.current_sec += 1
    else:
        self.current_min += 1
        self.current_sec = 0
        self.time_calculating_crazy_method()
        self.current_sec += 1
```

⑰ 编写函数 time_display()检查播放状态,并检查是否播放到了末尾。具体实现代码如下。

```
def wait_for_end(self):
    pygame.display.init()
    SONG_END = pygame.USEREVENT + 1
    mus.set_endevent(SONG_END)
    for event in pygame.event.get():
        #if event.type != SONG_END:
            #self.wait_for_end()
        if event.type == SONG_END:
            pygame.display.quit()
            self.next()
    #self.tableWidget.selectRow(self.index)
    self.start_timer()
```

⑱ 编写函数 index_generat(),在播放列表中生成下一首歌曲的索引。具体实现代码如下。

```
def index_generate(self):
    if self.shuffle_box.isChecked():
        index_next = random.randint(0, self.playlist_len_for_random)
        if index_next == self.index:
            self.index_generate()
        else:
            self.index = index_next
    else:
        if self.index >= self.playlist_len_for_random:
            self.index = 0
        else:
            self.index += 1
```

⑲ 编写函数 next()播放下一首音乐。具体实现代码如下。

```
def next(self):
    if self.check_playlist():
        if (self.new_playlist and self.play_state and
                self.shuffle_box.isChecked() == False):
            self.index = 0
            self.play_music()
        else:
            self.index_generate()
            self.song_change()
```

⑳ 编写函数 prev()播放上一首音乐。具体实现代码如下。

```
def prev(self):
    if self.check_playlist():
        if self.index > 0:
            self.index_generate()
            self.index -= 2
            self.song_change()
        else:
            self.index_generate()
            self.index -= 1
            self.song_change()
```

到此,本实例的主要函数介绍完毕。执行后我们可以播放本地硬盘中的 MP3 文件。

11.8 使用 moviepy 库

moviepy 库是一个音乐理论和曲谱包,支持 MIDI 文件的播放和回放功能。可以使用 moviepy

11.8 使用 moviepy 库

库播放音乐理论，建立编辑、教育工具和其他应用程序。本节将详细讲解 moviepy 库的安装和使用方法。

11.8.1 安装 moviepy 库

安装 moviepy 库的命令如下。

```
pip install moviepy
```

在实现多媒体文件的剪切功能时，moviepy 库用到了另外一个第三方库 ImageMagick。读者需要登录 ImageMagick 官网下载适合自己计算机的安装包。

11.8.2 剪切一段视频

下面的实例文件 moviepy01.py 演示了，使用 moviepy 库剪切指定视频文件的过程。

源码路径：daima\11\11-8\moviepy01.py

```
from moviepy.editor import *

video = VideoFileClip("111.mp4").subclip(10,20)

#设置一个开头标题，分别设置文字内容、大小、颜色、居中、空白
txt_clip = ( TextClip("My Holidays 2013",fontsize=70,color='white')
             .set_position('center')
             .set_duration(10) )

result = CompositeVideoClip([video, txt_clip]) #显示开头标题
result.write_videofile("myHolidays_edited.webm",fps=25) #保存剪切的视频
```

执行后会输出：

```
[MoviePy] >>>> Building video myHolidays_edited.webm
[MoviePy] Writing audio in myHolidays_editedTEMP_MPY_wvf_snd.ogg
100%|██████████████| 221/221 [00:00<00:00, 319.15it/s]
[MoviePy] Done.
[MoviePy] Writing video myHolidays_edited.webm
100%|██████████████| 250/251 [00:56<00:00, 13.34it/s]
[MoviePy] Done.
[MoviePy] >>>> Video ready: myHolidays_edited.webm
```

通过上述代码，剪切了视频文件 111.mp4 中第 10～20s 的内容，并且在屏幕中心设置了一个文本标题，如图 11-4 所示。

图 11-4　设置的文本标题

11.8.3 视频合成

下面的实例文件 moviepy02.py 演示了使用 moviepy 库合成 3 个指定视频文件的过程。

源码路径：daima\11\11-8\moviepy02.py

```
from moviepy.editor import VideoFileClip, concatenate_videoclips
clip1 = VideoFileClip("111.mp4").subclip(10,20)
clip2 = VideoFileClip("222.mp4").subclip(50,60)
```

```
clip3 = VideoFileClip("333.mp4").subclip(20,30)
final_clip = concatenate_videoclips([clip1,clip2,clip3])
final_clip.write_videofile("my_concatenation.mp4")
```

通过上述代码，将 3 个视频文件 111.MP4、222.MP4、333.MP4 合成为 my_concatenation.mp4。执行后会输出：

```
[MoviePy] >>>> Building video my_concatenation.mp4
[MoviePy] Writing audio in my_concatenationTEMP_MPY_wvf_snd.mp3
100%|██████████████| 662/662 [00:02<00:00, 255.18it/s]
[MoviePy] Done.
[MoviePy] Writing video my_concatenation.mp4
100%|██████████████| 450/451 [00:35<00:00, 12.59it/s]
[MoviePy] Done.
[MoviePy] >>>> Video ready: my_concatenation.mp4
```

11.8.4 多屏显示

通过使用 moviepy 库可以将一个屏幕分割成多个区域，实现同一屏幕的双屏、4 屏、甚至更多屏显示效果。moviepy 库的左上角坐标为原点，具体说明如图 11-5 所示。

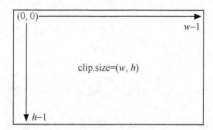

图 11-5　moviepy 库的坐标系统

通过下面的演示代码，可以设置视频在不同的坐标区域中显示。

```
clip2.set_pos((45,150)) #在指定坐标显示

clip2.set_pos("center") # 居中显示

#在顶部水平居中
clip2.set_pos(("center","top"))

#在左侧垂直居中
clip2.set_pos(("left","center"))

#在40%的宽度和70%的高度对应的位置显示
clip2.set_pos((0.4,0.7), relative=True)

#水平居中并向下移动
clip2.set_pos(lambda t: ('center', 50+t) )
```

下面的实例文件 moviepy03.py 演示了通过 moviepy 库用 4 屏显示某个指定视频文件的过程。

源码路径：daima\11\11-8\moviepy03.py

```
from moviepy.editor import VideoFileClip, clips_array, vfx
clip1 = VideoFileClip("111.mp4").margin(10) # add 10px contour
clip2 = clip1.fx( vfx.mirror_x)
clip3 = clip1.fx( vfx.mirror_y)
clip4 = clip1.resize(0.60) # downsize 60%
final_clip = clips_array([[clip1, clip2],
                          [clip3, clip4]])
final_clip.resize(width=480).write_videofile("my_stack.mp4")
```

执行后会以 4 屏显示视频 111.MP4，执行效果如图 11-6 所示。

11.8 使用 moviepy 库

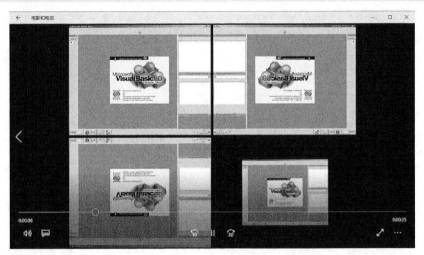

图 11-6　执行效果

11.8.5　设置视频属性

使用 moviepy 库可以重新设置视频文件的属性。下面的实例文件 moviepy04.py 演示了使用 moviepy 库重新设置视频文件属性的过程。

源码路径：daima\11\11-8\moviepy04.py

```
from moviepy.editor import *
clip = (VideoFileClip("111.MP4")
        .fx( vfx.resize, width=460)  #重设大小
        .fx( vfx.speedx, 2) #以双倍速度播放
        .fx( vfx.colorx, 0.5)) #变暗的颜色
clip.write_videofile("123.mp4")
```

通过上述代码，重新设置了视频文件 111.mp4 的大小、播放速度和颜色，并将新文件另存为 123.mp4。

下面的实例文件 moviepy05.py 演示了使用 moviepy 库提取视频文件中指定帧的过程。

源码路径：daima\11\11-8\moviepy05.py

```
from moviepy.editor import VideoFileClip
my_clip = VideoFileClip("111.MP4")
my_clip.save_frame("111.jpg")  #默认值，保存视频中的第一个帧
my_clip.save_frame("frame.png", t=2)  # 保存视频中t=2s时的帧
```

执行上述代码后，会分别提取视频文件 111.mp4 中的第一帧和第 2 秒时的帧，并分别另存为图片 111.jpg 和 frame.png，如图 11-7 所示。

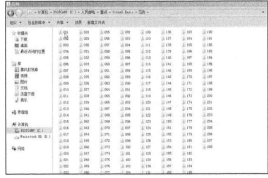

(a) 图片 111.jpg　　　　　　　　　　　(b) 图片 frame.png

图 11-7　提取的视频中的帧

另外，还可以使用 moviepy 库设置视频的显示方式。例如，通过下面的代码可以显示视频中指定的帧。

```
my_clip.show()    #默认值，显示第一帧
my_clip.show(10.5)    #显示 t=10.5s时的帧
my_clip.show(10.5, interactive = True)
```

设置可以预览指定的帧，例如下面的演示代码。

```
my_clip.preview()    #默认预览fps=15
my_clip.preview(fps=25)    #预览fps=25
my_clip.preview(fps=15, audio=False)    #不能生成/播放音频
```

可以从文件或直接从剪辑中嵌入视频、图像和声音，例如下面的演示代码。

```
my_clip = VideoFileClip("111.MP4")
ipython_display(my_clip)    #嵌入视频
ipython_display(my_clip)    #嵌入图片
ipython_display(my_clip)    #嵌入声音

ipython_display("111.jpeg")    #嵌入图片
ipython_display("my_video.mp4")    #嵌入视频
ipython_display("my_sound.mp3")    #嵌入声音
```

11.8.6 使用 moviepy 库和 Matplotlib 库实现数据的动态可视化

在 Python 中有多个数据可视化库，但极少能渲染 GIF 动画或视频动画。通过使用 moviepy 库，我们可以动态显示各种各样的数据。虽然 Matplotlib 库已经有了动画模块，但是使用 moviepy 库可以实现更轻量级、质量更好的视频，并且能达到两倍的显示速度。下面的实例文件 moviepy06.py 演示了联合使用 Matplotlib 库和 moviepy 库将数据转换为 GIF 动画的过程。

源码路径：daima\11\11-8\moviepy06.py

```python
import matplotlib.pyplot as plt
import numpy as np
from moviepy.video.io.bindings import mplfig_to_npimage
import moviepy.editor as mpy

#用Matplotlib画一个图
duration = 2
fig_mpl, ax = plt.subplots(1, figsize=(5, 3), facecolor='white')
xx = np.linspace(-2, 2, 200)    #向量
zz = lambda d: np.sinc(xx ** 2) + np.sin(xx + d)    # z轴的变化
ax.set_title("Elevation in y=0")
ax.set_ylim(-1.5, 2.5)
line, = ax.plot(xx, zz(0), lw=3)

#使用moviepy制作GIF动画
def make_frame_mpl(t):
    line.set_ydata(zz(2 * np.pi * t / duration))    #用于更新的曲线
    return mplfig_to_npimage(fig_mpl)    #图形的RGB图像

animation = mpy.VideoClip(make_frame_mpl, duration=duration)
animation.write_gif("sinc_mpl.gif", fps=20)
```

执行后会生成一个动画文件 sinc_mpl.gif，如图 11-8 所示。

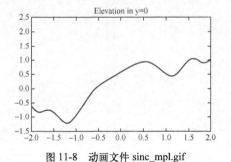

图 11-8 动画文件 sinc_mpl.gif

另外，在 Matplotlib 中有很多漂亮的主题，而且可以很好地与数字模块 Pandas 或 Scikit-Learn 配合使用。下面的实例文件 moviepy07.py 演示了，联合使用 Matplotlib 库和 moviepy 库将数据转换为 GIF 动画的过程。

源码路径：daima\11\11-8\moviepy07.py

```python
import numpy as np
import matplotlib.pyplot as plt
from sklearn import svm  #sklearn = scikit-learn
from sklearn.datasets import make_moons
from moviepy.editor import VideoClip
from moviepy.video.io.bindings import mplfig_to_npimage

X, Y = make_moons(50, noise=0.1, random_state=2)    #半随机数据

fig, ax = plt.subplots(1, figsize=(4, 4), facecolor=(1, 1, 1))
fig.subplots_adjust(left=0, right=1, bottom=0)
xx, yy = np.meshgrid(np.linspace(-2, 3, 500), np.linspace(-1, 2, 500))


def make_frame(t):
    ax.clear()
    ax.axis('off')
    ax.set_title("SVC classification", fontsize=16)

    classifier = svm.SVC(gamma=2, C=1)
    #使点一个接一个地出现
    weights = np.minimum(1, np.maximum(0, t ** 2 + 10 - np.arange(50)))
    classifier.fit(X, Y, sample_weight=weights)
    Z = classifier.decision_function(np.c_[xx.ravel(), yy.ravel()])
    Z = Z.reshape(xx.shape)
    ax.contourf(xx, yy, Z, cmap=plt.cm.bone, alpha=0.8,
                vmin=-2.5, vmax=2.5, levels=np.linspace(-2, 2, 20))
    ax.scatter(X[:, 0], X[:, 1], c=Y, s=50 * weights, cmap=plt.cm.bone)

    return mplfig_to_npimage(fig)


animation = VideoClip(make_frame, duration=7)
animation.write_gif("svm.gif", fps=15)
```

上述代码实现了一个 SVM 分类器，这样可以通过生成的动画 svm.gif，以更加直观的效果展示训练点数量在地图中增加的过程，如图 11-9 所示。

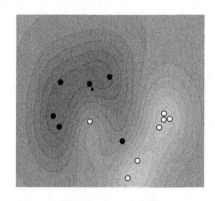

图 11-9　SVM 分类器

11.8.7　动画合成

下面的实例文件 moviepy08.py 演示了使用 moviepy 实现视频合成的过程。

源码路径：daima\11\11-8\moviepy08.py

```python
import moviepy.editor as mpy
```

```
clip_mayavi = mpy.VideoFileClip("sinc_mpl.gif")
clip_mpl = mpy.VideoFileClip("svm.gif").resize(height=clip_mayavi.h)
animation = mpy.clips_array([[clip_mpl, clip_mayavi]])
animation.write_gif("sinc_plot.gif", fps=20)
```

执行后会将两个动画文件 sinc_mpl.gif 和 svm.gif 合成一个动画文件 sinc_plot.gif，如图 11-10 所示。

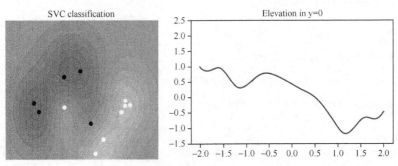

图 11-10 动画文件 sinc_plot.gif

11.8.8 使用 moviepy 库和 numpy 库实现文本动态化

下面的实例文件 moviepy09.py 演示了，联合使用 numpy 库和 moviepy 库实现文本动态化的过程。

源码路径：daima\11\11-8\moviepy09.py

```python
import numpy as np
from moviepy.editor import *
from moviepy.video.tools.segmenting import findObjects

#创建居中显示的移动文本
screensize = (720, 460)
txtClip = TextClip('Cool effect', color='white', font="Amiri-Bold",
                   kerning=5, fontsize=100)
cvc = CompositeVideoClip([txtClip.set_pos('center')],
                          size=screensize)

#接下来的4个函数定义了4种移动字母的方法

#辅助功能
rotMatrix = lambda a: np.array([[np.cos(a), np.sin(a)],
                                [-np.sin(a), np.cos(a)]])
#涡流函数
def vortex(screenpos, i, nletters):
    d = lambda t: 1.0 / (0.3 + t ** 8)
    a = i * np.pi / nletters
    v = rotMatrix(a).dot([-1, 0])
    if i % 2: v[1] = -v[1]
    return lambda t: screenpos + 400 * d(t) * rotMatrix(0.5 * d(t) * a).dot(v)

#瀑布函数
def cascade(screenpos, i, nletters):
    v = np.array([0, -1])
    d = lambda t: 1 if t < 0 else abs(np.sinc(t) / (1 + t ** 4))
    return lambda t: screenpos + v * 400 * d(t - 0.15 * i)

#到达
def arrive(screenpos, i, nletters):
    v = np.array([-1, 0])
    d = lambda t: max(0, 3 - 3 * t)
    return lambda t: screenpos - 400 * v * d(t - 0.2 * i)

#漩涡
def vortexout(screenpos, i, nletters):
    d = lambda t: max(0, t)
    a = i * np.pi / nletters
    v = rotMatrix(a).dot([-1, 0])
```

```
        if i % 2: v[1] = -v[1]
        return lambda t: screenpos + 400 * d(t - 0.1 * i) * rotMatrix(-0.2 * d(t) * a).dot(v)

#使用插件findobjects定位
letters = findObjects(cvc)    #ImageClips列表

#给字母做动画
def moveLetters(letters, funcpos):
    return [letter.set_pos(funcpos(letter.screenpos, i, len(letters)))
            for i, letter in enumerate(letters)]

clips = [CompositeVideoClip(moveLetters(letters, funcpos),
                            size=screensize).subclip(0, 5)
         for funcpos in [vortex, cascade, arrive, vortexout]]

#将一切写入文件
final_clip = concatenate_videoclips(clips)
final_clip.write_videofile('coolTextEffects.avi', fps=25, codec='mpeg4')
```

执行后会创建一个名为 coolTextEffects.avi 的视频文件，如图 11-11 所示。

图 11-11 生成的视频

11.9 使用 scikit-video 库

scikit-video 库是一个专门为 Python 推出的视频处理系统。本节将详细讲解 scikit-video 库的安装和使用方法。

11.9.1 安装并尝试使用 scikit-video 库

安装 scikit-video 库的命令如下。

```
pip install scikit-video
```

scikit-video 库的官方网站提供了视频素材文件，我们可以直接调用并操作这些视频，如图 11-12 所示。

图 11-12 官方提供的视频素材

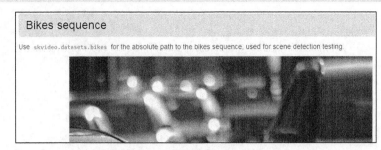

图 11-12　官方提供的视频素材（续）

在 scikit-video 库中，可以使用 skvideo.io.vread 加载任何视频（这里是 bigbuckbunny）到内存中。下面的实例文件 scikit-video01.py 演示了使用 scikit-video 库读取官方视频素材信息的过程。

源码路径：daima\11\11-9\scikit-video01.py

```
import skvideo.io
import skvideo.datasets
videodata = skvideo.io.vread(skvideo.datasets.bigbuckbunny())
print(videodata.shape)
```

在上述代码中，通过函数 skvideo.datasets.bigbuckbunny()读取了官方视频素材文件 bigbuckbunny.mp4 的信息，执行后会输出：

```
(132, 720, 1280, 3)
```

其中，132 代表包含的帧的数量，720 表示宽度（即有 720 行），1280 表示高度（即有 1280 列），3 表示通道数目。

下面的实例文件 scikit-video02.py 演示了，使用 scikit-video 库逐帧读取指定视频信息的过程。

源码路径：daima\11\11-9\scikit-video02.py

```
import skvideo.io
import skvideo.datasets
videogen = skvideo.io.vreader(skvideo.datasets.bigbuckbunny())
for frame in videogen:
        print(frame.shape)
```

通过上述代码，逐帧读取了官方视频素材文件 bigbuckbunny.mp4 的信息，执行后会输出：

```
(720, 1280, 3)
(720, 1280, 3)
(720, 1280, 3)
(720, 1280, 3)
#省略后面的好多帧
```

11.9.2　写入视频

在 scikit-video 库中，可以使用 skvideo.io.write 向视频中写入 ndarray。下面的实例文件 scikit-video03.py 演示了将随机生成的帧写入视频文件的过程。

源码路径：daima\11\11-9\scikit-video03.py

```
import skvideo.io
import numpy as np

outputdata = np.random.random(size=(5, 480, 680, 3)) * 255
outputdata = outputdata.astype(np.uint8)

skvideo.io.vwrite("outputvideo.mp4", outputdata)
```

通过上述代码，将随机生成的 ndarray 写入视频文件 outputvideo.mp4 中。另外，还可以使用 skvideo.io.ffprobe 获取视频的元数据信息。下面的实例文件 scikit-video04.py 演示了使用 skvideo.io.ffprobe 的过程。

源码路径：daima\11\11-9\scikit-video04.py

```
import skvideo.io
import skvideo.datasets
```

```
import json
metadata = skvideo.io.ffprobe(skvideo.datasets.bigbuckbunny())
print(metadata.keys())
print(json.dumps(metadata["video"], indent=4))
```

skvideo.io.ffprobe 会返回一个字典, 通过 json.dumps 美观地输出官方视频素材文件 bigbuckbunny.mp4 的元数据。

```
dict_keys(['video', 'audio'])
{
    "@index": "0",
    "@codec_name": "h264",
    "@codec_long_name": "H.264 / AVC / MPEG-4 AVC / MPEG-4 part 10",
    "@profile": "Main",
    "@codec_type": "video",
    "@codec_time_base": "1/50",
    "@codec_tag_string": "avc1",
    "@codec_tag": "0x31637661",
    "@width": "1280",
    "@height": "720",
    "@coded_width": "1280",
    "@coded_height": "720",
    "@has_b_frames": "0",
    "@sample_aspect_ratio": "1:1",
    "@display_aspect_ratio": "16:9",
    "@pix_fmt": "yuv420p",
    "@level": "31",
    "@chroma_location": "left",
    "@refs": "1",
    "@is_avc": "true",
    "@nal_length_size": "4",
    "@r_frame_rate": "25/1",
    "@avg_frame_rate": "25/1",
    "@time_base": "1/12800",
    "@start_pts": "0",
    "@start_time": "0.000000",
    "@duration_ts": "67584",
    "@duration": "5.280000",
    "@bit_rate": "1205959",
    "@bits_per_raw_sample": "8",
    "@nb_frames": "132",
    "disposition": {
        "@default": "1",
        "@dub": "0",
        "@original": "0",
        "@comment": "0",
        "@lyrics": "0",
        "@karaoke": "0",
        "@forced": "0",
        "@hearing_impaired": "0",
        "@visual_impaired": "0",
        "@clean_effects": "0",
        "@attached_pic": "0",
        "@timed_thumbnails": "0"
    },
    "tag": [
        {
            "@key": "creation_time",
            "@value": "1970-01-01T00:00:00.000000Z"
        },
        {
            "@key": "language",
            "@value": "und"
        },
        {
            "@key": "handler_name",
            "@value": "VideoHandler"
        }
    ]
}
```

11.9.3 视频基准测试

在 scikit-video 库中，可以使用函数 skvideo.motion.blockMotion()实现基于块的运动估算功能，并在给定的一系列帧之间返回运动矢量值。下面的实例文件 scikit-video05.py 演示了使用 blockMotion()返回运动矢量值的过程。

源码路径：daima\11\11-9\scikit-video05.py

```
import skvideo.io
import skvideo.motion
import skvideo.datasets

videodata = skvideo.io.vread(skvideo.datasets.bigbuckbunny())

motion = skvideo.motion.blockMotion(videodata)

print(videodata.shape)
print(motion.shape)
```

在默认情况下，blockmotion()使用 8×8 像素块和菱形搜索算法。执行后会输出：

```
(132, 720, 1280, 3)
(131, 90, 160, 2)
```

我们还可以使用 skvideo.motion.blockcomp 计算块运动矢量的运动补偿，下面的实例文件 scikit-video06.py 演示了这一用法。

源码路径：daima\11\11-9\scikit-video06.py

```
import skvideo.io
import skvideo.motion
import skvideo.datasets
# 计算向量
videodata = skvideo.io.vread(skvideo.datasets.bigbuckbunny())
motion = skvideo.motion.blockMotion(videodata)

# 视频补偿
compmotion = skvideo.motion.blockComp(videodata, motion)
```

11.9.4 图像的读取和写入

因为视频文件是由帧构成的，所以每个帧其实是一幅图像。scikit-video 库也能够实现对图像文件的解析。下面的实例文件 scikit-video07.py 演示了使用 skvideo.io.vread 读取指定图片信息的过程。

源码路径：daima\11\11-9\scikit-video07.py

```
import skvideo.io
vid = skvideo.io.vread("111.jpg")
T, M, N, C = vid.shape
print("Number of frames: %d" % (T,))
print("Number of rows: %d" % (M,))
print("Number of cols: %d" % (N,))
print("Number of channels: %d" % (C,))
```

通过上述代码，分别读取了图片 111.jpg 的帧数、行数、列数和通道数信息。执行后会输出：

```
Number of frames: 1
Number of rows: 588
Number of cols: 441
Number of channels: 3
```

另外，还可以使用 skvideo.io.vread 重设图像的分辨率。下面的实例文件 scikit-video08.py 演示了使用 skvideo.io.vread 重设指定图像行数和列数的过程。

源码路径：daima\11\11-9\scikit-video08.py

```
import skvideo.io
vid = skvideo.io.vread("111.jpg")
T, M, N, C = vid.shape
print("Number of frames: %d" % (T,))
print("Number of rows: %d" % (M,))
```

```
print("Number of cols: %d" % (N,))
print("Number of channels: %d" % (C,))
```
通过上述代码，将图片 111.jpg 的大小重设为 1440×2560。执行后会输出：
```
Number of frames: 1
Number of rows: 1440
Number of cols: 2560
Number of channels: 3
```
同样使用 skvideo.io.vwrite 可以实现写入图像功能。下面的实例文件 scikit-video09.py 演示了使用 skvideo.io.vwrite 写入并生成图像的过程。

源码路径：daima\11\11-9\scikit-video09.py
```
import skvideo.io
import numpy as np

#创建随机数据
image = np.random.random(size=(720, 1280))*255
print("Random image, shape (%d, %d)" % image.shape)
skvideo.io.vwrite("output.png", image)
```
执行后会根据随机数创建一个 1280×720 像素的图片 output.png，如图 11-13 所示，并输出以下内容。
```
Random image, shape (720, 1280)
```

图 11-13　图片 output.png

11.9.5　视频的读取和写入

在 scikit-video 库中，可以使用 skvideo.io.vread 设置只加载一个视频亮度通道。下面的实例文件 scikit-video10.py 演示了这一用法。

源码路径：daima\11\11-9\scikit-video10.py
```
import skvideo.utils

filename = skvideo.datasets.bigbuckbunny()

print("只加载luminance（亮度）通道")
vid = skvideo.io.vread(filename, outputdict={"-pix_fmt": "gray"})[:, :, :, 0]
print(vid.shape)
print("执行视频形状")
vid = skvideo.utils.vshape(vid)
print(vid.shape)
print("")

print("只加载前5个亮度通道帧")
vid = skvideo.io.vread(filename, num_frames=5, outputdict={"-pix_fmt": "gray"})[:, :, :, 0]
print(vid.shape)
print("执行视频形状")
vid = skvideo.utils.vshape(vid)
print(vid.shape)
print("")
```

执行后会输出:
只加载luminance(亮度)通道
(132, 720, 1280)
执行视频形状
(132, 720, 1280, 1)

只加载前5个亮度通道帧
(5, 720, 1280)
执行视频形状
(5, 720, 1280, 1)

在给出一个输入 YUV 视频时,必须设置其宽度、高度和格式。在默认情况下,scikit 视频的 pix_fmt 值是 yuvj444p。通过提供一致的存储和视频内容,在加载的同时可以保持信号的保真度。下面的实例文件 scikit-video11.py 演示了生成指定通道视频并提取视频中指定帧的过程。

源码路径:daima\11\11-9\scikit-video11.py

```
import skvideo.io
import skvideo.utils
import skvideo.datasets
from skimage import io, color

import skimage.io
import numpy as np

filename = skvideo.datasets.bigbuckbunny()
filename_yuv = "test.yuv"

#首先产生一个YUV视频
vid = skvideo.io.vread(filename)
T, M, N, C = vid.shape

#生成一个yuv文件,-pix_fmt值是yuvj444p
skvideo.io.vwrite(filename_yuv, vid)

#现在开始加载YUV视频

vid_luma = skvideo.io.vread(filename_yuv, height=M, width=N, outputdict={"-pix_fmt": "gray"})[:, :, :, 0]
vid_luma = skvideo.utils.vshape(vid_luma)

vid_rgb = skvideo.io.vread(filename_yuv, height=M, width=N)

#现在加载的YUV视频没有转换
vid_yuv444 = skvideo.io.vread(filename_yuv,height=M,width=N,outputdict={"-pix_fmt": "yuvj444p"})

#重新组织字节,因为在平面模式下输出的ffmpeg
vid_yuv444 = vid_yuv444.reshape((M * N * T * 3))
vid_yuv444 = vid_yuv444.reshape((T, 3, M, N))
vid_yuv444 = np.transpose(vid_yuv444, (0, 2, 3, 1))


#可视化
skvideo.io.vwrite("luma.mp4", vid_yuv444[:, :, :, 0])
skvideo.io.vwrite("chroma1.mp4", vid_yuv444[:, :, :, 1])
skvideo.io.vwrite("chroma2.mp4", vid_yuv444[:, :, :, 2])

#写出每个视频的第一帧
skimage.io.imsave("vid_rgb_frame1.png", vid_rgb[0])
skimage.io.imsave("vid_chroma1.png", vid_yuv444[0, :, :, 1])
skimage.io.imsave("vid_chroma2.png", vid_yuv444[0, :, :, 2])
```

在上述代码中,首先将官方视频素材 bigbuckbunny.mp4 加载到本地并另存为 test.yuv。然后创建 3 种不同格式的视频文件——chroma1.mp4、chroma2.mp4 和 luma.mp4。接下来,分别提取了视频中的指定帧,将提取的帧分别另存为 vid_chroma1.png、vid_chroma2.png 和 vid_rgb_frame1.png,如图 11-14 和图 11-15 所示。

11.9 使用 scikit-video 库

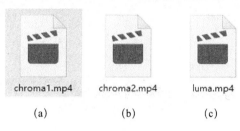

(a)   (b)   (c)

图 11-14 生成的 3 个视频

(a)   (b)   (c)

图 11-15 生成的 3 个帧图片

第 12 章

第三方网络开发库

Python 在网络通信方面的优点特别突出,远远领先其他语言。本章将详细讲解使用 Python 第三方库开发网络项目的知识,展示这些库在互联网应用中的强大功能。

12.1 处理 HTML 和 XML

因为 HTML 和 XML 是互联网应用中最常用的网页标记语言,所以用 Python 处理 HTML 和 XML 页面至关重要。本节将详细讲解常用的处理 HTML 和 XML 的第三方库。

12.1.1 使用 Beautiful Soup 库

Beautiful Soup 是一个可以从 HTML 或 XML 文件中提取数据的 Python 库。它能够将 HTML 和 XML 的标签文件解析成树状结构,然后方便地获取指定标签的对应属性。通过使用 Beautiful Soup 库,可以大大提高开发效率。

Beautiful Soup 3 目前已经停止开发,其官方推荐使用 Beautiful Soup 4,本书讲解的是 Beautiful Soup 4。可以使用两种命令安装 Beautiful Soup 库。

```
pip install beautifulsoup4
easy_install beautifulsoup4
```

接下来还需要安装解析器。Beautiful Soup 不但支持 Python 标准库中的 HTML 解析器,而且支持一些第三方的解析器,其中最常用的是 lxml。根据操作系统不同,可以使用如下命令来安装 lxml。

```
$ apt-get install Python-lxml
$ easy_install lxml
$ pip install lxml
```

下面的实例文件 bs01.py 演示了使用 Beautiful Soup 库解析 HTML 代码的过程。

源码路径:daima\12\12-1\bs01.py

```python
from bs4 import BeautifulSoup
html_doc = """
<html><head><title>The Dormouse's story</title></head>
<body>
<p class="title"><b>The Dormouse's story</b></p>

<p class="story">Once upon a time there were three little sisters; and their names were
<a href="http://example.com/elsie" class="sister" id="link1">Elsie</a>,
<a href="http://example.com/lacie" class="sister" id="link2">Lacie</a> and
<a href="http://example.com/tillie" class="sister" id="link3">Tillie</a>;
and they lived at the bottom of a well.</p>

<p class="story">...</p>
"""
soup = BeautifulSoup(html_doc,"lxml")
print(soup)
```

通过上述代码,解析了 html_doc 中的 HTML 代码。执行后会输出解析结果。

```
<html><head><title>The Dormouse's story</title></head>
<body>
<p class="title"><b>The Dormouse's story</b></p>
<p class="story">Once upon a time there were three little sisters; and their names were
<a class="sister" href="http://example.com/elsie" id="link1">Elsie</a>,
<a class="sister" href="http://example.com/lacie" id="link2">Lacie</a> and
<a class="sister" href="http://example.com/tillie" id="link3">Tillie</a>;
and they lived at the bottom of a well.</p>
<p class="story">...</p>
</body></html>
```

在使用 Beautiful Soup 库的过程中,还会涉及如下几个概念。

1. 标签处理

(1) 使用标签选择器获取标签信息。在解析 HTML 或 XML 文件时,可以使用标签选择器来获得某个具体的标签信息。例如,通过如下代码可以获取不同的标签信息。

```python
print(soup.title)
print(type(soup.title))
print(soup.head)
print(soup.p)
```

通过上述"soup.标签名"格式即可获得这个标签的内容。需要注意的是，如果文档中有多个这样的标签，返回的结果是第一个标签的内容。例如，通过 soup.p 可以获取 p 标签，通常在文档中有多个 p 标签，在执行 soup.p 后只会返回了第一个 p 标签的内容。

(2) 获取名称。例如，通过 soup.title.name 可以获得该 title 标签的名称，即 title。

(3) 获取属性。例如，通过下面两种方式可以获取标签 p 的 name 属性值。

```
print(soup.p.attrs['name'])
print(soup.p['name'])
```

(4) 获取内容。例如，通过如下代码可以获取第一个 p 标签的内容。

```
print(soup.p.string)
```

下面的实例文件 bs02.py 演示了使用 Beautiful Soup 库解析指定 HTML 标签的过程。

源码路径：daima\12\12-1\bs02.py

```
from bs4 import BeautifulSoup

html = '''
<html><head><title>The Dormouse's story</title></head>
<body>
<p class="title"><b>The Dormouse's story</b></p>

<p class="story">Once upon a time there were three little sisters; and their names were
<a href="http://example.com/elsie" class="sister" id="link1">Elsie</a>,
<a href="http://example.com/lacie" class="sister" id="link2">Lacie</a> and
<a href="http://example.com/tillie" class="sister" id="link3">Tillie</a>;
and they lived at the bottom of a well.</p>
<p class="story">...</p>
'''
soup = BeautifulSoup(html,'lxml')
print(soup.title)
print(soup.title.name)
print(soup.title.string)
print(soup.title.parent.name)
print(soup.p)
print(soup.p["class"])
print(soup.a)
print(soup.find_all('a'))
print(soup.find(id='link3'))
```

执行后将输出指定标签的信息：

```
<title>The Dormouse's story</title>
title
The Dormouse's story
head
<p class="title"><b>The Dormouse's story</b></p>
['title']
<a class="sister" href="http://example.com/elsie" id="link1">Elsie</a>
[<a class="sister" href="http://example.com/elsie" id="link1">Elsie</a>, <a class="sister" href="http://example.com/lacie" id="link2">Lacie</a>, <a class="sister" href="http://example.com/tillie" id="link3">Tillie</a>]
<a class="sister" href="http://example.com/tillie" id="link3">Tillie</a>
```

2. 节点处理

1) 基本节点

在 Beautiful Soup 库中，可以使用 contents 来处理节点。下面的实例文件 bs03.py 演示了将 p 标签下的所有子标签存入一个列表的过程。

源码路径：daima\12\12-1\bs03.py

```
html = """
<html>
    <head>
        <title>The Dormouse's story</title>
    </head>
    <body>
        <p class="story">
            Once upon a time there were three little sisters; and their names were
```

```
            <a href="http://example.com/elsie" class="sister" id="link1">
                <span>Elsie</span>
            </a>
            <a href="http://example.com/lacie" class="sister" id="link2">Lacie</a>
            and
            <a href="http://example.com/tillie" class="sister" id="link3">Tillie</a>
            and they lived at the bottom of a well.
        </p>
        <p class="story">...</p>
"""

from bs4 import BeautifulSoup

soup = BeautifulSoup(html,'lxml')
print(soup.p.contents)
```

执行后会输出：

```
['\n        Once upon a time there were three little sisters; and their names were\n ', 
   <a class="sister" href="http://example.com/elsie" id="link1">
<span>Elsie</span>
</a>, '\n', <a class="sister" href="http://example.com/lacie" id="link2">Lacie</a>,'\n
        and\n        ', <a class="sister" href="http://example.com/tillie" id="link3">
   Tillie</a>, '\n         and they lived at the bottom of a well.\n         ']
```

也就是说，在列表中会存入图 12-1 中框选的元素。

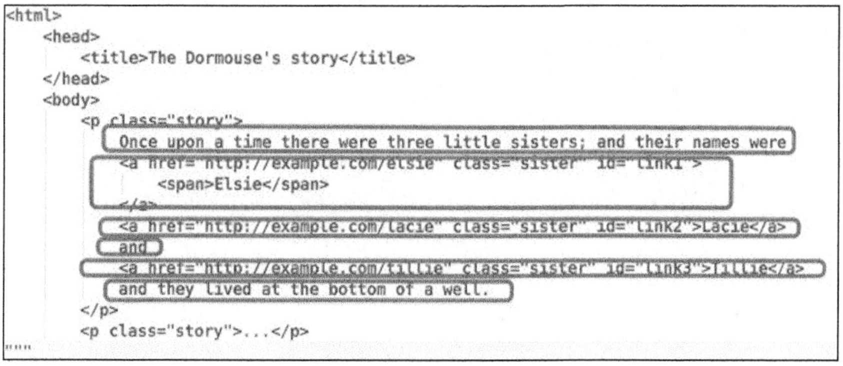

图 12-1　存入框选的元素

通过下面的实例文件 bs04.py，使用 children 也可以获取 p 标签下的所有子节点内容，其结果和通过 contents 获取的结果完全一样。不同的地方是 soup.p.children 是一个迭代对象，而不是列表，只能通过循环的方式获取所有的信息。

源码路径：daima\12\12-1\bs04.py

```
soup = BeautifulSoup(html,'lxml')
print(soup.p.children)
for i,child in enumerate(soup.p.children):
    print(i,child)
```

2）父节点和祖先节点

通过 soup.a.parent 可以获取父节点的信息，通过 list(enumerate(soup.a.parents))可以获取祖先节点。这个方法返回的结果是一个列表，会分别将 a 标签的父节点，以及父节点的父节点的信息存放到列表中，并且最后还会将整个文档放到列表中。所有列表的最后一个元素以及倒数第二个元素存的都是整个文档的信息。

3）兄弟节点

- soup.a.next_siblings：获取后面的兄弟节点。
- soup.a.previous_siblings：获取前面的兄弟节点。
- soup.a.next_sibling 获取下一个兄弟标签。
- soup.a.previous_sibling：获取上一个兄弟标签。

下面的实例文件 bs05.py 演示了处理标签兄弟节点和父节点的过程。

源码路径：daima\12\12-1\bs05.py

```
html = """
<html><head><title>The Dormouse's story</title></head>
<body>
<p class="title"><b>The Dormouse's story</b></p>

<p class="story">Once upon a time there were three little sisters; and their names were
<a href="http://example.com/elsie" class="sister" id="link1">Elsie</a>,
<a href="http://example.com/lacie" class="sister" id="link2">Lacie</a> and
<a href="http://example.com/tillie" class="sister" id="link3">Tillie</a>;
and they lived at the bottom of a well.</p>

<p class="story">...</p>
"""
from bs4 import BeautifulSoup
soup = BeautifulSoup(html, 'lxml')
title_tag = soup.title
print(title_tag)
print(title_tag.parent)
#在下面的代码中，因为<b>标签和<c>标签同一层——它们是同一个元素的子节点
#所以<b>和<c>可以被称为兄弟节点。当一段文档以标准格式输出时
#兄弟节点有相同的缩进级别。在代码中也可以使用这种关系
sibling_soup = BeautifulSoup("<a><b>text1</b><c>text2</c></b></a>", 'lxml')
print(sibling_soup.prettify())
#<b>标签有 .next_sibling 属性，但是没有 .previous_sibling 属性,因为<b>标签在同级节点中是第一个
#同理,<c>标签有.previous_sibling属性,却没有.next_sibling属性
print(sibling_soup.b.next_sibling)
print(sibling_soup.c.previous_sibling)

for sibling in soup.a.next_siblings:
    print(repr(sibling))

for sibling in soup.find(id="link3").previous_siblings:
    print(repr(sibling))
```

执行后会输出：

```
<title>The Dormouse's story</title>
<head><title>The Dormouse's story</title></head>
<html>
 <body>
  <a>
   <b>
    text1
   </b>
   <c>
    text2
   </c>
  </a>
 </body>
</html>
<c>text2</c>
```

3．标准选择器

1）find_all

通过使用函数 find_all(name,attrs,recursive,text,**kwargs)，可以根据标签名、属性和内容查找文档。下面的实例文件 bs06.py 演示了根据标签名查找文档的过程。

源码路径：daima\12\12-1\bs06.py

```
html='''
####省略部分代码
'''
from bs4 import BeautifulSoup
soup = BeautifulSoup(html, 'lxml')
print(soup.find_all('ul'))
print(type(soup.find_all('ul')[0]))
for ul in soup.find_all('ul'):
    print(ul.find_all('li'))
```

执行后会返回一个列表,并且在最后两行的 for 语句中,针对前面的解析结果再次执行 find_all()操作,从而获取所有 li 标签的信息。

```
[<ul class="list" id="list-1">
<li class="element">Foo</li>
<li class="element">Bar</li>
<li class="element">Jay</li>
</ul>, <ul class="list list-small" id="list-2">
<li class="element">Foo</li>
<li class="element">Bar</li>
</ul>]
<class 'bs4.element.Tag'>
[<li class="element">Foo</li>, <li class="element">Bar</li>, <li class="element">Jay</li>]
[<li class="element">Foo</li>, <li class="element">Bar</li>]
```

在使用函数 find_all()时,也可以根据属性 attrs 查找文档。attrs 通过传入字典的方式来查找标签,但是这里有个特殊情况。因为 class 在 Python 中是特殊的字段,所以如果想要查找 class 相关的信息,可以更改 attrs={'class_':'element'}或者 soup.find_all('',{"class":"element"}),特殊的标签属性可以不写 attrs,如 id。下面的实例文件 bs07.py 演示了使用函数 find_all()根据属性查找文档的过程。

源码路径:daima\12\12-1\bs07.py

```
html='''
####省略部分代码
'''
from bs4 import BeautifulSoup
soup = BeautifulSoup(html, 'lxml')
print(soup.find_all(attrs={'id': 'list-1'}))
print(soup.find_all(attrs={'name': 'elements'}))
```

执行后会输出:

```
[<ul class="list" id="list-1" name="elements">
<li class="element">Foo</li>
<li class="element">Bar</li>
<li class="element">Jay</li>
</ul>]
[<ul class="list" id="list-1" name="elements">
<li class="element">Foo</li>
<li class="element">Bar</li>
<li class="element">Jay</li>
</ul>]
```

下面的实例文件 bs08.py 演示了使用函数 find_all()根据 text 查找文档的过程。

源码路径:daima\12\12-1\bs08.py

```
html='''
####省略部分代码
'''
soup = BeautifulSoup(html, 'lxml')
print(soup.find_all(text='Foo'))
```

执行后会输出:

```
['Foo', 'Foo']
```

2)其他标准选择器
- find(name,attrs,recursive,text,**kwargs):返回匹配结果的第一个元素。
- find_parents():返回所有祖先节点。
- find_parent():返回直接父节点。
- find_next_siblings():返回后面的所有兄弟节点。
- find_next_sibling():返回后面的第一个兄弟节点。
- find_previous_siblings():返回前面的所有兄弟节点。
- find_previous_sibling():返回前面的第一个兄弟节点。
- find_all_next():返回节点后面所有符合条件的节点。

- find_next()：返回下一个符合条件的节点。
- find_all_previous()：返回节点前面所有符合条件的节点。
- find_previous()：返回前一个符合条件的节点。

下面的实例文件 bs09.py 演示了使用上述其他标准选择器的过程。

源码路径：daima\12\12-1\bs09.py

```
html = """
#省略部分代码
"""
from bs4 import BeautifulSoup
soup = BeautifulSoup(html, 'lxml')
first_link = soup.a
print(first_link)
print(first_link.find_next_siblings("a"))
first_story_paragraph = soup.find("p", "story")
print(first_story_paragraph.find_next_sibling("p"))
first_link = soup.a
print(first_link)
print(first_link.find_all_next(text=True))
print(first_link.find_next("p"))
first_link = soup.a
print(first_link)
print(first_link.find_all_previous("p"))
print(first_link.find_previous("title"))
```

执行后会输出：

```
<a class="sister" href="http://example.com/elsie" id="link1">Elsie</a>
[<a class="sister" href="http://example.com/lacie" id="link2">Lacie</a>, <a class="sister"
    href="http://example.com/tillie" id="link3">Tillie</a>]
<p class="story">...</p>
<a class="sister" href="http://example.com/elsie" id="link1">Elsie</a>
['Elsie', ',\n', 'Lacie', ' and\n', 'Tillie', ';\nand they lived at the bottom of a well.', '\
    n', '...', '\n']
<p class="story">...</p>
<a class="sister" href="http://example.com/elsie" id="link1">Elsie</a>
[<p class="story">Once upon a time there were three little sisters; and their names were
<a class="sister" href="http://example.com/elsie" id="link1">Elsie</a>,
<a class="sister" href="http://example.com/lacie" id="link2">Lacie</a> and
<a class="sister" href="http://example.com/tillie" id="link3">Tillie</a>;
and they lived at the bottom of a well.</p>, <p class="title"><b>The Dormouse's story</b></p>]
<title>The Dormouse's story</title>
```

4. CSS 选择器

Beautiful Soup 支持大部分的 CSS 选择器，通过在 Tag 或 BeautifulSoup 对象的 select()方法中传入字符串参数，可以使用 CSS 选择器的语法找到标签。下面的实例文件 bs10.py 演示了，使用 select()直接传入 CSS 选择器的方式完成标签选择的过程。

源码路径：daima\12\12-1\bs10.py

```
html='''
#省略部分代码
'''
from bs4 import BeautifulSoup
soup = BeautifulSoup(html, 'lxml')
print(soup.select('.panel .panel-heading'))
print(soup.select('ul li'))
print(soup.select('#list-2 .element'))
print(type(soup.select('ul')[0]))
```

执行后会输出：

```
[<div class="panel-heading">
<h4>Hello</h4>
</div>]
[<li class="element">Foo</li>, <li class="element">Bar</li>, <li class="element">Jay</
    li>, <li class="element">Foo</li>, <li class="element">Bar</li>]
[<li class="element">Foo</li>, <li class="element">Bar</li>]
<class 'bs4.element.Tag'>
```

下面的实例文件 bs11.py 演示了通过 get_text() 获取文本内容的过程。

源码路径：daima\12\12-1\bs11.py

```
html='''
#省略部分代码
'''
from bs4 import BeautifulSoup
soup = BeautifulSoup(html, 'lxml')
for li in soup.select('li'):
    print(li.get_text())
```

执行后会输出：

```
Foo
Bar
Jay
Foo
Bar
```

下面的实例文件 bs12.py 演示了通过[属性名]或者 attrs[属性名]方式获取属性的过程。

源码路径：daima\12\12-1\bs12.py

```
html='''
#省略部分代码
'''
from bs4 import BeautifulSoup
soup = BeautifulSoup(html, 'lxml')
for ul in soup.select('ul'):
    print(ul['id'])
    print(ul.attrs['id'])
```

执行后会输出：

```
list-1
list-1
list-2
list-2
```

12.1.2 使用 bleach 库

在使用 Python 进行 Web 开发时，必须要防止用户的 XSS（Cross Site Scripting，跨站脚本）攻击。我们可以自己写一个白名单，然后通过 Beautiful Soup 等处理 HTML 的库来过滤标签和属性。bleach 便是一个实现上述功能的 Python 库，是一个基于白名单的 HTML 清理和文本链接模块。

可以使用两种命令安装 bleach 库。

```
pip install bleach
easy_install bleach
```

下面的实例文件 bs01.py 演示了使用 bleach 库过滤 HTML 代码的过程。

源码路径：daima\12\12-1\bs01.py

```
import bleach
print(bleach.clean('an <script>evil()</script> example'))
print(bleach.linkify('an http://example.com url'))
```

通过上述代码实现了最基本的 HTML 过滤，执行后会输出：

```
an &lt;script&gt;evil()&lt;/script&gt; example
an <a href="http://example.com" rel="nofollow">http://example.com</a> url
```

接下来将详细讲解 bleach 库中的重要内置方法。

1. bleach.clean()

bleach.clean()是用于对 HTML 片段进行过滤的方法。需要注意的是，该方法过滤的是片段而非整个 HTML 文档，当不传递任何参数时，该方法只用来过滤 HTML 标签，不包括属性、CSS、JSON、XHTML 和 SVG 等其他内容。正因为如此，当对一些存在风险的属性进行渲染时，需要使用模板进行转义。如果你正在清理大量的文本，并传递相同的参数值或者想要更多的可配置性，可以考虑使用 bleach.sanitizer.Cleaner 实例。各个参数和返回值的具体说明如下。

- text（str）。要过滤的文本，通常为 HTML 片段文本。
- tags（list）。标签白名单，默认使用 bleach.sanitizer.ALLOWED_TAGS（参数值为包含标签字符串的可迭代对象，不在 tags 中的标签都会被清除或转义）。
- attributes（dict or list）。属性白名单，可以是一个可调用对象、列表或字典；默认使用 bleach.sanitizer.ALLOWED_ATTRIBUTES（同 tags 一样，dict 以标签为键，以标签对应属性组成的列表为值当键为*时，表示所有标签；而当为 list 时，其中的属性过滤应用于所有标签）。
- styles（list）。CSS 白名单，默认使用 bleach.sanitizer.ALLOWED_STYLES，但因为这个列表是空的，如果不加此参数则会把写进来的 style 值过滤掉。Protocols（list）表示链接协议白名单；默认使用 bleach.sanitizer.ALLOWED_PROTOCOLS=[u'http', u'https', u'mailto']。当有带链接或者锚的标签时，比如，有 href 属性的标签，则需要加上允许的协议，否则会把 href 属性过滤掉。可以通过对 bleach.sanitizer.ALLOWED_PROTOCOLS 添加值来扩展支持的协议。
- strip（bool）。表示是否清除白名单之外的元素。默认情况下，当为 False 时不清除，只进行转义；当为 True 时，会把白名单以外的标签清除掉。
- strip_comments（bool）。表示是否清除 HTML 注释内容，默认清除（设置为 True）。
- 返回值。Unicode 格式的文本。

下面的实例文件 bs02.py 演示了使用方法 bleach.clean()对不同参数进行过滤的过程。

源码路径：daima\12\12-1\bs02.py

```python
import bleach
print(bleach.clean(
    u'<b><i>an example</i></b>',
    tags=['b'],
))

print(bleach.clean(
    u'<p class="foo" style="color: red; font-weight: bold;">blah blah blah</p>',
    tags=['p'],
    attributes=['style'],
    styles=['color'],
))
attrs = {
    '*': ['class'],
    'a': ['href', 'rel'],
    'img': ['alt'],
}
print(bleach.clean(
    u'<img alt="an example" width=500>',
    tags=['img'],
    attributes=attrs
))

def allow_h(tag, name, value):
    return name[0] == 'h'
print(bleach.clean(
    u'<a href="http://example.com" title="link">link</a>',
    tags=['a'],
    attributes=allow_h,
))


tags = ['p', 'em', 'strong']
attrs = {
    '*': ['style']
}
styles = ['color', 'font-weight']
print(bleach.clean(
```

```
            u'<p style="font-weight: heavy;">my html</p>',
            tags=tags,
            attributes=attrs,
            styles=styles
))
print(bleach.clean(
        '<a href="smb://more_text">allowed protocol</a>',
        protocols=['http', 'https', 'smb']
))
print(bleach.clean(
        '<a href="smb://more_text">allowed protocol</a>',
        protocols=bleach.ALLOWED_PROTOCOLS + ['smb']
))

print(bleach.clean('<span>is not allowed</span>'))
print(bleach.clean('<b><span>is not allowed</span></b>', tags=['b']))

print(bleach.clean('<span>is not allowed</span>', strip=True))
print(bleach.clean('<b><span>is not allowed</span></b>', tags=['b'], strip=True))

html = 'my<!-- commented --> html'
print(bleach.clean(html))
print(bleach.clean(html, strip_comments=False))
```

执行后会输出：

```
<b>&lt;i&gt;an example&lt;/i&gt;</b>
<p style="color: red;">blah blah blah</p>
<img alt="an example">
<a href="http://example.com">link</a>
<p style="font-weight: heavy;">my html</p>
<a href="smb://more_text">allowed protocol</a>
<a href="smb://more_text">allowed protocol</a>
&lt;span&gt;is not allowed&lt;/span&gt;
<b>&lt;span&gt;is not allowed&lt;/span&gt;</b>
is not allowed
<b>is not allowed</b>
my html
my<!-- commented --> html
```

2. bleach.sanitizer.Cleaner()

bleach.sanitizer.Cleaner()的参数与前面的 clean()方法类似，filter 参数传入的是一个由 html5lib Filter 类组成的用来传递流内容的列表。其返回值和前面的 clean()方法类似，但是当传入的值不为文本格式时，会引发 TypeError 异常。

注意：方法 bleach.clean()实例化了 bleach.sanitizer.Cleaner 对象，而 bleach.sanitizer.Cleaner 对象则实例化了 bleach.sanitizer.BleachSanitizerFilter 对象，真正实现过滤作用的是 bleach.sanitizer. BleachSanitizerFilter 对象。bleach.sanitizer.BleachSanitizerFilter 是一个 html5lib 过滤器，可以在任何使用 html5lib 过滤器的地方使用。

3. leach.sanitizer.BleachSanitizerFilter

leach.sanitizer.BleachSanitizerFilter的参数与前面的bleach.clean()方法一样，strip_disallowed_elements 相当于 bleach.clean()方法中的 strip 参数，strip_html_comments 相当于 bleach.clean()方法中的 strip_comments 参数。

4. bleach.linkify(text, callbacks=[<function nofollow>], skip_tags=None, parse_email=False)

各个参数的具体说明如下。

- text（str）：要转换的 HTML 文本。
- callbacks（list）：由回调函数组成的列表，用来调整标签属性，默认使用 lbleach.linkifier.DEFAULT_CALLBACKS，callbacks 参数中的 callback 函数必须遵循如下格式。

```
def my_callback(attrs, new=False):
```

在上述格式中，attrs 参数和 clean()方法中的参数类似，是包含标签及其属性的字典。callbacks 可以用来为链接化后的标签中加入、删除或修改属性。参数 new 表明 callback 操作的对象是新的链接化字符（即类 url 字符，还未转换为链接），或者已存在的链接（即已经是链接的字符）。

- skip_tags（list）。由标签名组成的列表，表示这些标签不进行链接化处理。例如，可以设置['pre']，这样就会在链接化时跳过 pre 标签。
- parse_email（bool）。表示是否链接化 email 地址。

方法 bleach.linkify()会将 HTML 文本中的 url 形式字符转换为链接。url 形式字符包括 url、域名、email 等，但在以下情况下不会进行转换：

- 已经以链接格式呈现在文本中。
- 该标签属性值包含 url 格式。
- email 地址。

总体来讲，方法 bleach.linkify()会尽可能多地将文本中的链接转换为 a 标签。下面的实例文件 bs03.py 演示了使用方法 bleach.linkify()添加指定属性的过程。

源码路径：daima\12\12-1\bs03.py

```python
from bleach.linkifier import Linker

def set_title(attrs, new=False):
    attrs[(None, u'title')] = u'link in user text'
    return attrs

linker = Linker(callbacks=[set_title])
print(linker.linkify('abc http://example.com def'))

#下面的代码将生成的链接设置为内部链接在当前页打开，外部链接在新建页打开
import urllib
from bleach.linkifier import Linker

def set_target(attrs, new=False):
    p = urllib.parse.urlparse(attrs[(None, u'href')])
    if p.netloc not in ['my-domain.com', 'other-domain.com']:
        attrs[(None, u'target')] = u'_blank'
        attrs[(None, u'class')] = u'external'
    else:
        attrs.pop((None, u'target'), None)
    return attrs

linker = Linker(callbacks=[set_target])
print(linker.linkify('abc http://example.com def'))
```

执行后会输出：

```
abc <a href="http://example.com" title="link in user text">http://example.com</a> def
abc <a class="external" href="http://example.com" target="_blank">http://example.com</a> def
```

通过方法 bleach.linkify()中的参数 callback，可以实现类似属性白名单的过滤操作。不但可以删除标签中已有属性，而且可以删除那些没有经过链接化的文本内容的标签属性，功能与前面的 clean()方法类似。下面的实例文件 bs04.py 演示了使用 callback 参数删除指定属性的过程。

源码路径：daima\12\12-1\bs04.py

```python
from bleach.linkifier import Linker

def allowed_attrs(attrs, new=False):
    """Only allow href, target, rel and title."""
    allowed = [
        (None, u'href'),
        (None, u'target'),
        (None, u'rel'),
        (None, u'title'),
        u'_text',
    ]
    return dict((k, v) for k, v in attrs.items() if k in allowed)
```

```
linker = Linker(callbacks=[allowed_attrs])
print(linker.linkify('<a style="font-weight: super bold;" href="http://example.com">link</a>'))

#除了删除白名单之外的属性，还可以删除指定属性
def remove_title(attrs, new=False):
    attrs.pop((None, u'title'), None)
    return attrs
linker = Linker(callbacks=[remove_title])
print(linker.linkify('<a href="http://example.com">link</a>'))
print(linker.linkify('<a title="bad title" href="http://example.com">link</a>'))
```

执行后会输出：

```
<a href="http://example.com">link</a>
<a href="http://example.com">link</a>
<a href="http://example.com">link</a>
```

5.bleach.linkifier.Linker

bleach.linkifier.Linker 的原型如下。

```
class bleach.linkifier.Linker(callbacks=[<function nofollow>],skip_tags=None,parse_email=False,
url_re=<_sre.SRE_Pattern object at 0x25b8e90>, email_re=<_sre.SRE_Pattern object at 0x258b5f0>)
```

各个参数的具体说明如下。

- callbacks（list）：同 linkify 函数。
- skip_tags（list）：同 linkify 函数。
- parse_email（bool）：同 linkify 函数。
- url_re（re）：匹配 url 的正则对象。
- email_re（re）：匹配 email 地址的正则对象。
- 返回值：链接化的 Unicode 字符。

当使用一套统一的规则进行文本链接化处理时，建议使用 bleach.linkifier.Linker 实例，因为 linkify 方法的本质就是调用此实例。下面的实例文件 bs05.py 演示了使用 bleach.linkifier.Linke 处理链接的过程。

源码路径：daima\12\12-1\bs05.py

```
from bleach.linkifier import Linker
linker = Linker(skip_tags=['pre'])
print(linker.linkify('a b c http://example.com d e f'))
```

执行后会输出：

```
a b c <a href="http://example.com" rel="nofollow">http://example.com</a> d e f
```

6.bleach.linkifier.LinkifyFilter

bleach.linkifier.LinkifyFilter 的原型如下。

```
class bleach.linkifier.LinkifyFilter(source, callbacks=None, skip_tags=None, parse_email=False,
url_re=<_sre.SRE_Pattern object at 0x25b8e90>, email_re=<_sre.SRE_Pattern object at 0x258b5f0>)
```

各个参数的具体说明如下。

- source（TreeWalker）：数据流。
- callbacks（list）：同 linkify 函数。
- skip_tags（list）：同 linkify 函数。
- parse_email（bool）：同 linkify 函数。
- url_re（re）：同 Linker 实例。
- email_re（re）：同 Linker 实例。

方法 bleach.linkify()是通过 bleach.linkifier.LinkifyFilter 实例进行链接化的，与前面讲解的 bleach.linkifier.Cleaner 一样，此实例也可以当作 html5lib 过滤器实例使用。例如，可以使用此实例把参数传入 bleach.linkifier.Cleaner 中，使得文本过滤和文本链接化同时进行。

下面的实例文件 bs06.py 演示了使用 bleach.linkifier.LinkifyFilter 处理链接的过程。

源码路径：daima\12\12-1\bs06.py

```python
from bleach import Cleaner
from bleach.linkifier import LinkifyFilter
#使用bleach.linkifier.LinkifyFilter的默认配置
cleaner = Cleaner(tags=['pre'])
print(cleaner.clean('<pre>http://example.com</pre>'))

cleaner = Cleaner(tags=['pre'], filters=[LinkifyFilter])
print(cleaner.clean('<pre>http://example.com</pre>'))
#下面演示传参后对比
from functools import partial
from bleach.sanitizer import Cleaner
from bleach.linkifier import LinkifyFilter

cleaner = Cleaner(
    tags=['pre'],
    filters=[partial(LinkifyFilter, skip_tags=['pre'])]
)
print(cleaner.clean('<pre>http://example.com</pre>'))
```

执行后会输出：

```
<pre>http://example.com</pre>
<pre><a href="http://example.com">http://example.com</a></pre>
<pre>http://example.com</pre>
```

12.1.3 使用 cssutils 库

cssutils 库是一个 Python 包，用于解析和构建 CSS。在 cssutils 库中只有 DOM，没有任何渲染设备。可以使用两种命令安装 cssutils 库。

```
pip install cssutils
easy_install cssutils
```

下面的实例文件 css01.py 演示了使用 cssutils 库处理 CSS 标记的过程。

源码路径：daima\12\12-1\css01.py

```python
import cssutils

css = u'''/* a comment with umlaut &auml; */
        @namespace html "http://www.w3.org/1999/xhtml";
        @variables { BG: #fff }
        html|a { color:red; background: var(BG) }'''
sheet = cssutils.parseString(css)

for rule in sheet:
    if rule.type == rule.STYLE_RULE:
            #遍历属性
            for property in rule.style:
                    if property.name == 'color':
                            property.value = 'green'
                            property.priority = 'IMPORTANT'
                            break
            #简易处理
            rule.style['margin'] = '01.0eM' # or: ('1em', 'important')

sheet.encoding = 'ascii'
sheet.namespaces['xhtml'] = 'http://www.w3.org/1999/xhtml'
sheet.namespaces['atom'] = 'http://www.w3.org/2005/Atom'
sheet.add('atom|title {color: #000000 !important}')
sheet.add('@import "sheets/import.css";')

# cssutils.ser.prefs.resolveVariables == True
print(sheet.cssText)
```

执行后会输出：

```
@charset "ascii";
@import "sheets/import.css";
/* a comment with umlaut \E4  */
@namespace xhtml "http://www.w3.org/1999/xhtml";
@namespace atom "http://www.w3.org/2005/Atom";
```

```
xhtml|a {
    color: green !important;
    background: #fff;
    margin: 1em
    }
atom|title {
    color: #000 !important
    }
```

12.1.4 使用 html5lib 库

html5lib 库是一个兼容标准 HTML 文档和片段解析及序列化的库。在本章前面讲解使用 Beautiful Soup 库从 HTML 或 XML 文件中提取数据时，使用的解析器是 lxml，而 html5lib 库也是 Beautiful Soup 支持的一种解析器。

html5lib 库是纯 Python 实现的，其解析方式与浏览器相同。可以使用两种命令安装 html5lib 库。

```
pip install html5lib
easy_install html5lib
```

在 12.1.1 节中的所有实例中，可以将解析器 lxml 修改为 html5lib，修改之后的代码都可以成功运行。下面的实例文件 ht501.py 演示了在实例文件 bs01.py 中使用 html5lib 解析 HTML 的过程。

源码路径：daima\12\12-1\ht501.py

```
from bs4 import BeautifulSoup
html_doc = """
<html><head><title>The Dormouse's story</title></head>
<body>
<p class="title"><b>The Dormouse's story</b></p>

<p class="story">Once upon a time there were three little sisters; and their names were
<a href="http://example.com/elsie" class="sister" id="link1">Elsie</a>,
<a href="http://example.com/lacie" class="sister" id="link2">Lacie</a> and
<a href="http://example.com/tillie" class="sister" id="link3">Tillie</a>;
and they lived at the bottom of a well.</p>

<p class="story">...</p>
"""
soup = BeautifulSoup(html_doc,"html5lib")
print(soup)
```

执行后会输出：

```
<html><head><title>The Dormouse's story</title></head>
<body>
<p class="title"><b>The Dormouse's story</b></p>

<p class="story">Once upon a time there were three little sisters; and their names were
<a class="sister" href="http://example.com/elsie" id="link1">Elsie</a>,
<a class="sister" href="http://example.com/lacie" id="link2">Lacie</a> and
<a class="sister" href="http://example.com/tillie" id="link3">Tillie</a>;
and they lived at the bottom of a well.</p>

<p class="story">...</p>
</body></html>
```

同理，也可以使用 html5lib 解析出 HTML 中的指定标签。下面的实例文件 ht502.py 演示了这一功能。

源码路径：daima\12\12-1\ht502.py

```
html = """
<html><head><title>The Dormouse's story</title></head>
<body>
<p class="title"><b>The Dormouse's story</b></p>

<p class="story">Once upon a time there were three little sisters; and their names were
<a href="http://example.com/elsie" class="sister" id="link1">Elsie</a>,
<a href="http://example.com/lacie" class="sister" id="link2">Lacie</a> and
<a href="http://example.com/tillie" class="sister" id="link3">Tillie</a>;
```

```
and they lived at the bottom of a well.</p>

<p class="story">...</p>
"""

from bs4 import BeautifulSoup

#添加一个解析器
soup = BeautifulSoup(html,'html5lib')
print(soup.title)
print(soup.title.name)
print(soup.title.text)
print(soup.body)

#从文档中找到所有<a>标签的内容
for link in soup.find_all('a'):
    print(link.get('href'))


#从文档中找到所有文字内容
print(soup.get_text())
```

执行后会输出：

```
<title>The Dormouse's story</title>
title
The Dormouse's story
<body>
<p class="title"><b>The Dormouse's story</b></p>

<p class="story">Once upon a time there were three little sisters; and their names were
<a class="sister" href="http://example.com/elsie" id="link1">Elsie</a>,
<a class="sister" href="http://example.com/lacie" id="link2">Lacie</a> and
<a class="sister" href="http://example.com/tillie" id="link3">Tillie</a>;
and they lived at the bottom of a well.</p>

<p class="story">...</p>
</body>
http://example.com/elsie
http://example.com/lacie
http://example.com/tillie
The Dormouse's story

The Dormouse's story

Once upon a time there were three little sisters; and their names were
Elsie,
Lacie and
Tillie;
and they lived at the bottom of a well.

...
```

12.1.5 使用 MarkupSafe 库

MarkupSafe 库为 XML/HTML/XHTML 标记提供了安全字符串。可以使用两种命令安装 MarkupSafe 库。

```
pip install MarkupSafe
easy_install MarkupSafe
```

下面的实例文件 mark01.py 演示了使用 MarkupSafe 库构建安全 HTML 的过程。

源码路径：daima\12\12-1\mark01.py

```
from markupsafe import Markup, escape
#实现支持HTML字符串的Unicode子类
print(escape("<script>alert(document.cookie);</script>"))
tmpl = Markup("<em>%s</em>")
print(tmpl % "Peter > Lustig")

#可以通过重写__html__自定义等效HTML标记
```

```python
class Foo(object):
    def __html__(self):
        return '<strong>Nice</strong>'

print(escape(Foo()))
print(Markup(Foo()))
```

执行后会输出：

```
&lt;script&gt;alert(document.cookie);&lt;/script&gt;
<em>Peter &gt; Lustig</em>
<strong>Nice</strong>
<strong>Nice</strong>
```

下面的实例文件 mark02.py 演示了使用 MarkupSafe 库实现自定义格式化的过程。

源码路径：daima\12\12-1\mark02.py

```python
from markupsafe import Markup, escape

class User(object):

        def __init__(self, id, username):
            self.id = id
            self.username = username

        def __html_format__(self, format_spec):
            if format_spec == 'link':
                return Markup('<a href="/user/{0}">{1}</a>').format(
                    self.id,
                    self.__html__(),
                )
            elif format_spec:
                raise ValueError('Invalid format spec')
            return self.__html__()

        def __html__(self):
            return Markup('<span class=user>{0}</span>').format(self.username)

user = User(1, 'foo')
print(Markup('<p>User: {0:link}').format(user))
```

执行后会输出：

```
<p>User: <a href="/user/1"><span class=user>foo</span></a>
```

12.1.6 使用 PyQuery 库

PyQuery 是一个解析 HTML 的库，是 Python 对 jQuery 的封装。如果读者拥有前端开发经验，那么应该接触过 jQuery。PyQuery 是 Python 仿照 jQuery 的严格实现。PyQuery 的语法与 jQuery 几乎完全相同。可以使用两种命令安装 PyQuery 库。

```
pip install pyquery
easy_install pyquery
```

下面的实例文件 pyq01.py 演示了使用 PyQuery 库实现字符串初始化的过程。

源码路径：daima\12\12-1\pyq01.py

```python
html = '''
<div>
    <ul>
        <li class="item-0">first item</li>
        <li class="item-1"><a href="link2.html">second item</a></li>
        <li class="item-0 active"><a href="link3.html"><span class="bold">third
            item</span></a></li>
        <li class="item-1 active"><a href="link4.html">fourth item</a></li>
        <li class="item-0"><a href="link5.html">fifth item</a></li>
    </ul>
</div>
'''

from pyquery import PyQuery as pq
doc = pq(html)
```

```
print(doc)
print(type(doc))
print(doc('li'))
```

在上述代码中,因为 PyQuery 写起来比较麻烦,所以在导入的时候添加了别名 pq。

```
from pyquery import PyQuery as pq
```

在上述代码中,doc 其实就是一个 PyQuery 对象,我们可以通过 doc 进行元素的选择。其实这里 doc 就是一个 css 选择器,所以 CSS 选择器的规则都可以用,直接使用 doc(标签名)就可以获取所有该标签的内容。如果要获取 class,则使用 doc('.class_name');如果要获取 id,则使用 doc('#id_name')。执行后会输出:

```
<div>
    <ul>
        <li class="item-0">first item</li>
        <li class="item-1"><a href="link2.html">second item</a></li>
        <li class="item-0 active"><a href="link3.html"><span class="bold">third
            item</span></a></li>
        <li class="item-1 active"><a href="link4.html">fourth item</a></li>
        <li class="item-0"><a href="link5.html">fifth item</a></li>
    </ul>
</div>
<class 'pyquery.pyquery.PyQuery'>
<li class="item-0">first item</li>
        <li class="item-1"><a href="link2.html">second item</a></li>
        <li class="item-0 active"><a href="link3.html"><span class="bold">third
            item</span></a></li>
        <li class="item-1 active"><a href="link4.html">fourth item</a></li>
        <li class="item-0"><a href="link5.html">fifth item</a></li>
```

下面的实例文件 pyq02.py 演示了使用 PyQuery 库解析 HTML 内容的过程。

源码路径:daima\12\12-1\pyq02.py

```
from pyquery import PyQuery as pyq

html = '''
<html>
    <title>这是标题</title>
<body>
    <p id="hi">Hello</p>
    <ul>
        <li>list1</li>
        <li>list2</li>
    </ul>
</body>
</html>
'''
jq = pyq(html)
print(jq('title'))    #获取title标签的源码
print(jq('title').text())    #获取title标签的内容
print(jq('#hi').text())    #获取id为hi的标签的内容

li = jq('li')    #处理多个元素
for i in li:
    print(pyq(i).text())
```

执行后会输出:

```
<title>这是标题</title>

这是标题
Hello
list1
list2
```

下面的实例文件 pyq03.py 演示了使用 PyQuery 库分别解析本地 HTML 文件和网络页面的过程。

源码路径:daima\12\12-1\pyq03.py

```
from pyquery import PyQuery as pq
doc = pq(filename='123.html')
```

```
print(doc)
print(doc('head'))

doc1 = pq(url="http://www.baidu.com",encoding='utf-8')
print(doc1)
print(doc1('head'))
```

在上述代码中，在 pq()里可以传入 url 参数，也可以传入文件参数。当然，这里的文件通常是一个本地 HTML 文件，如 pq(filename='index.html')。执行后会分别解析本地文件 123.html 和百度主页。

下面的实例文件 pyq04.py 演示了使用 PyQuery 库基于 CSS 选择器查找的过程。

源码路径：daima\12\12-1\pyq04.py

```
html = '''
<div id="container">
    <ul class="list">
         <li class="item-0">first item</li>
         <li class="item-1"><a href="link2.html">second item</a></li>
         <li class="item-0 active"><a href="link3.html"><span class="bold">third
             item</span></a></li>
         <li class="item-1 active"><a href="link4.html">fourth item</a></li>
         <li class="item-0"><a href="link5.html">fifth item</a></li>
    </ul>
 </div>
'''
from pyquery import PyQuery as pq
doc = pq(html)
print(doc('#container .list li'))
```

在上述代码中需要注意 doc('#container .list li')，小括号中的 3 项不必挨着，只要三者是层级关系即可。执行后会输出：

```
<li class="item-0">first item</li>
         <li class="item-1"><a href="link2.html">second item</a></li>
         <li class="item-0 active"><a href="link3.html"><span class="bold">third
             item</span></a></li>
         <li class="item-1 active"><a href="link4.html">fourth item</a></li>
         <li class="item-0"><a href="link5.html">fifth item</a></li>
```

在使用 PyQuery 库时，可以通过已经查找的标签来查找这个标签下的子标签或者父标签，而不用从头开始查找。下面的实例文件 pyq05.py 演示了使用 PyQuery 库查找子元素的过程。

源码路径：daima\12\12-1\pyq05.py

```
html = '''
<div id="container">
    <ul class="list">
         <li class="item-0">first item</li>
         <li class="item-1"><a href="link2.html">second item</a></li>
         <li class="item-0 active"><a href="link3.html"><span class="bold">third
             item</span></a></li>
         <li class="item-1 active"><a href="link4.html">fourth item</a></li>
         <li class="item-0"><a href="link5.html">fifth item</a></li>
    </ul>
 </div>
'''
from pyquery import PyQuery as pq
doc = pq(html)
items = doc('.list')
print(type(items))
print(items)
lis = items.find('li')
print(type(lis))
print(lis)
```

执行后会输出：

```
<class 'pyquery.pyquery.PyQuery'>
<ul class="list">
```

```
            <li class="item-0">first item</li>
            <li class="item-1"><a href="link2.html">second item</a></li>
            <li class="item-0 active"><a href="link3.html"><span class="bold">third
                item</span></a></li>
            <li class="item-1 active"><a href="link4.html">fourth item</a></li>
            <li class="item-0"><a href="link5.html">fifth item</a></li>
    </ul>
<class 'pyquery.pyquery.PyQuery'>
<li class="item-0">first item</li>
            <li class="item-1"><a href="link2.html">second item</a></li>
            <li class="item-0 active"><a href="link3.html"><span class="bold">third
                item</span></a></li>
            <li class="item-1 active"><a href="link4.html">fourth item</a></li>
            <li class="item-0"><a href="link5.html">fifth item</a></li>
```

从上述执行结果可以看出，通过 pyquery 找到的结果是一个 PyQuery 对象，我们可以继续查找，上述代码中的 items.find('li')则表示查找 ul 里的所有 li 标签。当然，这里通过 children 可以实现同样的效果，并且通过 children 方法得到的结果也是一个 PyQuery 对象。例如：

```
li = items.children()
print(type(li))
print(li)
```

同时在 children 里也可以用 CSS 选择器，例如：

```
li2 = items.children('.active') print(li2)
```

在使用 PyQuery 库时，通过.parent 可以找到父元素的内容。下面的实例文件 pyq06.py 演示了使用 PyQuery 库查找父元素的过程。

源码路径：daima\12\12-1\pyq06.py

```
html = '''
<div id="container">
    <ul class="list">
            <li class="item-0">first item</li>
            <li class="item-1"><a href="link2.html">second item</a></li>
            <li class="item-0 active"><a href="link3.html"><span class="bold">third
                item</span></a></li>
            <li class="item-1 active"><a href="link4.html">fourth item</a></li>
            <li class="item-0"><a href="link5.html">fifth item</a></li>
    </ul>
 </div>
'''
from pyquery import PyQuery as pq
doc = pq(html)
items = doc('.list')
container = items.parent()
print(type(container))
print(container)
```

执行后会输出下面的结果。从结果可以看出返回了两部分内容：一个是父节点的信息；另一个是父节点的父节点的信息，即祖先节点的信息。同理，在通过 parents 查找的时候也可以添加 CSS 选择器来进行内容的筛选。

```
<class 'pyquery.pyquery.PyQuery'>
<div id="container">
    <ul class="list">
            <li class="item-0">first item</li>
            <li class="item-1"><a href="link2.html">second item</a></li>
            <li class="item-0 active"><a href="link3.html"><span class="bold">third
                item</span></a></li>
            <li class="item-1 active"><a href="link4.html">fourth item</a></li>
            <li class="item-0"><a href="link5.html">fifth item</a></li>
    </ul>
 </div>
```

在使用 PyQuery 库时，通过 siblings()可以获取所有的兄弟标签。当然，这里不包括标签自身的。同理，在.siblings()中也可以通过 CSS 选择器进行筛选。下面的实例文件 pyq07.py 演示了使用 PyQuery 库获取兄弟标签的过程。

源码路径：daima\12\12-1\pyq07.py

```
html = '''
<div class="wrap">
     <div id="container">
          <ul class="list">
               <li class="item-0">first item</li>
               <li class="item-1"><a href="link2.html">second item</a></li>
               <li class="item-0 active"><a href="link3.html"><span class="bold">third
                   item</span></a></li>
               <li class="item-1 active"><a href="link4.html">fourth item</a></li>
               <li class="item-0"><a href="link5.html">fifth item</a></li>
          </ul>
     </div>
 </div>
'''
from pyquery import PyQuery as pq
doc = pq(html)
li = doc('.list .item-0.active')
print(li.siblings())
```

在上述代码中，doc('.list .item-0.active') 中的 item-0 和 active 是紧挨着的，这表示两者是"并"的关系，这样满足条件的就剩下一个了——third item 的那个标签。执行后会输出：

```
<li class="item-1"><a href="link2.html">second item</a></li>
               <li class="item-0">first item</li>
               <li class="item-1 active"><a href="link4.html">fourth item</a></li>
               <li class="item-0"><a href="link5.html">fifth item</a></li>
```

12.2 处理 HTTP

HTTP（HyperText Transfer Protocol，超文本传输协议）是互联网上应用最广泛的一种网络协议，所有的 WWW 文件都必须遵守这个标准。本节将详细讲解处理 HTTP 的常用第三方库。

12.2.1 使用 aiohttp 库

Python 的标准库 asyncio 实现了单线程并发 I/O 操作，以及 TCP、UDP、SSL 等协议，而 aiohttp 则是基于 asyncio 实现的 HTTP 框架，实现了异步 HTTP 网络处理。

可以使用两种命令安装 aiohttp 库。

```
pip install aiohttp
easy_install aiohttp
```

下面的实例演示了使用 aiohttp 库实现异步处理的过程。首先看客户端文件 aiocli01.py，其功能是从 Python 网站中检索信息。具体实现代码如下。

源码路径：daima\12\12-2\aiocli01.py

```
import aiohttp
import asyncio
import async_timeout

async def fetch(session, url):
    async with async_timeout.timeout(10):
        async with session.get(url) as response:
            return await response.text()

async def main():
    async with aiohttp.ClientSession() as session:
        html = await fetch(session, 'http://Python官网域名.org')
        print(html)

loop = asyncio.get_event_loop()
loop.run_until_complete(main())
```

然后看服务器端文件 aioser01.py，其功能是获取客户端访问者的信息并输出欢迎信息。具体实现代码如下。

源码路径：daima\12\12-2\aioser01.py

```python
from aiohttp import web

async def handle(request):
    name = request.match_info.get('name', "Anonymous")
    text = "Hello, " + name
    return web.Response(text=text)

app = web.Application()
app.add_routes([web.get('/', handle),
                web.get('/{name}', handle)])

web.run_app(app)
```

aiohttp 库最大的用处体现在服务器端。下面的实例文件 aio02.py 演示了使用 aiohttp 库创建一个指定地址和端口的服务器的过程。

源码路径：daima\12\12-2\aio02.py

```python
import asyncio

from aiohttp import web

async def index(request):
    await asyncio.sleep(0.5)
    return web.Response(body=b'<h1>Index</h1>')

async def hello(request):
    await asyncio.sleep(0.5)
    text = '<h1>hello, %s!</h1>' % request.match_info['name']
    return web.Response(body=text.encode('utf-8'))

async def init(loop):
    app = web.Application(loop=loop)
    app.router.add_route('GET', '/', index)
    app.router.add_route('GET', '/hello/{name}', hello)
    srv = await loop.create_server(app.make_handler(), '127.0.0.1', 8000)
    print('Server started at http://127.0.0.1:8000...')
    return srv

loop = asyncio.get_event_loop()
loop.run_until_complete(init(loop))
loop.run_forever()
```

执行后会启动这个创建的服务器。

```
Server started at http://127.0.0.1:8000...
```

下面的实例文件 aio03.py 演示了，使用 aiohttp 库爬取指定 CSDN 博客中技术文章的地址的过程。

源码路径：daima\12\12-2\aio03.py

```python
import urllib.request as request
from bs4 import BeautifulSoup as bs
import asyncio
import aiohttp

@asyncio.coroutine
async def getPage(url,res_list):
    print(url)
    headers = {'User-Agent':'Mozilla/4.0 (compatible; MSIE 5.5; Windows NT)'}
    # conn = aiohttp.ProxyConnector(proxy="http://127.0.0.1:8087")
    async with aiohttp.ClientSession() as session:
        async with session.get(url,headers=headers) as resp:
            assert resp.status==200
            res_list.append(await resp.text())

class parseListPage():
    def __init__(self,page_str):
        self.page_str = page_str
    def __enter__(self):
```

```
                    page_str = self.page_str
                    page = bs(page_str,'lxml')
                    #获取文章链接
                    articles = page.find_all('div',attrs={'class':'article_title'})
                    art_urls = []
                    for a in articles:
                            x = a.find('a')['href']
                            art_urls.append('http://blog.csdn.net'+x)
                    return art_urls
            def __exit__(self, exc_type, exc_val, exc_tb):
                    pass

page_num = 100
page_url_base = 'http://blog.csdn.net/u011475134/article/list/'
page_urls = [page_url_base + str(i+1) for i in range(page_num)]
loop = asyncio.get_event_loop()
ret_list = []
tasks = [getPage(host,ret_list) for host in page_urls]
loop.run_until_complete(asyncio.wait(tasks))

articles_url = []
for ret in ret_list:
    with parseListPage(ret) as tmp:
        articles_url += tmp
ret_list = []

tasks = [getPage(url, ret_list) for url in articles_url]
loop.run_until_complete(asyncio.wait(tasks))
loop.close()
```

执行后会输出：

```
http://blog.csdn.net/u011475134/article/list/29
http://blog.csdn.net/u011475134/article/list/30
http://blog.csdn.net/u011475134/article/list/73
http://blog.csdn.net/u011475134/article/list/2
#省略后的爬取结果
```

12.2.2 使用 requests 库

库 requests 是用 Python 基于 urllib 编写的，采用的是 Apache 2 Licensed 开源协议的 HTTP 库。Requests 比 urllib 更加方便，可以节约开发者大量的时间。

可以使用两种命令安装 requests 库。

```
pip install requests
easy_install requests
```

下面的实例文件 Requests01.py 演示了使用 requests 库返回指定 URL 地址请求的过程。

源码路径：daima\12\12-2\Requests01.py

```
import requests

r = requests.get(url='http://指定网址的域名')    #最基本的GET请求
print(r.status_code)    #获取返回的状态
r = requests.get(url='http://指定网址的域名', params={'wd': 'python'})    #带参数的GET请求
print(r.url)
print(r.text)    #输出解码后的返回数据
```

在上述代码中，创建了一个名为 r 的 Response 对象，可以从这个对象中获取所有想要的信息。执行后会输出：

```
200
http://指定网址的域名/?wd=python
<!DOCTYPE html PUBLIC "-//W3C//DTD XHTML 1.0 Transitional//EN" "http://www.w3.org/TR/
    xhtml1/DTD/xhtml1-transitional.dtd">
<html xmlns="http://www.w3.org/1999/xhtml">
<head>
<meta http-equiv="X-UA-Compatible" content="IE=edge">
<meta http-equiv="Content-Type" content="text/html; charset=gbk" />
<title>门户 -  Powered by Discuz!</title>
```

```
<meta name="keywords" content="门户" />
<meta name="description" content="门户 " />
<meta name="generator" content="Discuz! X3.2" />
#省略后面的结果
```

上述实例只演示了 get 接口的用法，其他接口的用法也十分简单。

```
requests.get('https://GitHub的域名.com/timeline.json') #GET请求
requests.post("http://HTTPBin域名.org/post") #POST请求
requests.put("http://HTTPBin域名.org/put") #PUT请求
requests.delete("http://HTTPBin域名.org/delete") #DELETE请求
requests.head("http://HTTPBin域名.org/get") #HEAD请求
requests.options("http://HTTPBin域名.org/get") #OPTIONS请求
```

例如，要查询 HTTPBin 网站页面的具体参数，需要在 url 里面加上这个参数。假如我们想看有没有 Host=HTTPBin 域名.org 这条数据，url 形式应该是 http://HTTPBin 域名.org/get?Host= HTTPBin 域名.org。在下面的实例文件 Requests02.py 中，提交的数据是向这个地址传送 data 的数据。

源码路径：daima\12\12-2\Requests02.py

```
import requests
url = 'http://HTTPBin域名.org/get'
data = {
    'name': 'zhangsan',
    'age': '25'
}
response = requests.get(url, params=data)
print(response.url)
print(response.text)
```

执行后会输出：

```
http://HTTPBin域名.org/get?name=zhangsan&age=25
{
  "args": {
    "age": "25",
    "name": "zhangsan"
  },
  "headers": {
    "Accept": "*/*",
    "Accept-Encoding": "gzip, deflate",
    "Connection": "close",
    "Host": "HTTPBin域名.org",
    "User-Agent": "python-requests/2.12.4"
  },
  "origin": "39.71.61.153",
  "url": "http://HTTPBin域名.org/get?name=zhangsan&age=25"
}
```

requests 库也能够处理 JSON 数据，其内置方法 response.json()等同于 json.loads（response.text）方法。下面的实例文件 Requests03.py 演示了分别使用 get 和 post 方式处理 JSON 数据的过程。

源码路径：daima\12\12-2\Requests03.py

```
import requests
import json
r = requests.post('https://api.GitHub域名.com/some/endpoint',data=json.dumps({'some':'data'}))
print(r.json())
response = requests.get("http://HTTPBin域名.org/get")
print(type(response.text))
print(response.json())
print(json.loads(response.text))
```

执行后会输出：

```
{'message': 'Not Found', 'documentation_url': 'https://developer.HTTPBin域名.com/v3'}
<class 'str'>
{'args': {}, 'headers': {'Accept': '*/*', 'Accept-Encoding': 'gzip, deflate', 'Connection':
    'close', 'Host': 'HTTPBin域名.org', 'User-Agent': 'python-requests/2.12.4'}, 'origin':
    '39.71.61.153', 'url': 'http://HTTPBin域名.org/get'}
{'args': {}, 'headers': {'Accept': '*/*', 'Accept-Encoding': 'gzip, deflate', 'Connection':
    'close', 'Host': 'HTTPBin域名.org', 'User-Agent': 'python-requests/2.12.4'}, 'origin':
    '39.71.61.153', 'url': 'http://HTTPBin域名.org/get'}
```

12.2 处理HTTP

在使用 requests 库获取网页信息时，很多验证网站需要我们提供 header（头部信息）。例如，要访问知乎页面（不是会员登录知乎，就看不到里面的内容），我们可以尝试不加 header 信息进行访问。下面的实例文件 Requests04.py 实现了此功能。

源码路径：daima\12\12-2\Requests04.py

```
import requests
url = 'https://www.zhihu.com/'
response = requests.get(url)
response.encoding = "utf-8"
print(response.text)
```

执行后会提示发生内部服务器错误，也就是说，连知乎登录页面的 html 都下载不下来。

```
<html><body><h1>500 Server Error</h1>
An internal server error occured.
</body></html>
```

要成功访问知乎页面，就必须加对应的 headers 信息。我们可以通过浏览器工具查看知乎页面的 headers 信息，如图 12-2 所示。

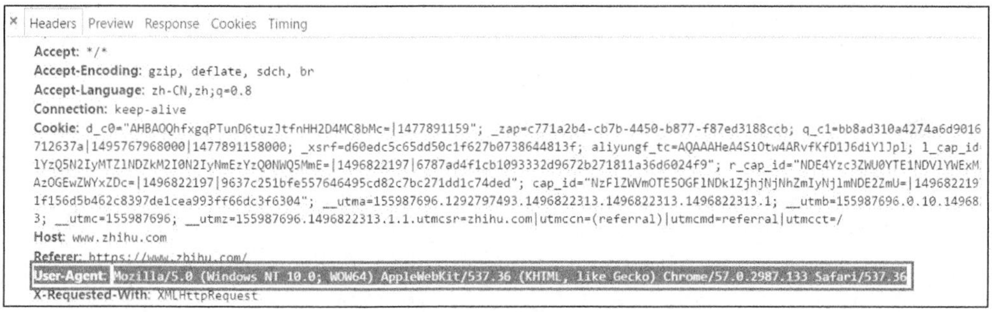

图 12-2　查看知乎页面的 headers 信息

下面的实例文件 Requests05.py 演示了添加 headers 后获取知乎页面信息的过程。

源码路径：daima\12\12-2\Requests05.py

```
import requests
url = 'https://www.zhihu.com/'
headers = {
    'User-Agent': 'Mozilla/5.0 (Windows NT 10.0; WOW64) AppleWebKit/537.36 (KHTML, like
        Gecko) Chrome/57.0.2987.133 Safari/537.36'
}
response = requests.get(url, headers=headers)
print(response.text)
```

执行后会成功获取知乎主页的信息。

```
<!doctype html>
<html lang="zh" data-hairline="true" data-theme="light"><head><meta charSet="utf-8"/><
    title data-react-helmet="true">知乎 - 发现更大的世界</title><meta name="viewport"
    content="width=device-width,initial-scale=1,maximum-scale=1"/><meta name="renderer"
    content="webkit"/><meta name="force-rendering" content="webkit"/><meta http-equiv=
    "X-UA-Compatible" content="IE=edge,chrome=1"/><meta name="google-site-verification"
    content="FTeR0c8arOPKh8c5DYh_9uu98_zJbaWw53J-Sch9MTg"/><link rel="shortcut icon"
#省略后面的结果
```

在使用 requests 内置方法后，会返回一个 response 对象，在里面存储了服务器响应的内容，例如，上述代码中的 response.text。在下面的实例文件 Requests06.py 中，当我们访问 r.text 时，会使用其响应的文本编码进行解码，并且可以修改其编码让 r.text 使用自定义的编码进行解码。

源码路径：daima\12\12-2\Requests06.py

```
import requests
r = requests.get('http://www.toppr.net')
print(r.text, '\n{}\n'.format('*'*79), r.encoding)
r.encoding = 'GBK'
print(r.text, '\n{}\n'.format('*'*79), r.encoding)
```

上述库 requests 还支持如下响应。
- r.status_code：响应状态码。
- r.raw：返回原始响应体，也就是 urllib 的 response 对象，使用 r.raw.read() 读取。
- r.content：字节方式的响应体，会自动解码和压缩。
- r.text：字符串方式的响应体，会自动根据响应头部的字符编码进行解码。
- r.headers：以字典对象存储服务器响应头，但是这个字典比较特殊，字典键不区分大小写，若键不存在则返回 None。
- r.json()：Requests 中内置的 JSON 解码器。
- r.raise_for_status()：请求失败（非 200 响应）时抛出异常。

在使用 requests 库时，我们可以通过代理或证书来访问并获取指定网站的信息。下面的实例文件 Requests07.py 演示了使用手工设置的证书来访问远程页面信息的过程。

源码路径：daima\12\12-2\Requests07.py

```
import requests
response = requests.get('https://www.12306.cn', cert=('/path/server.crt', '/path/key'))
print(response.status_code)
```

下面的实例文件 Requests08.py 演示了用设置的普通代理访问远程页面信息的过程。

源码路径：daima\12\12-2\Requests08.py

```
import requests
proxies = {
    "http": "http://127.0.0.1:9743",
    "https": "https://127.0.0.1:9743",
}
response = requests.get("https://www.taobao.com", proxies=proxies)
print(response.status_code)
```

下面的实例文件 Requests09.py 演示了，使用设置的用户名和密码代理访问远程页面信息的过程。

源码路径：daima\12\12-2\Requests09.py

```
import requests
proxies = {
    "http": "http://user:password@127.0.0.1:9743/",
}
response = requests.get("https://www.taobao.com", proxies=proxies)
print(response.status_code)
```

下面的实例文件 Requests10.py 演示了，使用设置的 socks 代理访问远程页面信息的过程。

源码路径：daima\12\12-2\Requests10.py

```
import requests
proxies = {
    'http': 'socks5://127.0.0.1:9742',
    'https': 'socks5://127.0.0.1:9742'
}
response = requests.get("https://www.taobao.com", proxies=proxies)
print(response.status_code)
```

如果要访问需要认证的网站，可以通过 requests.auth 模块进行认证。下面的实例文件 Requests11.py 演示了通过认证方法访问远程页面信息的过程。

源码路径：daima\12\12-2\Requests11.py

```
from requests.auth import HTTPBasicAuth
#方法一
r = requests.get('http://120.27.34.24:9001', auth=HTTPBasicAuth('user', '123'))
#方法二
<br>r = requests.get('http://120.27.34.24:9001', auth=('user', '123'))
print(r.status_code)
```

12.2.3 使用 httplib2 库

httplib2 库是一个第三方的开源库，比 Python 内置库 http.client 更完整地实现了 HTTP 协议，

12.2 处理 HTTP

同时比 urllib.request 提供了更好的抽象。

可以使用两种命令安装 httplib2 库。

```
pip install httplib2
easy_install httplib2
```

1. 获取内容

一旦拥有了 http 对象，将会非常简单地获取网页数据，只要以需要的数据的地址作为参数调用 request() 方法即可，这会对该 url 执行一个 http GET 请求。下面的实例文件 http201.py 演示了使用 httplib2 库获取网页数据的过程。

源码路径：daima\12\12-2\http201.py

```python
import httplib2
#获取HTTP对象
h =httplib2.Http()
#发出同步请求，并获取内容
resp, content = h.request("http://www.baidu.com/")
print(resp)
print(content)
```

方法 request() 返回两个值：第一个是一个 httplib2.Response 对象，其中包含了服务器返回的所有 http 头，比如，status 为 200 表示请求成功；第二个 content 变量包含了 http 服务器返回的实际数据。数据以 bytes 对象返回，而不是字符串。如果你需要一个字符串，则需要确定字符编码并自己进行转换。

2. 处理缓存

缓存是很多机制都必有的功能，httplib2 库与 Python 内置库相比的最大优势是可以处理缓存。下面的实例文件 http202.py 演示了使用 httplib2 库处理网页缓存数据的过程。

源码路径：daima\12\12-2\http202.py

```python
import httplib2
#获取HTTP对象
h =httplib2.Http('.cache')
#发出同步请求，并获取内容
resp, content = h.request("http://www.baidu.com")
print(resp)
print("......"*3)
httplib2.debuglevel = 1
h1 = httplib2.Http('.cache')
resp,content = h1.request('http://www.baidu.com')

print(resp)
print('debug',resp.fromcache)
```

执行后会输出网页的源码，获取带有缓存的 HTTP 对象 h1，并存储在当前环境的 ".cache" 目录下。

```
{'date': 'Fri, 20 Apr 2018 08:06:48 GMT', 'content-type': 'text/html; charset=utf-8',
   'transfer-encoding': 'chunked', 'connection': 'Keep-Alive', 'vary': 'Accept-Encoding',
   'set-cookie': 'BAIDUID=AAEB2299BC40EA4AFC6AEA020A6FC971:FG=1; expires=Thu, 31-Dec-
   37 23:55:55 GMT; max-age=2147483647; path=/; domain=.baidu.com, BIDUPSID=
   AAEB2299BC40EA4AFC6AEA020A6FC971; expires=Thu, 31-Dec-37 23:55:55 GMT; max-age=
   2147483647; path=/; domain=.baidu.com, PSTM=1524211608; expires=Thu, 31-Dec-
   37 23:55:55 GMT; max-age=2147483647; path=/; domain=.baidu.com, BDSVRTM=0; path=/,
   BD_HOME=0; path=/, H_PS_PSSID=26193_1469_21111_20928; path=/; domain=.baidu.com',
   'p3p': 'CP=" OTI DSP COR IVA OUR IND COM "', 'cache-control': 'private', 'cxy_all':
   'baidu+28c18a0926430a3884a32911dee55ed', 'expires': 'Fri, 20 Apr 2018 08:06:31
   GMT', 'x-powered-by': 'HPHP', 'server': 'BWS/1.1', 'x-ua-compatible': 'IE=Edge,
   chrome=1', 'bdpagetype': '1', 'bdqid': '0xab4be86f00008f3b', 'status': '200',
   'content-length': '115109', '-content-encoding': 'gzip', 'content-location':
   'http://www.baidu.com'}
.................
connect: (www.baidu.com, 80) ************
send: b'GET / HTTP/1.1\r\nHost: www.baidu.com\r\nuser-agent: Python-httplib2/0.11.3
   (gzip)\r\naccept-encoding: gzip, deflate\r\n\r\n'
reply: 'HTTP/1.1 200 OK\r\n'
```

```
header: Date header: Content-Type header: Transfer-Encoding header: Connection header:
Vary header: Set-Cookie header: Set-Cookie header: Set-Cookie header: Set-Cookie
header: Set-Cookie header: Set-Cookie header: P3P header: Cache-Control header:
Cxy_all header: Expires header: X-Powered-By header: Server header: X-UA-Compatible
header: BDPAGETYPE header: BDQID header: Content-Encoding {'date': 'Fri, 20 Apr
2018 08:06:49 GMT', 'content-type': 'text/html; charset=utf-8', 'transfer-encoding': '
chunked', 'connection': 'Keep-Alive', 'vary': 'Accept-Encoding', 'set-cookie':
'BAIDUID=166923F37969C306C0ABB71432D96615:FG=1; expires=Thu, 31-Dec-37 23:55:55 GMT;
max-age=2147483647; path=/; domain=.baidu.com, BIDUPSID=166923F37969C306C0ABB71432D96615;
expires=Thu, 31-Dec-37 23:55:55 GMT; max-age=2147483647; path=/; domain=.baidu.com,
PSTM=1524211609; expires=Thu, 31-Dec-37 23:55:55 GMT; max-age=2147483647; path=/;
domain=.baidu.com, BDSVRTM=0; path=/, BD_HOME=0; path=/, H_PS_PSSID=26254_1466_
13289_21095_26105; path=/; domain=.baidu.com', 'p3p': 'CP=" OTI DSP COR IVA OUR IND
COM "', 'cache-control': 'private', 'cxy_all': 'baidu+6b76b5a82a8a96ac81598ed88b6af038',
'expires': 'Fri, 20 Apr 2018 08:06:31 GMT', 'x-powered-by': 'HPHP', 'server':
'BWS/1.1', 'x-ua-compatible': 'IE=Edge,chrome=1', 'bdpagetype': '1', 'bdqid':
'0xbf58353700009f7f', 'status': '200', 'content-length': '114616', '-content-encoding':
'gzip', 'content-location': 'http://www.baidu.com'}
debug False
```

此时 debug 的值是 True，说明是从本地的.cache 进行读取的，没经过原网站。如果我们想不读取缓存，只需用如下代码实现即可。

```
resp,content = h1.request('http://www.baidu.com/sitemap.xml',headers={'cache-control':
'no-cache' })
```

httplib2 库允许我们添加任意的 http 头部到发出的请求中。为了跳过所有缓存（不仅包括本地的磁盘缓存，还包括任何处于你和远程服务器之间的缓存代理服务器），只需要在 headers 字典里面加入 no-cache 头即可。

3. 处理 Last-Modified 和 ETag 头

当服务器的资源发生改变但是还有本地缓存需要更新时，http 定义了 Last-Modified 和 Etag 头，这些头称为验证器。如果本地缓存已经不是较新的，客户端可以在下一个请求中发送验证器来检查数据实际上有没有改变。如果数据没有改变，服务器返回 304 状态码，但不返回数据。所以，虽然还会在网络上有一个来回，但是最终可以少下载一点字节。这里实际上有如下两个状态码。

- 304：服务器这次返回的状态码，导致 httplib2 查看它的缓存。
- 200：服务器上次返回的状态码，并和页面数据一起保存在 httplib2 的缓存里。

4. http2lib 处理压缩

每当 httplib2 发送一次请求时，它包含了 Accept-Encoding 头来告诉服务器它能够处理 deflate 或者 gzip 压缩。当 request()方法返回的时候，httplib2 就已经解压缩了响应体并将其放在 content 变量里。如果你想知道响应是否压缩过，你可以检查 response['-content-encoding']。

5. POST 发送构造数据

下面的实例文件 http203.py 演示了使用 POST 发送构造数据的过程。

源码路径：daima\12\12-2\http203.py

```
from urllib.parse import urlencode

import httplib2

httplib2.debuglevel = 1

h = httplib2.Http('.cache')

data = {'status': 'Test update from Python 3'}

h.add_credentials('diveintomark', 'MY_SECRET_PASSWORD', 'identi.ca')

resp, content = h.request('https://www.baidu.com',
    'POST',
```

```
        urlencode(data),
        headers={'Content-Type': 'application/x-www-form-urlencoded'})
```
执行后会输出：
```
connect: (www.baidu.com, 443)
send: b'POST / HTTP/1.1\r\nHost: www.baidu.com\r\nContent-Length: 32\r\ncontent-type:
    application/x-www-form-urlencoded\r\nuser-agent: Python-httplib2/0.11.3 (gzip)\r\
    naccept-encoding: gzip, deflate\r\n\r\n'
send: b'status=Test+update+from+Python+3'
reply: 'HTTP/1.1 302 Found\r\n'
header: Bdpagetype header: Connection header: Content-Length header: Content-Type
    header: Date header: Location header: Server header: Set-Cookie header: X-Ua-Compatible
```
在上述代码中，add_credentials()方法的第三个参数是该证书有效的域名。我们应该总是指定这个参数，如果省略了这个参数，并且之后重用这个 httplib2.Http 对象访问另一个需要认证的站点，可能会导致 httplib2 将一个站点的用户名、密码泄露给其他站点。httplib2 返回的数据总是字节串，而不是字符串。为了将其转化为字符串，你需要用合适的字符编码进行解码，比如：
```
print(content.decode('utf-8'))
```

12.2.4 使用 urllib3 库

urllib3 库是一个具有线程安全连接池、支持文件 post 的 HTTP 库。可以使用两种命令安装 urllib3 库。
```
pip install urllib3
easy_install urllib3
```
下面的实例文件 urllib301.py 演示了，使用 urllib3 库中的 request()方法创建请求的过程。

源码路径：daima\12\12-2\urllib301.py
```
import urllib3
import requests
#忽略InsecureRequestWarning
requests.packages.urllib3.disable_warnings()
# 一个PoolManager实例生成请求，由该实例对象处理与线程池的连接以及线程安全的所有细节
http = urllib3.PoolManager()
#通过request()方法创建一个请求
r = http.request('GET', 'http://www.baidu.com/')
print(r.status) # 200
#获得html源码和utf-8解码
print(r.data.decode)
```
通过上述代码获取了百度主页的信息。
```
200
<!DOCTYPE html><!--STATUS OK-->
<html>
<head>
    <meta http-equiv="content-type" content="text/html;charset=utf-8">
    <meta http-equiv="X-UA-Compatible" content="IE=Edge">
    <link rel="dns-prefetch" href="//s1.指定URL.com"/>
    <link rel="dns-prefetch" href="//t1.baidu.com"/>
    <link rel="dns-prefetch" href="//t2.baidu.com"/>
    <link rel="dns-prefetch" href="//t3.baidu.com"/>
    <link rel="dns-prefetch" href="//t10.baidu.com"/>
#省略后面的效果
```
下面的实例文件 urllib302.py 演示了在 request()方法中添加 header 创建请求的过程。

源码路径：daima\12\12-2\urllib302.py
```
import urllib3
header = {
        'User-Agent': 'Mozilla/5.0 (Windows NT 6.1; Win64; x64) AppleWebKit/537.36
    (KHTML, like Gecko) Chrome/63.0.3239.108 Safari/537.36'
    }
http = urllib3.PoolManager()
r = http.request('GET',
                'https://www.baidu.com/s?',
                fields={'wd': 'hello'},
```

```
                    headers=header)
print(r.status) # 200
print(r.data.decode())
```

下面的实例文件 urllib303.py 演示了使用 urllib3 库中的 post()方法创建请求的过程。

源码路径：daima\12\12-2\urllib303.py

```
import urllib3
http = urllib3.PoolManager()
#还可以通过request()方法向请求中添加一些其他信息
header = {
            'User-Agent': 'Mozilla/5.0 (Windows NT 6.1; Win64; x64) AppleWebKit/537.36
    (KHTML, like Gecko) Chrome/63.0.3239.108 Safari/537.36'
}
r = http.request('POST',
                            'http://HTTPBin域名.org/post',
                            fields={'hello':'world'},
                            headers=header)
print(r.data.decode())

#对于POST和PUT请求需要手动对传入的数据进行编码，并添加在URL之后
encode_arg = urllib.parse.urlencode({'arg': '我的'})
print(encode_arg.encode())
r = http.request('POST',
                            'http://HTTPBin域名.org/post?'+encode_arg,
                            headers=header)
# unicode解码
print(r.data.decode('unicode_escape'))
```

执行后会输出：

```
{
  "args": {},
  "data": "",
  "files": {},
  "form": {
    "hello": "world"
  },
  "headers": {
    "Accept-Encoding": "identity",
    "Connection": "close",
    "Content-Length": "129",
    "Content-Type": "multipart/form-data; boundary=b33b20053e6444ee947a6b7b3f4572b2",
    "Host": "HTTPBin域名.org",
    "User-Agent": "Mozilla/5.0 (Windows NT 6.1; Win64; x64) AppleWebKit/537.36 (KHTML,
        like Gecko) Chrome/63.0.3239.108 Safari/537.36"
  },
  "json": null,
  "origin": "39.71.61.153",
  "url": "http://HTTPBin域名.org/post"
}

b'arg=%E6%88%91%E7%9A%84'
{
  "args": {
    "arg": "我的"
  },
  "data": "",
  "files": {},
  "form": {},
  "headers": {
    "Accept-Encoding": "identity",
    "Connection": "close",
    "Content-Length": "0",
    "Host": "HTTPBin域名.org",
    "User-Agent": "Mozilla/5.0 (Windows NT 6.1; Win64; x64) AppleWebKit/537.36 (KHTML,
        like Gecko) Chrome/63.0.3239.108 Safari/537.36"
  },
  "json": null,
  "origin": "39.71.61.153",
  "url": "http://HTTPBin域名.org/post?arg=我的"
}
```

下面的实例文件 urllib304.py 演示了使用 urllib3 库发送 json 数据的过程。

源码路径：daima\12\12-2\urllib304.py

```python
import urllib3
http = urllib3.PoolManager()
data={'attribute':'value'}
encode_data= json.dumps(data).encode()

r = http.request('POST',
                    'http://HTTPBin域名.org/post',
                    body=encode_data,
                    headers={'Content-Type':'application/json'}
                )
print(r.data.decode('unicode_escape'))
```

执行后会输出：

```
{
  "args": {},
  "data": "{\"attribute\": \"value\"}",
  "files": {},
  "form": {},
  "headers": {
    "Accept-Encoding": "identity",
    "Connection": "close",
    "Content-Length": "22",
    "Content-Type": "application/json",
    "Host": "HTTPBin域名.org"
  },
  "json": {
    "attribute": "value"
  },
  "origin": "39.71.61.153",
  "url": "http://HTTPBin域名.org/post"
}
```

下面的实例文件 urllib305.py 演示了使用 urllib3 库获取远程 CSV 数据的过程。

源码路径：daima\12\12-2\urllib305.py

```python
import urllib3
#两个文件的源url
url1 = 'http://指定网址/all_week.csv'
url2 = 'http://指定网址/all_month.csv'
#开始创建一个HTTP连接池
http = urllib3.PoolManager()
#请求第一个文件并写入文件
response = http.request('GET', url1)
with open('all_week.csv', 'wb') as f:
    f.write(response.data)
#请求第二个文件
response = http.request('GET', url2)
with open('all_month.csv', 'wb') as f:
    f.write(response.data)

#最后释放这个HTTP连接
response.release_conn()
```

执行后会将这两个远程 CSV 文件下载并保存到本地，如图 12-3 所示。

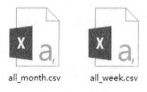

图 12-3　下载并保存到本地的 CSV 文件

下面的实例文件 urllib306.py 演示了使用 urllib3 库抓取并显示百度新闻的过程。

源码路径：daima\12\12-2\urllib306.py

```
from bs4 import BeautifulSoup
import urllib3

def get_html(url):
    try:
        userAgent = 'Mozilla/5.0 (Windows; U; Windows NT 6.1; en-US; rv:1.9.1.6)
            Gecko/20091201 Firefox/3.5.6'
        http = urllib3.PoolManager(timeout=2)
        response = http.request('get', url, headers={'User_Agent': userAgent})
        html = response.data
        return html
    except Exception as e:
        print(e)
        return None

def get_soup(url):
    if not url:
        return None
    try:
        soup = BeautifulSoup(get_html(url))
    except Exception as e:
        print(e)
        return None
    return soup

def get_ele(soup, selector):
    try:
        ele = soup.select(selector)
        return ele
    except Exception as e:
        print(e)
        return None

def main():
    url = 'http://www.baidu.com/'
    soup = get_soup(url)
    ele = get_ele(soup, '#headLineDefault > ul > ul:nth-of-type(1) > li.topNews > h1 > a')
    headline = ele[0].text.strip()
    print(headline)

if __name__ == '__main__':
    main()
```

因为新闻内容是随着时间的推移发生变化的，所以每次的执行效果会不一样。

12.3 电子邮件

使用 Python 可以开发出功能强大的邮件系统，本节将详细讲解使用 Python 第三方库开发邮件系统的过程。

12.3.1 使用 envelopes 库

envelopes 库是 Python 处理邮件的一个第三方模块，是对 Python 内置模块 email 和 smtplib 的封装。对于 Linux 系统，直接使用如下命令即可安装 envelopes 库。

```
pip install envelopes
```

如果是 Windows 系统，需要先下载 envelopes 库的源码文件压缩包，解压之后使用如下命令安装。

```
python setup.py install
```

下面的实例文件 envelopes01.py 演示了使用 envelopes 库向指定邮箱发送邮件的过程。

源码路径：daima\12\12-3\envelopes01.py

```
from envelopes import Envelope, GMailSMTP

envelope = Envelope(    # 实例化Envelope
    from_addr=(u'from@example.com', u'From Example'),    #必选参数，表示发件人信息，前面是发件
```

```
                    #箱，后面是发送人；只有发件箱也可以
                    to_addr=(u'to@example.com', u'To Example'),   #必选参数,可以直接发送给多人(u'user1@example.
                    com', u'user2@example.com')
                    subject=u'Envelopes demo',        #必选参数，邮件标题
                    html_body=u'<h1>活着之上</h1>'#可选参数，带HTML的邮件正文
                    text_body=u"I'm a helicopter!",   #可选参数，文本格式的邮件正文
                    cc_addr=u'boss1@example.com',     #可选参数，抄送人，也可以是列表形式
                    bcc_addr=u'boss2@example.com',    #可选参数，隐藏抄送人，也可以是列表
                    headers=u'',   #可选参数，邮件头部内容，字典形式
                    charset=u'',   #可选参数，邮件字符集
)
envelope.add_attachment('/Users/bilbo/Pictures/helicopter.jpg')   #增加附件，注意，文件是
#完整路径，也可以加入多个附件

#使用临时连接发送邮件
envelope.send('smtp.163.com', login='from@example.com',
              password='password', tls=True)   #发送邮件
...
gmail = GMailSMTP('from@example.com', 'password')
gmail.send(envelope)
```

下面的实例文件 envelopes02.py 演示了，使用 envelopes 库构建 Flask Web 邮件发送程序的过程。

源码路径：daima\12\12-3\envelopes02.py

```
from envelopes import Envelope, SMTP
import envelopes.connstack
from flask import Flask, jsonify
import os


app = Flask(__name__)
app.config['DEBUG'] = True
conn = SMTP('127.0.0.1', 1025)


@app.before_request
def app_before_request():
    envelopes.connstack.push_connection(conn)


@app.after_request
def app_after_request(response):
    envelopes.connstack.pop_connection()
    return response


@app.route('/mail', methods=['POST'])
def post_mail():
    envelope = Envelope(
        from_addr='%s@localhost' % os.getlogin(),
        to_addr='%s@localhost' % os.getlogin(),
        subject='Envelopes in Flask demo',
        text_body="I'm a helicopter!"
    )

    smtp = envelopes.connstack.get_current_connection()
    smtp.send(envelope)
    return jsonify(dict(status='ok'))

if __name__ == '__main__':
    app.run()
```

12.3.2 使用 Inbox 库

Inbox.py 是一个用 Python 实现的简单 SMTP 服务器，它是异步的。单个实例处理邮件的速度超过 1000 封/秒。可以使用如下命令安装 Inbox 库。

```
pip install Inbox
```

在下面的实例文件 Inbox01.py 中，使用 Inbox 库创建一个带有 HTTP REST 接口的 SMTP 服务器，用于访问邮件信息。

源码路径：daima\12\12-3\Inbox.py

```
class StorageHandler(object):
    def __init__(self):
        self.messages = {}
        self.attachments = {}
```

```python
        self.inboxes = {}

    def store_mail(self, message):
        recps = [recp.lower() for recp in message['to']]

        message['id'] = hashlib.md5(recp + str(time.time())).hexdigest()

        message['content'] = zlib.compress(message['content'], 9)

        self.messages[message['id']] = message

        for recp in recps:
            if recp not in self.inboxes:
                self.inboxes[recp] = []

            self.inboxes[recp].append(message['id'])

    def store_attachment(self, payload, mimetype):
        id = hashlib.sha1(payload).hexdigest()
        self.attachments[id] = (zlib.compress(payload, 9), mimetype)
        return id

    def get_mail(self, recp, *args, **kwargs):
        out = []
        if recp not in self.inboxes:
            return out

        for msg_id in self.inboxes[recp]:
            found = True
            msg2 = copy.deepcopy(self.messages[msg_id])
            msg2['content'] = zlib.decompress(msg2['content'])

            for (key, value) in kwargs.items():
                if value not in msg2[key]:
                    found = False
            if found:
                out.append(msg2)
        return out

    def get_attachment(self, id):
        if id not in self.attachments:
            return ("", "text/plain")
        else:
            return (zlib.decompress(self.attachments[id][0]), self.attachments[id][1])

class HttpHandler(asyncore.dispatcher_with_send):
    def __init__(self, sock, store):
        asyncore.dispatcher_with_send.__init__(self, sock)
        self.store = store
        self.inbuffer = ""
        self.outbuffer = ""
        self.headers = {}
        self.headers_received = False
        self.body = ""

    def parse_headers(self, data):
        lines = data.splitlines()
        (method, path, version) = lines[0].split()
        for line in lines:
            if ':' in line:
                (key, value) = line.split(':', 1)
                self.headers[key.strip().lower()] = value.strip()

        self.headers['http'] = (method, path, version)
        self.headers_received = True

    def generate_http_response(self, payload, mimetype):
        out = "HTTP/1.1 200 Ok.\r\n"
        out += "Content-Type: %s\r\n" % mimetype
        out += "Content-Length: %i\r\n" % (len(payload))
        out += "Connection: close\r\n"
        out += "\r\n"
```

```python
            out += payload
        return out

    def handle_http_request(self):
        (method, path, version) = self.headers['http']
        content = ""
        mime = 'application/json'
        if method == "GET":
            parts = path.split("/")[1:]
            if parts[0] == "inbox":
                if len(parts) > 1:
                    recp = parts[1]
                    if '@' in recp:
                        mails = self.store.get_mail(recp)
                        content = json.dumps(mails, indent=2)
            if parts[0] == "files":
                if len(parts) > 1:
                    (content, mime) = self.store.get_attachment(parts[1])
        self.outbuffer += self.generate_http_response(content, mime)

        self.header_received = False
        self.headers = {}
        self.body = ""

    def writable(self):
        return len(self.outbuffer) >= 0

    def handle_write(self):
        sent = self.send(self.outbuffer)
        self.outbuffer = self.outbuffer[:sent]

    def handle_close(self):
        self.close()

    def handle_read(self):
        chunk = self.recv(8192)
        if chunk:
            self.inbuffer += chunk

        if '\r\n\r\n' in self.inbuffer:
            (header, body) = self.inbuffer.split('\r\n\r\n', 1)
            self.parse_headers(header)
            self.body = body
            self.inbuffer = ""

        if self.headers_received:
            if 'http' in self.headers:
                if self.headers['http'][0] == "GET":
                    self.handle_http_request()
                else:
                    self.body += self.inbuffer
                    self.inbuffer = ""
                    if len(self.body) >= self.headers['content-length']:
                        self.handle_http_request()

class HttpServer(asyncore.dispatcher):
    def __init__(self, host, port, storage):
        asyncore.dispatcher.__init__(self)
        self.create_socket(socket.AF_INET, socket.SOCK_STREAM)
        self.set_reuse_addr()
        self.bind((host, port))
        self.listen(5)
        self.storage = storage

    def handle_accept(self):
        pair = self.accept()
        if pair is None:
            pass
        else:
            sock, addr = pair
            handler = HttpHandler(sock, self.storage)
```

```
inbox = Inbox()
storage = StorageHandler()


@inbox.collate
def handle(to, sender, body):
    parser = email.parser.Parser()
    mail = parser.parsestr(body)
    message = {}
    message['to'] = to
    message['sender'] = sender
    message['subject'] = mail['subject']
    message['received'] = time.strftime("%Y-%m-%d %H:%M:%S")
    message['content'] = ""
    message['attachments'] = []

    for part in mail.walk():
        if part.get_content_maintype() == "multipart":
            continue

        if not part.get_filename():
            if part.get_content_maintype() == "text":
                message['content'] += part.get_payload(decode=False)
        else:
            attachment = {}
            attachment['filename'] = part.get_filename()
            attachment['type'] = part.get_content_type()
            payload = part.get_payload(decode=True)
            attachment['payload-id'] = storage.store_attachment(payload, part.get_content_type())
            message['attachments'].append(attachment)

    storage.store_mail(message)

server = HttpServer('0.0.0.0', 8123, storage)

handle = open('123.txt', 'wb')
sys.stderr = handle

inbox.serve(address='0.0.0.0', port=4467)
```

12.4 处理 URL

URL 是 Uniform Resource Locator 的缩写，其中文含义是统一资源定位器，也就是我们平常所说的 WWW 网址。本节将详细讲解使用 Python 第三方库开发 URL 程序的过程。

12.4.1 使用 furl 库

furl 库是一个用来处理 URL 的 Python 库，Python 可以用它更加优雅地操作 URL 地址。可以使用如下命令安装 furl 库。

```
pip install furl
```

下面的实例文件 url01.py 演示了使用 furl 库处理 URL 分页的过程。

源码路径：daima\12\12-4\url01.py

```
from furl import furl
f = furl('http://www.baidu.com/?page=1')
f.scheme, f.host, f.port, f.path, f.query, f.fragment
print(f.url)
f.args
f.args['page']
f.args['page'] = 2
print(f.url)
```

执行后会输出：

```
http://www.baidu.com/?page=1
```

```
http://www.baidu.com/?page=2
```

通过使用 furl 库，我们可以灵活地处理 URL 中的参数。下面的实例文件 url02.py 演示了，使用 furl 库处理 URL 参数的过程。

源码路径：daima\12\12-4\url02.py

```python
from furl import furl
f= furl('http://www.baidu.com/?bid=12331')
#输出参数
print(f.args)
#增加参数
f.args['haha']='123'
print(f.args)
#修改参数
f.args['haha']='124'
print(f.args)
#删除参数
del f.args['haha']
print(f.args)
```

执行后会输出：

```
{'bid': '12331'}
{'bid': '12331', 'haha': '123'}
{'bid': '12331', 'haha': '124'}
{'bid': '12331'}
```

在使用 furl 库时，可以通过使用内联方法来处理 URL 中的参数。下面的实例文件 url03.py 演示了，使用内联方法处理 URL 参数的过程。

源码路径：daima\12\12-4\url03.py

```python
from furl import furl
f= furl('http://www.baidu.com/?bid=12331')
#增加参数
print(furl('http://www.baidu.com/?bid=12331').add({'haha':'123'}).url)
#设置参数（只保留设置的参数）
print(furl('http://www.baidu.com/?bid=12331').set({'haha':'123'}).url)
#移除参数
print(furl('http://www.baidu.com/?bid=12331').remove(['bid']).url)
```

执行后会输出：

```
http://www.baidu.com/?bid=12331&haha=123
http://www.baidu.com/?haha=123
http://www.baidu.com/
```

12.4.2　使用 purl 库

purl 库是一个简单的、不可变的 URL 类，提供了简洁的 API 来询问和处理 URL。可以使用如下命令安装 purl 库。

```
pip install purl
```

下面的实例文件 purl01.py 演示了使用 purl 库处理 3 种构造类型 URL 的过程。

源码路径：daima\12\12-4\purl01.py

```python
from purl import URL
#字符串构造函数
from_str = URL('https://www.baidu.com/search?q=testing')
print(from_str)
#关键字构造器
from_kwargs = URL(scheme='https', host='www.baidu.com', path='/search', query='q=testing')
print(from_kwargs)
#联合使用
from_combo = URL('https://www. baidu.com').path('search').query_param('q', 'testing')
print(from_combo)
```

执行后会输出：

```
https://www.baidu.com/search?q=testing
https://www.baidu.com/search?q=testing
https://www.baidu.com/search?q=testing
```

在 purl 库中，URL 对象是不可变的，所有的修改器方法都返回一个新的实例。下面的实例文件 purl02.py 中，演示了使用 purl 库返回各个 URL 对象值的过程。

源码路径：daima\12\12-4\purl02.py

```
from purl import URL
u = URL('https://www.baidu.com/search?q=testing')
print(u.scheme())
print(u.host())
print(u.domain())
print(u.username())
u.password()
print(u.netloc())
u.port()
print(u.path())
print(u.query())
print(u.fragment())
print(u.path_segment(0))
print(u.path_segments())
print(u.query_param('q'))
print(u.query_param('q', as_list=True))
print(u.query_param('lang', default='GB'))
print(u.query_params())
print(u.has_query_param('q'))
print(u.has_query_params(('q', 'r')))
print( u.subdomains())
print(u.subdomain(0))
```

执行后会输出：

```
https
www.baidu.com
www.baidu.com
None
www.baidu.com
/search
q=testing

search
('search',)
testing
['testing']
GB
{'q': ['testing']}
True
False
['www', 'baidu', 'com']
www
```

在 purl 库中，每个访问器方法被重载为一个类似于 jQuery API 的修改器方法。下面的实例文件 purl03.py 演示了，使用 purl 库修改 URL 参数值的过程。

源码路径：daima\12\12-4\purl03.py

```
from purl import URL
u = URL.from_string('https://GitHub的域名.com/codeinthehole')
#输出
print(u.path_segment(0))
#修改器（创建一个新实例）
new_url = u.path_segment(0, 'tangentlabs')
print(new_url)
print(new_url is u)
print(new_url.path_segment(0))
```

执行后会输出：

```
codeinthehole
https://GitHub的域名.com/tangentlabs
False
tangentlabs
```

除了用上面介绍的重载方法外，可以使用方法 add_path_segment()在当前路径的末尾添加一个字段值。下面的实例文件 purl04.py 演示了，使用方法 add_path_segment()在当前路径末尾添加字段值的过程。

源码路径：daima\12\12-4\purl04.py

```
from purl import URL
u = URL().scheme('http').domain('www.example.com').path('/some/path').query_param('q', 'search term')
```

```
print(u)
new_url = u.add_path_segment('here')
print( new_url.as_string())
```

执行后会输出：

```
http://www.example.com/some/path?q=search+term
http://www.example.com/some/path/here?q=search+term
```

12.4.3 使用 webargs 库

webargs 库是一个解析 HTTP 请求参数的 Python 框架，内置支持流行的 Web 框架，包括 Flask、Django、Bottle、Tornado、Pyramid、webapp2、Falcon 和 aiohttp。可以使用如下命令安装 webargs 库。

```
pip install webargs
```

下面的实例文件 webargs01.py 演示了，在 Flask 程序中使用 webargs 库处理 URL 参数的过程。

源码路径：daima\12\12-4\webargs01.py

```
from flask import Flask
from webargs import fields
from webargs.flaskparser import use_args

app = Flask(__name__)

hello_args = {
    'name': fields.Str(required=True)
}

@app.route('/')
@use_args(hello_args)
def index(args):
    return 'Hello ' + args['name']

if __name__ == '__main__':
    app.run()
```

在浏览器地址栏中输入 http://127.0.0.1:5000/?name='World'后会显示执行效果，如图 12-4 所示。

图 12-4 执行效果

下面的实例文件 webargs02.py 演示了在 aiohttp 程序中使用 webargs 库的过程。

源码路径：daima\12\12-4\webargs02.py

```
import asyncio
import datetime as dt

from aiohttp import web
from aiohttp.web import json_response
from webargs import fields, validate
from webargs.aiohttpparser import use_args, use_kwargs

hello_args = {
    'name': fields.Str(missing='Friend')
}
@asyncio.coroutine
@use_args(hello_args)
def index(request, args):
    """A welcome page.
    """
    return json_response({'message': 'Welcome, {}!'.format(args['name'])})

add_args = {
    'x': fields.Float(required=True),
    'y': fields.Float(required=True),
}
@asyncio.coroutine
@use_kwargs(add_args)
def add(request, x, y):
```

```python
        """An addition endpoint."""
        return json_response({'result': x + y})

dateadd_args = {
    'value': fields.Date(required=False),
    'addend': fields.Int(required=True, validate=validate.Range(min=1)),
    'unit': fields.Str(missing='days', validate=validate.OneOf(['minutes', 'days']))
}
@asyncio.coroutine
@use_kwargs(dateadd_args)
def dateadd(request, value, addend, unit):
    """A datetime adder endpoint."""
    value = value or dt.datetime.utcnow()
    if unit == 'minutes':
        delta = dt.timedelta(minutes=addend)
    else:
        delta = dt.timedelta(days=addend)
    result = value + delta
    return json_response({'result': result.isoformat()})


def create_app():
    app = web.Application()
    app.router.add_route('GET', '/', index)
    app.router.add_route('POST', '/add', add)
    app.router.add_route('POST', '/dateadd', dateadd)
    return app

def run(app, port=5001):
    loop = asyncio.get_event_loop()
    handler = app.make_handler()
    f = loop.create_server(handler, '0.0.0.0', port)
    srv = loop.run_until_complete(f)
    print('serving on', srv.sockets[0].getsockname())
    try:
        loop.run_forever()
    except KeyboardInterrupt:
        pass
    finally:
        loop.run_until_complete(handler.finish_connections(1.0))
        srv.close()
        loop.run_until_complete(srv.wait_closed())
        loop.run_until_complete(app.finish())
    loop.close()
```

下面的实例文件 webargs03.py 演示了在 Tornado 程序中使用 webargs 库的过程。

源码路径：daima\12\12-4\webargs03.py

```python
import tornado.ioloop
import tornado.web

from webargs import fields
from webargs.tornadoparser import parser

class HelloHandler(tornado.web.RequestHandler):

    hello_args = {
        'name': fields.Str()
    }

    def post(self, id):
        reqargs = parser.parse(self.hello_args, self.request)
        response = {
            'message': 'Hello {}'.format(reqargs['name'])
        }
        self.write(response)

application = tornado.web.Application([
    (r"/hello/([0-9]+)", HelloHandler),
], debug=True)

if __name__ == "__main__":
    application.listen(8888)
    tornado.ioloop.IOLoop.instance().start()
```